The Handbook of Environmental Chemistry

Volume 135

Founding Editor: Otto Hutzinger

Series Editors: Damià Barceló • Andrey G. Kostianoy

In over four decades, *The Handbook of Environmental Chemistry* has established itself as the premier reference source, providing sound and solid knowledge about environmental topics from a chemical perspective. Written by leading experts with practical experience in the field, the series continues to be essential reading for environmental scientists, environmental managers, and decision-makers in industry, government, agencies and public-interest groups. The series also supports and advances the UN's sustainable development goals, mainly SDG14 (Life below water), SDG6 (Clean Water and Sanitation), SDG13 (Climate Change) and SDG15 (Life on Land).

Two distinguished Series Editors, internationally renowned volume editors as well as a prestigious Editorial Board safeguard the publication of volumes according to high scientific standards.

Presenting a wide spectrum of viewpoints and approaches in topical volumes, the scope of the series covers topics such as

- local and global changes of natural environment and climate
- anthropogenic impact on the environment
- water, air and soil pollution
- remediation and waste characterization
- environmental contaminants
- biogeochemistry and geoecology
- chemical reactions and processes
- chemical and biological transformations as well as physical transport of chemicals in the environment
- environmental modeling

A particular focus of the series lies in methodological advances in environmental analytical chemistry.

The Handbook of Environmental Chemistry is available both in print and online via http://link.springer.com/bookseries/698. Articles are published online as soon as they have been reviewed and approved for publication.

Meeting the needs of the scientific community, publication of volumes in subseries has been discontinued to achieve a broader scope for the series as a whole.

Disinfection By-Products in Water

Volume Editors: Chao Chen · Susan Andrews · Yuefeng Xie

With contributions by

L. M. F. Amiri · S. Andrews · H. Chen · B. Chen · C. Chen · A. Chow · A. T.-S. Chow · W. Chu · H. Du · P. Du · Z. Du · M. B. Emelko · M. Fan · E. H. Goslan · J. Guan · G. He · L. He · R. Hofmann · Q. Kong · J. Li · Y. Li · Y. Liu · W. Ma · R. Meng · J. Ni · Y. Pan · Z. Sheng · M. Stone · H. L. Tang · L. Wang · W. Wang · Y. Wang · D. Wu · Q. Wu · X. Xiao · Y. F. Xie · Y. Xie · B. Xu · M. Yang · X. Yang · K. Ye · Q. Zeng · D. Zhang · L. Zhang · M. Zhang · T.-Y. Zhang · M. Zhou

Springer

Editors
Chao Chen
Tsinghua University
Beijing, China

Susan Andrews
University of Toronto
Toronto, ON, Canada

Yuefeng Xie
Pennsylvania State University
Pennsylvania, USA

ISSN 1867-979X ISSN 1616-864X (electronic)
The Handbook of Environmental Chemistry
ISBN 978-3-031-74259-0 ISBN 978-3-031-74260-6 (eBook)
https://doi.org/10.1007/978-3-031-74260-6

This Springer imprint is published by the registered company Springer Nature Switzerland AG
The registered company address is: Gewerbestrasse 11, 6330 Cham, Switzerland

Preface

Disinfection of drinking water is a critical tool for maintaining public health. Disinfectants inactivate pathogens—microorganisms that cause human disease including bacteria, viruses, and protozoa effectively. Disinfection has saved millions of lives from waterborne epidemics since the nineteenth century. However, disinfectants also have their side effects. They can react with some organic matter or other precursors to yield disinfection byproducts (DBPs). Fifty years ago, Dr. Johannes Rook, a Dutch chemist, published a groundbreaking paper on the formation of trihalomethanes in chlorinated drinking water, effectively founding the field of DBPs. Since then, hundreds of DBPs have been identified in water treated with chlorine, chloramines, chlorine dioxide, ozone, etc.

In the United States, Canada, China, and many other countries, regulations have been established for several organic DBPs, such as trihalomethanes (THMs) and haloacetic acids (HAAs), as well as inorganic DBPs such as chlorite and bromate. Emerging DBPs, including nitrogenous, iodinated, aromatic, and heterocyclic compounds, have also been detected and studied. Recent research continues to explore the formation, detection, removal, and control of DBPs, while climate change and energy development introduce new complexities to the field.

This book represents the latest advancements and offers a comprehensive review of DBPs. It also commemorates the 50th anniversary of Dr. Rook's landmark paper from 1974. The book is divided into four parts: the first introduces DBPs, highlighting their significance and impact on water quality. It also covers detection methods, offering insights into cutting-edge techniques and emerging technologies, and explores the sources and characteristics of DBP precursors. Part II features contributions from experts who overview various DBPs, including halogenated aliphatic, iodinated, aromatic, and heterocyclic types, and discusses their properties and implications. The section also addresses nitrogenous DBPs, such as haloacetonitriles and nitrosamines, and inorganic DBPs. Part III examines the interplay between energy consumption, climate change, and DBP formation and considers DBPs in the context of water treatment plants, water reuse facilities, swimming pools, and daily life. The final section focuses on the health effects of

DBPs, with detailed analysis of their toxicity and associated health risks, and reviews current research on risk assessment and management strategies.

In an era marked by water scarcity and increasing demand for safe drinking water, the presence of DBPs remains a critical public health concern. This book serves as an invaluable resource for researchers, water professionals, public health officials, and policymakers seeking a thorough understanding of DBPs.

This book is the result of a collaboration among esteemed researchers from around the world invited by the three co-editors to share their expertise. All three editors have been intimately involved in DBP research through their entire academic career. As a PhD student in 2002–2005, Dr. Chen Chao investigated the control of THMs and HAAs by enhancing precursor removal and optimizing disinfection processes. Dr. Chen investigated the formation and control of N-nitrosodimethylamine (NDMA) after graduation. Dr. Chen also served as the Secretary and Chair of International Water Association (IWA) Specialist Group of Disinfection and co-organized the IWA Disinfection and Disinfection Byproduct conference in Beijing, China. As a PhD student in the early 1990s, Dr. Andrews investigated organic DBP formation in ozonated, chlorinated, and/or chloraminated water. She has investigated the formation, analysis, and control of many DBPs, especially on NDMA, aldehydes, organic acids, and other emerging DBPs after graduation. Her current efforts include evaluating a variety of disinfection technologies including using light (ultraviolet, solar) as an alternative to traditional chemical disinfectants (chlorine). As a PhD student in the late 1980s, Dr. Yuefeng Xie investigated the formation and removal of mutagenic activity and mutagen X (MX) during the water treatment process. He has investigated the formation, analysis, and control of various DBPs, especially on cyanogen halides, ketoacids, and trihaloaldehydes after graduation. Recently, he has become more involved in DBP regulations and DBP compliance technologies, especially for small water systems.

We extend our gratitude to the chapter authors for their invaluable contributions and to the Springer team, particularly Ms. Neelofar Yasmeen and Dr. Sofia Costa, for their assistance and patience throughout the publication process.

Beijing, China — Chao Chen
Toronto, ON, Canada — Susan Andrews
Middletown, PA, USA — Yuefeng Xie

Contents

Part I General Knowledge for DBPs

History of Disinfection Byproducts 3
Yuefeng Xie, Chao Chen, and Susan Andrews

Detection Methods for DBPs 31
Wenhai Chu, Di Zhang, Baiyang Chen, Wei Ma, Guiying He, Mengting Yang, Yi Li, and Susan Andrews

DBP Precursors 59
Penghui Du, Alex Tat-Shing Chow, and Chao Chen

Part II Individual DBPs

Halogenated Aliphatic DBPs 91
Emma H. Goslan

Iodinated DBPs 113
Tian-Yang Zhang and Bin Xu

Formation, Analysis, and Removal of Halogenated Cyclic DBPs 131
Ruixue Meng, Meijiao Zhou, Yuefeng Xie, and Wei Wang

Nitrogenous DBPs 169
Wenhai Chu, Di Zhang, and Zhenqi Du

Nitrosamines 195
Yu Wang, Chao Chen, and Susan Andrews

Inorganic Disinfection By-Products (DBPs) 231
Qingqing Kong, Mengge Fan, Jingmeng Guan, Liuqing Zhang, and Xin Yang

Part III DBPs in our Lives

DBPs in Drinking Water Treatment Plants . . . 257
Emma H. Goslan and Chao Chen

DBP Formation and Control in Water Reuse . . . 273
Qianyuan Wu, Liu He, Xiao Xiao, DeXiu Wu, and Ron Hofmann

Disinfection Byproducts in Daily Life . . . 303
Lei Wang, H. Larry Tang, and Baiyang Chen

DBPs in Food and Cooking . . . 333
Yan Liu, Jia Ni, Kai Ye, Guiying He, Zan Sheng, Yang Pan, and Mengting Yang

Climate Change Implications for DBP Occurrence in Municipal Drinking Water . . . 357
Monica B. Emelko, Alex Chow, Fariba Amiri, Huan Chen, and Micheal Stone

Disinfection Byproducts in the Energy Sector . . . 377
H. Larry Tang and Yuefeng Xie

Part IV Health effect of DBPs

Toxicity of DBPs . . . 395
Haiying Du and Jinhua Li

Health Risk of DBPs . . . 445
Min Zhang and Qiang Zeng

Future Perspectives . . . 473
Susan Andrews, Chao Chen, and Yuefeng Xie

Correction to: Climate Change Implications for DBP Occurrence in Municipal Drinking Water . . . C1
Monica B. Emelko, Alex Chow, Fariba Amiri, Huan Chen, and Micheal Stone

Part I
General Knowledge for DBPs

History of Disinfection Byproducts

Yuefeng Xie, Chao Chen, and Susan Andrews

Contents

1 Waterborne Disease and Water Chlorination 5
1.1 Waterborne Disease, Chloroform, and Dr. John Snow 5
1.2 Water Chlorination and Dr. John Leal 6
1.3 Disinfection and Chick's Law 7
2 Haloforms and Dr. Johannes J. Rook 8
2.1 Landmark Paper on Trihalomethanes 8
2.2 Trihalomethane Control and Other DBPs 10
3 Formation of Haloacetic Acids in Chlorinated Water 11
4 Emerging or Remerging Disinfection Byproducts 12
4.1 Chloral and Three Brominated Trihaloacetaldehydes 12
4.2 Haloacetonitriles, Chloropicrin, and Cyanogen Halides 12
4.3 Nitrosamines 13
4.4 MX 14
5 Ozonation Disinfection Byproducts 14
5.1 Aldehydes 14
5.2 Aldoacids and Ketoacids 15
5.3 Carboxylic Acids 15
6 Other Contaminants and Analytical Artifacts 16
6.1 Contaminants from Chlorine 16
6.2 Artifacts from Extraction Salt, Sodium Chloride 17
6.3 Artifacts from Extraction Solvents 17
7 Conferences and Books on Disinfection Byproducts 17
7.1 Water Chlorination 18

Y. Xie (✉)
Department of Civil, Construction and Environmental Engineering, The Pennsylvania State University, Middletown, PA, USA
e-mail: yxx4@psu.edu

C. Chen
School of Environment, Tsinghua University, Beijing, P. R. China
e-mail: chen_water@tsinghua.edu.cn

S. Andrews
Department of Civil and Mineral Engineering, University of Toronto, Toronto, ON, Canada
e-mail: susan.andrews@utoronto.ca

Chao Chen, Susan Andrews, and Yuefeng Xie (eds.), *Disinfection By-Products in Water*,
Hdb Env Chem (2024) 135: 3–30, DOI 10.1007/698_2024_1152,

Published online: 26 September 2024

7.2 American Chemical Symposium Series on Disinfection Byproducts 18
7.3 Gordon Research Conferences 18
7.4 IWA Conferences 19
7.5 Chinese DBP Conferences 19
7.6 Books on Disinfection Byproducts 19
8 Global DBP Regulations 20
8.1 General Information of Global DBP Regulations 20
8.2 The United States 20
8.3 WHO 23
8.4 Canada 24
8.5 China 25
8.6 European Union 25
9 Summary 26
References 26

Abstract This chapter briefly reviews the history of water chlorination, early disinfection byproduct (DBP) studies, and global DBP regulations. The evolution of research concerning several classes of DBPs, including trihalomethanes, haloacetic acids, emerging DBPs, ozonation DBPs, and their artifacts are summarized. Selected information associated with conferences and books related to DBPs are also summarized for the readers' reference. Also summarized are the maximum acceptable DBP concentrations reported by the World Health Organization (WHO) and in DBP regulations of the United States, the European Union, Canada, Japan, and China.

Keywords Chick Harriette, Disinfection byproducts (DBPs), Emerging DBPs, Haloacetic acids, History, Johannes Rook, John Leal, John Snow, Regulation, Trihalomethanes

It has been half a century since Dr. Johannes Rook's landmark paper on trihalomethane (THM) formation in chlorinated drinking water. Since then, other disinfection byproducts (DBPs), including haloacetic acids (HAAs), nitrogenous DBPs, chloral hydrate, and 3-chloro-4-(dichloromethyl)-5-hydroxy-2(5H)-furanone (MX) have also been identified in chlorinated and/or chloraminated water by other researchers. In addition, non-halogenated DBPs, including aldehydes, ketoacids, and carboxylic acids, have been identified as byproducts of treatment processes such as ozonation. This chapter briefly reviews the history of water chlorination, early DBP research, and DBP regulations.

1 Waterborne Disease and Water Chlorination

1.1 Waterborne Disease, Chloroform, and Dr. John Snow

When modern science and technology on epidemiology and water supply were not developed in the nineteenth century, large cities commonly discharged their wastewater into local sewers without additional treatment and disinfection. Many of the water bodies that received sewage also served as drinking water sources, and water suppliers generally piped water from those drinking water sources to residential neighborhoods without treatment and disinfection. Unfortunately, many waterborne diseases, including cholera and typhoid fever, were often transmitted by drinking water. However, it was difficult to connect the sewers, drinking water, and disease because of the prominent miasma theory that mistakenly associated the outbreak of many of these diseases with bad air emanating from rotting organic matter [1].

It was Dr. John Snow who correctly linked contaminated drinking water sources with cholera, a waterborne disease. Dr. Snow was an English physician who practiced in the Soho District of London. Today Dr. Snow is well recognized as the pioneer of disease mapping and the founding father of epidemiology. He was also a pioneering anesthesiologist.

In 1854, a cholera epidemic broke out in the Soho district, London, and killed 600 people in a 10-day period. With firsthand cholera death information, Dr. Snow mapped out the locations of over 10 public water pumps (wells) and the residential locations of deaths from cholera in the Soho District. Based on visual inspections of the patterns on this cholera death map, Dr. Snow correctly linked these deaths to the drinking water pump on Broad Street. He also considered the much lower cholera death figures from a nearby workhouse and a brewery to further establish this exclusive linkage [1]. The workhouse, with 535 people and surrounded by houses in which deaths from cholera occurred, only registered five deaths. This death rate was 20 times lower than for adjacent residential houses. The brewery, with 75 people, had no deaths from cholera. Both places had their own wells and never used the drinking water pump on Broad Street. These results also refuted the miasma theory.

With this information, Dr. Snow convinced the authorities to remove the handle of the Broad Street pump. The cholera cases decreased in the neighborhood following the removal of the pump handle. It was also reported that Dr. Snow used chlorine (chloride of lime) for water disinfection at the Broad Street pump. However, there was no evidence to support this [2].

During the next cholera outbreak in London in 1854, Dr. Snow conducted a "Grand Experiment" in neighborhoods that received water from two suppliers, the Lambeth Company and the Southwark and Vauxhall Water Company [1]. Both companies draw their water from the polluted Thames River; however, the Lambeth Company had its water intake upstream at a less polluted part of the river. Dr. Snow tabulated the cholera deaths and calculated cholera deaths per 10,000 houses for seven weeks for both areas. Neighborhoods that received water from the Southwark and Vauxhall Water Company, with its water source heavily polluted with sewage, had 319 cholera deaths per 10,000 houses. Neighborhoods that received water from the Lambeth Company, with less polluted water, had only 37 cholera deaths per 10,000 houses. These results supported Dr. Snow's hypothesis that the source of the cholera epidemic was contaminated water.

Before these pioneering epidemiologic studies, Dr. Snow was an established anesthetist. He published a practical guide, "*On the Inhalation of Ether*" in 1847. When chloroform was introduced as a much more potent anesthetic, Dr. Snow investigated a death due to anesthesia with chloroform [3]. He concluded that the uncontrolled administration of chloroform was to blame. He proposed and published the use of scientifically engineered vaporizers for chloroform administration. In 1853 Dr. Snow administered chloroform to Queen Victoria to assist with the delivery of her daughter. A book *On Chloroform and Other Anaesthetics: Their Action and Administration* written by Dr. Snow was published shortly after he died in 1858. This book is regarded as one of the first textbooks in the field of anesthesiology.

In 1954, the 100th anniversary of Dr. Snow's cholera epidemic study with the Broad Street pump, a drinking house, John Snow pub, was established in his honor. The irony is that Dr. Snow signed a pledge promising to abstain from alcohol in 1835. Dr. Snow was also a vegetarian and drank only distilled water. Lastly, Dr. Snow surely did not know that chloroform, the anesthetic he promoted extensively, is a water chlorination byproduct and is now regulated in drinking water as a contaminant.

1.2 Water Chlorination and Dr. John Leal

John Leal was an American physician, public health officer, and water treatment expert. In addition to his career as a physician, he was also active in public health protection. His passion for public health protection is thought to have been influenced by his father's death from chronic dysentery caused by the contaminated water [2].

It had become well known that Dr. Snow had the handle of the Broad Street pump removed during London's cholera outbreak. In North America, Dr. Leal was also a furious fighter for public health protection, and as a member of the Paterson Board of Health he convinced the city to remove the handles from the pumps of four shallow wells contaminated by sewage [2]. After these pump handles were reinstalled under political pressure, Dr. Leal removed these handles again in a direct exercise of his authority on the Board of Health.

While it is not entirely clear how the decision to use chlorine for drinking water disinfection came about, perhaps in response to reports surrounding Dr. Snow's experiences, the first continuous use of chlorine in a water supply started in Middelkerke, Belgium, in 1902 [2]. Chlorine, as calcium hypochlorite, was added to the influent of a rapid sand filter with a treatment capacity of 100,000 gal per day (378 m^3/d). Around the same time, chlorine was also used to disinfect the slow sand filter influent in Lincoln, England, in 1905.

In North America, between 1905 and 1908, there was a court battle concerning the possible contamination of the Boonton Water Supply System between Jersey City, New Jersey, USA, and its water supply contractor, the Jersey City Water Supply Company (JCWSC) [2]. On June 4, 1908, the judge gave JCWSC 90 days to "present other plans or devices for maintaining the purity of the water delivered" or lose a big part of a US$7.6 million construction payment. As the sanitary advisor to JCWSC, Dr. Leal implemented a chlorination process with calcium hypochlorite at Boonton on September 16, 1908, designed by George Fuller, a sanitary engineer [2]. The capacity of the chlorination system was 40,000,000 gal per day (about 151,000 m^3/day). Since then, the adoption of chlorination for water supplies has become an essential drinking water treatment process worldwide.

1.3 Disinfection and Chick's Law

Dr. Harriette Chick received her doctorate degree in bacteriology in 1904 for her research on microorganisms in polluted water. In 1905, Dr. Chick became the first woman to obtain a position at the Lister Institute of Preventive Medicine in London. As a research assistant, Dr. Chick first investigated the relative biological potency of several disinfectants. In 1908, she published a paper entitled "An Investigation of the Laws of Disinfection" [4], which discussed the relationship between the inactivation efficiency of organisms and contact time with a disinfectant, later named as "the Chick's Law."

This fundamental study set up the most basic principle for various disinfection practices on various microorganisms. Dr. Chick discovered a very complete analogy between a chemical reaction and the disinfection process according to her experiments with two microorganisms (*B. paratyphosus* and spores of *B. anthracis*) and disinfectants including metallic salts ($HgCl_2$ and $AgNO_3$), phenol and their emulsified disinfectants. She noticed that the disinfection kinetics followed a pseudo first order model if the disinfectants were employed at sufficiently high concentrations. The resulting Chick's Law was often presented using Eqs. 1 and 2 as follows:

$$-\frac{dn}{dt} = Kn \text{ or on integration } K = \frac{1}{t2 - t1} \log \frac{n1}{n2} \quad (1)$$

where n is the number of surviving bacteria, t is the inactivation time and K is a constant, and

$$\frac{1}{C_0 - C_t} \log \frac{Ct}{C_0 t_0} = \text{constant} \quad (2)$$

where C is the concentration of disinfectant, t is the inactivation time.

Chick noticed that the disinfection experiments with *B. paratyphosus* did not fit Eq. 1 very well. She attributed that to the differences in resistance to disinfectants among individual organisms. She also gave values of 1.0, 1.5, and 2.0 to the parameter n to better fit the results. She also noticed that temperature influenced disinfection kinetics in a reproducible manner and applied the Arrhenius Equation to describe its influence.

Eight months later, Herbert E. Watson modified the equations to the form shown below, which helped with the understanding of the disinfectant principle [5].

$$n \log C + \log t = \text{constant} \tag{3}$$

The new Eq. 3 also established the *Ct* value ($C^n t$, $n = 1$ generally) as the basic parameter to describe the inactivation capability of specific disinfectants on specific microorganisms under certain conditions. This principle of disinfection is also known as the Chick-Watson's Law.

2 Haloforms and Dr. Johannes J. Rook

Dr. Johannes Rook received his PhD in biochemistry in 1949. Before his tenure at the Rotterdam Water Works between 1963 and 1984 he worked several jobs, including one at the Amstel Brewery. At the time, water sample preparation and concentration prior to gas chromatography analysis were done by two methods [6]. The first method involved adsorption onto granular activated carbon and desorption by chloroform and the second method involved liquid-liquid extraction followed by evaporation. Both methods had problems when used to analyze volatile organic compounds (e.g., chloroform) in water. Built on his experience in the brewery industry, likely on taste and odor analysis, Dr. Rook developed a new water sample concentration technique, headspace analysis, to analyze volatile organic contaminants in treated water. When applying this method for water quality control at the Berenplaat Water Treatment Plant, Spijkenisse, Netherlands, Dr. Rook observed four new peaks which were identified later as chloroform or trichloromethane (TCM), bromodichloromethane (BDCM), chlorodibromomethane (CDBM), and bromoform or tribromomethane (TBM).

2.1 *Landmark Paper on Trihalomethanes*

Dr. Rook published his landmark paper "Formation of Halomethanes during Chlorination of Natural Waters" in 1974 [7]. Even though this journal, *Water Treatment and Examination*, ceased publication shortly thereafter, researchers have well cited this paper since it was published. Citing this paper as the first research on chloroform detection in chlorinated drinking water, however, does not give a full justification for this landmark paper. This paper is actually a comprehensive investigation on DBP formation in chlorinated drinking water and correctly identifies humic substances as the DBP precursors.

While the identification of THMs was not the focus of Dr. Rook's 1974 paper, he did summarize his discovery of THMs in the introduction of this paper. Using a newly developed headspace gas chromatography method [7], the biggest of four new peaks was identified as TCM by comparing gas chromatography (GC) retention times using three GC columns with different stationary phases. The identification of other peaks, BDCM, CDBM, and TBM, was made by mass spectrometry.

Dr. Rook first investigated the potential of THM contamination from chlorine gas. By dissolving chlorine gas into reagent water at various dosages, Dr. Rook found that although chlorine gas contained tiny impurities of THMs, their peaks were over 100 times smaller than those found in chlorinated water. The impurity of bromine in gas chlorine was also investigated. He concluded that chlorine gas was not a significant source of chloroform or brominated THMs in chlorinated water. Carbon tetrachloride was also identified as an impurity in gaseous chlorine; however, it was found not to be a DBP in chlorinated water.

Dr. Rook also investigated the potential for acetone to be a haloform precursor in river water. Although the chlorination of acetone can produce THMs, an excessively high and improbable level of acetone, 2–9 mg/L, needs to be present in water to form the THMs detected in chlorinated water.

The Berenplaat Water Treatment Plant had two raw water sources, the Rhine River and the Meuse River. Either one river or a blend of the two rivers was used seasonally. Dr. Rook reported the THMs formation for 11 months between 1973 and 1974, along with several water quality parameters, including chlorine dose, temperature, and levels of inorganic bromide, ammonia, pH, and color. He observed a good correlation between chloroform formation and the color of water. He then studied the formation of THMs in chlorinated peat extract and natural water from a peaty region. The detection of chloroform at 200 μg/L in chlorinated non-polluted natural water supported Dr. Rook's hypothesis that natural humic substances serve as precursors to THMs.

Dr. Rook also studied the effect of bromide on the formation of brominated THMs and noticed the incorporation of bromine into THMs was higher than chlorine. In this 1974 paper, Dr. Rook concluded that "There is strong evidence that haloforms are produced by chlorination of humic substances in natural waters."

At approximately the same time in North America, the formation of THMs in chlorinated drinking water was also detected, studied, and reported by Bellar et al. [8]. With a newly developed purge-and-trap gas chromatography method, they detected chloroform, BDCM, and CDBM in chlorinated drinking water at the Cincinnati Water Treatment Plant, Cincinnati, USA. Mass spectrometry was used to verify these compounds. They also investigated the effects of treatment processes, chlorination time, chlorine addition, and source water on these THMs and concluded that increasing chlorination time and adding higher chlorine doses resulted in a substantial increase in THM concentrations. Well water samples with low organic content had much lower THM levels than surface water with high organic content. However, Bellar et al. would be shown later to have incorrectly attributed ethyl alcohol, detected in finished water, as the possible THM precursor [8]. They also incorrectly attributed the bromine impurity in chlorine gas as the main precursor of brominated THMs.

Chloroform was also detected in chlorinated drinking water by other researchers using headspace or carbon adsorption method sampling methods. However, these studies did not fully investigate the sources of THMs and assumed that the THMs were simply contaminants in the source water [6].

2.2 Trihalomethane Control and Other DBPs

In addition to the landmark paper in 1974, Dr. Rook published an equally important paper on THM control in the *Journal of American Water Works Association* [9]. This paper reported on his study of the THM precursors, formation, and control. He affirmed that the main precursor for THMs in chlorinated water is the group of natural, colored, organic materials known as humic substances. He reported that increasing total organic carbon (TOC), chlorine dose, pH, and chlorination time all increased the formation of THMs. More importantly, Dr. Rook reported a substantial reduction of THM formation potential (>50% for chloroform) with increased coagulant doses. This higher dose application of coagulation technology was later adopted as Enhanced Coagulation to control THM formation under the United States Environmental Protection Agency (USEPA) Stage 1 Disinfectants and Disinfection By-products Rule [10]. He also investigated THM removal by other treatment processes and reported a complete removal of chloroform by granular activated carbon for several weeks and a 50% removal of chloroform by aeration. Activated carbon adsorption has been adopted in many point-of-use water treatment devices and aeration has also been adopted by water systems in recent years as a localized treatment for THM control in water distribution systems [11].

Another important paper by Dr. Rook was published in *Environmental Science and Technology* [12]. This paper reported that the most reactive sites in organic precursors for THM formation are meta-dihydroxylated benzene moieties. The degradation pathways of fulvic acid and resorcinol were proposed as well. Dr. Rook also identified 20 chlorinated compounds in heavily chlorinated water from the Rotterdam Storage Reservoir by mass spectrometry. One noticeable compound was dichloroacetic acid. However, Dr. Rook believed that the formation of these nonvolatile compounds typically was below μg/L level in drinking water chlorinated at doses of 1–10 mg/L and so was not significant. This misconception was likely due to the lack of analytical methods for these organic acids at that time.

Additional information concerning THMs can be found in the chapter "Halogenated Aliphatic DBPs."

3 Formation of Haloacetic Acids in Chlorinated Water

The investigation of nonvolatile organic DBPs lagged a few years behind THMs due to the difficulty in analyzing organic acids by gas chromatography. However, using chemical derivatization with diazomethane, trichloroacetic acid (TCAA) was identified by Dr. Uden's group at the University of Massachusetts, Amherst, Massachusetts [13]. With the same analytical method, they detected high levels of dichloroacetic acid (DCAA) and TCAA in both chlorinated humic acid solution and in chlorinated drinking water [14, 15]. DCAA and TCAA were detected at 79 μg/L and 73 μg/L, and 133 μg/L and 160 μg/L, respectively, in two chlorinated drinking waters. These levels were comparable to the levels of chloroform (87 and 190 μg/L) in these waters. Dr. Uden's group also investigated the effects of chlorination time, chlorine dose, and pH on these two HAAs and chloroform in chlorinated fulvic acid solutions [14]. Their results indicated that the effects of chlorination time and chlorine dose on the formation of chloroform and DCAA and TCAA are similar. However, while increasing pH increased the formation of chloroform, the formation of DCAA and TCAA was reduced. Although the group credited the microwave-induced-and-sustained helium plasma as the new analytical technique for HAA detection, the chemical derivatization with diazomethane to convert acids to methyl esters they developed was the true breakthrough.

Peters et al. [16] developed an HAA analytical method using chemical derivatization and gas chromatography with a more common electron capture detection technique and analyzed water samples from 20 Dutch drinking water plants. They also synthesized three mixed chloro/bromo acetic acids: bromochloroacetic acid, chlorodibromoacetic acid, and bromodichloroacetic acid. In addition to DCAA and TCAA, six brominated and mixed chloro-bromoacetic acids were also detected in Dutch drinking water, including monobromoacetic acid, dibromoacetic acid, and tribromoacetic acid, bromochloroacetic acid, chlorodibromoacetic acid, and bromodichloroacetic acid. Monochloroacetic acid was not analyzed because it co-eluted with the solvent and THM peaks. All these HAAs are referred to as HAA9 today. The detection of these mixed bromochloroacetic acids was delayed by the lack of chemical standards and their methyl ester mass spectra at the time. The formation of these mixed bromochloroacetic acids in chlorinated drinking water was predicted by Ireland et al. [17] several years earlier. However, inconsistencies in their mass spectra may have hampered their efforts in detecting these compounds in drinking water [18].

The fact that HAAs could be formed at levels comparable to chloroform, and that they escaped notice in early studies, was largely due to the difficulties in HAA measurement by gas chromatography. Chemical derivatization with diazomethane to convert organic acids to their methyl esters was a critical sample preparation method for HAA analysis. However, diazomethane needs to be freshly prepared from a precursor compound, most often the potent carcinogen 1-methyl-3-nitro-1-nitrosoguanidine (MNNG). Because of the hazardous nature of diazomethane and the toxicity of its precursor compound, there was some concern about using these chemicals for HAA analysis, especially in water utility laboratories. A more sustainable analytical method, liquid-liquid microextraction-acidic methanol derivatization as developed by Xie et al. [19], has become a method of choice for many laboratories.

Additional information concerning HAAs can be found in the chapter "Halogenated Aliphatic DBPs."

4 Emerging or Remerging Disinfection Byproducts

In addition to THMs and HAAs, other DBPs have also been identified and quantified in chlorinated drinking water. Due to their low occurrence levels, many of these compounds initially did not attract sustained research interest. However, interest in some of these DBPs was regained because of their toxicity. These DBPs have often been labeled as "emerging" DBPs. However, for many researchers who participated in the early studies on these DBPs, "re-emerging" may be a better description.

4.1 Chloral and Three Brominated Trihaloacetaldehydes

Chloral, or trichloroacetaldehyde, was first reported in chlorinated drinking water by Uden and Miller [15]. Because it readily forms trichloroacetaldehyde hydrate, the name chloral hydrate is commonly used. Xie and Reckhow [20] detected and identified three brominated trihaloacetaldehydes, including bromochloroacetaldehyde, chlorodibromoacetaldehyde, and tribromoacetaldehyde, in ozonated and chlorinated fulvic acid solutions containing inorganic bromide. Their study on chloral hydrate also indicated that preozonation increased its formation potential and GAC adsorption and biofiltration reduced its formation potential [21]. This indicated that some ozonation byproducts (e.g., acetaldehyde) might serve as the precursors of chloral hydrate.

4.2 Haloacetonitriles, Chloropicrin, and Cyanogen Halides

McKinney et al. [22] reported the detection of dichloroacetonitrile in two drinking water samples but did not investigate its origin. Trehy and Bieber [23] detected and identified three dihaloacetonitriles, i.e., dichloroacetonitrile, bromochloroacetonitrile, and dibromoacetonitrile, in chlorinated groundwater and surface water samples and confirmed that dihaloacetonitriles were produced by the chlorination process.

Trehy and Bieber [23] also reported that dihaloacetonitriles were not found in water supplies that had been lime-softened. Further studies indicated that hydrolysis of dihaloacetontriles at neutral and basic pHs leads to the formation of dihaloacetic acids. This also explains some of the difficulties involved in analyzing dihaloacetonitriles, i.e., that they can decompose during storage and analysis.

Sayato et al. [24] identified chloropicrin in chlorinated humic acid solutions in the presence of nitrite. They also investigated formation mechanisms for chloropicrin by studying the chlorination of alanine and resorcinol. Merlet et al. [25] reported the formation of chloropicrin in drinking water. Thibaud in the same group [26] also identified brominated trihalonitromethanes during the chlorination of nitromethane and of 3-nitrophenol in the presence of bromide. They also reported a significant decrease in chloropicrin formation in chlorinated fulvic acids and fractions of dissolved organic matter present in two surface waters with increasing bromide concentrations.

The formation of cyanogen chloride by chlorination was extensively investigated by Ohya and Kanno [27]. They detected and proposed the formation of cyanogen chloride in chlorinated humic acid solutions and chlorinated amino acid solutions in the presence of ammonium ion [27, 28]. Cyanogen chloride is commonly detected in chloraminated water. It is also a highly toxic chemical that is used as a chemical weapon agent.

Additional information concerning nitrogen-containing DBPs can be found in the chapter "Nitrogenous DBPs."

4.3 Nitrosamines

Nitrosamines are a family of about 200 species with the general formula of $R_1N(R_2)$-N=O. They are widely detected in food, tobacco, water, wastewater, sludge, soil, air, and some reagents in certain industries [29]. Researchers have shown that over 90% of nitrosamines and nitrosamides, another family of N-nitroso chemicals with about 100 species, exhibit some level of toxicity, especially strong carcinogenicity. The most common nitrosamine, N,N-nitrosodimethylamine (NDMA), was initially confirmed as a carcinogen in the dry-cleaning industry by Magee and Barnes in 1956 [30]. Other widely detected volatile nitrosamines include N-nitrosodiethylamine (NDEA), N-nitrosodi-n-propylamine (NDPA), N-nitrosopiperidine (NPIP), N-nitrosopyrrolidine (NPYR) and N-nitrosomorpholine (NMOR). NDMA and other nitrosamines have been widely detected in food as processing and preservation byproducts [31], as contaminants in some ground waters in California polluted by rocket fuel in the late 1980s [32, 33], and identified as DBPs in disinfected drinking water since 1994 in Canada [34] and later in the U.S., China, and all over the world [35, 36].

Nitrosamine DBPs are formed by the reactions of several types of disinfectants with appropriate nitrosamine precursors. These precursors generally have a dialkylamine functional group which is the target of attack of the disinfectants or oxidants. Dimethylamine was the first, simplest, and most widely studied NDMA precursor [37, 38]. Chloramine disinfectants produce the highest yields of NDMA and other nitrosamines in the water industry. However, free chlorine, chlorine dioxide, and ozone can also react with certain precursors to form NDMA and other nitrosamines. For example, ozone can react with one fungicide, i.e., N, N-dimethylsulfamide to form NDMA [39].

Additional information concerning nitrosamines can be found in the chapter "Nitrosamines."

4.4 MX

In the early 1980s, the mutagenic activity of chlorinated water was also investigated using the Ames Test, a biological assay named after its inventor, Dr. Bruce Ames. MX, 3-chloro-4-(dichloromethyl)-5-hydroxy-2(5H)-furanone, is a nonvolatile chlorinated organic compound initially identified in chlorinated pulp-bleaching waters [40, 41]. MX, or Mutagen X, was actually the code name for a compound with high mutagenicity but at that time not chemically identified. A geometric isomer of MX was also identified, (E)-2-chloro-3-(dichloromethyl)-4-oxobutenoic acid, or simply E-MX.

Extensive studies on MX and E-MX in drinking water were conducted by Kronberg and Holmbom [42, 43]. Their studies indicated that although the concentrations of MX were low, at 26, 15, and 67 ng/L in three drinking waters, MX contributed 19%, 33%, and 46% of mutagenic activities, respectively. Similar levels of MX were also reported in Canada and China [44, 45]. Their studies also indicated that granular activated carbon is effective in MX removal.

Because of the high percentage of contribution of MX to mutagenic activity in chlorinated water, research interest in MX re-emerged several times over the past 30 years. In recent years, other toxicity tests such as microplate CHO cell cytotoxicity and genotoxicity assays have been extensively used to evaluate the toxicity of DBPs. Plewa et al. [46] reported that bromoacetic acid was 31 times more cytotoxic and 14 times more genotoxic in CHO cells than MX. Many other nitrogenous or iodinated DBPs also showed much higher cytotoxicity and/or genotoxicity than MX.

Additional information concerning MX can be found in the chapter "Halogenated Aromatic & Heterocyclic DBPs."

5 Ozonation Disinfection Byproducts

5.1 Aldehydes

Yamada and Somiya [47] developed an analytical method for low molecular weight aldehydes. This method uses O-(2,3,4,5,6-pentafluorobenzyl) hydroxylamine (PFBOA) to derivatize aldehydes into volatile derivatives that are sensitive to electron capture detection (ECD). In an ozonated surface water, they detected formaldehyde, glyoxal, and methylglyoxal at levels up to 38, 1.8, and 1.7 μg/L, respectively. They also detected higher levels of these aldehydes at pH 10.4 than at pH 6.6. Their study indicated that these three aldehydes can be effectively removed by granular activated carbon in an ozone demonstration plant. This is likely due to the biological activity on the carbon surface.

Glaze et al. [48] improved the PFBOA derivatization method and investigated the formation of low molecular weight aldehydes in three full-scale drinking water treatment plants and a pilot plant. Their results showed a much higher formation of formaldehyde, glyoxal, and methylglyoxal and a proportional relationship between the formation of formaldehyde and the influent TOC.

In addition to potential health risks, these low molecular weight aldehydes could also cause biological regrowth in drinking water distribution systems. Acetaldehyde is also a precursor of trihaloacetaldehydes, including chloral hydrate, and THMs.

5.2 *Aldoacids and Ketoacids*

Xiong et al. [49] identified an aldoacid, glyoxylic acid, at 130–170 μg/L in an ozonated fulvic solution using a double derivatization method, PFBHA as a derivatizing agent for the carbonyl function in aqueous solution, and diazomethane as a methylating agent for the carboxyl function in organic solution. Using a similar procedure, Xie and Reckhow [50] identified two additional ketoacids, pyruvic acid and ketomalonic acid, as well as glyoxylic acid in ozonated fulvic acid solutions and ozonated drinking water. The combined concentration of these three ketoacids, 208 μg/L, was almost 8 times higher than that of four major aldehydes, 27 μg/L. Increasing ozone dosage increased the formation of ketoacids. Further study indicated that ketoacids formed during preozonation can be effectively removed by biologically active carbon.

Gloxylic acid is a known mutagen. Pyruvic acid is a known THM and HAA precursor. There is also a strong correlation between ketoacids and assimilable organic carbon (AOC). High levels of ketoacids could result in high potential of microbial regrowth in distribution systems.

5.3 *Carboxylic Acids*

The likely endpoint of strongly oxidative processes like ozonation is the formation of carboxylic acids, stopping short of complete mineralization of the oxidation of organic matter to carbon dioxide. Carboxylic acids are generally considered to be even less toxic than their aldehyde or ketone counterparts, which was good news initially. However, much like their aldo/keto counterparts, it has been found that they also contribute organic matter for the regrowth of microorganisms that escape disinfection. As such, knowledge of their presence and quantity has become of interest, and means to measure some of them directly as well as measure the potential for regrowth indirectly have been developed.

An excellent review on the occurrence, analysis, and toxicity of carboxylic acid DBPs was prepared by Pulicharla et al. [51]. It summarizes the formation of carboxylic acids in several water sources that have been treated and/or disinfected with ozone. Operational parameters such as ozone dose, temperature, and reaction time are discussed. Also summarized are a variety of analytical methods available for detecting and measuring them. Several methods are needed for this class of DBPs because they have a wide range of molecular weights and polarities, making it very difficult to develop a single accurate analytical method for quantification. Possibilities for their toxicity in drinking water are also explored. While carboxylic acids are relatively non-toxic themselves, they can contribute to the overall toxicity of drinking water in other ways. For example, similar to the aldehydes, aldoacids, and ketoacids mentioned in previous sections, they can promote microbial activity in treated and distributed water. Those organisms that regrow may produce toxic microbial products. Finally, some carboxylic acids can also be precursors to halogenated DBPs such as THMs and HAAs if chlorine is also present.

6 Other Contaminants and Analytical Artifacts

During the times when new DBPs were being identified, a few contaminants and analytical artifacts were also incorrectly identified as new DBPs. These included contaminants in chlorine-dosing chemicals and artifacts from extraction salts and extraction solvents.

6.1 Contaminants from Chlorine

Carbon tetrachloride is often detected in chlorinated water and is sometimes reported as a disinfection byproduct. As reported by Dr. Rook [7] in his landmark paper, carbon tetrachloride is a contaminant that can be introduced by gas chlorine. Water samples could also be contaminated with carbon tetrachloride through industrial discharges.

Due to the risks involved with liquified chlorine gas (often simply referred to as gaseous chlorine), many water utilities switched from liquified chlorine gas to hypochlorite solutions. Perchlorate, chlorate, and bromate, although true inorganic DBPs, are often detected in hypochlorite solutions. The levels of perchlorate, chlorate, and bromate can be minimized with proper manufacture, storage conditions, and handling of hypochlorite solutions.

More information concerning these inorganic DBPs can be found in the chapter "Inorganic DBPs."

6.2 Artifacts from Extraction Salt, Sodium Chloride

For DBP analysis, it is not unusual for high concentrations of simple inorganic salts to be added to samples to encourage organic DBPs to be more fully extractable into organic solvents for subsequent analysis (generally known as the "salting out effect"). Sodium chloride was often used as an extraction salt in the early days because it is generally inert, dissolves completely, and does not solidify during the extraction process. Sodium chloride in reagent grade often contains 0.01% of bromide. Xie and Reckhow [52] reported that using sodium chloride during extraction could add 30 mg/L of bromide to water samples and result in the formation of brominated DBPs in unquenched or incompletely quenched samples. When sodium chloride is used as the extraction salt, many studies may have reported much higher than expected concentrations of brominated DBPs in water that is low in bromide. Many of these brominated DBPs are artifacts formed during preparations for sample extraction.

6.3 Artifacts from Extraction Solvents

Liquid-liquid extraction with organic solvents is the most common separation technique for the analysis of DBPs in water, and some of these solvents can be contaminated with the DBPs being measured. For example, chloroform is a common contaminant in methyl tert-butyl ether (MTBE).

Unique artifacts can also form in these analytical solvents when exposed to free chlorine, chloramines, and free bromine. Xie and Reckhow [53] reported the formation of two bromohydrins in brominated MTBE or brominated pentane, and three chlorinated acetals in chloraminated diethyl ether. These phenomena have caused a certain amount of re-evaluation of the results of some studies. For example, 3-bromo-2-methyl-2-butanol and other bromohydrins were reported as novel disinfection byproducts in ozonated water high in bromide [54], and three chlorinated acetals were also reported as novel chloramination byproducts [55]. This shows the importance of proper quenching of disinfectant/oxidant residuals and proper preparation of natural organic matter free blanks during DBP analysis. Although effective ozone residual quenching minimized the formation of these bromohydrins as reported, it was mistakenly assumed as the degradation reactions between bromohydrins and quenching reagents [56].

7 Conferences and Books on Disinfection Byproducts

The formation of THMs in chlorinated drinking water caught many researchers' attention. The inaugural Water Chlorination conference was held at Oak Ridge National Laboratory, Oak Ridge, Tennessee, USA, in 1975, shortly after the discovery of THM formation in chlorinated water. After the last water chlorination conference in 1987, several conference series, including the American Chemical Society (ACS) symposia, Gordon research conferences, and IWA conferences have been held regularly.

7.1 Water Chlorination

Robert Jolly and Carl Gehrs at Oak Ridge National Laboratory conceived the idea of organizing a water chlorination conference to facilitate information exchange and gain a better understanding of the environmental impact and health effects of water chlorination [57]. They invited researchers in the field and held the conference in October 1975. The three-day conference attracted nearly 300 registrants. After the inaugural Water Chlorination conference, five additional Water Chlorination conferences were held in various US locations in 1977, 1979, 1981, 1984, and 1987. These conferences also resulted in a book series, Water Chlorination, which were proceedings from these conferences [57–61]. They served as an important information source for many researchers in the early days of DBP research.

7.2 American Chemical Symposium Series on Disinfection Byproducts

An ACS conference on Disinfection Byproducts was organized by Dr. Roger Minear and Gary Amy and held in Chicago, USA, in 1993 to continue the tradition of the Water Chlorination conferences [62]. The second ACS symposium was organized by Sylvia Barrett, Stuart Krasner, and Gary Amy and held in 1999 [63]. The third and fourth symposia were organized by Tanju Karanfil, Stuart Krasner, William Mitch, Paul Westerhoff, and Yuefeng Xie and held in 2007 and 2014 [64, 65]. These ACS symposia also resulted in four disinfection byproduct books, proceedings from these symposia.

7.3 Gordon Research Conferences

The inaugural Drinking Water Disinfection By-Products Gordon Research Conference was organized by Dr. Susan Richardson and Dr. Mark Nieuwenhuijsen and held at Mount Holyoke College, Massachusetts, USA in 2006. Six more Gordon Research Conferences were chaired by Benjamin Blount, Manolis Kogevinas, Michael J Plewa, William A Mitch, Xingfang Li, and Urs von Gunten and held in 2009, 2012, 2015, 2017, 2019, and 2023, respectively. In addition to DBP formation and control, these Gordon Research Conferences were more focused on DBP exposure, toxicity, and epidemiology. Because of the Gordon Research Conferences' "No Publication Policy," no books or proceedings have been published from these conferences.

7.4 IWA Conferences

The International Water Association (IWA) Disinfection Specialist Group also organized three Disinfection and Disinfection Byproduct conferences, including one held in Mexico (chaired by Blanca Jimenez-Cisneros), one in China (chaired by Wenjun Liu and Chao Chen), and one in Italy (chaired by Manuela Antonelli and Andrea Turolla), in 2012, 2018, and 2022, respectively. The fourth conference will be held in Spain, chaired by José Antonio Sánchez Pérez and Inmaculada Polo López, in 2024.

7.5 Chinese DBP Conferences

DBP researchers in China also initiated the Water Disinfection and Disinfection Byproduct National Conference series. The inaugural Water Disinfection and Disinfection Byproduct National Conference of China was held in 2013 at Tongji University, Shanghai, China. Eight more conferences were held at Tsinghua University, 2015; Tongji University, 2016; Harbin Institute of Technology, 2017; Tsinghua University, 2018; Xiamen University, 2019; Xi'an University of Architecture Science and Technology, 2020; Sun-Yat-Sen University, 2022; Research Center for Eco-Environmental Sciences, 2023; respectively. The tenth conference will be held at Nanjing University in 2024.

No books or proceedings were published for these IWA and Chinese national conferences.

7.6 Books on Disinfection Byproducts

In addition to the edited books above, there are several other edited books based on conference proceedings. These are:

I. Formation and Control of Disinfection By-Products in Drinking Water, edited by Singer [66].
II. Disinfection By-Products in Drinking Water: Current Issues, edited by Fielding and Farrimond [67].

III. Disinfection By-Products: Relevance to Human Health, edited by Hrudey and Charrois [68].
IV. Disinfection By-Products in Drinking Water, edited by Thompson, Gillespie, and Goslan [69].
V. Disinfection By-Products in Drinking Water: Detection and Treatment, edited by Prasad [70].

Notably, there is also one individually written book, *Disinfection Byproducts in Drinking Water: Formation, Analysis, and Control* by Yuefeng Xie [71].

8 Global DBP Regulations

8.1 General Information of Global DBP Regulations

Many countries which have established drinking water standards regulate at least one DBP. The drinking water standards for DBP in different countries and regions generally follow those in the Drinking Water Standards and Health Advisories of the United States and those in the Guidelines for Drinking Water Quality of the World Health Organization (WHO). Many of the basic DBP regulations are summarized in Table 1.

The regulation process has shown a very prominent effect on reducing DBP concentrations in drinking water. One study estimated that global mean THM concentrations in drinking water were significantly decreased from 78 μg/L (1973–1983) to 52 μg/L (1984–2004) and 22 μg/L (after 2004) [72].

8.2 The United States

The United States has been actively regulating disinfection byproducts in drinking water since the discovery of THM formation in chlorinated water. The USEPA set up four DBP regulations. They are the Total Trihalomethane Rule, the Information Collection Rule, and the Stage 1 and Stage 2 Disinfectants and Disinfection By-products Rules.

In 1979, five years after Dr. Rook's landmark paper on THM formation in chlorinated drinking water, the USEPA [73] promulgated the Total Trihalomethane (TTHM) Rule in drinking water. This TTHM rule was an interim regulation and applied to large community water systems, serving at least 10,000 people, and using a disinfectant during the treatment process.

To better support future DBP regulations, the USEPA promulgated the Information Collection Rule (ICR) in May 1996 [74]. This rule applied to large public water systems and mandated drinking water surveillance for 18 months. There were three requirements: monitoring for DBP and related parameters, monitoring for disease-

Table 1 Global regulations on DBPs in drinking water (unit: μg/L)

Category	DBP species	IARC category	Conc. of 10^{-6} cancer risk	WHO 2014	US 2018	EU 2020	CAN 2014	JP 2014	CN 2022
Trihalomethanes (THMs)	TCM	2B	18	30				60	60
	DCBM	2B	3	60				30	60
	DBCM	3		100				100	100
	TBM	3		60				90	100
	TTHM[a]				80	100	100	100	1[a]
Haloacetic acids (HAAs)	MCAA				60	60	80	20	
	DCAA	2B	0.38	50				30	50
	TCAA	2B	0.18	200				30	100
	BCAA	2B							
	DBAA	2B							
	IAA								20[b]
Inorganic DBPs	ClO^-	3		700	1,000	250	1,000	600	700
	ClO_3^-			700		250	1,000	20	700
	BrO_3^-	2B		10	10	10	10	10	10
	CNCl as CN^-							10	70
Aldehyde	HCHO	1					N.Q.	80	900[b]
Haloaldehyde	CH	2A							100[b]
N-nitrosamines (NAs)	NDMA	2A	0.0007	0.1			0.04		0.1[b]
	NDEA	2A	0.001						
	NDPA	2B	0.005						
	NDBA	2B	0.003						
	NMEA	2B	0.0015						
	NPIP	2B	0.0035						
	NPYR	2B	0.005						
	NMOR	2B	0.015						
	NDPhA	3	7						

(continued)

Table 1 (continued)

Category	DBP species	IARC category	Conc. of 10^{-6} cancer risk	WHO 2014	US 2018	EU 2020	CAN 2014	JP 2014	CN 2022
Haloacetonitriles (HANs)	MCAN	3							
	DCAN	3						10	
	DBAN	2B							
	BCAN	3							
	TCAN	3							

Note: The most recent regulations on DBPs of WHO, United States (US), European Union (EU), Canada (CAN), Japan (JP), and China (CN) are summarized with the limit and issue year

[a] TTHM has two means of limit: That for US, EU and JP is the total sum of four THMs. That for CN is the sum of ratios of each THM species to their individual limit

[b] IAA and NDMA have been included in the Appendix of Standards for Drinking Water Quality in China, which currently is not mandatory

causing microorganisms and microbial indicators, and treatment studies. A total of 296 water systems were involved in this DBP data collection national study. An ICR database was developed based on the reported occurrence of DBPs and microbial pathogens. DBP and microbial pathogen occurrence information for the nation, as well as for each participating water system and plant, were available to researchers, regulators, as well as the general public.

The USEPA then promulgated the Stage 1 Disinfectants and Disinfection Byproducts Rule in December 1998 [75]. This rule applies to nearly all sizes of public water systems. The rule established a maximum contaminant level of 4 mg/L for chlorine and chloramine, and 0.8 mg/L for chlorine dioxide. The rule also established maximum contaminant levels of 80, 60, 0.1, and 1.0 mg/L for THMs, HAA, bromate, and chlorite, respectively. The rule also established the best available technologies for disinfectants and DBP control, including enhanced coagulation and granular activated carbon adsorption for DBP precursor removal. However, the compliance calculation for THMs and HAAs was based on a running annual average, the arithmetic average of analytical results for samples taken at all monitoring locations during the current quarter and previous three calendar quarters.

The Stage 2 Disinfectants and Disinfection Byproducts Rule was promulgated in January 2006 to address the spatial and temporal variation of DBP levels in the water distribution systems and further reduce peak THM and HAA levels [10]. The Stage 2 rule did not change the maximum contaminant levels of disinfectants and DBPs but specified the locational running annual averages for THM and HAA compliance. The running annual average specified in the Stage 1 Rule used the average of all monitoring locations and could result in inequality in DBP protection in some DBP monitoring locations. The Stage 2 rule also required water systems to conduct an Initial Distribution System Evaluation to identify high THM and HAA locations in their distribution systems and use these locations for DBP compliance monitoring. The Stage 2 rule also specifies the operational evaluation levels calculation and requires water systems to conduct operational evaluation and further reduce DBP levels due to the seasonal variation.

8.3 WHO

WHO included only chloroform in the first version of their Guidelines for Drinking Water Quality in 1984, with the guideline value of 30 μg/L [76]. In the second edition, issued in 1996, the guideline values for each species of THMs were fully established (TCM: 200 μg/L, TBM: 100 μg/L, BDCM: 60 μg/L, DBCM: 100 μg/L) [77]. The third edition of the Guidelines pointed out that the total THM concentrations should not go beyond 100 μg/L in finished or chlorinated water and THMs were not expected to be found in raw water (unless near a pollution source) [78]. In 2017, the guideline values for each THM species were further revised in the fourth edition (TCM: 300 μg/L, TBM: 100 μg/L, BDCM: 60 μg/L, DBCM: 100 μg/L), and required the sum of the ratio of the concentration of each to its respective guideline value to not exceed 1 [79].

WHO also included HAAs (DCAA and TCAA), CNCl, chloral hydrate, 3 haloacetonitriles (i.e., dichloroacetonitrles, dibromoacetonitriles, and trichloroacetonitriles), and chloropicrin in the second, third, and fourth version of the Guidelines for Drinking Water Quality. MCAA, haloacetones, and bromochloroacetonitrile were mentioned in the second version but were not given a guideline value due to low toxicity or insufficient database.

8.4 Canada

In Canada, disinfection byproducts in drinking water are controlled through a 2-tiered advisory and regulatory system. Overarching health-based Maximum Acceptable Concentrations (MACs) are determined for each byproduct by literature reviews and/or studies conducted at the national level by Health Canada. As their name suggests, these MACs are the maximum concentrations of DBPs that are expected to be safe to drink for all Canadians. They are regularly updated by Health Canada and form the basis for guidelines published as the Guidelines for Canadian Drinking Water Quality [80].

Formal and enforceable regulation of DBPs in Canada generally occurs at the provincial level. The 'guidelines' are used by provincial governments to set maximum acceptable levels of DBPs for regulations that are appropriate to their regions. Provinces cannot set regulations higher than the MACs, but they can set them at lower levels than the MACs where warranted and when the feasibility and economics of available treatment technologies are acceptable.

In terms of limits for specific DBPs, while the Guidelines in Canada are generally similar to those in the United States, they are not exactly the same [81]. For example, at the time of writing the MACs for TTHM and HAA5 were 100 μg/L and 80 μg/L, respectively, and the MACs for chlorite and chlorate were 1 mg/L. Canada is also among the few countries where nitrosamines are monitored and regulated. The MAC for NDMA in Canada is 40 ng/L [80], but the province of Ontario has set its regulatory limit to be 9 ng/L [82]. NDMA is one of the most potent carcinogenic DBPs detected to date. It was discovered in drinking water in Ontario in the early 1990s [83].

As with other jurisdictions, documentation that accompanies the DBP guidelines and regulations also includes information to support decision-making in various areas related to providing safe drinking water. These areas include, but are not necessarily limited to, impacts of the following: water quality characteristics, treatment processes, distribution systems, health effects, and monitoring. As new information concerning these and other factors becomes known, the MACs are re-evaluated.

8.5 China

China issued the first national Sanitary Standard for Drinking Water (GB5749-1985) in 1985 which included 35 water quality parameters [84]. Only one DBP parameter, i.e., chloroform, was included in this standard with a limit of 0.06 mg/L.

In 2006, China updated the standard greatly and named it the "Standards for Drinking Water Quality (GB5749-2006)" to match international drinking water standards. This standard had a total of 106 parameters, including microbiology, esthetics, chemicals (metals, pesticides, DBPs, etc.), radioactivity, and disinfectants [85]. Four individual THMs, total THM, two individual HAAs (i.e., DCAA and TCAA), chloral hydrate, cyanogen chloride, aldehyde, bromate, chlorite, and chlorate were regulated in this standard. Among these 13 DBP items, chloroform was listed as a conventional parameter that must be monitored at least once per month for all water companies. Aldehyde, chlorite, and chlorate are conventional parameters for facilities that use ozone, chlorine dioxide, and mixtures of chlorine and chlorine dioxide as disinfectants, respectively. Other DBPs are listed as non-conventional parameters which must be monitored at least twice per year.

In 2022, China updated the standard again as "Standards for Drinking Water Quality (GB5749-2022)" [86]. The improvements concerning DBPs were of two parts: (1) increasing the monitoring frequency of 4 THMs and 2 HAAs to once per month; and (2) adding NDMA and iodinated HAAs into the appendix of this standard which can be used as the reference for future regulation.

8.6 European Union

The European Union has included THMs and HAAs in the Council Directive 98/83/EC, 1998 [87] concerning the quality of water intended for human consumption. This document was updated in 2022 with some modifications [88]. At the time of writing, THMs, HAAs, chlorite, chlorate, and bromate have been given the limits of 100, 60, 250, 250, and 10 μg/L in this document. Each member country of the European Union should follow this Directive at a minimum and may set up their own drinking water standards with more stringent DBP regulations. For example, the Netherlands and Germany added NDMA to their own drinking water standards with a limit of 12 ng/L.

The UK is no longer part of the EU and England and Wales do not follow the 2020 revised EU directive although it retains the standards set under previous EU directives. Scotland is following the 2020 revised EU directive. England and Wales are currently going through a process to set new drinking water standards.

9 Summary

Since the discovery of chloroform formation in chlorinated water, there has been a global effort to identify it and other DBPs in drinking water, including THMs, HAAs, chloral hydrate, MX, nitrogenous DBPs, their brominated analogs, and ozonation DBPs. Using innovative analytical methods is the key to discovering these DBPs. Good designs for comprehensive studies have also been critical in identifying the origin and formation mechanisms for DBPs. As the sensitivity of analytical methods and sample concentration procedures improve, it is anticipated that more and more DBPs will be identified in disinfected drinking water.

As for many other science and engineering disciplines, multidisciplinary partnerships are needed to address future DBPs research. Scientists and engineers in public health, policy, and water treatment need to work together to reduce both the microbial risks and chemical risks of drinking water. Chemists and toxicologists need to collaborate to identify DBPs that are toxicology significant. Scientists and engineers need to better control the formation of DBPs, especially those that are toxicologically significant.

References

1. Tulchinsky TH (2018) John Snow, Cholera, the Broad Street Pump; waterborne diseases then and now case studies in public health. Elsevier, pp 77–99. https://doi.org/10.1016/B978-0-12-804571-8.00017-2
2. McGuire MJ (2013) The chlorine revolution: water disinfection and the fight to save lives. American Water Works Association. ISBN: 9781583219133
3. Ramsay MAE (2006) John snow, MD: anaesthetist to the queen of England and Pioneer epidemiologist. Baylor Univ Med Cent Proc 19(1):24–28. https://doi.org/10.1080/08998280.2006.11928120
4. Chick H (1908) An investigation of the Laws of disinfection. J Hyg 8(1):92–158. https://doi.org/10.1017/S0022172400006987
5. Waston HE (1908) A note on the variation of the rate of disinfection with change in the concentration of the disinfectant. J Hyg 8(4):536–542. https://doi.org/10.1017/S0022172400015928
6. Symons J (1999) Disinfection by-products: a historical perspective. Formation and control of disinfection by-products in drinking water. American Water Works Association, pp 1–25
7. Rook JJ (1974) Formation of Haloforms during chlorination of natural waters. J Water Treat Exam 23:234–243
8. Bellar TA et al (1974) The occurrence of organohalides in chlorinated drinking waters. J AWWA 66(12):703–706. https://doi.org/10.1002/j.1551-8833.1974.tb02129.x
9. Rook JJ (1976) Haloforms in drinking water. J AWWA 68(3):168–172. https://doi.org/10.1002/j.1551-8833.1976.tb02376.x
10. US EPA (1998) National Primary Drinking Water Regulations; disinfectants and disinfection byproducts; final rule; Fed Regist, vol. 63, p. 69392
11. McGuire MJ et al (2014) Not your Granddad's disinfection by-product problems and solutions. J AWWA 106(8):54–73. https://doi.org/10.5942/jawwa.2014.106.0128
12. Rook JJ (1977) Chlorination reactions of fulvic acids in natural waters. Environ Sci Technol 11(5):478–482. https://doi.org/10.1021/es60128a014

13. Quimby BD et al (1980) Determination of the aqueous chlorination products of humic substances by gas chromatography with microwave plasma emission detection. Anal Chem 52(2): 259–263. https://doi.org/10.1021/ac50052a010
14. Miller JW, Uden PC (1983) Characterization of nonvolatile aqueous chlorination products of humic substances. Environ Sci Technol 17(3):150–157. https://doi.org/10.1021/es00109a006
15. Uden PC, Miller JW (1983) Chlorinated acids and chloral in drinking water. J AWWA 75(10): 524–527. https://doi.org/10.1002/j.1551-8833.1983.tb05213.x
16. Peters R et al (1991) The analysis of halogenated acetic acids in Dutch drinking water. Water Res 25(4):473–477. https://doi.org/10.1016/0043-1354(91)90084-4
17. Ireland JC et al (1988) Letter to the editors. Biol Mass Spectrom 17(6):483–486. https://doi.org/10.1002/bms.1200170612
18. Xie YF et al (1992) Mass spectra of synthesized mixed bromochloroacetic acids. Org Mass Spectrom 27(7):807–810. https://doi.org/10.1002/oms.1210270708
19. Xie YF et al (1998) Analyzing HAAs and Ketoacids without diazomethane. J AWWA 90(4): 131–138. https://doi.org/10.1002/j.1551-8833.1998.tb08416.x
20. Xie YF, Reckhow D (1993) Identification of trihaloacetaldehydes in ozonated and chlorinated fulvic acid solutions. Analyst 118(1):71. https://doi.org/10.1039/an9931800071
21. Xie YF, Reckhow D (1992) Formation and control of trihaloacetaldehydes in drinking water. Proceedings of AWWA water quality technology conference
22. McKinney JD et al (1976) Possible factors in the drinking water of laboratory animals causing reproductive failure. In: Keith LH (ed) Identification and analysis of organic pollutants in water, vol 1. Ann Arbor Science, pp 417–432
23. Trehy ML, Bieber TI (1980) Effects of commonly used water treatment processes on the formation of THMs and DHANs. Proceedings of the American Water Works Association 1980 annual conference, American Water Works Association
24. Sayato Y et al (1982) Studies on mechanism of volatile chlorinated organic compound formation (III) mechanism of formation of chloroform and chloropicrin by chlorination of humic acid. Jpn J Water Pollut Res 5(3):127–134
25. Merlet N et al (1985) Chloropicrin formation during oxidative treatments in the preparation of drinking water. Sci Total Environ 47:223–228. https://doi.org/10.1016/0048-9697(85)90332-8
26. Thibaud H et al (1988) Influence de la Concentration en Bromures sur la Formation de Chloropicrine lors de la Chloration des Eaux. Mise en Evidence de Trihalonitromethanes Bromes. Water Res 22(3):381–390. https://doi.org/10.1016/S0043-1354(88)90300-4
27. Ohya T, Kanno S (1985) Formation of cyanide ion or cyanogen chloride through the cleavage of aromatic rings by nitrous acid or chlorine. VIII. On the reaction of humic acid with hypochlorous acid in the presence of ammonium ion. Chemosphere 14(11–12):1717–1722. https://doi.org/10.1016/0045-6535(85)90112-2
28. Hirose Y et al (1988) Formation of cyanogen chloride by the reaction of amino acids with hypochlorous acid in the presence of ammonium ion. Chemosphere 17(5):865–873. https://doi.org/10.1016/0045-6535(88)90059-8
29. Krasner SW et al (2013) Formation, precursors, control, and occurrence of nitrosamines in drinking water: a review. Water Res 47:4433–4450. https://doi.org/10.1016/j.watres.2013.04.050
30. Magee PN, Barnes JM (1956) The production of malignant primary hepatic tumours in the rat by feeding dimethylnitrosamine. Br J Cancer 10:114–122. https://doi.org/10.1038/bjc.1956.15
31. Li X et al (2021) Intake of volatile nitrosamines by Chinese residents in different provinces via food and drinking water. Sci Total Environ 754:142121. https://doi.org/10.1016/j.scitotenv.2020.142121
32. Li X et al (2024) Intake of tobacco nitrosamines of smokers in various provinces of China and their cancer risk: a meta-analysis. J Environ Sci 141:249–260. https://doi.org/10.1016/j.jes.2023.05.018
33. SWRCB CA, U.S. (2017) Groundwater fact sheet N-nitrosodimethylamine (NDMA). https://www.waterboards.ca.gov/gama/docs/ndma.pdf

34. Jobb DB, Hunsinger RB, Meresz O (1994) Removal of N-nitrosodimethylamine from the ohsweken (six nations) water supply. Ontario Ministry of Environment and Energy, Science and Technology Branch, Toronto
35. Najm I, Trussell RR (2001) NDMA formation in water and wastewater. J AWWA 93(2):92–99
36. Bei E et al (2016) Occurrence of nitrosamines and their precursors in drinking water systems around mainland China. Water Res 98:168–175. https://doi.org/10.1016/j.watres.2016.04.013
37. Choi J, Valentine RL (2002) Formation of N-nitrosodimethylamine (NDMA) from reaction of monochloramine: a new disinfection by-product. Water Res 36:817–824. https://doi.org/10.1016/S0043-1354(01)00303-7
38. Mitch WA, Sedlak DL (2002) Formation of N-nitrosodimethylamine (NDMA) from dimethylamine during chlorination. Environ Sci Technol 36(4):588–595
39. Schmidt CK, Brauch HJ (2008) N,N-dimethosulfamide as precursor for N-nitrosodimethylamine (NDMA) formation upon ozonation and its fate during drinking water treatment. Environ Sci Technol 42(17):6340–6346
40. Holmbom BR et al (1981) Isolation and identification of an Ames-mutagenic compound present in Kraft chlorination effluents. Tappi 64(3):172–174
41. Holmbom BR et al (1984) Fractionation, isolation and characterization of Ames-mutagenic compounds in Kraft chlorination effluents. Environ Sci Technol 18(5):333–337. https://doi.org/10.1021/es00123a009
42. Kronberg L et al (1985) Fractionation of mutagenic compounds formed during chlorination of humic water. Sci Total Environ 47:343–347. https://doi.org/10.1016/0048-9697(85)90340-7
43. Kronberg L et al (1988) Identification and quantification of the Ames mutagenic compound 3-Chloro-4-(dichloromethyl)-5-hydroxy-2(5H)-furanone and of its geometric isomer (E)-2-Chloro-3-(Dichloromethyl)-4-oxobutenoic acid in chlorine-treated humic water and drinking water extracts. Environ Sci Technol 22(9):1097–1103. https://doi.org/10.1021/es00174a016
44. Andrews RC et al (1990) Occurrence of the mutagenic compound 'MX' in drinking water and its removal by activated carbon. Environ Technol 11(7):685–694. https://doi.org/10.1080/09593339009384913
45. Xie YF, Wang Z (1990) Effect of water treatment on mutagenic activity and concentration of MX in water. Proceedings of AWWA water quality technology conference
46. Plewa MJ et al (2002) Mammalian cell cytotoxicity and genotoxicity analysis of drinking water disinfection by-products. Environ Mol Mutagen 40(2):134–142. https://doi.org/10.1002/em.10092
47. Yamada H, Isao S (1989) The determination of carbonyl compounds in ozonated water by the PFBOA method. Ozone Sci Eng 11(2):127–141. https://doi.org/10.1080/01919518908552431
48. Glaze WH et al (1989) Ozonation byproducts. 2. Improvement of an aqueous-phase derivatization method for the detection of formaldehyde and other carbonyl compounds formed by the ozonation of drinking water. Environ Sci Technol 23(7):838–847. https://doi.org/10.1021/es00065a013
49. Xiong F et al (1992) Long-term ozone consumption by aquatic fulvic acids acting as precursors of radical chain reactions. Environ Sci Technol 26(5):1059–1064. https://doi.org/10.1021/es00029a029
50. Xie YF, Recknow DA (1992) Formation of ketoacids in ozonated drinking water. Ozone Sci Eng 14(3):269–275. https://doi.org/10.1080/01919519208552480
51. Pulicharla R et al (2020) Trends in ozonation disinfection by-products – occurrence, analysis and toxicity of carboxylic acids. Water 12(3):756. https://doi.org/10.3390/w12030756
52. Xie YF, Recknow DA (1996) Comment on 'Identification of Halogenated Compounds in Chlorinated Seawater and Drinking Water Produced Offshore Using n-Pentane Extraction and Open-Loop Stripping Technique'. Environ Sci Technol 30(2):720–720. https://doi.org/10.1021/es940776y
53. Xie YF, Recknow DA (1994) Formation of halogenated artifacts in brominated, chloraminated, and chlorinated solvents. Environ Sci Technol 28(7):1357–1360. https://doi.org/10.1021/es00056a026

54. Cavanagh JE et al (1992) Ozonation byproducts: identification of bromohydrins from the ozonation of natural waters with enhanced bromide levels. Environ Sci Technol 26(8): 1658–1662. https://doi.org/10.1021/es00032a027
55. Kanniganti et al. (1992) Identification of compounds in mutagenic extracts of aqueous monochloraminated fulvic acid. Environ Sci Technol 26(10): 1998–2004. https://doi.org/10.1021/es00034a018
56. Glaze WH et al (1993) Evaluating the formation of brominated DBPs during ozonation. J AWWA 85(1):96–103. https://doi.org/10.1002/j.1551-8833.1993.tb05926.x
57. Jolley RL (ed) (1978) Water chlorination: environmental impact and health effects, vol 1. Ann Arbor Science
58. Jolley RL (ed) (1978) Water chlorination: environmental impact and health effects, vol 2. Ann Arbor Science
59. Jolley RL (ed) (1980) Water chlorination: environmental impact and health effects, vol 3. Ann Arbor Science
60. Jolley RL (ed) (1983) Water chlorination: environmental impact and health effects, vol 4. Ann Arbor Science
61. Jolley RL (ed) (1990) Water chlorination: chemistry, environmental impact and health effects, vol 6. Lewis Publishing
62. Minear RA, Amy GL (eds) (1996) Disinfection by-products in water treatment: the chemistry of their formation and control. Lewis Publishing
63. Barrett SE et al (eds) (2000) Natural organic matter and disinfection by-products characterization and control in drinking water. American Chemical Society
64. Karanfil T, Krasner S et al (eds) (2008) Disinfection by-products in drinking water: occurrence, formation, health effects, and control. American Chemical Society; Distributed by Oxford University Press. https://doi.org/10.1021/bk-2008-0995
65. Karanfil T, Mitch B et al (eds) (2015) Recent advances in disinfection by-products. American Chemical Society
66. Singer PC (ed) (1999) Formation and control of disinfection by-products in drinking water. American Water Works Association
67. Fielding M, Farrimond M (eds) (1999) Disinfection by-products in drinking water: current issues. Royal Society of Chemistry
68. Hrudey SE, Charrois JWA (2012) Disinfection by-products: relevance to human health. IWA Publishing
69. Thompson KC et al (eds) (2016) Disinfection by-products in drinking water. Royal Society of Chemistry
70. Prasad MNV (ed) (2020) Disinfection by-products in drinking water: detection and treatment. Butterworth-Heinemann. an imprint of Elsevier, Pulicharla R
71. Xie YF (2004) Disinfection byproducts in drinking water: formation, analysis, and control. Lewis Publishers
72. Shi YJ, Xia W et al (2024) Trihalomethanes in global drinking water: distributions, risk assessments, and attributable disease burden of bladder cancer. J Hazard Mater 469:133760
73. USEPA (1979) National interim primary drinking water regulations; control of trihalomethanes in drinking water. Fed Regist 44:68624
74. USEPA (1996) National Primary Drinking Water Regulations; monitoring requirements for public drinking water supplies; final rule. Fed Regist 61:24354
75. USEPA (2006) National Primary Drinking Water Regulations: stage 2 disinfectants and disinfection byproducts rule. Fed Regist 71(2006):388
76. WHO (1984) Guidelines for drinking-water quality, 1st edn. https://www.who.int/publications/i/item/9241541709
77. WHO (1996) Guidelines for drinking-water quality, 2nd edn. https://www.who.int/publications/i/item/WHO-EOS-98.1
78. WHO (2008) Guidelines for drinking-water quality, 3rd edn. https://www.who.int/publications/i/item/9789241547611

79. WHO (2022) Guidelines for drinking-water quality. 4th edition incorporating the first and second addenda. https://www.who.int/publications/i/item/9789240045064
80. Health Canada (2024) Guidelines for Canadian drinking water quality. June. https://www.canada.ca/en/health-canada/services/environmental-workplace-health/reports-publications/water-quality/guidelines-canadian-drinking-water-quality-summary-table.html
81. Wang X, Mao Y et al (2015) Disinfection byproducts in drinking water and regulatory compliance: a critical review. Front Environ Sci Eng 1:3–15. https://doi.org/10.1007/s11783-014-0734-1
82. Ontario DWQS (2020) Ontario drinking water quality standards. https://www.ontario.ca/laws/regulation/030169
83. Jobb D, Hunsinger R, Meresz O, Taguchi V (1993) A study of the occurrence and inhibition of formation of N-nitrosodimethyl amine (NDMA) in the Ohsweken water supply. In: Proceedings of the 1992 water quality technology conference. American Water Works Association, Denver
84. Ministry of Health (1985) Standards for drinking water quality (GB 5749–1985). http://down.foodmate.net/standard/yulan.php?itemid=3467
85. National Health Commission of the People's Republic of China (NHC) (2006) Standards for drinking water quality (GB 5749-2006). http://www.nhc.gov.cn/cmsresources/zwgkzt/wsbz/new/20070628143525.pdf
86. National Health Commission of the People's Republic of China (NHC) (2022) Standards for drinking water quality (GB 5749–2022). http://c.gb688.cn/bzgk/gb/showGb?type=online&hcno=99E9C17E3547A3C0CE2FD1FFD9F2F7BE
87. Council Directive 98/83/EC (1998) Quality of water intended for human consumption. Off J Eur Commun 41, 5 Dec 1998, L 330, 3 November 1998
88. Directive (EU) 2020/2184 of the European Parliament and of the Council of 16 December 2020 on the quality of water intended for human consumption. Off J Eur Union, L 435, 23 Dec 2020

Detection Methods for DBPs

Wenhai Chu, Di Zhang, Baiyang Chen, Wei Ma, Guiying He, Mengting Yang, Yi Li, and Susan Andrews

Contents

1 Gas Chromatography (GC) for Volatile and Semi-volatile DBPs 32
 1.1 Sample Preservation 32
 1.2 Sample Pretreatment 34
 1.3 Chromatographic Separation and Detection of Compounds 35
2 Liquid Chromatography (LC) for Semi-volatile and Emerging DBPs 36
 2.1 Fundamentals of LC in DBP Analysis 36
 2.2 Application of LC in Detecting DBPs 37
 2.3 Summary 39
3 Ion Chromatography (IC) for Haloacetic Acids 39
 3.1 Standard IC Methods 39
 3.2 Sample Pretreatment and Online Treatment Methods 43
 3.3 IC Separation Columns 43
 3.4 IC Detectors 46
4 Precursor Ion Scan (PIS) Method for Screening New DBPs 46
 4.1 Principles of the PIS Method 46
 4.2 Screening and Identification of New Polar Halogenated DBPs 47
 4.3 Evaluation of Formation Profiles of Polar Halogenated DBPs 48
References 51

W. Chu and D. Zhang
Tongji University, Shanghai, People's Republic of China
e-mail: 1world1water@tongji.edu.cn; Tongji_Zhangdi@tongji.edu.cn

B. Chen and W. Ma
Harbin Institute of Technology (Shenzhen), Shenzhen, People's Republic of China

G. He and M. Yang
Shenzhen University, Shenzhen, People's Republic of China
e-mail: heguiying2023@email.szu.edu.cn; yangmt@szu.edu.cn

Y. Li and S. Andrews (✉)
University of Toronto, Toronto, ON, Canada
e-mail: livia.li@mail.utoronto.ca; susan.andrews@utoronto.ca

Chao Chen, Susan Andrews, and Yuefeng Xie (eds.), *Disinfection By-Products in Water*, Hdb Env Chem (2024) 135: 31–58, DOI 10.1007/698_2024_1148,

Published online: 26 September 2024

Abstract In the 50 years since the discovery of chloroform and other haloforms as drinking water disinfection products (DBPs), significant efforts have been made to develop sensitive and selective analytical methods for them and to determine the possible existence of other DBPs. Hundreds of DBPs have been identified and quantified, and yet this is still understood to be a small fraction of the total. The analytical advances that have been made are really too numerous to describe in detail in this chapter, so we do not provide an exhaustive review herein. Instead, we have provided overviews of the currently used procedures for measuring major classes of organic DBPs (e.g., gas, liquid, and ion chromatography) and inorganic or non-volatile DBPs (e.g., liquid chromatography, ion chromatography) with various detectors (e.g., electron capture, mass spectrometry, conductivity). Much greater detail concerning current analytical methods for all DBPs can be found in the documented standard methods that have been adopted by the scientific and regulatory organizations that are active in virtually every nation that produces drinking water. Also, with a view toward the future of this topic, the chapter ends with a look at an emerging non-targeted screening method, precursor ion scanning or PIS, which holds promise for providing a more comprehensive accounting of the chemical byproducts of water disinfection processes.

Keywords Analysis, Detection, Electron capture, Gas chromatography, Ion chromatography, Liquid chromatography, Mass spectrometry, Preparation, Quantification, Screening, Separation

1 Gas Chromatography (GC) for Volatile and Semi-volatile DBPs

DBP assay usually includes sample preservation, sample extraction, chromatographic separation of compounds, and detector identification. Because DBPs are usually present in drinking water at μg/L or even ng/L concentrations, it is necessary to isolate DBPs of interest from complex substrates and concentrate them for analysis. Sample pretreatment includes sample preservation, concentration, and extraction to stabilize DBPs prior to analysis, eliminate or reduce potential interferences, improve the sensitivity of analytical method, and convert the analytes to a more suitable form that can be easily characterized by the instrument used. Gas chromatography (GC) equipped with electron capture detection (ECD) or mass spectrometry detection is widely used in environmental labs for the analysis of volatile and semi-volatile DBPs (e.g., THMs, HAAs, HANs, and HALs).

1.1 Sample Preservation

Water treatment plants typically lack on-site testing capability. After sampling, water samples must be transported to a laboratory for the determination of DBPs. Hence, it is crucial to ensure that DBPs remain stable during sample preservation to achieve accurate quantification. Two key factors influencing this stability are the choice of quenching agents and maintaining the appropriate pH.

Quenching Agents Residual disinfectant is usually maintained in distribution systems to suppress microbial regrowth. The residual organic matter remaining after water treatment continues to react with the disinfectant, giving rise to the formation of target DBPs. Additionally, previously formed DBPs may also undergo reactions with disinfectant, transforming into other substances, thus affecting their determination in distribution systems. A key aspect of DBP analysis that differs from other environmental pollutant analyses is the need for a quencher to extinguish remaining disinfectants (such as chlorine or chloramines).

Except for ammonium chloride, which serves as a specialized quenching agent for chlorine and reacts with hypochlorous acid or hypochlorite anion to form chloramines, quenching agents are typically reducing agents (e.g., ascorbic acid, sodium thiosulfate, sodium sulfite). These agents participate in redox reactions with disinfectants (oxidants), preventing further reactions between disinfectants and organic/inorganic substances and DBPs in the water sample. Generally, excess quenching agent is added to the water sample for use, usually 120% of the stoichiometric ratio of the reaction with the disinfectant [1].

The properties of an ideal quenching agent include rapid and complete reaction with residual disinfectants, chemically inert toward the DBPs or other analytes of interest, negligible interfering effects in the analysis of the DBPs or other analytes of interest, and non-detectable signal for the compound or its reaction products. Commonly used quenching agents are sodium sulfite, sodium arsenite, sodium borohydride, ascorbic acid, ammonium chloride, etc. It has been reported that some quenchers, especially sulfites, and thiosulfates, can lead to the decomposition of some DBPs. In Ina's study [1], the measured concentrations of HAAs in both the quenched and the non-quenched samples were found to decrease over the experimental period (up to 50% and 30% decrease in the quenched and non-quenched samples, respectively), so the influence of different quenchers on the stability of different types of DBPs should be clearly defined when choosing them.

Ina's study has also shown that all the above-mentioned quenchers do not have adverse effects on the concentration of trihalomethanes (THMs) and haloacetic acids (HAAs) [1]. Ascorbic acid is suitable for the analysis of haloacetonitriles (HANs) and haloketones (HKs), but not suitable for the analysis of chlorite and halonitromethanes (HNMs). It was found that ascorbic acid reduced BDCNM, DBCNM, TBNM, and TBAN to undetectable concentrations within hours of contact in quenched chlorinated and chloraminated waters. Sodium arsenite, sodium borohydride, and ascorbic acid can be used for halogenated acetaldehyde (HALs) analysis. All quenchers have an adverse effect on the concentration of chloroform. Therefore, organic DBPs (i.e., THMs, HAAs, HANs, HKs, and HALs) can be quenched with ascorbic acid and inorganic DBPs can be analyzed with sodium sulfite. It is important to note that the sample still needs to be analyzed as soon as possible after the quenching agent is used to minimize degradation and hydrolysis.

In future studies, we expect to find a quencher that does not destroy any of the compounds of interest and does not produce by-products.

pH DBPs may undergo hydrolysis or decomposition under pH conditions that are representative of drinking water. The investigation of their stability is of great significance to understand their fate in drinking water. Meanwhile, the loss caused by hydrolysis during sample preservation can be reduced by adjusting the pH of water samples. In fact, many DBPs are intermediates that can be converted to other end products by hydrolysis (base or acid catalysis) or chlorination (hypochlorous acid or hypochlorite ion). For example, haloacetamides (HAMs) can be decomposed directly into their corresponding haloacetic acids (HAAs) through a nucleophilic reaction between carbonyl carbon and hydroxide, with both the hydrolysis rate and the chlorination rate increasing with the pH and the number of halogens substituted on the methyl group [2]. In a hydrolysis experiment to study HAMs, the stability of 10 HAMs (BCAM, DBAM, CIAM, BIAM, DIAM, BDCAM, DBCAM, TBAM, DCAM, and TCAM) was monitored for 336 h (14 days) at pH 5 to 9. The results showed that less than 25% of the HAMs was hydrolyzed at pH 5.0, and that the results at pH 6.0 and 7.0 were similar to those of pH 5.0. Under weakly alkaline conditions (pH 8.0), the hydrolysis rate of HAMs was higher than that under weakly acidic and neutral conditions. However, at pH 9.0, the degradation rate was significantly higher than that of pH 8.0. The experiment fully showed that acidic conditions are more conducive to the stability of HAMs than alkaline conditions [2]. The stability of seven mono-, di-, and trihaloacetonitriles was also examined at different pH levels. All seven HANs were most stable at pH 6, and the rate of loss increased with both increasing pH and the number of halogens. In addition, it is known that haloacetonitriles (HANs) produce haloacetamide (HAMs) and haloacetic acid (HAAs) by the nucleophilic attack of hydroxide and hypochlorite on its nitrile carbon [3]. Therefore, in addition to the use of quenchers to eliminate residual chlorine, adjusting the pH of the sample solution is equally critical to stabilizing the analytes themselves prior to analysis. The pH of water samples can be adjusted to a range of 3–5 to sustain most of the DBPs.

1.2 Sample Pretreatment

Liquid-liquid extraction (LLE), also known as solvent extraction and fractionation, is a method of separating compounds based on their relative solubility in two different immiscible solutions, usually water and organic solvents. It is the extraction of a substance from one liquid phase to another liquid phase. It has been shown that organic solvents such as methyl tert-butyl ether (MTBE), n-hexane, and ethyl acetate can be used as extraction solvents for the detection and identification of THMs and HAMs. LLE is a conventional DBP monitoring method, which has the advantages of low cost and simple operation but also has the possible disadvantages of long sample preparation times, low recovery rates, and high detection limits.

In recent years, solid phase extraction (SPE) has been widely used for determining trace target analytes. The SPE uses the affinity of solutes dissolved or suspended in a liquid (called the mobile phase) to the solid through which the sample passes (called the stationary phase) to separate the mixture into desired and unwanted components. SPE can be used to separate analytes of interest from a wide variety of substrates, facilitating the increased concentration and purification (relative to LLE) of samples for analysis. SPE pretreatment has been used in many studies to detect and identify HAMs, NAs, and other DBPs in water samples.

In addition to sample extraction, certain samples also require derivatization. As polar acids, HAAs undergo almost complete ionization in water, and cannot be directly analyzed by GC. HAAs have been generally determined based on EPA method 552.3 which involves a derivatization step that converts HAAs into corresponding haloacetic acid methyl ester (HAAMEs), neutralization of excess acid, and then analysis with GC-ECD or GC-MS. HAMs can interfere with the quantification of HAAs through by generating the same HAAMEs during derivatization. The concentrations of HAAs in water in the presence of HAMs need to be corrected, such that the accurate concentration of an HAA species (abbreviated as HAA_R) is equal to the difference between the measured concentration of that specific HAA by GC-MS (abbreviated as $HAA_{M\text{-}GC}$) and the concentration of the corresponding HAM (abbreviated as HAM_M) and its transformation rate (TR). For example, for the determination of TCAA in the presence of trichloroacetamide (TCAM), $TCAA_R = TCAA_M - (TCAM_M \times TR)$ [4]. The TR should be determined with HAM standards of the same concentration of HAMs in the sample. As such, convenient and precise methods for HAA determination continue to be developed.

1.3 Chromatographic Separation and Detection of Compounds

In gas chromatography (GC), a sample is propelled through a column containing a moving carrier gas, usually an inert gas such as helium or a non-reactive gas such as nitrogen. The column is packed with a stationary phase, usually consisting of microscopic layers of liquid or polymer on an inert solid carrier, to complete a specific type of separation. Gas chromatography has been widely used for the separation of low molecular weight DBPs with volatile or semi-volatile heat stability. The most common currently used detectors coupled to gas chromatography for the detection of DBPs are electron capture detectors (ECD) and mass spectrometry detectors (especially mass selection detectors). However, there is still a lack of analytical methods to quantify the full classes of DBPs at sufficiently low limits of quantification (ng/L), limiting the completeness of most DBP toxicological studies. In addition to optimizing sample extraction methods, better quantitative analysis of DBPs can be performed in terms of optimizing instrument conditions and temperature programming [5], to take advantage of the diversity of DBPs' different polarities, volatilities, and thermal stabilities. For example, although when using an Rtx-5 column to analyze haloacetamide, there will be a significant peak tail in the chromatogram, and the peak width can reach 40s, when using Rtx-624 and Rtx-200 columns there will be a significant improvement in peak shape, forming a typical 5–10s wide GC peak [5]. For high-boiling compounds prone to thermal degradation at the GC inlet, such as halonitromethanes, if the system contains a non-programmable inlet, then the sample can be run continuously at two different temperature programs, one at a higher temperature and one at a lower temperature, with the lower inlet temperature and transmission line temperature used to calibrate the curve on a separate GC method. If the system can program the inlet temperature through programmable temperature evaporation (PTV) or multi-mode inlet (MMI), the sample can be analyzed through temperature programming.

2 Liquid Chromatography (LC) for Semi-volatile and Emerging DBPs

2.1 Fundamentals of LC in DBP Analysis

Liquid chromatography (LC) is a technique that separates molecules based on their relative affinities for a liquid mobile phase and a solid stationary phase. LC can be classified into different types depending on the properties of the molecules that are exploited for separation, e.g., size exclusion, ion exchange, hydrophobic interaction, affinity, and multimodal chromatography. The affinity of the molecules for the mobile phase determines their elution time from the column, which can be adjusted by changing the mobile phase composition.

Liquid chromatography is widely used for monitoring pesticides, pharmaceuticals, personal care products, and surfactants in water. However, its application in disinfection byproduct (DBP) analysis is limited compared to gas chromatography (GC). GC with mass spectrometry detection (GC-MS) and GC with electron capture detection (GC-ECD) are currently the most common analytical methods for certain classes of disinfection by-products (DBPs) in water. These GC methods generally require extraction and concentration of DBPs using solid phase extraction (SPE) or liquid-liquid extraction before analysis. Therefore, they are more suitable for low molecular weight, volatile, and semi-volatile DBPs. Ion chromatography (IC), which can be considered as a sub-type of LC utilizing conductivity detection, is widely used for the detection of oxyhalide inorganic disinfection byproducts (DBPs) such as chlorate (ClO_3^-), chlorite (ClO_2^-), perchlorate (ClO_4^-), and bromate (BrO_3^-). Section 3 provides detailed insights into the use of IC for HAA detection.

Liquid chromatographic separation has become more often used in combination with mass spectrometry (MS) detection for DBP analysis. Conductivity and UV-Vis absorbance detectors, although employed for disinfection byproduct (DBP) identification, are generally less selective and less sensitive than MS. The ionization source, which converts analytes into ions for MS detection, is a crucial factor that determines the types of DBPs that can be analyzed. Among the various ionization techniques, electrospray ionization (ESI), atmospheric pressure ionization (API), and atmospheric pressure chemical ionization (APCI) are the most common ones for LC-MS, as they offer high sensitivity and wide applicability. In recent years, ESI has been increasingly employed in combination with LC-MS to identify new DBPs that have large, polar, and/or thermally unstable structures that are not amenable to GC-MS analysis. LC-MS methods for DBP analysis usually employ reverse-phase LC (RPLC) with ESI. RPLC columns with different types of alkyl chains as the stationary phase, such as butyl (C4), octyl (C8), or octadecyl (C18), can effectively separate a wide range of DBPs that are non-polar or have low polarity [6–11].

2.2 *Application of LC in Detecting DBPs*

2.2.1 LC as an Alternative for Semi-volatile DBPs

Among the commonly known DBPs, LC methods offer an alternative to GC-MS for analyzing semi-volatile DBPs such as HAAs [12–15], HAMs [16, 17], and nitrosamines [15, 18–20].

The analysis of HAAs is challenging due to their high acidity and hydrophilicity, which require derivatization prior to GC-ECD or GC-MS methods. These techniques involve liquid-liquid extraction (LLE) and derivatization steps that are time-consuming and labor-intensive. Given the detection challenges, HAAs are well suited for analysis by LC-ESI-MS combined with solid phase extraction (SPE) at μg/L concentrations. Previous literature reported complete LC separation of HAAs using volatile ion-pairing agents, such as 5 mM tributylamine or 5 mM triethylamine, which produced both [M - H]- and [M -COOH]- ions in negative ion ESI-MS [8, 12]. In addition to SPE-LCMS methods, reverse-phase LC-MSMS methods have been developed for HAA detection [13–15]. These methods involve direct aqueous injection, offering highly specific detection without the need for derivatization. They efficiently separate HAAs from common anionic interferences, ensuring accurate results [13–15]. Additional insights on the analysis of HAAs using methods alternative to GC-ECD or GC-MS will be provided in Sect. 3.

Similar to HAAs, haloacetamides (HAMs) are a group of compounds that pose analytical difficulties due to their low volatility and varying polarity depending on the degree of halogenation. A novel method developed by Chu et al. (2012) enabled simultaneous detection of 13 HAMs in drinking water by integrating SPE enrichment and high-performance LC separation with triple quadrupole MS (SPE-HPLC/tqMS) with atmospheric pressure chemical ionization (APCI) using selective reaction monitoring (SRM) in positive mode [16]. Zhou et al. (2023) also reported their determination of 13 HAMs using ultra-performance LC-ESI-MS/MS (UPLC-ESI-MS/MS) using multiple reaction monitoring information-dependent acquisition-enhanced product ion (MRM-IDA-EPI) mode without pretreatment for real water samples [17].

Nitrosamines are a class of DBPs that pose high carcinogenic risks to humans even at ng/L levels in water. Most analytical methods for nitrosamines, such as

GC/MS, are only suitable for (semi) volatile and thermally stable compounds and lack structural information for unknown DBPs. SPE-LC-MSMS methods have been developed to detect N-nitrosodimethylamine (NDMA) – a well-known semi-volatile nitrosamine in a wide range of water matrices [18, 19]. Zhao et al. (2006) developed an LC-MS-MS method that can analyze a wide range of nitrosamines which are both GC-detectable and non-detectable in drinking water at low ng/L levels [20]. The method involves SPE enrichment, LC separation, and a tandem quadrupole linear ion trap mass spectrometer (MS/MS) with positive ESI and multiple reaction monitoring (MRM) mode to achieve high sensitivity and selectivity. Alternatively, a rapid method for analyzing nitrosamines and haloacetic acids using LC-MS with positive-negative switching and MRM has been developed, enabling analysis without sample enrichment [15].

2.2.2 LC for Emerging DBPs

LC-MS methods are not only useful for GC-detectable DBPs as an alternative, but also for identifying known and unknown emerging DBPs that pose potential health risks, such as haloquinones (HQs), halogenated aromatic DBPs, and polar and thermally unstable brominated and iodinated DBPs.

Halobenzoquinones (HBQs) have been detected as DBPs at ng/L levels in drinking water and swimming pool water [21]. HBQs are predicted to be carcinogenic and genotoxic, potentially presenting a risk to human health. Analytical methods based on SPE-LC-MS have been developed for determining HBQs. An SPE-UHPLC-MS/MS method developed by Li's group was able to measure both HBQs and their transformation products, halo-hydroxyl-benzoquinones (OH-HBQs), which were also found to be DBPs in drinking water [22]. A unique property of HBQ compounds is their ability to generate $(M + H)^-$ ions through a reduction process when subjected to negative ion-ESI, instead of the conventional $(M - H)^-$ ions that are usually observed with negative ionization, while OH-HBQ compounds produce $(M - H)^-$, $(M)^-$, and $(M + H)^-$ as the major ions under negative ESI [22].

Another class of emerging DBPs that were detected and identified by UPLC/ESI-tqMS at low ng/L are halogenated aromatic DBPs. These compounds have similar total organic halogen (TOX) levels as aliphatic DBPs in chlorinated water, but they are more toxic and contribute significantly to the overall toxicity of the water. Some of the most common halogenated aromatic DBPs in drinking water are halonitrophenols (HNPs), halophenols (HPs), halohydroxybenzoic acids (HHBAs), and halohydroxybenzaldehydes (HHBDs) [23, 24]. In a recent study, seven halohydroxybenzonitriles (HHBNs) have also been identified as a new group of halogenated aromatic DBPs in drinking water by a SPE-UPLC/ESI-hybrid quadrupole time-of-flight mass spectrometer (ESI-QTOF-MS) with concentrations in the range of 0.1–36 ng/L [24].

Compared to their chlorinated analogs, brominated and iodinated DBPs exhibit higher toxicity. However, these DBPs also pose analytical challenges due to their high polarity, low thermal stability, and large molecular weight, which hinder their detection by GC-related methods. An SPE-HPLC-MS/MS method developed by Hu et al. (2018) with negative ESI and multiple reaction monitoring (MRM) mode enabled the simultaneous detection of iodinated haloacetic acids (HAAs) and aromatic iodinated DBPs, including monoiodoacetic acid (MIAA) and diiodoacetic acid (DIAA) 3,5-diiodo-4-hydroxybenzaldehyde, 3,5-diiodosalicylic acid, 2,6-diiodo-4-nitrophenol and 2,4,6-triiodophenol with a detection limit <0.2 ng/L [25]. Zhang's group proposed a novel method for the rapid and selective detection of new polar brominated and iodinated DBPs using ESI-tqMS. The method involves performing precursor ion scan (PIS) of bromide at m/z 79 and 81 and iodide at m/z 126.9, respectively [26, 27]. The application of PIS for screening new DBPs will be elaborated in Sect. 4.

2.3 Summary

LC-MS methods have been extensively employed for detecting DBPs, extending beyond the limited examples discussed earlier. By using soft ionization techniques, such as ESI, LC-MS methods can minimize the fragmentation of the analytes and preserve their molecular structure. The high resolution and sensitivity of modern MS instruments enable the identification and quantification of many known and unknown emerging DBPs. Table 1 provides a summary of the LC-MS methods detailed above for DBP detection. LC-MS methods have also been successfully applied to identify and quantify different types of DBPs, such as halogenated acids and transformation products of humic and fulvic acid fractions, estrogens, endocrine disruptors, pesticides, pharmaceuticals, and microcystins. In addition, LC-MS methods are essential for toxicological assessment of DBPs, as they provide accurate and reliable information on the chemical composition and concentration of these compounds.

3 Ion Chromatography (IC) for Haloacetic Acids

3.1 Standard IC Methods

For ionic DBPs like HAAs, ion chromatography (IC) is a suitable alternative for analysis. In 2007 and 2019, USEPA issued two standard methods, namely Methods 557.0 and 557.1 [28, 29], for HAAs analysis based on IC principles to help comply with disinfection/DBP rules (Table 2). Method 557.0 couples IC and ESI-MS/MS (IC-ESI-MS/MS) and the Method 557.1 employs two-dimensional (2D) IC and traditional suppressed conductivity detection (2D IC-CD) [28, 29]. These methods have almost no pretreatment requirements (except for filtering), so they are by far the most convenient and direct HAA detection methods. However, their main

Table 1 A summary of LC-MS methods for DBP detection

Analytes	Matrix	Sample pretreatment	Separation column	Detection method	Detection limit	Recovery	Reference
HAAs	MilliQ water, ground-water, tap water	SPE	250 × 4 mm LiChrospher RP-C_{18} column	ion-pair RPLC-ESI-MS	0.1–1.6 μg/L	18–74%	[12]
HAAs	Treated water from groundwater and surface water sources	–	100 × 2.1 mm HSS T3 column	RP-UHPLC-ESI-MS/MS	<0.8 μg/L	90–110%	[13]
HAAs	Treated surface water	–	100 × 4.6 mm Luna C18 column and 125 × 4 mm Purospher C8 column	LC-ESI-HRMS	<6 μg/L	91–120%	[14]
HAAs	Recycled and environmental water	–	10 × 2.1 mm Raptor ARC-18 column	UPLC-ESI-MS	0.05–0.1 μg/L	22–76%	[15]
HAMs	MilliiQ water, drinking water	SPE	100 × 2.1 mm Hypersil GOLD C18 packed column	SPE-LC-APCI-tqMS	7.6–19.7 ng/L	62–92%	[16]
HAMs	MilliQ water, finished water, tap water	–	150 × 2.1 mm Atlantis T3 column	UPLC-ESI-MSMS	1–100 ng/L	–	[17]
Nitrosamines	Recycled and environmental water		10 × 2.1 mm Raptor ARC-18 column	UPLC-ESI-MS	0.075–0.1 μg/L	64–92%	[15]
Nitrosamines	MilliQ water, drinking water	SPE	150 × 0.32 mm C_8 column	SPE-LC-ESI-MSMS	0.1–10.6 ng/L	41–111%	[20]
Halobenzoquinones (HBQs)	MiiliQ water, source, and treated water	SPE	100 × 2.0 mm Luna C_{18} column	LC–ESI-MS/MS	3–8.7 ng/L	64–116%	[21]
HBQs and halo-hydroxyl-benzoquinones (OH-HBQs)	MilliQ water, tap water	SPE	100 × 2.0 mm Luna C_{18} column	SPE – UHPLC–ESI-MS/MS	0.02–0.8 ng/L	68–96%	[22]

Halogenated aromatic DBPs	Simulated drinking water	LLE	50 × 2.1 mm HSS T3 column	UPLC-ESI-tqMS	–	–	[23]
Halohydroxybenzonitriles (HHBNs)	MilliQ water, drinking water	SPE	100 × 2.1 mm HSS T3 column	UPLC / ESI-QTOFMS	0.01–0.02 ng/L	87–112%	[24]
Iodinated haloacetic acids (HAAs) and aromatic iodinated DBPs	MilliQ water, tap water	SPE	100 × 2.1 mm XSelect HSS T3 column	SPE-HPLC-MSMS	0.02–0.15 ng/L	71–100%	[25]
Polar iodinated DBPs	Simulated drinking water	LLE	50 × 2.1 mm HSS T3 column	UPLC/ESI-tqMS	–	–	[26]
Polar brominated DBPs	Simulated drinking water	LLE	50 × 2.1 mm HSS T3 column	UPLC/ESI-tqMS	–	–	[27]

Table 2 Regulatory limits of HAAs in drinking water around the globe (μg/L)

	MCAA	DCAA	TCAA	HAA_5	IAA
China	–	50	100	–	20[a]
Japan	20	30	30	–	–
Korea	–	–	–	100	–
Malaysia	–	50	100	–	–
Singapore	20	50	200	–	–
Philippines	20	50	200	–	–
USA	–	–	–	60	–
Scotland				60	
Canada	–	–	–	80	–
Australia	150	100	100	–	–
New Zealand	20	50	200	–	–
Russia	50	–	5,000	–	–
WHO	20	50	200	–	–
EU	–	–	–	60	–

[a] Reference limits; –: no regulatory limits

disadvantages are that the relevant equipment is very expensive and they have no or low HAA enrichment effects, so they might not be sensitive enough to detect emerging HAAs at trace levels (e.g., monoiodoacetic acid at low ng/L levels) [30].

To overcome the cost burdens, the Chinese government promulgated a standard method (GB/T 5750-2022) that uses conventional IC tandem CD for the analysis of chlorinated and brominated HAAs [31]. Only when interfering compounds overlap with HAAs peaks (e.g., chloride and sulfate interfere with DCAA), solid phrase desalination cartridges are required for sample pretreatment. Now that the regulated HAAs in China are only DCAA and TCAA [32], which can be separated well from other coexisting anions, most ion exchange columns (IECs) used for IC are already robust enough to separate them from coexisting anions, thus satisfying contemporary regulatory needs well. When required, the desalination procedure is designed to pass samples through silver, barium, or hydrogen columns, or a combo so that interfering bicarbonate, chloride, and sulfate can be eliminated from water. However, it cannot effectively remove other anions (such as nitrate) from water samples [33]. In addition, the simple IC-CD method cannot separate mono-HAAs (e.g., MCAA, MBAA, and MIAA) well nor enrich HAAs, therefore its applicability is limited and its MDLs for HAAs are high (e.g., 1.9–8.3 μg/L) [31]. Notably, H columns should be always placed after Ba/Ag columns to remove residual Ag^+ or Ba^{2+} cations, which could foul IECs by forming precipitates with eluent (e.g., $Na_2CO_3/NaHCO_3$) under alkaline conditions [34].

Novel and simple IC methods have been proposed in recent decades. For example, to enrich HAAs and separate HAAs from coexisting anions at affordable cost, a modified 2D-IC method employing a one-pump column-switching IC technique was developed [35]. Like Method 557.0 and 557.1, this method also allows direct sample injection without requiring extraction and derivatization pretreatments, giving operational convenience. But unlike conventional 2D-IC Method 557.1, which uses two pumps and two parallel sets of IC columns, this method applies only one pump and one IC column to remove interfering anions and enrich HAAs simultaneously. The key process or technology is that the column and pump are used repeatedly by switching valve ports. In addition, IC might also be combined with capillary electrophoresis (CE) to form a hybrid 2D method [36].

3.2 Sample Pretreatment and Online Treatment Methods

Eliminating anions and enriching HAAs can also be achieved by conventional sample pretreatment techniques like SPE and LLE (Table 3). Similar to HAA extraction in GC methods, SPE and LLE techniques used for IC also need to protonate HAAs first to convert ionic, hydrophilic HAAs into non-charged hydrophobic HAAs to increase sorbent and extractant interactions and maximize the extraction efficiency. Interestingly, by following the same pK_a principle but reversing the pH direction, one study applied not only forward-extraction to extract HAAs from water to MTBE (a typical extractant), but also back-extraction to transfer HAAs from MTBE back to water with neutral or higher pH [37], which successfully enriched the HAAs two-fold and enabled low MDLs at a range of 0.033–0.246 μg/L. Given that different HAAs have different pK_a and hydrophobicity characteristics, the enrichment degrees vary significantly. Although robust, the multiple LLE method has not yet been automated which might hinder its widespread acceptance in the future.

In addition, after IEC separation HAAs might undergo derivatization with nicotinamide in a post-column reactor (PCR) to form signals detectable with a fluorescence detector (FD) [38–40]. Because of the need for a special detector and derivatization equipment, this method is not widely used nowadays although it was developed 10 years ago.

3.3 IC Separation Columns

Among the IC methods that have been tested (Table 3), most have employed hydroxide-driven IECs (made of alkanol quaternary ammonium ions), and eight studies used carbonate-driven IECs (made of alkyl or polyvinyl alcohol quaternary ammonium ions). Compared to hydroxide-driven IECs, carbonate-driven IECs usually feature relatively higher baseline noise but more rapid elution of anions [41, 42]. To minimize baseline noise, researchers optimized the types of suppressors used, and helped reduce the MDLs for HAAs from 1.53–53.6 μg/L to 0.37–31.64 μg/L for carbonate-driven IECs, thus reaching similar MDLs (0.6–9.32 μg/L) as for hydroxide-driven IEC [33]. In addition, a carbonate removal device can be installed between suppressor and detector to remove carbonate/bicarbonate and reduce the baseline of eluents [43, 44]. In addition, certain columns

Table 3 A summary of IC methods used for HAAs analysis in water

Analytes	Separation equipment	Detector	Pretreatment	Detection limits (μg/L)	% Recovery (±RSD%)	Reference
HAA_5 & TFAA	IC	CD	No	8–80	/	[56]
HAA_5 & TFAA	IC	UV	No	5.1–130	/	[56]
HAA_9	IC	CD	LLE-back-extraction	0.045–1.1	48–115 (<9.8)	[57]
HAA_9	IC	CD	SPE- Ba, Ag and H cartridges	0.09–21.5	13–87 (<29.2)	[42]
HAA_9	IC	CD	No	0.5–8	/ (<9.2)	[58]
HAA_8	IC	CD	SPE	0.02–2.46	/ (<9.2)	[58]
HAA_9	IC	CD	Ag cartridge	1.11–9.32	96.6–99.1 (<3.22)	[33]
HAA_3	IC	CD	No	29.5–321.3	/ (<2.37)	[59]
HAA_3	IC	CD	Microwave-based evaporative concentration	0.06–0.85	91–93 (<4.96)	[47]
HAA_6	IC	CD	SPE- Ba, Ag and H cartridges	5.08–32.5	62–88 (<29)	[34]
HAA_5	IC	CD	SPE-Ba, Ag, H cartridges	0.48–2.61	99.6–103.2 (<2.01)	[60]
HAA_3	IC	CD	Adding Ba $(OH)_2{\cdot}4H_2O$ and Ag_2O	3.6–35.4	92–106	[61]
HAA_9	IC	CD	SPE-Guard-Ba, -Ag, and -H cartridges	8–210	22–106 (<11)	[62]
HAA_3	IC	CD	H-column, Ba-column, and Ag-H-column	3.39–6.81	60.4–100.4 (<2.7)	[63]
HAA_3	IC	CD	SPE	50–360	83–107 (<3.8)	[64]
HAA_5	IC	CD	Multi-step LLE	0.033–0.246	70–110 (<9.91)	[37]
HAA_{10}	IC	ICP-MS	No	0.36–3.28	67.4–123.1 (<23.8)	[55]
HAA_{10}	IC	ESI-MS	No	19–1,440	/ (<13)	[58]
HAA_9	IC	ESI-MS	SPE	2–138	/ (<13)	[58]
HAA_9	IC	ESI-MS	No	0.08–0.51	56–125 (<9.2)	[65]
HAA_{10}	IC	ESI-MS	SPE	0.16–0.61	83.1–106.7 (<5.2)	[66]
HAA_{13}	IC	ESI-MS	No	0.01–0.14	77–125 (<6.9)	[67]

(continued)

Table 3 (continued)

Analytes	Separation equipment	Detector	Pretreatment	Detection limits (μg/L)	% Recovery (±RSD%)	Reference
HAA_{12}	IC	ESI-MS	No	0.026–0.134	88.1–134 (<30)	[68]
HAA_9	IC	ESI-MS	No	0.009–0.099	86.7–110 (<7.4)	[69]
HAA_{11}	IC	ESI-MS	No	0.090–0.216	83.9–110 (<8.4)	[45]
Br-HAA_6	IC	ICP-MS	Ag and H cartridges	0.13–0.36	80.8–113.8	[70]
HAA_9	IC	ICP-MS	Ag cartridge	0.34–23.6	91.7–103.8 (<5.27)	[54]
D/TCAA	IC	ESI-MS/MS	No	0.27–0.28	74.2–116.3 (<9.9)	[46]
HAA_9	IC	ESI-HRMS	No	0.0011–0.18	70–130 (<8.04)	[71]
HAA_5 and IAA	IC	MS	No	0.009–0.022	/ (<4.4)	[41]
HAA_9	PCR-IC	FD	No	0.6–10.1	58–161 (<32)	[38]
HAA_9	PCR-IC	FD	Ba and two Ag cartridges	1.4–7.8	75.9–112 (<34.6)	[40]
HAA_9	SIA-PCR-IC	FD	No	0.4–0.9	90–114 (<9.2)	[39]
HAA_5	2D-IC	CD	No	0.3–0.64	90.5–115.4 (<5.69)	[44]
HAA_9	2D-IC	CD	No	0.7–1.7	60–119	[43]
HAA_{10}[1]	IC	HRMS	SPE	0.51–2.30	/ (<1.38)	[72]
HAA_9	ICxCE	C^4D	No	/	69–117 (<17.4)	[36]

PCR post-column reaction with nicotinamide, *XSD* halogen-specific detector, *ESI-MS* electrospray ionization-mass spectrometry, *ICP-MS* inductively coupled plasma mass spectrometry, *FD* fluorescence detector, *CD* conductimetry detector, *SIA* sequential injection analysis, *HRMS* high-resolution mass spectrometry, *C^4D* contactless conductivity detection

(e.g., Dionex IonPac AS24 analytical column, 2 mm × 250 mm) with high capacity (e.g., 140 μeq) can separate HAA_9 well within 25 min [45], and even a simple guard column (Dionex IonPac AG18 guard column, 2 mm × 50 mm) with low capacity (e.g., 2.5 μeq) can elute DCAA and TCAA well within 3 min [46]. Therefore, the choice of hydroxide-driven or carbonate-driven IECs is dependent on not only separation efficiency of HAAs but also the type of target compounds.

3.4 IC Detectors

The most commonly used detector for IC is the conductivity detector (CD), followed by UV, FD, ICP-MS, ESI-MS, and HRMS (Table 3). The mass-based signals of HAAs in CD depend highly on atomic masses, which can be ranked as mono > di > tri-HAAs for HAAs with identical halogen types [44, 47]. However, the molarity-based signals of HAAs in CD do not correlate well with the numbers or types of halogens. In comparison, when using UV detection at 210 nm, the signals of HAAs on a molarity basis were well ranked as I > Br > Cl for HAAs with identical numbers of halogens, and tri > di > mono-HAAs for HAAs with identical halogen types [48–52]. A similar trend is observed for FD [39]. Similar to UV and FD, ICP-MS exhibited poorer sensitivity to chlorine than that to iodine [53]. Therefore, analytical methods employing ICP-MS often obtained higher MDLs for MCAA, DCAA, and TCAA (15.6–23.6 μg/L [54]) than those for I-HAAs (0.36–0.72 μg/L [55]).

4 Precursor Ion Scan (PIS) Method for Screening New DBPs

4.1 Principles of the PIS Method

Both chlorine and bromine have two natural isotopes, i.e., $^{35}Cl/^{37}Cl$ and $^{79}Br/^{81}Br$, thus DBPs containing varying numbers of chlorine/bromine atoms present different halogen patterns during mass spectrometry analyses [73]. Coupled with the PIS mode, a triple-quadrupole mass spectrometer can be adapted for nontargeted screening for halogenated DBPs with high sample throughput. Specifically, by setting a PIS of *m/z* 79/81, compounds in the sample undergo ionization via ESI, and their molecular ions sequentially pass through the first mass analyzer. Then, each ion collides with the collision gas in the collision cell, generating fragment ions. The second mass analyzer is configured to allow only the specific fragment ion to pass, enabling detection exclusively of compounds capable of producing the pre-selected fragment ion (*m/z* 79/81). Bromine-free compounds are less likely to concurrently generate fragment ions at *m/z* 79 and 81. Even if fragment ions may be produced by chance, only actual bromine-containing compounds can exhibit the correct isotopic pattern of the ion/ion cluster peaks observed in PISs of *m/z* 79 and 81. The isotopic pattern (meeting a specific abundance ratio) can be regarded as a distinguishing "fingerprint".

The theoretical isotopic abundance ratios displayed for PISs of *m/z* 35/37 and 79/81 have been previously derived and reported [26, 74]. For a DBP containing a Br atoms and b Cl atoms, it is expected to exhibit a total of $a + b + 1$ isotopic peaks in full scan. The abundance of the Nth isotopic peak for PISs of *m/z* 79 ($A_{N,\text{pre79}}$) and 81 ($A_{N,\text{pre79}}$) can be represented as follows:

$$A_{N,\ \text{pre79}} = \sum\nolimits_{k=\max(N-b-1,0)}^{\min(N-1,a)} \left[C_a^k C_b^{N-k-1} 3^{-(N-k-1)} (a-k) \right] \quad (1)$$

$$A_{N,\ \text{pre81}} = \sum\nolimits_{k=\max(N-b-1,0)}^{\min(N-1,a)} \left[C_a^k C_b^{N-k-1} 3^{-(N-k-1)} k \right] \quad (2)$$

where k is any integer between max $(N - a - 1, 0)$ and min $(N - 1, a)$; and C_a^k and C_b^{N-k-1} are the binomial coefficients.

Upon acquiring the abundances of the isotopic peaks in PIS spectra, the theoretical abundance ratio can be determined. The observed ratio can indicate the quantities of bromine and chlorine atoms present within a halogenated DBP. It should be mentioned that although the number of iodine atoms cannot be determined when the PIS method setting *m/z* 126.9, due to iodine having only one naturally occurring isotope, the PIS method remains a valuable choice for selectively detecting electrospray-ionizable iodinated compounds in chlorinated water samples [26].

4.2 *Screening and Identification of New Polar Halogenated DBPs*

One major application of the PIS method is screening and identification of new polar halogenated DBPs. Through the combination of ion/ion cluster pairs in the PIS spectra of *m/z* 79 and 81, all polar brominated DBPs can be specifically screened in samples. Likewise, through PISs of *m/z* 35/37 and 126.9, in turn, enables the screening of polar chlorinated and iodinated DBPs. Using this powerful method, Zhang et al. (2008) have found many polar brominated DBPs in U.S. drinking water samples (comprised of finished water collected from a drinking water treatment plant and corresponding tap water), typical ion clusters including *m/z* 127/129/131, 171/173/175, 205/207/209, 215/217/219, 249/251/253, 395/397/399, 397/399/401, 410/412/414/416, 428/430, and others [74]. Some of them were tentatively identified as new DBPs for the first time. For instance, the observed ion cluster *m/z* 410/412/414/416 with an abundance ratio of 1:3:3:1 was reported to be 1-bromoamino-1,2-dibromo-1,2,2-tricarboxylethane. Pan et al. (2016) characterized the formation profiles of iodinated DBPs in simulated chlorinated and chloraminated tap water during the cooking process using iodized table salt, and nine emerging iodinated DBPs that were initially detected by PIS of *m/z* 126.9 were initially identified to be diiodoacetic acid, 3,5-diiodo-4-hydroxybenzoic acid, 3-iodo-4-hydroxybenzoic acid, 3-iodo-4-hydroxy-5-methylbenzoic acid, 3-iodo-4-hydroxybenzaldehyde, 3,5-diiodo-4-hydroxybenzaldehyde, 2,4,6-triiodophenol, 2,4-diiodo-6-nitrophenol, and 2,6-diiodo-4-nitrophenol, identifications which were subsequently confirmed [75].

In general, further structural elucidation of the ion clusters of interest is achieved through product ion scan. Characteristic fragment information is useful for speculating the detailed structures [76]. For example, the loss of 44 (CO_2) from the molecular ion indicates that the compound should contain a carboxyl group. The loss of 79/81 ($^{79}Br/^{81}Br$) instead of 80/82 ($H^{79}Br/H^{81}Br$) seems to suggest the absence of hydrogen in the molecular ions. Additional techniques that can be utilized to finalize the identification process include ultra-performance liquid chromatography (UPLC), multiple reaction monitoring (MRM), HRMS, confirmation with commercially available or synthesized standard compounds, Fourier transform infrared spectroscopy (FTIR), etc. MRM mode can also be used to confirm bromine/chlorine counts and iodine presence via ion transitions. Halogenated compounds exhibit specific retention times (RTs) in UPLC: aliphatic acids (0–2.5 min), benzoic acids (2.5–6.0 min), and phenols (6.0–8.0 min) [27]. Specific RTs under MRM mode help classify analytes, while proposed structures match RTs, isotopic ratios, and fragment ions with authentic compounds for validation. Pan et al. (2016) reported a new group of DBPs trihalo-hydroxy-cyclopentene-diones in drinking water [77]. Identifying the structure of this new set of compounds posed a challenge due to the absence of readily available standard compounds, so they synthesized a standard by reacting 2,4,6-trihydroxybenzaldehyde with bromine, followed by purification using HPLC and lyophilization, and characterized using FTIR.

For detected compounds lacking standard references, HRMS can determine accurate formulas, aiding in the validation of proposed structures. Generally, Orbitrap mass spectrometers exhibit mass measurement errors below 10 ppm, thus a mass accuracy of 5 ppm, derived from the comparison between theoretical and measured accurate *m/z* values, indicates clear and unequivocal formula identification [78]. With the assistance of HRMS analyses, 26 and 40 emerging halogenated compounds were identified in chlorinated and chloraminated tannic acid samples, respectively, all reported as DBPs for the first time [76, 79]. Until now, utilizing the robust PIS method has unveiled more than 200 previously unknown DBPs across various water matrices. Table 4 lists representative ion clusters that have been initially observed through the PIS method and further confirmed using the aforementioned techniques.

4.3 Evaluation of Formation Profiles of Polar Halogenated DBPs

Another important aspect of the application of the PIS method involves utilizing it to assess the comprehensive formation profiles of polar halogenated DBPs in tested samples. However, it should be noted that this particular method can only be used for semi-quantitative comparisons. Analysis for total organic halogen (TOX) serves as a prevalent method for gauging the collective level of halogenated compounds in tested samples, encompassing both identified and unidentified halogenated compounds [80]. The halogen-specific fractions include total organic chlorine (TOCl), bromine (TOBr), and iodine (TOI). Similarly, the total ion intensities (TII) detected in the PIS spectra at *m/z* 35 (TII_{PIS35}), *m/z* 79 (TII_{PIS79}), and *m/z* 126.9 ($TII_{PIS126.9}$)

Table 4 Representative polar halogenated DBPs initially detected by the PIS method

Matrix	Compounds	Formula	*m/z*	Additional techniques	Reference
Tap drinking water	Chlorobromopropanedioic acid	$C_3H_2ClBrO_4$	215/217/219	Product ion scan	[74]
Finished drinking water	1-Bromoamino-1,2-dibromo-1,2,2-tricarboxylethane[a]	$C_5H_4Br_3NO_6$	410/412/414/416	Product ion scan	[74]
Simulated drinking water	2-Chloro-2-(4-iodo-2,5-dicarboxyl-2,5-dihydro-3-furyl)propanoic acid	$C_9H_8ClIO_7$	389	UPLC, SIR, product ion scan	[26]
Simulated raw water	2,4-Dibromo-6-nitro-benzenediol	$C_6H_3Br_2NO_4$	310/312/314	UPLC, full scan, product ion scan	[96]
Simulated drinking water	Bromo-4-(dichloro-methyl)-5-hydroxy-2(5H)-furanone[a]	$C_5H_3BrCl_2O_3$	259/261/263/265	n.a.	[97]
Simulated drinking water	2,6-Dibromo-1,4-hydro-Quinone	$C_6H_4Br_2O_2$	265/267/269	UPLC, MRM, product ion scan, comparison with standard compound	[27]
Simulated drinking water	3,5-Dichloro-4-hydroxy-benzoic acid	$C_7H_4Cl_2O_3$	205/207/209	UPLC, MRM, product ion scan, comparison with standard compound	[88]
Simulated drinking water	2,4,6-Tribromophenol	$C_6H_3Br_3O$	327/329/331/333	UPLC, MRM, product ion scan, comparison with standard compound	[84]
Simulated tap water	2,4,6-Tribromo-5-hydroxyisophthalic acid	$C_8H_3Br_3O_5$	415/417/419/421	Proposed from its transformation product, SIR, comparison with synthesized compound	[85]
Real tap water	2,6-Dichloro-4-bromophenol	$C_6H_3Cl_2BrO$	239/241/243/245	n.a.	[86]
Tap water	Tribromo-hydroxy-cyclopen-tene-dione	$C_5HBr_3O_3$	345/347/349/351	Product ion scan, HRMS, FTIR, comparison with synthesized compound	[77]
Simulated drinking water	3,5-Diiodosalicylic acid	$C_7H_4I_2O_3$	389	Product ion scan, comparison with standard compound	[87]
ClO_2-treated drinking water	Trichloromethanol	$CHCl_3O$	133/135/137/139	Product ion scan, comparison with synthesized compound	[88]
Simulated drinking water	3,5,6-Trichloro-2-pyridinol	$C_5H_2NOCl_3$	196/198/200/202	HRMS, product ion scan, comparison with standard compound	[89]

(continued)

Table 4 (continued)

Matrix	Compounds	Formula	*m/z*	Additional techniques	Reference
Simulated cooking water	3-Iodo-4-hydroxybenzoic acid	$C_7H_5IO_3$	263	UPLC, product ion scan, comparison with standard compound	[75]
Saline sewage effluent	2,6-Dibromo-4-nitrophenol	$C_6H_3Br_2NO_3$	294/296/298	SIR, product ion scan, comparison with standard compound	[90]
Saline secondary effluent	2,6-Diiodo-4-nitrophenol	$C_6H_3I_2NO_3$	390	MRM, comparison with standard compound	[91]
Saline wastewater effluent	Tetrabromopyrrole	C_4HBr_4N	378/380/382/384/386	Product ion scan, comparison with standard compound	[92]
Saline wastewater effluent	2-Bromo-6-iodo-3,4,5-trihydroxybenzenesulfonic acid	$C_6H_4BrISO_6$	409/411	MRM, product ion scan comparison with standard compound	[81]
Simulated wastewater effluent	3,5-Dichloro-4-dodecyl-benzenesulfonic acid	$C_{18}H_{28}Cl_2SO_3$	393/395/397	MRM	[98]
Simulated saline wastewater in receiving seawater	2,6-Diiodo-1,4-hydro-quinone	$C_6H_4I_2O_2$	361	n.a.	[83]
Swimming pool water	2-Bromophenol	C_6H_5BrO	171/173	Full scan, comparison with standard compound	[93]
Chlorinated tannic acid samples	Dichloro-bromo-oxo-hexenedioic acid	$C_6H_3BrCl_2O_5$	303/305/307/309	MRM, product ion scan, HRMS	[76]
Chloraminated tannic acid samples	Bromo-3,4-dihydroxy-5-[(3,4,5-trihydroxybenzoyl)oxy]-benzoic acid	$C_{14}H_9BrO_9$	399/341	MRM, product ion scan, HRMS	[79]
Chlorinated acetaminophen solutions	Dichloro-4-acetamidophenol	$C_8H_7NCl_2O_2$	218/220/222	n.a.	[94]
Chlorinated sugar samples	n.a.	$C_{12}H_{17}O8Br$	367/369	n.a.	[95]

[a] Or its isomers

can be considered as parameters indicating the relative prevalence of polar chlorine-, bromine-, and iodine-containing compounds respectively, across diverse water samples.

In general, halogen-specific TIIs show a strong correlation with their respective TOX values. Gong and Zhang (2015) investigated the DBP formation characteristics in disinfected wastewater effluents [81]; they observed that the $TII_{PIS126.9}$ value demonstrated a consistent pattern comparable to the TOI level across various disinfection conditions. Yang et al. (2019) examined the DBP formation traits resulting from chlorinating tannic acid (a typical precursor of NOM) samples with varying biodegradation times [76]. Both of the observed TII and TOX results suggest that overall polar halogenated DBPs share a similar trend as the overall DBP mixtures in the samples. The consistent increase in the TOBr/TOX or TII_{PIS79}/(TII_{PIS79} + TII_{PIS35}) ratio during biodegradation signals a need for heightened attention toward brominated DBPs.

Moreover, Bu et al. (2018) developed an innovative and environmentally friendly pre-treatment approach aimed at facilitating TOX analysis [82]. The PIS method was employed to assess the efficiency of ultraviolet photolysis in transforming TOX into halides. In comparison with the simulated unpretreated drinking water sample, the TII in the pretreated sample decreased by 96.4%, further validating the efficacy of the new pretreatment approach.

TII may also serve as a potential indicator for the overall toxicity present in a given sample. The PIS method was utilized in investigating the photoconversion of representative DBPs and DBP mixtures derived from disinfected sewage effluents in receiving water [83]. Samples exhibiting elevated developmental toxicity also displayed increased intensities of ion clusters in the PIS spectra (*m/z* 79/81, 35/37, and 126.9). The observed trend variations among different samples suggest that DBP mixtures were dehalogenated and thus their photoconversion could reduce the overall toxicity of receiving water.

References

1. Kristiana I, Lethorn A, Joll C, Heitz A (2014) To add or not to add: the use of quenching agents for the analysis of disinfection by-products in water samples. Water Res 59:90–98. https://doi.org/10.1016/j.watres.2014.04.006
2. Ding S, Chu W, Krasner SW, Yu Y, Fang C, Xu B, Gao N (2018) The stability of chlorinated, brominated, and iodinated haloacetamides in drinking water. Water Res 142:490–500. https://doi.org/10.1016/j.watres.2018.06.024
3. Yu Y, Reckhow DA (2015) Kinetic analysis of haloacetonitrile stability in drinking waters. Environ Sci Technol 49:11028–11036. https://doi.org/10.1021/acs.est.5b02772
4. Zhang A, Chu W, Bond T, Wang F, Pan Y, Tong J, Zhu H (2020) Interference from haloacetamides during the determination of haloacetic acids using gas chromatography. J Chromatogr A 1612:460652. https://doi.org/10.1016/j.chroma.2019.460652
5. Cuthbertson AA, Liberatore HK, Kimura SY, Allen JM, Bensussan AV, Richardson SD (2020) Trace analysis of 61 emerging Br-, Cl-, and I-DBPs: new methods to achieve part-per-trillion

quantification in drinking water. Anal Chem 92:3058–3068. https://doi.org/10.1021/acs.analchem.9b04377
6. Zwiener C, Richardson SD (2005) Analysis of disinfection by-products in drinking water by LC-MS and related MS techniques. Trends Anal Chem 24:613–621. https://doi.org/10.1016/j.trac.2005.03.014
7. Yang M, Zhang X (2016) Current trends in the analysis and identification of emerging disinfection byproducts. Trends Environ Anal Chem 10:24–34. https://doi.org/10.1016/j.teac.2016.03.002
8. Richardson SD (2002) The role of GC-MS and LC-MS in the discovery of drinking water disinfection by-products. J Environ Monit 4:1–9. https://doi.org/10.1039/b105578j
9. Richardson SD, Ternes TA (2011) Water analysis: emerging contaminants and current issues. Anal Chem 92:4614–4648. https://doi.org/10.1021/ac200915r
10. Craven CB, Tang Y, Carroll K, An L, Chen B, Li XF (2022) Closing the gap of known and unknown halogenated nitrogenous disinfection byproducts in water: Advanced mass spectrometry techniques. Trends Anal Chem 153:116624. https://doi.org/10.1016/j.trac.2022.116624
11. Wawryk NJP, Craven CB, Blackstock LKJ, Li XF (2021) New methods for identification of disinfection byproducts of toxicological relevance: progress and future directions. J Environ Sci (China) 99:151–159. https://doi.org/10.1016/j.jes.2020.06.020
12. Loos R, Barceló D (2001) Determination of haloacetic acids in aqueous environments by solid-phase extraction followed by ion-pair liquid chromatography-electrospray ionization mass spectrometric detection. J Chromatogr A 938:45–55. https://doi.org/10.1016/S0021-9673(01)01092-5
13. Grundy PL, Jarvis PR, Jefferson B, Fawell J, Haley JA, Goslan EH (2024) A validated reverse-phase LC-MS/MS method for the analysis of haloacetic acids in drinking water: supporting the transition from HAA5 to HAA9. H2Open J 7(3):272–285. https://doi.org/10.2166/h2oj.2024.008
14. Postigo C, Emiliano P, Valero F (2020) High-throughput and reliable determination of 13 haloacetic acids and dalapon in water and evaluation of control strategies. Environ Sci Water Res Technol 6:2499–2509. https://doi.org/10.1039/d0ew00296h
15. Alexandrou L, Bowen C, Jones OAH (2019) Fast analysis of multiple haloacetic acids and nitrosamines in recycled and environmental waters using liquid chromatography-mass spectrometry with positive-negative switching and multiple reaction monitoring. Anal Methods 11:3793–3799. https://doi.org/10.1039/c9ay01276a
16. Chu W, Gao N, Yin D, Krasner SW, Templeton MR (2012) Trace determination of 13 haloacetamides in drinking water using liquid chromatography triple quadrupole mass spectrometry with atmospheric pressure chemical ionization. J Chromatogr A 1235:178–181. https://doi.org/10.1016/j.chroma.2012.02.074
17. Zhou R, Xu Z, Wu Y, Yang L, Li X, Meng Y, Zhu P, Kong L, Ding X (2023) Analysis of 13 haloacetamide DBPs in drinking water using a highly sensitive LC-MS/MS method. Environ Sci Water Res Technol 9:1231–1237. https://doi.org/10.1039/d2ew00781a
18. Wang W, Hu J, Yu J, Yang M (2010) Determination of N-nitrosodimethylamine in drinking water by UPLC-MS/MS. J Environ Sci 22:1508–1512. https://doi.org/10.1016/S1001-0742(09)60281-3
19. Topuz E, Aydin E, Pehlivanoglu-Mantas E (2012) A practical LC-MS/MS method for the detection of NDMA at nanogram per liter concentrations in multiple water matrices. Water Air Soil Pollut 223:5793–5802. https://doi.org/10.1007/s11270-012-1315-1
20. Zhao YY, Boyd J, Hrudey SE, Li XF (2006) Characterization of new nitrosamines in drinking water using liquid chromatography tandem mass spectrometry. Environ Sci Technol 40:7636–7641. https://doi.org/10.1021/es061332s
21. Qin F, Zhao YY, Zhao Y, Boyd JM, Zhou W, Li XF (2010) A toxic disinfection by-product, 2,6-dichloro-1,4-benzoquinone, identified in drinking water. Angew Chem Int Ed 49:790–792. https://doi.org/10.1002/anie.200904934

22. Wang W, Qian Y, Li J, Moe B, Huang R, Zhang H, Hrudey SE, Li XF (2014) Analytical and toxicity characterization of halo-hydroxyl-benzoquinones as stable halobenzoquinone disinfection byproducts in treated water. Anal Chem 86:4982–4988. https://doi.org/10.1021/ac5007238
23. Pan Y, Zhang X (2013) Four groups of new aromatic halogenated disinfection byproducts: effect of bromide concentration on their formation and speciation in chlorinated drinking water. Environ Sci Technol 47:1265–1273. https://doi.org/10.1021/es303729n
24. Hu S, Kaw HY, Zhu L, Wang W (2022) Halohydroxybenzonitriles as a new group of halogenated aromatic DBPs in drinking water: are they of comparable risk to halonitrophenols? Water Res 219:118547. https://doi.org/10.1016/j.watres.2022.118547
25. Hu S, Gong T, Ma J, Tao Y, Xian Q (2018) Simultaneous determination of iodinated haloacetic acids and aromatic iodinated disinfection byproducts in waters with a new SPE-HPLC-MS/MS method. Chemosphere 198:147–153. https://doi.org/10.1016/j.chemosphere.2018.01.124
26. Ding G, Zhang X (2009) A picture of polar iodinated disinfection byproducts in drinking water by (UPLC/)ESI-tqMS. Environ Sci Technol 43:9287–9293. https://doi.org/10.1021/es901821a
27. Zhai H, Zhang X (2011) Formation and decomposition of new and unknown polar brominated disinfection byproducts during chlorination. Environ Sci Technol 45:2194–2201. https://doi.org/10.1021/es1034427
28. Zaffiro AD, Zimmerman M, Pepich BV, Slingsby RW, Jack RF, Pohl CA, Munch DJ (2009) Determination of haloacetic acids, bromate, and dalapon in drinking water by ion chromatography electrospray ionization tandem mass spectrometry (IC-ESI-MS/MS), EPA method 557, version 1.0
29. Wagner HP, Lin R, Srinivasan K, Fisher C, Bahten K, Pohl C, Jack RF (2017) Thermofisher method 557.1: determination of haloacetic acids in drinking water using two-dimensional ion chromatography with suppressed conductivity detection, version 1.0
30. Richardson SD, Fasano F, Ellington JJ, Crumley FG, Buettner KM, Evans JJ, Blount BC, Silva LK, Waite TJ, Luther GW, McKague AB, Miltner RJ, Wagner ED, Plewa MJ (2008) Occurrence and Mammalian cell toxicity of iodinated disinfection byproducts in drinking water. Environ Sci Technol 42:8330–8338. https://doi.org/10.1021/es801169k
31. Standard examination methods for drinking water—part 10: disinfection by-products indices (GB/T 5750.10–2022) (2022) State Administration for Market Regulation & Standardization Administration of the People's Republic of China, Beijing
32. Standards for drinking water quality (GB 5749-2022) (2022) State Administration for Market Regulation & Standardization Administration of the People's Republic of China, Beijing
33. Liu YJ, Mou SF (2003) Determination of trace levels of haloacetic acids and perchlorate in drinking water by ion chromatography with direct injection. J Chromatogr A 997:225–235. https://doi.org/10.1016/s0021-9673(03)00628-9
34. Barron L, Paull B (2004) Determination of haloacetic acids in drinking water using suppressed micro-bore ion chromatography with solid phase extraction. Anal Chim Acta 522:153–161. https://doi.org/10.1016/j.aca.2004.06.052
35. Yang Y, Ma W, Chen B, Peng C, Luo W, He H (2022) Determining haloacetic acids in drinking water by one-pump column-switching ion chromatography: An online and cost-effective tool for matrix removal and sample enrichment. Microchem J 183:107997. https://doi.org/10.1016/j.microc.2022.107997
36. Ranjbar L, Gaudry AJ, Breadmore MC, Shellie RA (2015) Online comprehensive two-dimensional ion chromatography x capillary electrophoresis. Anal Chem 87:8673–8678. https://doi.org/10.1021/acs.analchem.5b01130
37. Ma W, Li WY, Yang Y, Yang J, Chen BY, Xie YF (2022) Derivatization-free multi-step extraction for trace haloacetic acids analysis with ion chromatography: performance and mechanisms. J Hazard Mater 436:10. https://doi.org/10.1016/j.jhazmat.2022.129166
38. Simone PS, Ranaivo PL, Geme G, Brown MA, Emmert GL (2009) On-line monitoring of nine haloacetic acid species at the mu g L-1 level using post-column reaction-ion chromatography with nicotinamide fluorescence. Anal Chim Acta 654:133–140. https://doi.org/10.1016/j.aca.2009.09.021

39. Henson CM, Emmert GL, Simone PS (2014) A fully-automated analyzer for determining haloacetic acid concentrations in drinking water. Chemosphere 117:586–595. https://doi.org/10.1016/j.chemosphere.2014.09.018
40. Ranaivo PL, Henson CM, Simone PS, Emmert GL (2011) Analysis of haloacetic acids in drinking water using post-column reaction-ion chromatography with on-line internal standardization. Anal Methods 3:2873–2880. https://doi.org/10.1039/c1ay05398a
41. Mathew J, McMillin R, Gandhi J, Mohsin S, Czyborra S (2009) Trace level haloacetic acids in drinking water by direct injection ion chromatography and single quadrupole mass spectrometry. J Chromatogr Sci 47:505–509. https://doi.org/10.1093/chromsci/47.7.505
42. Barron L, Paull B (2004) Direct detection of trace haloacetates in drinking water using microbore ion chromatography—improved detector sensitivity using a hydroxide gradient and a monolithic ion-exchange type suppressor. J Chromatogr A 1047:205–212. https://doi.org/10.1016/j.chroma.2004.06.100
43. Verrey D, Louyer MV, Thomas O, Baures E (2013) Direct determination of trace-level haloacetic acids in drinking water by two-dimensional ion chromatography with suppressed conductivity. Microchem J 110:608–613. https://doi.org/10.1016/j.microc.2013.07.012
44. Teh HB, Li SFY (2015) Simultaneous determination of bromate, chlorite and haloacetic acids by two-dimensional matrix elimination ion chromatography with coupled conventional and capillary columns. J Chromatogr A 1383:112–120. https://doi.org/10.1016/j.chroma.2015.01.037
45. Cheng S, Wu YP, Young TR, Dodd MC, Wu J, Zhang H, Huo ZL, Qian YT, Li Y, Li WT, Li AM (2021) Rapid determination of trace haloacetic acids in water and wastewater using non-suppressed ion chromatography with electrospray ionization-tandem mass spectrometry. Sci Total Environ 754:9. https://doi.org/10.1016/j.scitotenv.2020.142297
46. Zuo YT, Cheng S, Jiang HH, Zhang H, Wu J, Sun HF, Zhu F, Li AM, Huo ZL, Li WT (2023) Rapid and simultaneous determination of dichloroacetic acid, trichloroacetic acid, ClO3-, ClO4-, and BrO3-using short-column ion chromatography-tandem electrospray mass spectrometry. Talanta 253:7. https://doi.org/10.1016/j.talanta.2022.124022
47. Liu YJ, Mou S (2004) Determination of bromate and chlorinated haloacetic acids in bottled drinking water with chromatographic methods. Chemosphere 55:1253–1258. https://doi.org/10.1016/j.chemosphere.2003.12.023
48. Nsubuga H, Basheer C (2013) Determination of haloacetic acids in swimming pool waters by membrane-protected micro-solid phase extraction. J Chromatogr A 1315:47–52. https://doi.org/10.1016/j.chroma.2013.09.050
49. Kou DW, Wang XY, Mitra S (2004) Supported liquid membrane microextraction with high-performance liquid chromatography-UV detection for monitoring trace haloacetic acids in water. J Chromatogr A 1055:63–69. https://doi.org/10.1016/j.chroma.2004.09.022
50. Wang XY, Kou DW, Mitra S (2005) Continuous, on-line monitoring of haloacetic acids via membrane extraction. J Chromatogr A 1089:39–44. https://doi.org/10.1016/j.chroma.2005.06.075
51. Wang XY, Saridara C, Mitra S (2005) Microfluidic supported liquid membrane extraction. Anal Chim Acta 543:92–98. https://doi.org/10.1016/j.aca.2005.04.033
52. Rocio-Bautista P, Pino V, Delgado B, Afonso AM, Ayala JH (2014) Vortex-assisted emulsification microextraction followed by in-syringe ultrasound-assisted back-microextraction to determine haloacetic acids in waters. Anal Methods 6:4115–4123. https://doi.org/10.1039/c4ay00432a
53. Chen BY, Bu YA, Yang J, Nian WM, Hao SY (2020) Methods for total organic halogen (TOX) analysis in water: past, present, and future. Chem Eng J 399:30. https://doi.org/10.1016/j.cej.2020.125675
54. Liu YJ, Mou SF, Chen DY (2004) Determination of trace-level haloacetic acids in drinking water by ion chromatography-inductively coupled plasma mass spectrometry. J Chromatogr A 1039:89–95. https://doi.org/10.1016/j.chroma.2003.12.069

55. Shi HL, Adams C (2009) Rapid IC-ICP/MS method for simultaneous analysis of iodoacetic acids, bromoacetic acids, bromate, and other related halogenated compounds in water. Talanta 79:523–527. https://doi.org/10.1016/j.talanta.2009.04.037
56. Nair LM, Saari-Nordhaus R, Anderson JM (1994) Determination of haloacetic acids by ion chromatography. J Chromatogr A 671:309–313. https://doi.org/10.1016/0021-9673(94)80255-6
57. Lopez-Avila V, Liu Y, Charan C (1999) Determination of haloacetic acids in water by ion chromatography – method development. J AOAC Int 82:689–704
58. Barron L, Paull B (2006) Simultaneous determination of trace oxyhalides and haloacetic acids using suppressed ion chromatography-electrospray mass spectrometry. Talanta 69:621–630. https://doi.org/10.1016/j.talanta.2005.10.032
59. Helaleh MIH, Tanaka K, Mori M, Xu Q, Taoda H, Ding MY, Hu WZ, Hasebe K, Haddad PR (2003) Vacancy ion-exclusion chromatography of haloacetic acids on a weakly acidic cation-exchange resin. J Chromatogr A 997:133–138. https://doi.org/10.1016/s0021-9673(03)00546-6
60. Sun YX, Gu P (2007) Determination of haloacetic acids in hospital effluent after chlorination by ion chromatography. J Environ Sci 19:885–891. https://doi.org/10.1016/s1001-0742(07)60147-8
61. Fu HT, Zhao LM, Luo M, Zhang HY, Zhang J (2008) Determination of chloracetic acids in drinking water by ion chromatography using silver oxide as precipitant eliminating interference of chlorate in matrix. Chin J Anal Chem 36:1407–1410. https://doi.org/10.1016/s1872-2040(08)60076-6
62. Bruzzoniti MC, De Carlo RM, Horvath K, Perrachon D, Prelle A, Tofalvi R, Sarzanini C, Hajos P (2008) High performance ion chromatography acids on macrocyclic cryptand anion of haloacetic exchanger. J Chromatogr A 1187:188–196. https://doi.org/10.1016/j.chroma.2008.02.028
63. Luo XM, Chen L, Zhao YQ (2015) Simultaneous determination of three chloroacetic acids, three herbicides, and 12 anions in water by ion chromatography. J Sep Sci 38:3096–3102. https://doi.org/10.1002/jssc.201401456
64. Yoshikawa K, Soda Y, Sakuragawa A (2009) Determination of chloroacetic acids in drinking water using suppressed ion chromatography with solid-phase extraction. Anal Sci 25:1491–1494. https://doi.org/10.2116/analsci.25.1491
65. Slingsby R, Saini C, Pohl C (2009) The determination of haloacetic acids in real world samples using IC-ESI-MS-MS. J Chromatogr Sci 47:523–528. https://doi.org/10.1093/chromsci/47.7.523
66. Zhong ZX, Li GK, Shao YJ, Zhu BH, Liu Z, Deng JC, Mo JT (2017) Amino-functionalized graphene oxide/neutral alumina nanocomposite based solid-phase extraction coupled with ion chromatography-mass spectrometry for the determination of trace haloacetic acids in drinking water. Anal Methods 9:2425–2432. https://doi.org/10.1039/c7ay00583k
67. Wu S, Anumol T, Gandhi J, Snyder SA (2017) Analysis of haloacetic acids, bromate, and dalapon in natural waters by ion chromatography-tandem mass spectrometry. J Chromatogr A 1487:100–107. https://doi.org/10.1016/j.chroma.2017.01.006
68. Bruzzoniti MC, Rivoira L, Meucci L, Fungi M, Bocina M, Binetti R, Castiglioni M (2019) Towards the revision of the drinking water directive 98/83/EC. Development of a direct injection ion chromatographic-tandem mass spectrometric method for the monitoring of fifteen common and emerging disinfection by-products along the drinking water supply chain. J Chromatogr A 1605:8. https://doi.org/10.1016/j.chroma.2019.07.004
69. Zhang X, Saini C, Pohl C, Liu Y (2020) Fast determination of nine haloacetic acids, bromate and dalapon in drinking water samples using ion chromatography-electrospray tandem mass spectrometry. J Chromatogr A 1621:8. https://doi.org/10.1016/j.chroma.2020.461052
70. Guo ZX, Cai QT, Yu CH, Yang ZG (2003) Determination of bromate and bromoacetic acids in water by ion chromatography-inductively coupled plasma mass spectrometry. J Anal At Spectrom 18:1396–1399. https://doi.org/10.1039/b308751d

71. Huang BB, Rohrer J (2020) Targeted quantitation mode comparison of haloacetic acids, bromate, and Dalapon in drinking water using ion chromatography coupled to high-resolution (orbitrap) mass spectrometry. J Chromatogr A 1630:8. https://doi.org/10.1016/j.chroma.2020.461538
72. Gallidabino MD, Hamdan L, Murphy B, Barron LP (2018) Suspect screening of halogenated carboxylic acids in drinking water using ion exchange chromatography – high resolution (Orbitrap) mass spectrometry (IC-HRMS). Talanta 178:57–68. https://doi.org/10.1016/j.talanta.2017.08.092
73. Rosenbaum JM, Cliff RA, Coleman ML (2000) Chlorine stable isotopes: a comparison of dual inlet and thermal ionization mass spectrometric measurements. Anal Chem 72(10):2261–2264
74. Zhang X, Talley JW, Boggess B, Ding G, Birdsell D (2008) Fast selective detection of polar brominated disinfection byproducts in drinking water using precursor ion scans. Environ Sci Technol 42(17):6598–6603
75. Pan Y, Zhang X, Li Y (2016c) Identification, toxicity and control of iodinated disinfection byproducts in cooking with simulated chlor (am) inated tap water and iodized table salt. Water Res 88:60–68
76. Yang M, Zhang X, Liang Q, Yang B (2019) Application of (LC/) MS/MS precursor ion scan for evaluating the occurrence, formation and control of polar halogenated DBPs in disinfected waters: a review. Water Res 158:322–337
77. Pan Y, Li W, Li A, Zhou Q, Shi P, Wang Y (2016a) A new group of disinfection byproducts in drinking water: trihalo-hydroxy-cyclopentene-diones. Environ Sci Technol 50(14):7344–7352
78. Liu Y, Gao J, Zhu Q, Zhou X, Chu W, Huang J, Liu C, Yang B, Yang M (2023) Zerovalent Iron/Cu combined degradation of halogenated disinfection byproducts and quantitative structure–activity relationship modeling. Environ Sci Technol 57(30):11241–11250
79. Liang Q, Gao J, Guo D, Huang J, Zhang J, Li J, Yang B, Chen B, Wu Q, Yang M (2021) Species and formation characteristics of halogenated DBPs in chloramination of tannic acid after biodegradation. Sci Total Environ 781:146690
80. Chen B, Bu Y, Yang J, Nian W, Hao S (2020) Methods for total organic halogen (TOX) analysis in water: past, present, and future. Chem Eng J 399:125675
81. Gong T, Zhang X (2015) Detection, identification and formation of new iodinated disinfection byproducts in chlorinated saline wastewater effluents. Water Res 68:77–86
82. Bu Y, Song M, Han J, Zhang Z, Chen B, Zhang X, Yang M (2018) A facile and green pretreatment method for nonionic total organic halogen (NTOX) analysis in water – Step II. Using photolysis to convert NTOX completely into halides. Water Res 145:579–587
83. Liu J, Zhang X, Li Y (2017) Photoconversion of chlorinated saline wastewater DBPs in receiving seawater is overall a detoxification process. Environ Sci Technol 51(1):58–67
84. Zhai H, Zhang X, Zhu X, Liu J, Ji M (2014) Formation of brominated disinfection byproducts during chloramination of drinking water: new polar species and overall kinetics. Environ Sci Technol 48(5):2579–2588
85. Pan Y, Zhang X, Wagner ED, Osiol J, Plewa MJ (2014) Boiling of simulated tap water: effect on polar brominated disinfection byproducts, halogen speciation, and cytotoxicity. Environ Sci Technol 48(1):149–156
86. Liu J, Zhang X, Li Y (2015) Effect of boiling on halogenated DBPs and their developmental toxicity in real tap waters. In: Recent advances in disinfection by-products. American Chemical Society, pp 45–60
87. Pan Y, Li W, An H, Cui H, Wang Y (2016b) Formation and occurrence of new polar iodinated disinfection byproducts in drinking water. Chemosphere 144:2312–2320
88. Han J, Zhang X, Liu J, Zhu X, Gong T (2017) Characterization of halogenated DBPs and identification of new DBPs trihalomethanols in chlorine dioxide treated drinking water with multiple extractions. J Environ Sci 58:83–92
89. Tang H, Zhong H, Pan Y, Zhou Q, Huo Z, Chu W, Xu B (2021) A new group of heterocyclic nitrogenous disinfection byproducts (DBPs) in drinking water: role of extraction pH in unknown DBP exploration. Environ Sci Technol 55(10):6764–6772

90. Ding G, Zhang X, Yang M, Pan Y (2013) Formation of new brominated disinfection byproducts during chlorination of saline sewage effluents. Water Res 47(8):2710–2718
91. Yang M, Zhang X (2013) Comparative developmental toxicity of new aromatic halogenated DBPs in a chlorinated saline sewage effluent to the marine polychaete *Platynereis dumerilii*. Environ Sci Technol 47(19):10868–10876
92. Yang M, Zhang X (2014) Halopyrroles: a new group of highly toxic disinfection byproducts formed in chlorinated saline wastewater. Environ Sci Technol 48(20):11846–11852
93. Xiao F, Zhang X, Zhai H, Lo IM, Tipoe GL, Yang M, Pan Y, Chen G (2012) New halogenated disinfection byproducts in swimming pool water and their permeability across skin. Environ Sci Technol 46(13):7112–7119
94. Li W, Zhang X, Han J (2022) Formation of larger molecular weight disinfection byproducts from acetaminophen in chlorine disinfection. Environ Sci Technol 56(23):16929–16939
95. Qiu C, He W, Li Y, Jiang F, Pan Y, Zhang M, Lin D, Zhang K, Yang Y, Wang W, Hua P (2022) Formation of halogenated disinfection byproducts in chlorinated real water during making hot beverage: effect of sugar addition. Chemosphere 305:135417
96. Xiao F, Zhang X, Zhai H, Yang M, Lo IM (2010) Effects of enhanced coagulation on polar halogenated disinfection byproducts in drinking water. Sep Purif Technol 76(1):26–32
97. Li Y, Zhang X, Krasner SW, Shang C, Zhai H, Liu J, Yang M (2011) Penetration of polar brominated DBPs through the activated carbon columns during total organic bromine analysis. J Environ Monit 13(10):2851–2857
98. Gong T, Zhang X, Li Y, Xian Q (2016) Formation and toxicity of halogenated disinfection byproducts resulting from linear alkylbenzene sulfonates. Chemosphere 149:70–75

DBP Precursors

Penghui Du, Alex Tat-Shing Chow, and Chao Chen

Contents

1 Introduction 60
2 Natural Organic Matter 61
2.1 Sources 61
2.2 Occurrence 62
3 Anthropogenic Chemicals 63
3.1 Pesticides 63
3.2 Pharmaceuticals and Personal Care Products (PPCPs) 64
3.3 Industrial Chemicals 64
3.4 Microplastics 65
3.5 Other Substances 65
4 Inorganic DBP Precursors 65
5 Evaluation of DBP Formation Potential 66
5.1 Formation Potential (FP) Test 66
5.2 Simulated Distribution System (SDS) Test 67
5.3 Uniform Formation Conditions (UFC) Test 67
6 Characterization Methods for DBP Precursors 68
6.1 Optical Properties 69
6.2 Molecular Composition 71
6.3 Other Physiochemical Properties 76
7 Spatial and Temporal Distribution of DBP Precursors 78
7.1 Spatial Distribution of DBP Precursors 78
7.2 Temporal Variation of DBP Precursors 79
8 Summary and Outlook 80
References 80

P. Du and A. T.-S. Chow (✉)
Department of Earth and Environmental Sciences, Faculty of Science, The Chinese University of Hong Kong, Hong Kong, China
e-mail: achow@cuhk.edu.hk

C. Chen
School of Environment, Tsinghua University, Beijing, China
e-mail: chen_water@tsinghua.edu.cn

Chao Chen, Susan Andrews, and Yuefeng Xie (eds.), *Disinfection By-Products in Water*, Hdb Env Chem (2024) 135: 59–88, DOI 10.1007/698_2024_1145,

Published online: 12 September 2024

Abstract Understanding disinfection byproduct (DBP) precursors is crucial for effectively controlling the harmful DBPs in water treatment. This chapter serves to introduce DBP precursors and outline approaches to assess their DBP formation potentials. Additionally, it delves into specific methodologies for characterizing these precursors, emphasizing their critical properties that influence DBP generation. In general, there exists a positive correlation between the formation of trihalomethanes (THMs) and haloacetic acids (HAAs) and the optical parameters and chemical properties associated with aromatic components in dissolved organic matter (DOM). Furthermore, DOM with lower molecular weight and higher hydrophobicity is more prone to reacting with chlorine, leading to the generation of DBPs. Moreover, this chapter provides an overview of the spatial-temporal distribution of DBP precursors. Overall, this chapter offers comprehensive knowledge of DBP precursors, which is essential for optimizing water treatment strategies to minimize DBP formation.

Keywords Anthropogenic chemicals, Disinfection byproduct formation potential, Dissolved organic matter, Molecular composition, Spatial-temporal distribution

1 Introduction

DBP precursors are a diverse group of substances present in water that, when exposed to oxidative disinfection processes such as chlorination, lead to the formation of DBPs (Fig. 1). DBP precursors primarily consist of dissolved organic matter (DOM), anthropogenic chemicals, and inorganic compounds. Understanding the diverse nature of DBP precursors is critical for developing more effective water treatment strategies. This includes not only optimizing the disinfection process to minimize DBP formation but also implementing pre-treatment measures to reduce the concentration of precursors in the water.

In this chapter primarily, we will delve into the general properties of DBP precursors and their potential in generating DBPs. Additionally, we will introduce the characterization methods of DBP precursors as well as their spatial-temporal

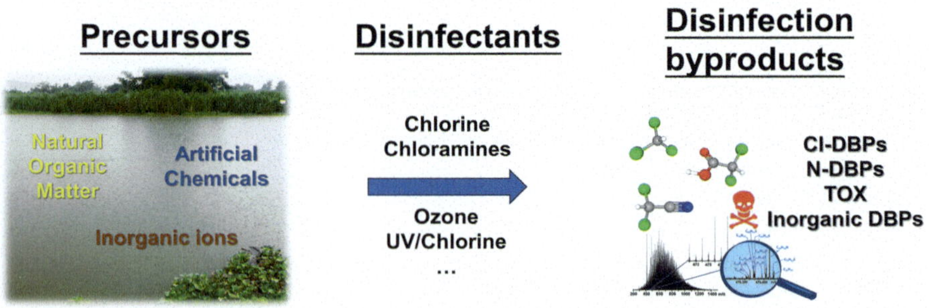

Fig. 1 The formation of disinfection byproducts (DBPs) from precursors

distribution, offering a comprehensive overview of these critical components in water treatment processes.

2 Natural Organic Matter

Natural organic matter (NOM) is a complex mixture of organic compounds derived from the decomposition of plant and animal residues and microbial activity, predominantly found in water bodies. Within NOM, the dissolved organic matter (DOM) is considered the most primary (~90%) and reactive component, playing a crucial role in the biogeochemical processes in aquatic ecosystems [1]. DOM is operationally defined as organic matter that can pass through a filter (the pore size of filters generally ranges in size between 0.70 and 0.22 μm). It typically consists of a complex mixture of organic compounds, including carbohydrates, proteins, lipids, and a variety of other undefined organic molecules [2]. This complexity and diversity render DOM highly active and central to a variety of chemical reactions in aquatic environments. When it comes to water treatment, particularly the disinfection of drinking water, the interaction between DOM and disinfectants like chlorine becomes critically important. Chlorine and similar disinfectants can react with DOM to form DBPs [3]. Importantly, not all DOM constituents are equally reactive with chlorine, and the chemical structures can affect the formation of a variety of DBP species. The source and composition of DOM can significantly influence the types and quantities of DBPs formed during disinfection. This aspect underscores the need for careful consideration of DOM characteristics in water treatment protocols to manage and mitigate potential risks associated with DBP formation.

2.1 Sources

The composition and character of aquatic DOM vary according to their sources, which fall broadly into two categories: the autochthonous pool, which is derived from within the aquatic ecosystem [4]; and the allochthonous pool, which comes from terrestrial vegetation and soil sources [5, 6]. These two DOM pools have fundamental differences in their optical and chemical characteristics, which in turn influence their responses during drinking water disinfection.

Autochthonous DOM refers to the DOM that is derived from the aquatic ecosystem through biological processes like phytoplankton exudation, bacterial production, and zooplankton activity. Phytoplankton exudation involves the release of various organic compounds by phytoplankton, which is a significant process in the aquatic ecosystem. In recent years, eutrophication and cyanobacterial bloom have led to the increasing release of autochthonous DOM from the biomass of algae or macrophytes [7, 8]. Bacteria activities are also viewed as the main source of DOM, through their metabolic processes, they decompose complex organic compounds, releasing simpler molecules that dissolve in water to form DOM [9]. Zooplankton activity involves the release of organic compounds by zooplankton, including fecal pellets and excretory products, into the water column, which are rich in organic carbon and form a substantial part of the autochthonous DOM [10]. Studies have shown how zooplankton activity contributes significantly to the autochthonous DOM pool, particularly in nutrient-rich environments [11, 12]. Generally, autochthonous DOM was more aliphatic and less oxidized compared with allochthonous DOM, which was marked by increases in lipid compounds and decreases in lignin components as compared to the allochthonous DOM-dominated freshwaters.

Allochthonous DOM is terrestrially derived, consisting of organic matter transported from the surrounding land into aquatic environments [13]. Common sources include leaves, branches, and trees that fall or are washed into water bodies like streams and estuaries. This material becomes dissolved or suspended in the water, forming a significant part of the aquatic DOM [14, 15]. Allochthonous DOM shows significant diversity in terms of level, composition, export flux, and biogeochemical activity. The composition of allochthonous DOM can vary widely based on its terrestrial origin, which includes various plant and soil materials, as well as human activities [16, 17]. Allochthonous DOM, particularly terrigenous DOM, often predominates in fluxes [18]. For example, during periods of higher water flow, terrigenous DOM forms the majority of DOM, and about 70% of the annual DOM fluxes to the coastal areas are of terrestrial origin [13]. It is also noted that allochthonous DOM is commonly more aged, refractory, and complex due to prior microbial and chemical processing on land. Terrestrially derived DOM is typically more recalcitrant, exhibiting structural complexity with high molecular weight and aromaticity. In contrast, DOM produced or transformed in-stream may be more bio-available, presenting as a simpler molecule with low molecular weight and aromaticity [19].

Moreover, the ongoing expansion of human activities has led to an increase in anthropogenic DOM being released into aquatic systems [20, 21]. This influx originates from various sources such as wastewater effluent, urban runoff, agricultural runoff, and industrial effluent [22]. Human activities may alter DOM composition by increasing the production of internally derived DOM and the processing of terrestrial DOM inputs, therefore potentially influencing the biogeochemical processes [23].

2.2 Occurrence

Generally, the levels of formed DBP have positive correlations with the concentration of DOM in water samples [24]. The parameter of dissolved organic carbon (DOC) is commonly used to characterize DOM concentration, and measurement is usually conducted in the laboratory using benchtop analyzers that oxidize organic carbon in the water sample to form carbon dioxide [25].

DOC typically accounts for 90% of the total aquatic organic carbon [1]. The DOC concentration varied largely across the global rivers (0.8–159 mg/L), with an average concentration of 10.4 mg/L [26]. In comparison, the DOC concentration in lakes generally ranged from 0.1 to >300 mg/L, and the median was 5.7 mg/L [27]. Fang et al. [28] reported that the DOC concentration gradually increased from upstream to downstream of the Yangtze River, ranging from 0.8 to 4.5 mg/L. DOC concentrations are nearly always lower in oceans than in the river, the DOC concentrations in open-ocean waters typically range from approximately 0.4 mg/L in the deep ocean to greater than 0.8 mg/L in DOC accumulating surface waters [29].

3 Anthropogenic Chemicals

Anthropogenic chemicals, also known as synthetic chemicals or man-made chemicals, are compounds that are not naturally occurring and are produced through human industrial processes. Anthropogenic chemicals generally include pesticides, pharmaceuticals and personal care products (PPCPs), microplastics, and various other synthetic compounds. These chemicals often enter water sources through agricultural runoff, industrial discharges, wastewater treatment plant effluents, and urban runoff. They can also react with disinfectants like chlorine or chloramine, contributing to the formation of various DBPs.

3.1 Pesticides

Agriculture is vital to the survival and development of humanity [30]. In activities closely related to agriculture, the use of pesticides, to control pests and weeds and enhance food yield, is an important one [31]. However, these chemicals also affect the soil and water through flow, leaching, and pulverization processes [32]. In this way, various pesticides have been detected in high concentrations in surface water worldwide [33]. Global studies on pesticide levels in surface waters indicate a significant presence of atrazine, metolachlor, chlorpyrifos, and tebuconazole [33–35]. Additionally, high levels and occurrences of other chemicals have also been detected in these waters. For instance, diuron concentrations ranged from 0.03 to 22,800 ng/L [36], dimethoate varied from 0.6 to 61,200 ng/L [33, 36], and the fungicide carbendazim was found in amounts between 0.04 and 2,240 ng/L [36, 37]. The occurrence of pesticides in aquatic environments can cause adverse effects on non-target organisms, including humans [33].

3.2 Pharmaceuticals and Personal Care Products (PPCPs)

PPCPs encompass a wide array of chemical substances, including all types of drugs (human and veterinary drugs, as well as their subsequent metabolites and conjugates) and non-medicinal consumer chemicals, such as sunscreen, soaps, moisturizers, fragrances, and shampoo. In recent years, the persistent presence of PPCPs has raised significant concerns about their potential threat to the ecological environment and human health [38–40]. PPCPs predominantly enter aquatic ecosystems through wastewater effluents from sewage treatment plants and large animal farms, with concentrations ranging from ng/L to μg/L detected in surface waters worldwide. Notably, antibiotics account for about half of this contamination [41]. A study by Spongberg et al. [42] revealed that in 86 water samples from Costa Rica, the most frequently detected PPCPs were doxycycline (77%), sulfadimethoxine (43%), salicylic acid (41%), and triclosan (34%). The highest concentrations were observed for doxycycline (74 μg/L), followed by ibuprofen (37 μg/L), gemfibrozil (17 μg/L), acetaminophen (13 μg/L), and ketoprofen (10 μg/L). Another study by Wang et al. [43] assessed 36 PPCPs in urban river water in Beijing, Changzhou, and Shenzhen, China. The most prevalent compounds included sulfadimethoxine (164 ng/L), sulpiride (77 ng/L), atenolol (53 ng/L), and indomethacin (51 ng/L). It is important to note that PPCPs are structurally designed to be biologically active at low concentrations and to have prolonged effects. These characteristics underscore the risks posed by the unintentional presence of PPCPs in the environment [44]. Studies also showed that PPCPs can serve as nitrosamine precursors during free chlorine, chlorine dioxide, and chloramine disinfection, and initial PPCP concentrations and contact time were key factors influencing nitrosamine formation [45, 46].

3.3 Industrial Chemicals

Industrial chemicals are defined as substances that originate from various industrial processes and products, such as surfactants, volatile organic compounds, phenols, polycyclic aromatic hydrocarbons, flame retardants, and per- and poly-fluoroalkyl substances. The growing presence of industrial chemicals in surface water is a global concern, originating mainly from industrial discharges, leakages, spills, and atmospheric deposition [47, 48]. The distribution and impact of industrial chemicals in freshwater systems are particularly alarming due to their toxicity and persistence in biota, sediment, and water [47, 49]. These industrial chemicals pose significant challenges to water and wastewater treatment, water reclamation, site remediation, and the management of landfill leachate due to the absence of effective and practical treatment technologies [50].

3.4 Microplastics

Microplastics, as defined by the U.S. National Oceanic and Atmospheric Administration, are plastic fragments less than 5 mm in length [51, 52]. They enter natural ecosystems from various sources, including cosmetics, clothing, food packaging, and industrial processes. It was estimated that the world's oceans contain between 15 and 51 trillion individual pieces of microplastic, weighing approximately 93,000 to 236,000 metric tons by 2014 [53, 54]. The impact of microplastics on aquatic ecosystems is profound, including ingestion by marine animals, interference with their reproductive systems, and even death [55]. Research indicates that microplastics can lead to the leaching of DOM and the formation of DBPs in water undergoing disinfection. The variety and amount of DBPs formed are influenced by factors such as the type and concentration of microplastics, their age, water salinity, and disinfection conditions [56, 57].

3.5 Other Substances

Other anthropogenic chemicals in water bodies may also serve as important precursors of DBPs, such as tire wear particles, rubber-derived tire chemicals, food additives, and nanoparticles. For instance, the friction between vehicle tires and road surfaces results in the generation of tire wear particles and leachable chemicals. Studies have found these particles in river sediments at concentrations ranging from 400 to 1,100 mg/kg, predominantly from surface runoffs [56–58]. Selbes et al. [58] reported leaching of DOC and dissolved nitrogen from scrap tires, with levels reaching up to 30.6 and 2.3 mg/L, respectively. Furthermore, Beita-Sandí et al. [59] demonstrated that chloramination of tire wear particle leachates led to the formation of nitrosamines, including *N*-nitrosodimethylamine, *N*-nitrosomorpholine, and N-nitrosopiperidine up to 126, 81, and 6.8 nmol/g-tire, respectively. Hence, tire wear particles and derived chemicals can be considered a major precursor of nitrosamines. Moreover, the proliferation of nanotechnology applications has also resulted in the release of nanoparticles into the environment. Research indicates that these nanoparticles can enhance the formation of trihalomethanes (THMs) during water disinfection processes [60, 61].

4 Inorganic DBP Precursors

Besides organic precursors, inorganic precursors of DBPs (e.g., bromide and iodide) also contribute to the formation of various DBPs during the treatment process [62]. Generally, bromide/iodide ions can be oxidized by oxidative disinfectants to hypobromous/hypoiodous acid, which may further react with DOM and other organic compounds to form brominated/iodinated DBPs [63, 64]. The bromated and iodinated DBPs are identified to be more toxic than their chlorinated analogs [65–67]. Moreover, bromate, a possible human carcinogen with a strict drinking water standard of 10 μg/L, can be generated from bromide during disinfection by ozone or other advance oxidation processes [68, 69]. Bromide and iodide in surface water primarily result from geogenic sources and anthropogenic sources, such as industrial point discharges, municipal waste incinerators, and hydraulic fracturing [68]. Based on a U.S. survey, median bromide levels in drinking water sources (groundwaters and surface waters) were in the range of 30–80 μg/L [70]. Meanwhile, the iodine levels in major rivers in the U.S., Canada, and Europe ranged from 0.5 to 212 $\mu g\ L^{-1}$, with a median and mean of 10 and 20 $\mu g\ L^{-1}$, respectively [71]. Overall, bromide and iodide ions in source water react with oxidants and NOM to produce regulated and emerging DBPs, posing adverse health risks.

5 Evaluation of DBP Formation Potential

The source and quality of water samples significantly affect DBP formation, so it is necessary to measure their potentials to generate DBP [72]. By evaluating the corresponding DBP formation potential, treatment plants can optimize their disinfection processes to reduce the formation of harmful byproducts while still effectively eliminating pathogens. This balancing act is crucial for maintaining both the safety and quality of drinking water. Three standardized tests are commonly used: formation potential (FP) test, simulated distribution system (SDS) test, and uniform formation conditions (UFC) test. Moreover, the index of specific DBP formation potential, calculated by dividing DBP formation potential by the DOC concentration, can be used to assess the capability of DOM samples to generate DBPs. This index is commonly employed to characterize DOM and its related properties [73, 74].

5.1 Formation Potential (FP) Test

FP test is a procedure to evaluate the potential of DBP precursors to form DBPs under standardized conditions. USEPA Standard Method 5710B outlines the reaction conditions for the DBP-FP test [75]. Theoretically, this test measures all possible DBPs formed using excess chlorine dosage to determine DBP-FP [76].

In the FP test, sample chlorination is typically conducted at pH ~8.2 and 25°C (Table 1). The chlorine dosage accounts for both organic (i.e., DOC) and inorganic (i.e., NH_3) demands: NaOCl (mg/L as Cl_2) = 3 × DOC (mg C/L) + 8 × NH_3–N (mg N/L) + 10, which required breakpoint chlorination (7.6 mg Cl_2/L per mg NH_3-N/L) before a free-chlorine residual could be achieved. Due to the high chlorine dose and extended incubation, chlorinated DBP levels in the test are generally much higher than those in finished water.

Table 1 Standard reaction conditions for trihalomethanes (THM) formation potential [75]

Incubation time	7 days
Incubation temperature	25 ± 2°C
Chlorine residual$_{7\ days}$	3–5 mg/L
pH	7.0 ± 0.2
pH buffer	Phosphate

Overall, the DBP-FP test is effective for evaluating DBP precursors and the impact of water treatment processes on their removal. It allows for comparison across different systems and laboratories due to standardized reaction conditions. However, it is not suitable for estimating actual DBP formation under real chlorination conditions or assessing the impact of changes in treatment processes on DBP formation in practical scenarios.

5.2 Simulated Distribution System (SDS) Test

The SDS test is designed to mirror DBP formation in actual water distribution systems. The SDS tests simulate conditions in a full-scale distribution system, focusing on aspects like materials, wall effects, bulk reactions, and water movement, which all influence DBP formation [77]. These tests are particularly effective in predicting the formation of specific DBPs such as THMs and haloacetic acids (HAAs) [78]. The final DBP concentrations measured in this test include both the initial DBPs and those formed during incubation, providing a comprehensive view of the SDS DBPs.

These SDS test results closely align with the DBPs found in actual distribution systems, both in terms of concentrations and types of DBPs [79]. SDS also considers various parameters such as chlorine dosage, incubation temperature, and holding time to predict the amounts of DBPs that would form in a distribution system, offering insights into the effective management of water treatment and distribution processes [80]. Water utilities can use this data to assess regulatory compliance of the finished water and the impact of any treatment process changes on this compliance [81].

Overall, the SDS is a valuable tool for assessing and predicting the formation potential of DBPs in drinking water systems, enabling better management of water treatment processes to ensure safe and high-quality drinking water.

5.3 Uniform Formation Conditions (UFC) Test

Contrasting with the FP test and SDS test, the UFC test employs standardized conditions, facilitating comparisons across different systems, laboratories, and even across seasons [82]. This test provides a consistent and reproducible way to

Table 2 Conditions for the uniform formation conditions test

Incubation time	24 ± 1 h
Incubation temperature	20 ± 1°C
Chlorine residual$_{24\ h}$	1.0 ± 0.4 mg/L
pH	8.0 ± 0.2
pH buffer	Borate

assess how different water treatment conditions affect the generation of DBPs [83]. These standard conditions, which are essential for the UFC test, are detailed in Table 2.

The UFC test is designed to facilitate comparisons of DBP formation across different water samples under similar conditions. It is particularly useful for contrasting DBP formation resulting from various water treatment processes and for assessing the seasonal variability of DBP precursors. In the UFC test, the DBP speciation and concentration are more reflective of actual chlorination conditions, with limited variation due to standardized chlorination conditions [84, 85].

To assess chlorine demand in water samples for the UFC test, a sodium hypochlorite (NaOCl) solution is introduced in different proportions. The goal is to maintain a target chlorine residual of 1.0 ± 0.4 mg Cl_2/L after 24 h at room temperature. The samples are buffered to a pH of around 8.0 and incubated in the dark at a controlled temperature. The chlorine residual is then measured, and samples outside the desired range are retested with adjusted chlorine doses. Once within range, the residual is neutralized for DBP extraction and analysis.

The uniformity of the test parameters in the UFC test ensures that results are consistent and comparable regardless of where or when the test is conducted, making it a valuable tool for uniform assessment and comparison of DBP formation potential in various water treatment scenarios [86].

6 Characterization Methods for DBP Precursors

The types and levels of DBPs are determined by the concentrations and properties of DBP precursors in source water. Understanding the characteristics and behavior of DBP precursors is critical for effectively managing these DBPs. While analyzing precursors with clear structures, such as anthropogenic chemicals and inorganic ions, is straightforward, DOM presents a more complex challenge. DOM often exhibits a diverse range of chemical properties that influence its behavior and interactions in aquatic systems. This complexity stems from the varied sources and formation processes of DOM. Extensive literature have documented how the types and levels of DBPs are dependent on the properties of DOM in water. Therefore, it is essential to introduce typical methods for characterizing DOM properties and their general relationships with DBP formation.

6.1 Optical Properties

6.1.1 Colored DOM

The colored fraction of the DOM (CDOM) pool absorbs ultraviolet (UV) and visible light. This absorption is primarily due to the presence of conjugated double bonds and aromatic structures within the organic molecules [87]. The intensity and wavelength of the absorption can provide information on the nature of the DOM. The absorbance in the UV range generally indicates the content of aromatic compounds, for example, the specific UV absorbance at 254 nm ($SUVA_{254}$) is commonly used to characterize the aromaticity of a DOM sample. The visible light absorption can impart the color of water, influencing its quality and perception. Moreover, the spectral slope of the absorbance curve in the UV range can provide information about the molecular weight (MW) and source of DOM.

As the exact chemical nature of CDOM is unknown, neither can be quantified in classical units of moles (or grams) per volume. Instead, they are quantified by the strength of their optical signals (Table 3). Due to their cost-efficiency and speedy analysis capabilities, optical methods for measuring absorbance are becoming more popular for monitoring the composition of DOM and inferring its sources. Furthermore, CDOM significantly influences the optical characteristics of natural waters, impacting various physical, chemical, and biological activities in aquatic ecosystems [87].

6.1.2 Fluorescent DOM

It is also important to note that many CDOM components, particularly with aromatic structures, are fluorescent (FDOM) [88]. They absorb light at certain wavelengths and re-emit it at longer wavelengths. The fluorescence of DOM is mainly attributed to two broad classes of compounds: humic-like and protein-like substances. Specifically, humic-like substances exhibit broad fluorescence emissions and are typically more abundant in terrestrial DOM. In comparison, protein-like substances show narrower fluorescence peaks, often linked to microbial or algal sources in aquatic environments [87].

Measurements of fluorescence in DOM are often conducted using three-dimensional excitation-emission matrices (3D-EEMs) [89–91]. These matrices are generated from a series of emission spectra obtained at incrementally increasing excitation wavelengths. Within a 3D-EEM, distinct peaks or areas are associated with specific types of DOM components, which provide insights into the DOM's origin, chemical composition, diagenetic status, and reactivity (Fig. 2). Broadly, these peaks are categorized into two types: humic-like fluorophores ($FDOM_H$) and amino acid-like fluorophores ($FDOM_A$) [92]. $FDOM_H$ typically shows broad emission peaks at wavelengths above 350 nm, reflecting humic-like substances formed through microbial transformation of biomolecules. In contrast, $FDOM_A$ peaks, usually narrower and found at wavelengths below 350 nm, are believed to signify

Table 3 Ecological definitions for optical parameters of DOM (abbreviations are em = emission and ex = excitation)

Parameters	Calculated	Description
A_x	The absorbance at a wavelength of x nm	
Specific UV absorbance at 254 nm ($SUVA_{254}$)	The ratio of the absorbance at 254 nm (A_{254}) to the DOC concentration	Indicator of the aromaticity of DOM
Slope ratio (S_R)	The ratio of the spectral slope of the 275–295 nm band to that of the 350–400 nm band	Indicator that has a negative correlation with the molecular weight (MW) of DOM
E2/E3	The absorbance ratio at 254 nm to that at 366 nm	Indicator that has a negative correlation with the MW of DOM
Fluorescence index (FI)	The ratio of Em at 470 nm to Em at 520 nm, obtained at Ex 370 nm	Determine the source of DOM, which is either: microbial (high FI ~ 1.8, derived from extracellular release and leachate from bacteria and algae) or terrestrially derived (low FI ~ 1.2, terrestrial plant and soil organic matter)
Freshness index (β:α)	The ratio of the fluorescence signal at Em 380 nm to the maximum signal observed between Em 420 and Em 435 nm, obtained at Ex 310 nm	Indicator of the contribution of recently produced DOM, where β represents more recently derived DOM and α represents more decomposed DOM
Humification index (HIX)	The peak area at Em 435–480 nm divided by the peak area at Em 300–345 nm + that at Em 435–480 nm, obtained at Ex 254 nm	Indicator of humic substance content or extent of humification. Higher values indicate an increasing degree of humification
Protein-like components	Tyrosine-like and tryptophan-like fluorescence components	Amino acids, free, bound in proteins, or associated with high-molecular-weight DOM
Humic-like components	Humic-like fluorescence components	Components that exhibit emission at long wavelengths are thought to be aromatic, contain many conjugated fluorescent molecules, and are red-shifted

aromatic amino acids like tryptophan and tyrosine, given the similarity in their fluorescence spectra [91, 93]. These characteristics help in understanding the nature and behavior of DOM in various environments.

Consequently, optical properties, particularly UV-vis absorbance and fluorescence, can be used to indicate water quality and assess the efficiency of water treatment processes [94]. Changes in the optical properties can indicate alterations in DOM sources and composition, which is important for understanding environmental processes [73, 95]. Table 3 outlines various standard parameters and indices derived from optical measurements, which encompass a range of data points. These

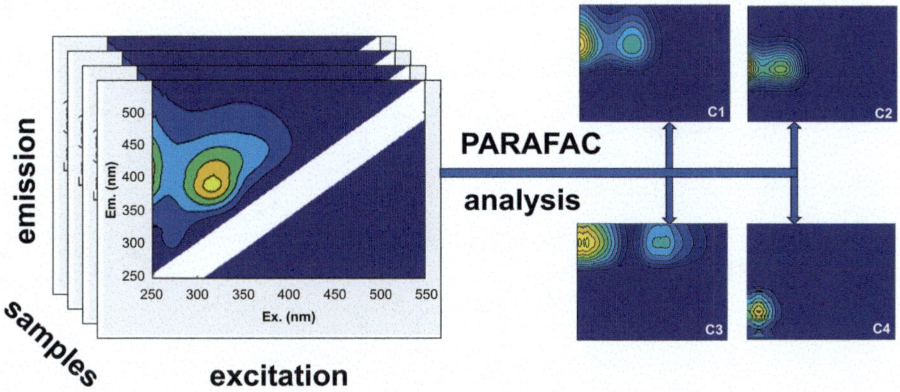

Fig. 2 EEM dataset arranged in a three-way structure and decomposed into PARAFAC components

include the absolute absorbance or fluorescence intensity at particular wavelengths, ratios calculated across different wavelengths, optical properties normalized to carbon content, and slope values found within certain spectral regions. These metrics are essential for interpreting and analyzing optical data effectively.

6.1.3 Key Optical Indices Related to DBP Formation

The optical indices can also be used to predict the DBP formation of target water samples. The DBP formation potential of the DOM sample is generally linked with several factors, including DOC, $SUVA_{254}$, specific absorbance, and fluorescence characteristics [96]. Extensive studies have reported that the formation of THMs and HAAs has a positive relationship with the parameters of $SUVA_{254}$ and/or HIX, indicating the importance of aromatic DOM structures for being DBP precursors [73, 97–99]. Aromatic structures substituted with oxygen- and nitrogen-containing functional groups readily react with chlorines to produce THMs [98]. Wang et al. [100] also reported that increased cumulative precipitation after a wildfire leads to increased DOM aromaticity, impacting the formation of various DBPs like THMs and HAAs. Moreover, a recent review investigated 45 peer-reviewed articles and concluded that the humic- and fulvic-like FDOM components exhibited strong relationships with multiple C-DBP classes including THMs, HAAs, haloacetaldehydes, and haloketones. In contrast, the formation of N-DBP classes demonstrated significant linear relationships with all fluorophore regions [101].

6.2 *Molecular Composition*

DOM is a complex mixture of organic compounds in water, having a diverse range of chemical properties that influence its behavior and interactions in aquatic systems.

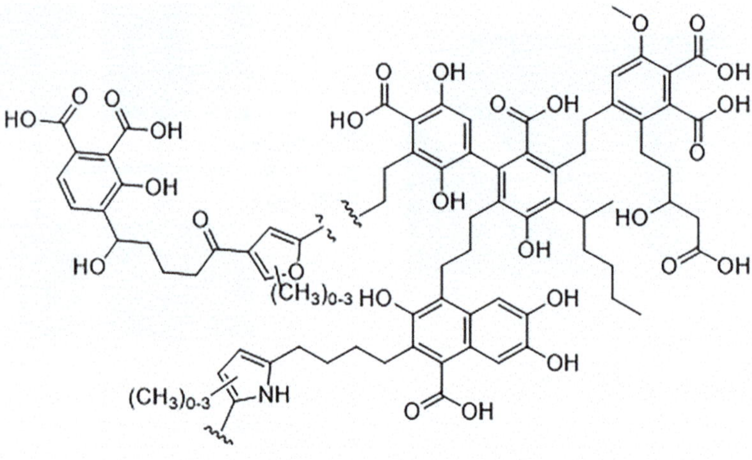

Fig. 3 Representative structure of dissolved organic matter (DOM)

This complexity arises from the diverse sources and formation processes of DOM. Understanding these properties is key to managing DOM in water treatment and assessing its environmental impact.

As shown in Fig. 3, DOM generally contains a variety of functional groups, including carboxyls, phenols, hydroxyls, carbonyls, and amines. These groups define the chemical reactivity and interactions of DOM with other substances in the environment. Many of these functional groups, especially carboxyls and phenols, contribute to the acid-base properties of DOM, affecting its charge and solubility at different pH levels. DOM's ability to form complexes with metals and other cations is a critical aspect, influencing the mobility and bioavailability of these metals in aquatic environments. DOM typically includes humic substances (humic and fulvic acids), carbohydrates, amino acids, lipids, and various other organic compounds. The composition of DOM can vary significantly depending on its source, with terrestrial DOM often being richer in humic substances, while aquatic sources might have more protein-like compounds. Moreover, DOM can act as both an electron donor and acceptor due to the redox-active components, playing a significant role in redox reactions in natural waters. The redox properties of DOM can impact microbial activity, influencing processes like denitrification and methanogenesis in aquatic and soil environments.

Gaining a comprehensive understanding of the molecular composition of DOM is vital for grasping its role in the global carbon cycle. DOM represents one of the most complex mixtures known to science, and to date, the structural details of only a handful of its components have been decoded, leaving much of its molecular makeup unknown. A core issue, and perhaps an insurmountable obstacle, in the molecular characterization of DOM lies in its nature as a virtually inseparable mixture. While modern chromatographic methods can partition this intricate DOM blend into groups based on characteristics like polarity or molecular size, the molecular intricacy within these isolated groups still exceeds the analytical capabilities of existing technologies [102, 103]. Only a few well-defined small biomolecules, such as monomeric carbohydrates, fatty acids, or amino acids, can be isolated and are then available for conventional analytical determination.

6.2.1 Nuclear Magnetic Resonance (NMR)

NMR spectroscopy is an effective tool for analyzing complex mixtures like DOM [104]. NMR spectra of DOM, which include thousands of compounds with varied molecular structures, differ significantly from those of simple organic compounds. Due to this complexity, it is often unfeasible to determine the molecular structure of each component in such mixtures.

NMR spectroscopy has been used historically to analyze DOM, particularly in studying the humic fraction of oceanic DOM. These studies using one-dimensional (1D) ^{1}H- and ^{13}C-NMR spectra typically reveal consistent spectra across depths, dominated by aliphatic structures with minor aromatic contributions, indicative of a largely autochthonous refractory material. Note that the 1D NMR spectra of DOM are characterized by broad, overlapping peaks due to its complex molecular mix. To achieve better structural resolution, multidimensional and advanced NMR techniques are being applied, such as 2D spectra combining ^{1}H- and ^{13}C-NMR data. Generally, 2D NMR offers more comprehensive information on mutual interactions of various nuclei, thereby facilitating a more detailed structural analysis of DOM constituents compared to 1D NMR. It is also important to note that the main challenges with NMR spectroscopy have been its limited sensitivity and insufficient spectral resolution.

6.2.2 Liquid Chromatography-Mass Spectrometry (LC-MS)

The introduction of soft ionization techniques like electrospray ionization (ESI) has revolutionized the molecular analysis of DOM. ESI allows for the transfer of whole organic molecules from an aqueous solution into a mass spectrometer, facilitating the molecular characterization of DOM. ESI–MS can map the molecular mass distribution of DOM across thousands of Daltons in approximately 1 Da increments. This provides a crucial structural overview and enables detailed molecular comparisons between different DOM samples. Coupling liquid chromatographic separation with mass spectrometry (LC-MS) adds an extra layer of analysis, significantly enhancing the depth of molecular information obtained (Fig. 4). LC/MS generates a highly complex molecular fingerprint for each DOM sample, consisting of thousands of data points. While targeted LC/MS analysis is useful for identifying specific compounds and their roles in microbial interactions in marine environments [105], the scope of molecules within targeted analytical windows is relatively small in the context of DOM [87]. Due to the vast molecular diversity of DOM, chromatographic

Fig. 4 Example outcome plot from LC-MS analysis

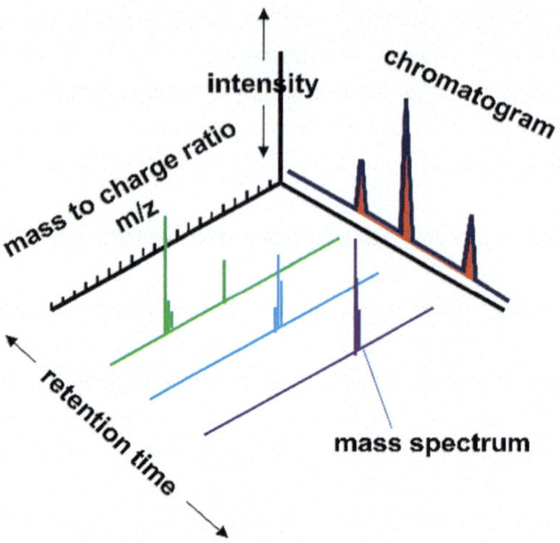

separation of individual compounds from the uncharacterized portion is challenging. Conversely, untargeted analysis encompasses a broader molecular spectrum, allowing for the assessment of molecular diversity in DOM samples.

6.2.3 Fourier-Transform Ion Cyclotron Resonance Mass Spectrometry (FT-ICR MS)

Ultrahigh-resolution mass spectrometry, particularly FT-ICR MS, has significantly advanced our ability to decipher the complex molecular structure of DOM. FT-ICR MS can measure the mass of intact molecules with extraordinary accuracy, exceeding 0.1 mDa, which is notably finer than the mass of a single electron (0.5 mDa). When combined with soft ionization methods like ESI, FT-ICR MS can detect the exact masses of tens of thousands of compounds simultaneously, obviating the need for chromatographic separation or other techniques to simplify DOM samples before analysis. In addition, the van Krevelen plot was commonly used for elemental mapping to DOM analysis [106], which can identify where relevant compound groups fall within a van Krevelen diagram (Fig. 5).

This capability has made FT-ICR MS an analytical cornerstone in studying the molecular composition of marine DOM across various ecosystems [103, 107, 108]. With its accuracy of less than 0.1 mDa, FT-ICR MS enables the highly sensitive distinction of thousands of individual molecular masses, allowing for the determination of their respective molecular formulas. However, the method does have limitations, such as the selectivity inherent in ionization techniques and the inability to distinguish between isomers.

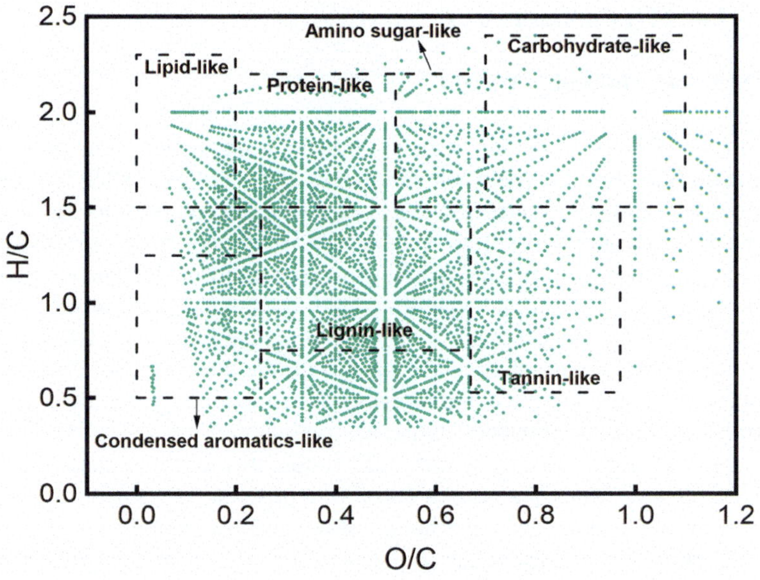

Fig. 5 van Krevelen diagram for different biochemical classifications in a typical DOM sample

6.2.4 Key DOM Constituents Influencing DBP Formation

The reactivity of chlorine with DOM to form organic DBPs is heavily dependent on the composition of DOM, which also affects the type and amount of DBPs formed. Some components in DOM could be more reactive than others in forming DBPs.

It is important to note that chlorine preferentially removes aromatic DOM formulas, suggesting that these compounds have a higher chlorine demand than other constituents [109]. This agrees well with the well-known fact that optical properties related to DOM aromaticity, such as $SUVA_{254}$, correlated positively with the formation of known DBPs, such as THMs [110]. Chlorine demand often has positive correlations with THM formation [111, 112]. Specifically, the lignin- and tannin-like compounds were found to positively correlate with the formation of THMs and haloacetonitriles (HANs) [109]. Moreover, additional strong positive correlations between HANs and more aliphatic formulas were also observed, suggesting HANs may have additional precursors. Collectively, aromatic DOM is highly reactive with free available chlorine and is more likely to form both targeted DBPs (e.g., THMs) and novel high molecular-weight DBPs compared to less aromatic fractions of DOM. In addition, organic acids in DOM are also identified as dominant precursors of THMs due to the C–O groups and unsaturated structures [113].

6.3 Other Physiochemical Properties

6.3.1 Size and Molecular Weight

DOM can be separated according to their size using fractionation techniques such as ultrafiltration (UF) and size exclusion chromatography (SEC) [24]. UF is a pressure-driven process where a porous membrane is used to retain components based on their size. It operates in either cross-flow mode or dead-end mode, with cross-flow mode being more common in full-scale drinking water treatment. UF membranes have small pore sizes (2–100 nm) and can effectively fractionate DOM at the molecular level. SEC, also known as gel permeation chromatography, separates DOM based on their size using a stationary phase with pores of different sizes [114]. Principally, as the sample passes through the column, the small compounds permeate the pores of the matrix to a larger extent than the larger compounds and are retained longer within the column, resulting in the largest molecules eluting first, and the smallest last. Both techniques allow for the separation of DOM into different-size fractions, providing valuable information about the distribution of organic compounds within the DOM pool [115]. However, it is important to note that the size fractions obtained through these techniques do not necessarily correspond to discrete molecular entities and should not be directly related to DBP precursors.

DOM is characterized by a broad range of molecular sizes and weights, extending from small, simple molecules like amino acids and sugars to large, complex macromolecules like humic and fulvic acids. This diversity in size and molecular weight reflects the varied origins and compositions of DOM, with different sources contributing different types of organic compounds. The smaller, lower molecular weight fractions of DOM typically consist of compounds that are more readily biodegradable and often more soluble in water. The larger, higher molecular weight fractions of DOM typically exhibit greater complexity, characterized by a higher concentration of aromatic and hydrophobic compounds.

The molecular weight of the DOM could significantly influence the formation of DBPs. Zhang. et al. [116] found that DBPs are mainly generated from DOM with molecular weights of 1 kDa and 3–10 kDa, with lower molecular weight DOM having a higher generation ability. This is also supported by an early study that DOM fraction with an apparent molecular weight of 1–10 kDa has been implicated as the primary source of THM precursors [24]. DOM compounds with lower molecular weight, having faster diffusion rates, are more readily available for reactions with disinfectants like chlorine. Additionally, their lower tendency to adsorb onto particulate matter in water increases their interaction with disinfectants, further enhancing DBP formation. Dong et al. [117] investigated the DBP toxicities of different molecular weight fractions (i.e., <1 kD, 1–3 kD, 3–5 kD, and > 5 kD), and revealed that the <1 kD fraction for both chloraminated and chlorinated waters exhibited the highest levels of predictive cytotoxicity and direct genotoxicity.

6.3.2 Hydrophobicity

DOM in water can be separated into different fractions depending on the compound's hydrophobicity. The separation of DOM based on hydrophobicity is usually conducted by using adsorbent resins [118]. This method involves passing the sample through a column packed with resin like XAD-8, where hydrophobic DOM adsorbs onto the resin while non-hydrophobic fractions pass through. The non-hydrophobic fractions can be further fractionated into hydrophilic and transphilic fractions using another resin like XAD-4. The hydrophobic fraction adsorbed on the XAD-8 resin can be eluted using methanol, a strong base, and a strong acid in sequence, and further fractionated into hydrophobic neutral, hydrophobic acid, and hydrophobic base fractions [119].

Hydrophobic DOM, often composed of larger, aromatic, and more complex molecules such as certain humic substances, tends to be less soluble in water. In contrast, the hydrophilic fraction, including more polar and smaller molecules like sugars, amino acids, and low-molecular-weight acids, is more readily soluble in water. The balance between these hydrophobic and hydrophilic components plays a crucial role in DOM's behavior in the environment.

Hydrophobic fractions of DOM have been implicated as the primary source of THM precursors [24, 99, 118, 120]. Hydrophobic DOM generally leads to increased and potentially more problematic DBP formation [113]. This increased reactivity is due to the abundance of reactive functional groups in hydrophobic compounds, leading to more DBP formation. For instance, Fan et al. [121] reported that hydrophobic acids and hydrophobic neutrals are key contributors to the formation of mutagenic DBPs in both river and reservoir water. Moreover, it is also noted that chlorination of DOM fractions in filtered river water yields more THMs and HAAs than chloramination, with hydrophobic acid being the dominant contributor [122].

6.3.3 Polarity

Separating DOM by polarity involves fractionation methods like Solid Phase Extraction (SPE), where DOM components are separated based on their affinity to water and non-polar solvents [123, 124]. Initially, a filtered DOM sample is prepared, then polar compounds (like acids, bases, and neutrals) are eluted using water in SPE, while non-polar, hydrophobic compounds are later eluted with non-polar solvents like methanol. For more detailed separation, techniques such as gel permeation chromatography or high-performance liquid chromatography (HPLC) can be used, further fractionating DOM based on size or specific interactions. The resulting fractions are analyzed using methods like mass spectrometry, NMR, and UV-visible spectroscopy, providing insight into the composition and properties of DOM, which is essential for environmental and water treatment research.

7 Spatial and Temporal Distribution of DBP Precursors

The spatial and temporal distribution of DOM in aquatic environments is a critical aspect of water chemistry, significantly influencing the quality and treatment of water resources. Spatially, DOM concentrations and compositions vary significantly across different water bodies, influenced by factors such as local geology, land use, and climatic conditions. Temporally, DOM exhibits dynamic changes, which are driven by biological activity, weather patterns, and hydrological cycles. Exploring the spatial-temporal distribution of DOM is thus essential for predicting and managing the formation of harmful DBPs during water treatment processes.

7.1 Spatial Distribution of DBP Precursors

The spatial distribution of DOM in aquatic systems is a multifaceted subject, encompassing a range of environmental, chemical, and geographical factors. DOM's spatial variability is crucial for understanding ecosystem dynamics, water quality, and the challenges in water treatment. Clear and distinct differences in the quality and quantity of DOM were evident in streams and rivers draining different land use types. For instance, forests and wetlands can be a significant source of high molecular weight, aromatic DOM while cropland and pasture can contribute amino acid-like, microbial-derived DOM with a low molecular weight [125, 126]. In forested watersheds, upper soil horizons and leaf debris tend to contribute low molecular weight, and reactive DOM molecules to the streams [127]. In urban and agricultural watersheds, impervious surfaces, crop detritus, and upper soils might contribute terrestrial, humic, slightly higher molecular weight, yet reactive DOM that can also promote in-stream DOM production and microbial processing [128, 129]. Conversely, in wetlands, organic matter-rich soils and sub-surface water pools contribute humic, complex, and recalcitrant terrestrial DOM that may limit in-stream productivity [130, 131]. The influence of geographical factors, such as proximity to coastlines or altitude, also plays a significant role, as discussed by Ruchi et al. [132], who observed distinct variations in DOM composition in rivers and coastal areas. Literature also reported that Riverine DOC concentration was 46.8% higher in the north than in the south of China, which is likely due to increased anthropogenic activities combined with soil organic matter via soil erosion/leaching through precipitation and agricultural runoff/discharge from the north [133, 134]. Table 4 presents the optical properties of DOM samples collected from different sites along with Maozhou River, China, suggesting the significant discrepancy of DOM in spatial distribution [135].

As DOM is an important DBP precursor, and thus spatial variations can have a significant impact on water treatability and drinking water safety. A study by Fang et al. [28] revealed that there was considerable diversity in DOM characteristics and chlorine reactivity along the Yangtze River, China. The upstream and midstream

Table 4 Averaged spectroscopic indices of DOM-containing water samples from different sites along with Maozhou River, China [135][a]

Sites	SUV_{254} (L/mgC/m)	E2/E3	FI	HIX	BIX
Mainstream	1.87	6.26	2.08	2.74	1.03
Tributary	2.00	5.42	1.98	2.39	0.98
Pond	1.39	6.20	1.85	3.16	0.91
Reservoir	2.01	5.34	1.71	3.98	0.90

[a] $SUVA_{254}$: specific ultraviolet absorbance at 254 nm; FI: fluorescence index; HIX: humification index; BIX: biological index

waters were mainly characterized by humic substances with terrestrial origins, while high levels of low-molecular-weight organics and microbial-derived protein-like biopolymers were present in downstream water, which significantly contributed to the formation potentials of haloacetaldehydes and HANs that dominated DBP-associated mammalian cell cytotoxicity.

7.2 *Temporal Variation of DBP Precursors*

Temporal variation in DBP precursors at a specific location might be influenced by external factors, including seasonal changes, and both natural or anthropogenic disturbances, leading to changes in substrates and water inputs or losses, and alterations in on-site geochemical processes. For example, a study revealed the seasonal changes in DOM of Tai Lake, China that considerable biosignatures were present in summer DOM, yet with a near absence of algal extract compounds. The time-dependent molecular signature of Tai Lake DOM was likely dominated by microbial metabolism rather than abiotic chemical transformations [136]. Furthermore, long-term temporal variations, possibly driven by climate change and anthropogenic activities, have been noted. It was also suggested that global shifts in climate patterns are altering the hydrological cycle, leading to changes in DOM characteristics over time. Chow et al. [137] conducted a one-year monitoring study in a wetland in South Carolina, results showed that the DOC and DON concentrations were significantly higher during the peak litterfall season (October to December) compared to spring (March to May). Moreover, Majidzadeh et al. [138] found a 22%–99% increase in DOC load in the outflow of 18 sub-watersheds during flooding compared to baseflow conditions, while the DOM during flooding was observed with increased aromaticity and humification degree. Correspondingly, the formation potential of THMs was observed to be 238% (±117%) higher under the flooding condition compared with the baseflow ($p < 0.001$) using chlorine as a disinfectant. These shifts in DOM composition and concentration over time have implications for water treatment strategies, as they might lead to an increase in the formation of DBPs, posing a challenge for maintaining water quality standards.

8 Summary and Outlook

The presence of DBPs in water is garnering increased concern due to their potential harm to both aquatic life and humans. It is crucial to comprehend the nature of DBP precursors to mitigate risks in water treatment processes. This chapter delves into the primary categories of DBP precursors, including DOM, anthropogenic chemicals, and inorganic substances, and explores prevalent methods for assessing the DBP formation potential in each sample. Moreover, it highlights the fundamental characteristics of DBP precursors, such as the aromaticity, molecular weight, and hydrophobicity, that significantly influence DBP formation, and the techniques for analyzing these characteristics. Additionally, the chapter discusses the spatial and temporal variations of DOM in aquatic environments, offering insights that could enhance the management of DBP formation in water treatment processes.

Due to the complex nature of water matrices, research on DBP precursors still faces challenges in effective detection and characterization, as well as relationships between DBP generation and properties of DBP precursors. Innovations in water treatment technology are essential for removing these precursors and ensuring water safety. Additionally, it is vital to study the impact of environmental factors such as climate change and urbanization on the distribution of DBP precursors for future research and water management strategies.

Acknowledgements The authors would like to thank the Research Assistant Professorship Scheme of The Chinese University of Hong Kong (no. 3130207), the General Research Fund of the Research Grants Council of Hong Kong (no. 14300824), and the Environmental System Science Program of the U.S. Department of Energy (no. DE-SC00023311) for their financial support.

References

1. Wetzel RG (2001) Limnology: lake and river ecosystems. Gulf Professional Publishing
2. Nebbioso A, Piccolo A (2013) Molecular characterization of dissolved organic matter (DOM): a critical review. Anal Bioanal Chem 405(1):109–124. https://doi.org/10.1007/s00216-012-6363-2
3. Kimura SY, Ortega-Hernandez A (2019) Formation mechanisms of disinfection byproducts: recent developments. Curr Opin Environ Sci Health 7:61–68. https://doi.org/10.1016/j.coesh.2018.11.002
4. Guillemette F, del Giorgio PA (2012) Simultaneous consumption and production of fluorescent dissolved organic matter by lake bacterioplankton. Environ Microbiol 14(6):1432–1443. https://doi.org/10.1111/j.1462-2920.2012.02728.x
5. Kieber RJ, Whitehead RF, Skrabal SA (2006) Photochemical production of dissolved organic carbon from resuspended sediments. Limnol Oceanogr 51(5):2187–2195. https://doi.org/10.4319/lo.2006.51.5.2187
6. Miller C, Gordon KG, Kieber RJ, Willey JD, Seaton P (2009) Chemical characteristics of chromophoric dissolved organic matter in rainwater. Atmos Environ 43(15):2497–2502. https://doi.org/10.1016/j.atmosenv.2009.01.056
7. Liu S, He Z, Tang Z, Liu L, Hou J, Li T et al (2020) Linking the molecular composition of autochthonous dissolved organic matter to source identification for freshwater lake ecosystems

by combination of optical spectroscopy and FT-ICR-MS analysis. Sci Total Environ 703: 134764. https://doi.org/10.1016/j.scitotenv.2019.134764
8. Zhou Y, Davidson TA, Yao X, Zhang Y, Jeppesen E, de Souza JG et al (2018) How autochthonous dissolved organic matter responds to eutrophication and climate warming: evidence from a cross-continental data analysis and experiments. Earth Sci Rev 185:928–937. https://doi.org/10.1016/j.earscirev.2018.08.013
9. Broman E, Asmala E, Carstensen J, Pinhassi J, Dopson M (2019) Distinct coastal microbiome populations associated with autochthonous- and allochthonous-like dissolved organic matter. Front Microbiol 10. https://doi.org/10.3389/fmicb.2019.02579
10. Popa CL, Dontu SI, Carstea EM, Ioja I-C, Florescu LI, Dumitrache AC et al (2023) Land use impact on the levels of fluorescent dissolved organic matter, phytoplankton and zooplankton in urban lakes. Limnologica 99:126062. https://doi.org/10.1016/j.limno.2023.126062
11. Urban-Rich J, McCarty JT, Fernández D, Acuña JL (2006) Larvaceans and copepods excrete fluorescent dissolved organic matter (FDOM). J Exp Mar Biol Ecol 332(1):96–105. https://doi.org/10.1016/j.jembe.2005.11.023
12. Johnston SE, Finlay K, Spencer RGM, Butman DE, Metz M, Striegl R et al (2022) Zooplankton release complex dissolved organic matter to aquatic environments. Biogeochemistry 157(3):313–325. https://doi.org/10.1007/s10533-021-00876-7
13. Hosen JD, Aho KS, Fair JH, Kyzivat ED, Matt S, Morrison J et al (2021) Source switching maintains dissolved organic matter Chemostasis across discharge levels in a large temperate river network. Ecosystems 24(2):227–247. https://doi.org/10.1007/s10021-020-00514-7
14. Raymond PA, Saiers JE, Sobczak WV (2016) Hydrological and biogeochemical controls on watershed dissolved organic matter transport: pulse-shunt concept. Ecology 97(1):5–16. https://doi.org/10.1890/14-1684.1
15. Singh NK, Reyes WM, Bernhardt ES, Bhattacharya R, Meyer JL, Knoepp JD et al (2016) Hydro-climatological influences on long-term dissolved organic carbon in a mountain stream of the southeastern United States. J Environ Qual 45(4):1286–1295. https://doi.org/10.2134/jeq2015.10.0537
16. Aitkenhead-Peterson JA, McDowell WH, Neff JC (2003) 2 - Sources, production, and regulation of Allochthonous dissolved organic matter inputs to surface waters. In: Findlay SEG, Sinsabaugh RL (eds) Aquatic ecosystems. Academic Press, Burlington, pp 25–70
17. Dong Y, Li Y, Kong F, Zhang J, Xi M (2020) Source, structural characteristics and ecological indication of dissolved organic matter extracted from sediments in the primary tributaries of the Dagu River. Ecol Indic 109:105776. https://doi.org/10.1016/j.ecolind.2019.105776
18. Yates CA, Johnes PJ, Brailsford FL, Evans CD, Evershed RP, Glanville HC et al (2023) Determining patterns in the composition of dissolved organic matter in fresh waters according to land use and management. Biogeochemistry 164(1):143–162. https://doi.org/10.1007/s10533-022-00964-2
19. Fellman JB, Spencer RGM, Hernes PJ, Edwards RT, D'Amore DV, Hood E (2010) The impact of glacier runoff on the biodegradability and biochemical composition of terrigenous dissolved organic matter in near-shore marine ecosystems. Mar Chem 121(1):112–122. https://doi.org/10.1016/j.marchem.2010.03.009
20. Li P, Zhao C, Liu K, Xiao X, Wang Y, Wang Y et al (2021) Anthropogenic influences on dissolved organic matter in three coastal bays, North China. Front Earth Sci:9. https://doi.org/10.3389/feart.2021.697758
21. Harjung A, Attermeyer K, Aigner V, Krlovic N, Steniczka G, Švecová H et al (2020) High anthropogenic organic matter inputs during a festival increase river heterotrophy and refractory carbon load. Environ Sci Technol 54(16):10039–10048. https://doi.org/10.1021/acs.est.0c02259
22. Xenopoulos MA, Barnes RT, Boodoo KS, Butman D, Catalán N, D'Amario SC et al (2021) How humans alter dissolved organic matter composition in freshwater: relevance for the earth's biogeochemistry. Biogeochemistry 154(2):323–348. https://doi.org/10.1007/s10533-021-00753-3

23. Williams CJ, Frost PC, Morales-Williams AM, Larson JH, Richardson WB, Chiandet AS et al (2016) Human activities cause distinct dissolved organic matter composition across freshwater ecosystems. Glob Chang Biol 22(2):613–626. https://doi.org/10.1111/gcb.13094
24. Chow AT, Gao S, Dahlgren RA (2005) Physical and chemical fractionation of dissolved organic matter and trihalomethane precursors: a review. J Water Supply Res Technol AQUA 54(8):475–507. https://doi.org/10.2166/aqua.2005.0044
25. Kolka R, Weishampel P, Fröberg M (2008) Measurement and importance of dissolved organic carbon. J Field Measure For Carbon Monit:171–176. https://doi.org/10.1007/978-1-4020-8506-2_13
26. Liu F, Wang D (2022) Dissolved organic carbon concentration and biodegradability across the global rivers: a meta-analysis. Sci Total Environ 818:151828. https://doi.org/10.1016/j.scitotenv.2021.151828
27. Sobek S, Tranvik LJ, Prairie YT, Kortelainen P, Cole JJ (2007) Patterns and regulation of dissolved organic carbon: an analysis of 7,500 widely distributed lakes. Limnol Oceanogr 52(3):1208–1219. https://doi.org/10.4319/lo.2007.52.3.1208
28. Fang C, Yang X, Ding S, Luan X, Xiao R, Du Z et al (2021) Characterization of dissolved organic matter and its derived disinfection byproduct formation along the Yangtze River. Environ Sci Technol 55(18):12326–12336. https://doi.org/10.1021/acs.est.1c02378
29. Hansell DA, Carlson CA, Repeta DJ, Schlitzer R (2009) Dissolved organic matter in the ocean: a controversy stimulates new insights. Oceanography 22(4):202–211. https://doi.org/10.5670/oceanog.2009.109
30. Sadowski A, Baer-Nawrocka A (2018) Food and environmental function in world agriculture – interdependence or competition? Land Use Policy 71:578–583. https://doi.org/10.1016/j.landusepol.2017.11.005
31. Alletto L, Coquet Y, Benoit P, Heddadj D, Barriuso E (2010) Tillage management effects on pesticide fate in soils, A review. Agron Sustain Dev 30(2):367–400. https://doi.org/10.1051/agro/2009018
32. Nsibande SA, Forbes PBC (2016) Fluorescence detection of pesticides using quantum dot materials – a review. Anal Chim Acta 945:9–22. https://doi.org/10.1016/j.aca.2016.10.002
33. de Souza RM, Seibert D, Quesada HB, de Jesus BF, Fagundes-Klen MR, Bergamasco R (2020) Occurrence, impacts and general aspects of pesticides in surface water: a review. Process Saf Environ Prot 135:22–37. https://doi.org/10.1016/j.psep.2019.12.035
34. Battaglin WA, Bradley PM, Iwanowicz L, Journey CA, Walsh HL, Blazer VS (2018) Pharmaceuticals, hormones, pesticides, and other bioactive contaminants in water, sediment, and tissue from Rocky Mountain National Park, 2012–2013. Sci Total Environ 643:651–673. https://doi.org/10.1016/j.scitotenv.2018.06.150
35. Köck-Schulmeyer M, Postigo C, Farré M, Barceló D, López de Alda M (2019) Medium to highly polar pesticides in seawater: analysis and fate in coastal areas of Catalonia (NE Spain). Chemosphere 215:515–523. https://doi.org/10.1016/j.chemosphere.2018.10.049
36. Carazo-Rojas E, Pérez-Rojas G, Pérez-Villanueva M, Chinchilla-Soto C, Chin-Pampillo JS, Aguilar-Mora P et al (2018) Pesticide monitoring and ecotoxicological risk assessment in surface water bodies and sediments of a tropical agro-ecosystem. Environ Pollut 241:800–809. https://doi.org/10.1016/j.envpol.2018.06.020
37. Metcalfe CD, Helm P, Paterson G, Kaltenecker G, Murray C, Nowierski M et al (2019) Pesticides related to land use in watersheds of the Great Lakes basin. Sci Total Environ 648: 681–692. https://doi.org/10.1016/j.scitotenv.2018.08.169
38. Cizmas L, Sharma VK, Gray CM, McDonald TJ (2015) Pharmaceuticals and personal care products in waters: occurrence, toxicity, and risk. Environ Chem Lett 13:381–394. https://doi.org/10.1007/s10311-015-0524-4
39. Nozaki K, Tanoue R, Kunisue T, Tue NM, Fujii S, Sudo N et al (2023) Pharmaceuticals and personal care products (PPCPs) in surface water and fish from three Asian countries: species-specific bioaccumulation and potential ecological risks. Sci Total Environ 866:161258. https://doi.org/10.1016/j.scitotenv.2022.161258

40. Dey S, Bano F, Malik A (2019) Pharmaceuticals and personal care product (PPCP) contamination – a global discharge inventory. Pharmaceuticals and personal care products: waste management and treatment technology. Elsevier, pp 1–26
41. Yang Y, Ok YS, Kim K-H, Kwon EE, Tsang YF (2017) Occurrences and removal of pharmaceuticals and personal care products (PPCPs) in drinking water and water/sewage treatment plants: a review. Sci Total Environ 596–597:303–320. https://doi.org/10.1016/j.scitotenv.2017.04.102
42. Spongberg AL, Witter JD, Acuña J, Vargas J, Murillo M, Umaña G et al (2011) Reconnaissance of selected PPCP compounds in Costa Rican surface waters. Water Res 45(20): 6709–6717. https://doi.org/10.1016/j.watres.2011.10.004
43. Wang Z, Zhang X-H, Huang Y, Wang H (2015) Comprehensive evaluation of pharmaceuticals and personal care products (PPCPs) in typical highly urbanized regions across China. Environ Pollut 204:223–232. https://doi.org/10.1016/j.envpol.2015.04.021
44. Ebele AJ, Abou-Elwafa Abdallah M, Harrad S (2017) Pharmaceuticals and personal care products (PPCPs) in the freshwater aquatic environment. Emerg Contam 3(1):1–16. https://doi.org/10.1016/j.emcon.2016.12.004
45. Zhang A, Li Y, Song Y, Lv J, Yang J (2014) Characterization of pharmaceuticals and personal care products as N-nitrosodimethylamine precursors during disinfection processes using free chlorine and chlorine dioxide. J Hazard Mater 276:499–509. https://doi.org/10.1016/j.jhazmat.2014.05.069
46. Shen R, Andrews SA (2011) Demonstration of 20 pharmaceuticals and personal care products (PPCPs) as nitrosamine precursors during chloramine disinfection. Water Res 45(2):944–952. https://doi.org/10.1016/j.watres.2010.09.036
47. Choo G, Wang W, Cho H-S, Kim K, Park K, Oh J-E (2020) Legacy and emerging persistent organic pollutants in the freshwater system: relative distribution, contamination trends, and bioaccumulation. Environ Int 135:105377. https://doi.org/10.1016/j.envint.2019.105377
48. Morin-Crini N, Lichtfouse E, Liu G, Balaram V, Ribeiro ARL, Lu Z et al (2022) Worldwide cases of water pollution by emerging contaminants: a review. Environ Chem Lett 20(4): 2311–2338. https://doi.org/10.1007/s10311-022-01447-4
49. Jiang W, Huang T, Chen H, Lian L, Liang X, Jia C et al (2018) Contamination of short-chain chlorinated paraffins to the biotic and abiotic environments in the Bohai Sea. Environ Pollut 233:114–124. https://doi.org/10.1016/j.envpol.2017.10.034
50. Cui J, Gao P, Deng Y (2020) Destruction of per- and polyfluoroalkyl substances (PFAS) with advanced reduction processes (ARPs): a critical review. Environ Sci Technol 54(7): 3752–3766. https://doi.org/10.1021/acs.est.9b05565
51. Collignon A, Hecq J-H, Galgani F, Collard F, Goffart A (2014) Annual variation in neustonic micro- and meso-plastic particles and zooplankton in the bay of Calvi (Mediterranean–Corsica). Mar Pollut Bull 79(1):293–298. https://doi.org/10.1016/j.marpolbul.2013.11.023
52. Ghosh S, Sinha JK, Ghosh S, Vashisth K, Han S, Bhaskar R (2023) Microplastics as an emerging threat to the global environment and human health. Sustainability 15(14):10821. https://doi.org/10.3390/su151410821
53. Ioakeimidis C, Fotopoulou KN, Karapanagioti HK, Geraga M, Zeri C, Papathanassiou E et al (2016) The degradation potential of PET bottles in the marine environment: an ATR-FTIR based approach. Sci Rep 6:23501. https://doi.org/10.1038/srep23501
54. van Sebille E, Wilcox C, Lebreton L, Maximenko N, Hardesty BD, van Franeker JA et al (2015) A global inventory of small floating plastic debris. Environ Res Lett 10(12):124006. https://doi.org/10.1088/1748-9326/10/12/124006
55. Issac MN, Kandasubramanian B (2021) Effect of microplastics in water and aquatic systems. Environ Sci Pollut Res 28(16):19544–19562. https://doi.org/10.1007/s11356-021-13184-2
56. Ghanadi M, Kah M, Kookana RS, Padhye LP (2023) Formation of disinfection by-products from microplastics, tire wear particles, and other polymer-based materials. Water Res 230: 119528. https://doi.org/10.1016/j.watres.2022.119528

57. Ateia M, Kanan A, Karanfil T (2020) Microplastics release precursors of chlorinated and brominated disinfection byproducts in water. Chemosphere 251:126452. https://doi.org/10.1016/j.chemosphere.2020.126452
58. Selbes M, Yilmaz O, Khan AA, Karanfil T (2015) Leaching of DOC, DN, and inorganic constituents from scrap tires. Chemosphere 139:617–623. https://doi.org/10.1016/j.chemosphere.2015.01.042
59. Beita-Sandí W, Selbes M, Ersan MS, Karanfil T (2019) Release of nitrosamines and nitrosamine precursors from scrap tires. Environ Sci Technol Lett 6(4):251–256. https://doi.org/10.1021/acs.estlett.9b00172
60. Sharma VK, Yang X, Cizmas L, McDonald TJ, Luque R, Sayes CM et al (2017) Impact of metal ions, metal oxides, and nanoparticles on the formation of disinfection byproducts during chlorination. Chem Eng J 317:777–792. https://doi.org/10.1016/j.cej.2017.02.071
61. Metch JW, Ma Y, Pruden A, Vikesland PJ (2015) Enhanced disinfection by-product formation due to nanoparticles in wastewater treatment plant effluents. Environ Sci Water Res Technol 1(6):823–831. https://doi.org/10.1039/C5EW00114E
62. Dong H, Qiang Z, Richardson SD (2019) Formation of iodinated disinfection byproducts (I-DBPs) in drinking water: emerging concerns and current issues. Acc Chem Res 52(4): 896–905. https://doi.org/10.1021/acs.accounts.8b00641
63. Gong T, Zhang X (2015) Detection, identification and formation of new iodinated disinfection byproducts in chlorinated saline wastewater effluents. Water Res 68:77–86. https://doi.org/10.1016/j.watres.2014.09.041
64. Ding G, Zhang X, Yang M, Pan Y (2013) Formation of new brominated disinfection byproducts during chlorination of saline sewage effluents. Water Res 47(8):2710–2718. https://doi.org/10.1016/j.watres.2013.02.036
65. Yang Y, Komaki Y, Kimura SY, Hu H-Y, Wagner ED, Mariñas BJ et al (2014) Toxic impact of bromide and iodide on drinking water disinfected with chlorine or chloramines. Environ Sci Technol 48(20):12362–12369. https://doi.org/10.1021/es503621e
66. Richardson SD, Plewa MJ, Wagner ED, Schoeny R, DeMarini DM (2007) Occurrence, genotoxicity, and carcinogenicity of regulated and emerging disinfection by-products in drinking water: a review and roadmap for research. Mutat Res/Rev Mutat Res 636(1): 178–242. https://doi.org/10.1016/j.mrrev.2007.09.001
67. Wagner ED, Plewa MJ (2017) CHO cell cytotoxicity and genotoxicity analyses of disinfection by-products: an updated review. J Environ Sci 58:64–76. https://doi.org/10.1016/j.jes.2017.04.021
68. Morrison CM, Hogard S, Pearce R, Mohan A, Pisarenko AN, Dickenson ERV et al (2023) Critical review on bromate formation during ozonation and control options for its minimization. Environ Sci Technol 57(47):18393–18409. https://doi.org/10.1021/acs.est.3c00538
69. Fang J-Y, Shang C (2012) Bromate formation from bromide oxidation by the UV/persulfate process. Environ Sci Technol 46(16):8976–8983. https://doi.org/10.1021/es300658u
70. Westerhoff P, Sharma N, Zeng C, Karanfil T, Kim D, Ghosh A et al (2022) Occurrence survey of bromide and iodide in water supplies. Water Res Found Project:4711
71. Moran JE, Oktay SD, Santschi PH (2002) Sources of iodine and iodine 129 in rivers. Water Resour Res 38(8):24-1–24-10. https://doi.org/10.1029/2001WR000622
72. Lee M-H, Hur J (2014) Photodegradation-induced changes in the characteristics of dissolved organic matter with different sources and their effects on disinfection by-product formation potential. CLEAN-Soil Air Water 42(5):552–560. https://doi.org/10.1002/clen.201200685
73. Du P, Chen G, Zhang P, Yang B, Wang J (2023) Photo-transformation of wastewater effluent organic matter reduces the formation potential and toxicity of chlorinated disinfection byproducts. Ecotoxicol Environ Saf 265:115515. https://doi.org/10.1016/j.ecoenv.2023.115515
74. Wu Q-Y, Li C, Du Y, Wang W-L, Huang H, Hu H-Y (2016) Elimination of disinfection byproduct formation potential in reclaimed water during solar light irradiation. Water Res 95: 260–267. https://doi.org/10.1016/j.watres.2016.02.023

75. APHA (1926) Standard methods for the examination of water and wastewater. American Public Health Association
76. Xie Y (2003) Disinfection byproducts in drinking water: formation, analysis, and control. CRC Press
77. Poleneni SR, Inniss EC (2019) Array of prediction tools for understanding extent of wall effects on DBP formation in drinking water distribution systems. J Water Supply Res Technol AQUA 68(6):390–398. https://doi.org/10.2166/aqua.2019.002
78. Sfynia C, Bond T, Ganidi N, Kanda R, Templeton MR (2017) Predicting the formation of haloacetonitriles and haloacetamides by simulated distribution system tests. Procedia Engineering 186:186–192. https://doi.org/10.1016/j.proeng.2017.03.226
79. Masters S, Wang H, Pruden A, Edwards MA (2015) Redox gradients in distribution systems influence water quality, corrosion, and microbial ecology. Water Res 68:140–149. https://doi.org/10.1016/j.watres.2014.09.048
80. Koch B, Krasner SW, Sclimenti MJ, Schimpff WK (1991) Predicting the formation of DBPs by the simulated distribution system. Procedia Eng 83(10):62–70. https://doi.org/10.1002/j.1551-8833.1991.tb07233.x
81. Kennedy A, Flint L, Aligata A, Hoffman C, Arias-Paić M (2021) Regulated disinfection byproduct formation over long residence times. Water Res 188:116523. https://doi.org/10.1016/j.watres.2020.116523
82. Kanan A, Soyluoglu M, Karanfil T (2021) Removal of the precursors of regulated DBPs and TOX from surface waters and wastewater effluents using mixed anion exchange resins. Chemosphere 263:128094. https://doi.org/10.1016/j.chemosphere.2020.128094
83. Hui T, Using UFC (2006) Test to evaluate the removal of DBPs precursors. China Water Wastewater
84. Kralles ZT, Ikuma K, Dai N (2020) Assessing disinfection byproduct risks for algal impacted surface waters and the effects of peracetic acid pre-oxidation. Environ Sci Water Res Technol 6(9):2365–2381. https://doi.org/10.1039/D0EW00237B
85. Kanan A, Karanfil T (2020) Estimation of haloacetonitriles formation in water: uniform formation conditions versus formation potential tests. Sci Total Environ 744:140987. https://doi.org/10.1016/j.scitotenv.2020.140987
86. Summers RS, Hooper SM, Shukairy HM, Solarik G, Owen D (1996) Assessing DBP yield: uniform formation conditions. J Am Water Works Ass 88(6):80–93. https://doi.org/10.1002/j.1551-8833.1996.tb06573.x
87. Dittmar T, Stubbins A (2014) 12.6-Dissolved organic matter in aquatic systems. Treatise Geochem 2:125–156. https://doi.org/10.1016/B978-0-08-095975-7.01010-X
88. D'Andrilli J, Silverman V, Buckley S, Rosario-Ortiz FL (2022) Inferring ecosystem function from dissolved organic matter optical properties: a critical review. Environ Sci Technol 56(16):11146–11161. https://doi.org/10.1021/acs.est.2c04240
89. McKnight DM, Boyer EW, Westerhoff PK, Doran PT, Kulbe T, Andersen DT (2001) Spectrofluorometric characterization of dissolved organic matter for indication of precursor organic material and aromaticity. Limnol Oceanogr 46(1):38–48. https://doi.org/10.4319/lo.2001.46.1.0038
90. Birdwell JE, Valsaraj KT (2010) Characterization of dissolved organic matter in fogwater by excitation–emission matrix fluorescence spectroscopy. Atmos Environ 44(27):3246–3253. https://doi.org/10.1016/j.atmosenv.2010.05.055
91. Coble PG (1996) Characterization of marine and terrestrial DOM in seawater using excitation-emission matrix spectroscopy. Mar Chem 51(4):325–346. https://doi.org/10.1016/0304-4203(95)00062-3
92. Coble PG (2007) Marine optical biogeochemistry: the chemistry of ocean color. Chem Rev 107(2):402–418. https://doi.org/10.1021/cr050350+
93. Yamashita Y, Tanoue E (2003) Chemical characterization of protein-like fluorophores in DOM in relation to aromatic amino acids. Mar Chem 82(3):255–271. https://doi.org/10.1016/S0304-4203(03)00073-2

94. Zhang P, Cao C, Wang Y-H, Yu K, Liu C, He C et al (2021) Chemodiversity of water-extractable organic matter in sediment columns of a polluted urban river in South China. Sci Total Environ 777:146127. https://doi.org/10.1016/j.scitotenv.2021.146127
95. Yang B, Du P, Chen G, Zhang P, Zhang Q, Wang Z et al (2023) Dual role of soil-derived dissolved organic matter in the sulfamethoxazole oxidation by manganese dioxide. Water Res 235:119901. https://doi.org/10.1016/j.watres.2023.119901
96. Marhaba TF, Kochar IH (1999) Rapid prediction of disinfection by-product formation potential by fluorescence. Environ Eng Policy 2(1):29–36. https://doi.org/10.1007/BF03500904
97. Hur J, Lee M-H, Song H, Schlatman MA (2013) Microbial transformation of dissolved organic matter from different sources and its influence on disinfection byproduct formation potentials. Environ Sci Pollut Res 20(6):4176–4187. https://doi.org/10.1007/s11356-012-1384-0
98. Reckhow DA, Singer PC, Malcolm RL (1990) Chlorination of humic materials: byproduct formation and chemical interpretations. Environ Sci Technol 24(11):1655–1664. https://doi.org/10.1021/es00081a005
99. Chen C, Zhang X, Zhu L, He W, Han H (2011) Changes in different organic matter fractions during conventional treatment and advanced treatment. J Environ Sci 23(4):582–586. https://doi.org/10.1016/S1001-0742(10)60423-8
100. Wang J-J, Dahlgren RA, Erşan MS, Karanfil T, Chow AT (2016) Temporal variations of disinfection byproduct precursors in wildfire detritus. Water Res 99:66–73. https://doi.org/10.1016/j.watres.2016.04.030
101. Fernández-Pascual E, Droz B, O'Dwyer J, O'Driscoll C, Goslan EH, Harrison S et al (2023) Fluorescent dissolved organic matter components as surrogates for disinfection byproduct formation in drinking water: a critical review. ACS ES&T Water 3(8):1997–2008. https://doi.org/10.1021/acsestwater.2c00583
102. Woods GC, Simpson MJ, Pautler BG, Lamoureux SF, Lafrenière MJ, Simpson AJ (2011) Evidence for the enhanced lability of dissolved organic matter following permafrost slope disturbance in the Canadian High Arctic. Geochim Cosmochim Acta 75(22):7226–7241. https://doi.org/10.1016/j.gca.2011.08.013
103. Koch BP, Ludwichowski K-U, Kattner G, Dittmar T, Witt M (2008) Advanced characterization of marine dissolved organic matter by combining reversed-phase liquid chromatography and FT-ICR-MS. Mar Chem 111(3):233–241. https://doi.org/10.1016/j.marchem.2008.05.008
104. Mitschke N, Vemulapalli SPB, Dittmar T (2023) NMR spectroscopy of dissolved organic matter: a review. Environ Chem Lett 21(2):689–723. https://doi.org/10.1007/s10311-022-01528-4
105. Kujawinski EB (2011) The impact of microbial metabolism on marine dissolved organic matter. Ann Rev Mar Sci 3(1):567–599. https://doi.org/10.1146/annurev-marine-120308-081003
106. Kim S, Kramer RW, Hatcher PG (2003) Graphical method for analysis of ultrahigh-resolution broadband mass spectra of natural organic matter, the Van Krevelen diagram. Anal Chem 75(20):5336–5344. https://doi.org/10.1021/ac034415p
107. Cooper WT, Chanton JC, D'Andrilli J, Hodgkins SB, Podgorski DC, Stenson AC et al (2022) A history of molecular level analysis of natural organic matter by FTICR mass spectrometry and the paradigm shift in organic geochemistry. Mass Spectrom Rev 41(2):215–239. https://doi.org/10.1002/mas.21663
108. Stenson AC, Landing WM, Marshall AG, Cooper WT (2002) Ionization and fragmentation of humic substances in electrospray ionization Fourier transform-ion cyclotron resonance mass spectrometry. Anal Chem 74(17):4397–4409. https://doi.org/10.1021/ac020019f
109. Milstead RP, Remucal CK (2021) Molecular-level insights into the formation of traditional and novel halogenated disinfection byproducts. ACS ES&T Water 1(8):1966–1974. https://doi.org/10.1021/acsestwater.1c00161
110. Sadiq R, Rodriguez MJ (2004) Disinfection by-products (DBPs) in drinking water and predictive models for their occurrence: a review. Sci Total Environ 321(1):21–46. https://doi.org/10.1016/j.scitotenv.2003.05.001

111. Ged EC, Chadik PA, Boyer TH (2015) Predictive capability of chlorination disinfection byproducts models. J Environ Manage 149:253–262. https://doi.org/10.1016/j.jenvman.2014.10.014
112. Chowdhury S, Champagne P, McLellan PJ (2009) Models for predicting disinfection byproduct (DBP) formation in drinking waters: a chronological review. Sci Total Environ 407(14):4189–4206. https://doi.org/10.1016/j.scitotenv.2009.04.006
113. Zhang H, Qu J, Liu H, Wei D (2009) Characterization of dissolved organic matter fractions and its relationship with the disinfection by-product formation. J Environ Sci 21(1):54–61. https://doi.org/10.1016/S1001-0742(09)60011-5
114. Aukes PJK, Schiff SL, Venkiteswaran JJ, Elgood RJ, Spoelstra J (2021) Size-based characterization of freshwater dissolved organic matter finds similarities within a waterbody type across different Canadian ecozones. Limnol Oceanogr Lett 6(2):85–95. https://doi.org/10.1002/lol2.10180
115. Asmala E, Massicotte P, Carstensen J (2021) Identification of dissolved organic matter size components in freshwater and marine environments. Limnol Oceanogr Lett 66(4):1381–1393. https://doi.org/10.1002/lno.11692
116. Zhang Y, Zhang N, Zhao P, Niu Z (2018) Characteristics of molecular weight distribution of dissolved organic matter in bromide-containing water and disinfection by-product formation properties during treatment processes. J Environ Sci 65:179–189. https://doi.org/10.1016/j.jes.2017.03.013
117. Dong H, Cuthbertson AA, Plewa MJ, Weisbrod CR, McKenna AM, Richardson SD (2023) Unravelling high-molecular-weight DBP toxicity drivers in chlorinated and chloraminated drinking water: effect-directed analysis of molecular weight fractions. Environ Sci Technol 57(47):18788–18800. https://doi.org/10.1021/acs.est.3c00771
118. Aiken GR, McKnight DM, Thorn KA, Thurman EM (1992) Isolation of hydrophilic organic acids from water using nonionic macroporous resins. Org Geochem 18(4):567–573. https://doi.org/10.1016/0146-6380(92)90119-I
119. Chen C, Zhang X-j, Zhu L-x, Liu J, He W-j (2008) Disinfection by-products and their precursors in a water treatment plant in North China: seasonal changes and fraction analysis. Sci Total Environ 397(1–3):140–147. https://doi.org/10.1016/j.scitotenv.2008.02.032
120. Fan Z, Yang H, Li S, Yu X (2020) Tracking and analysis of DBP precursors' properties by fluorescence spectrometry of dissolved organic matter. Chemosphere 239:124790. https://doi.org/10.1016/j.chemosphere.2019.124790
121. Fan Z, Gong S, Xu X, Zhang X, Zhang Y, Yu X (2014) Characterization, DBPs formation, and mutagenicity of different organic matter fractions in two source waters. Int J Hyg Environ Health 217(2):300–306. https://doi.org/10.1016/j.ijheh.2013.07.002
122. Lu J, Zhang T, Ma J, Chen Z (2009) Evaluation of disinfection by-products formation during chlorination and chloramination of dissolved natural organic matter fractions isolated from a filtered river water. J Hazard Mater 162(1):140–145. https://doi.org/10.1016/j.jhazmat.2008.05.058
123. Rosario-Ortiz FL, Snyder S, Suffet I (2007) Characterization of the polarity of natural organic matter under ambient conditions by the polarity rapid assessment method (PRAM). Environ Sci Technol 41(14):4895–4900. https://doi.org/10.1021/es062151t
124. Rosario-Ortiz FL, Gerringer FW, Suffet IH (2009) Application of a novel polarity method for the characterization of natural organic matter during water treatment. J Water Supply Res Technol AQUA 58(3):159–169. https://doi.org/10.2166/aqua.2009.066
125. Mulholland P (2003) Large-scale patterns in dissolved organic carbon concentration, flux, and sources. Aquatic ecosystems. Elsevier, pp 139–159
126. Graeber D, Boëchat IG, Encina-Montoya F, Esse C, Gelbrecht J, Goyenola G et al (2015) Global effects of agriculture on fluvial dissolved organic matter. Sci Rep 5(1):16328. https://doi.org/10.1038/srep16328

127. Inamdar S, Finger N, Singh S, Mitchell M, Levia D, Bais H et al (2012) Dissolved organic matter (DOM) concentration and quality in a forested mid-Atlantic watershed, USA. Biogeochemistry 108(1):55–76. https://doi.org/10.1007/s10533-011-9572-4
128. Kaushal SS, Mayer PM, Vidon PG, Smith RM, Pennino MJ, Newcomer TA et al (2014) Land use and climate variability amplify carbon, nutrient, and contaminant pulses: a review with management implications. J Am Water Resour Assoc 50(3):585–614. https://doi.org/10.1111/jawr.12204
129. Osburn CL, Handsel LT, Peierls BL, Paerl HW (2016) Predicting sources of dissolved organic nitrogen to an estuary from an agro-urban coastal watershed. Environ Sci Technol 50(16): 8473–8484. https://doi.org/10.1021/acs.est.6b00053
130. Yamashita Y, Scinto LJ, Maie N, Jaffé R (2010) Dissolved organic matter characteristics across a subtropical Wetland's landscape: application of optical properties in the assessment of environmental dynamics. Ecosystems 13(7):1006–1019. https://doi.org/10.1007/s10021-010-9370-1
131. Fellman JB, Hood E, Spencer RGM (2010) Fluorescence spectroscopy opens new windows into dissolved organic matter dynamics in freshwater ecosystems: a review. Limnol Oceanogr 55(6):2452–2462. https://doi.org/10.4319/lo.2010.55.6.2452
132. Bhattacharya R, Osburn CL (2020) Spatial patterns in dissolved organic matter composition controlled by watershed characteristics in a coastal river network: the Neuse River Basin, USA. Water Res 169:115248. https://doi.org/10.1016/j.watres.2019.115248
133. Li S-L, Zhang H, Yi Y, Zhang Y, Qi Y, Mostofa KMG et al (2023) Potential impacts of climate and anthropogenic-induced changes on DOM dynamics among the major Chinese rivers. Geogr Sustain 4(4):329–339. https://doi.org/10.1016/j.geosus.2023.07.003
134. Liu D, Du Y, Yu S, Luo J, Duan H (2020) Human activities determine quantity and composition of dissolved organic matter in lakes along the Yangtze River. Water Res 168: 115132. https://doi.org/10.1016/j.watres.2019.115132
135. Ye Q, Zhang Z-T, Liu Y-C, Wang Y-H, Zhang S, He C et al (2019) Spectroscopic and molecular-level characteristics of dissolved organic matter in a highly polluted Urban River in South China. ACS Earth Space Chem 3(9):2033–2044. https://doi.org/10.1021/acsearthspacechem.9b00151
136. Zhang F, Harir M, Moritz F, Zhang J, Witting M, Wu Y et al (2014) Molecular and structural characterization of dissolved organic matter during and post cyanobacterial bloom in Taihu by combination of NMR spectroscopy and FTICR mass spectrometry. Water Res 57:280–294. https://doi.org/10.1016/j.watres.2014.02.051
137. Chow AT, Dai J, Conner WH, Hitchcock DR, Wang J-J (2013) Dissolved organic matter and nutrient dynamics of a coastal freshwater forested wetland in Winyah Bay, South Carolina. Biogeochemistry 112(1):571–587. https://doi.org/10.1007/s10533-012-9750-z
138. Majidzadeh H, Uzun H, Chen H, Bao S, Tsui MT-K, Karanfil T et al (2020) Hurricane resulted in releasing more nitrogenous than carbonaceous disinfection byproduct precursors in coastal watersheds. Sci Total Environ 705:135785. https://doi.org/10.1016/j.scitotenv.2019.135785

Part II
Individual DBPs

Halogenated Aliphatic DBPs

Emma H. Goslan

Contents

1 THMs 92
1.1 Basic Properties 92
2 HAAs 93
2.1 Basic Properties 93
3 Precursors and Formation Mechanisms 95
4 Occurrence 96
4.1 THMs 96
4.2 HAAs 97
5 Regulation 97
6 HALs 100
6.1 HAL Precursors and Formation Mechanisms 101
6.2 HAL Occurrence 102
7 HKs 102
7.1 HK Precursors and Formation Mechanisms 102
7.2 HK Occurrence 105
References 105

Abstract In the 50 years since the discovery of the trihalomethane (THM), chloroform, as a disinfection by-product (DBP), numerous studies have been carried out on the measurement of THMs. These have determined their occurrence in drinking waters and investigated their precursors and formation mechanisms. THMs have been regulated in many countries around the world and are generally used as a surrogate to minimize DBP formation overall. In more recent years, haloacetic acids (HAAs) have also been regulated in some countries and have gained prominence in the literature. Although not regulated, other aliphatic DBPs that have been studied include haloacetaldehydes and haloketones. This chapter summarizes the up-to-date literature on the properties, precursors, and formation mechanisms as well as occurrence and regulation of aliphatic DBPs with a particular focus on THMs and HAAs.

E. H. Goslan (✉)
Cranfield University, Cranfield, UK
e-mail: e.h.goslan@cranfield.ac.uk

Chao Chen, Susan Andrews, and Yuefeng Xie (eds.), *Disinfection By-Products in Water*, Hdb Env Chem (2024) 135: 91–112, DOI 10.1007/698_2024_1160,

Published online: 8 October 2024

Keywords Chlorination, Disinfection by-products, Drinking water, Haloacetic acids, Trihalomethanes

1 THMs

1.1 Basic Properties

Trihalomethanes are halogen-substituted single-carbon compounds with the general formula CHX_3 where X may be chlorine, bromine, or iodine, or a combination of these. From the point of view of drinking water regulation, four members of the group are important – trichloromethane, bromodichloromethane, dibromochloromethane, and tribromomethane, with trichloromethane tending to dominate in chlorinated waters [1]. Trichloromethane, or chloroform, was the first DBP to be identified in chlorinated water in 1974 [2]. The properties of trichloromethane are its characteristic odor and a burning, sweet taste. Its odor threshold values are 2.4 mg/L in water and 420 mg/m^3 in air [3, 4]. The odor threshold for tribromomethane in water is 0.3 mg/L [4, 5]. The physico-chemical properties of THMs are reported in Table 1 and the structures are shown in Fig. 1.

Table 1 Physico-chemical properties of THMs

Identity	Trichloromethane	Bromodichloro-methane	Dibromochloro-methane	Tribromo-methane
CAS No	67-66-3	75-27-4	124-48-1	75-25-2
Molecular formula	$CHCl_3$	$CHBrCl_2$	$CHBr_2Cl$	$CHBr_3$
Boiling point at 101.3 kPa (°C)	61.3	90	119	149–150
Melting point (°C)	−63.2	−57.1	−22	8.3
Relative density (20°C, g/cm^3)	1.484	1.98	2.38	2.90
Water solubility (g/L)	7.5–9.3 (25°C)	3.32 (30°C)	1.05 (30°C)	3.19 (30°C)
Vapor pressure (kPa)	8.13 (0°C) 21.28 (20°C)	6.67 (20°C)	2.0 (10°C)	0.75 (25°C)
Log octanol–water partition coefficient	1.97	1.88	2.08	2.38

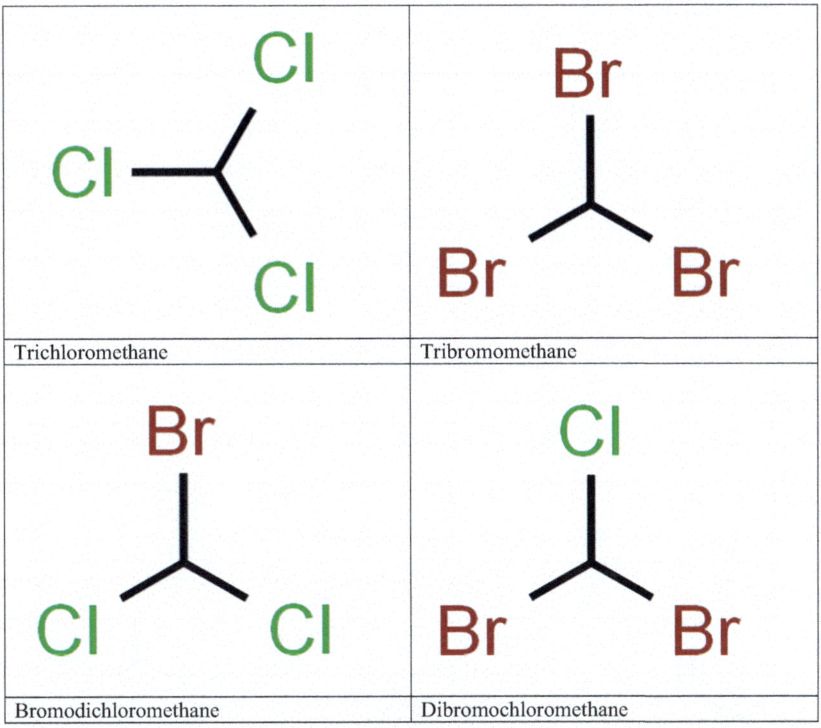

Fig. 1 Structures of THMs

2 HAAs

2.1 Basic Properties

There are nine chloro- and bromo-haloacetic acids (HAAs) in total [6]. They can be classified as:

- Monohalogenated acetic acids (known as MXAA or XHAA) comprising monochloroacetic acid (MCAA) and monobromoacetic acid (MBAA)
- Dihalogenated acetic acids (DXAA or X_2AA) comprising dichloroacetic acid (DCAA), dibromoacetic acid (DBAA), and bromochloroacetic acid (BCAA)
- Trihalogenated acetic acids (TXAA or X_3AA) comprising trichloroacetic acid (TCAA), bromodichloroacetic acid (BDCAA), dibromochloroacetic acid (DBCAA), and tribromoacetic acid (TBAA)

Regulation is currently in place for five of the HAAs in various parts of the world. These are MCAA, MBAA, DCAA, TCAA, and DBAA and are often referred as HAA5. The properties of all nine HAAs are given as well as their structures (Table 2 and Fig. 2). The boiling point of methyl esters is reported as HAAs are converted to their methyl esters prior to analysis by gas chromatography.

Table 2 Properties of HAAs

HAA	Molecular formula	Molecular weight (g/mol)	Boiling point (°C)	Boiling point of ester (°C)	CAS number
MCAA	$C_2H_3ClO_2$	94.45	189	130	79-11-8
MBAA	$C_2H_3BrO_2$	138.95	206–208	132	79-08-3
DCAA	$C_2H_2Cl_2O_2$	128.94	194	143	79-43-6
BCAA	$C_2H_2BrClO_2$	173.39	215	174	5589-96-8
TCAA	$C_2HCl_3O_2$	163.39	196	168	76-03-9
DBAA	$C_2H_2Br_2O_2$	217.84	128–130	NR[a]	631-64-1
BDCAA	$C_2HBrCl_2O_2$	207.84	NR[a]	NR[a]	71133-14-7
DBCAA	$C_2HBr_2ClO_2$	252.29	NR[a]	NR[a]	5278-95-5
TBAA	$C_2HBr_3O_2$	296.74	245	225	75-96-7

[a] Not reported

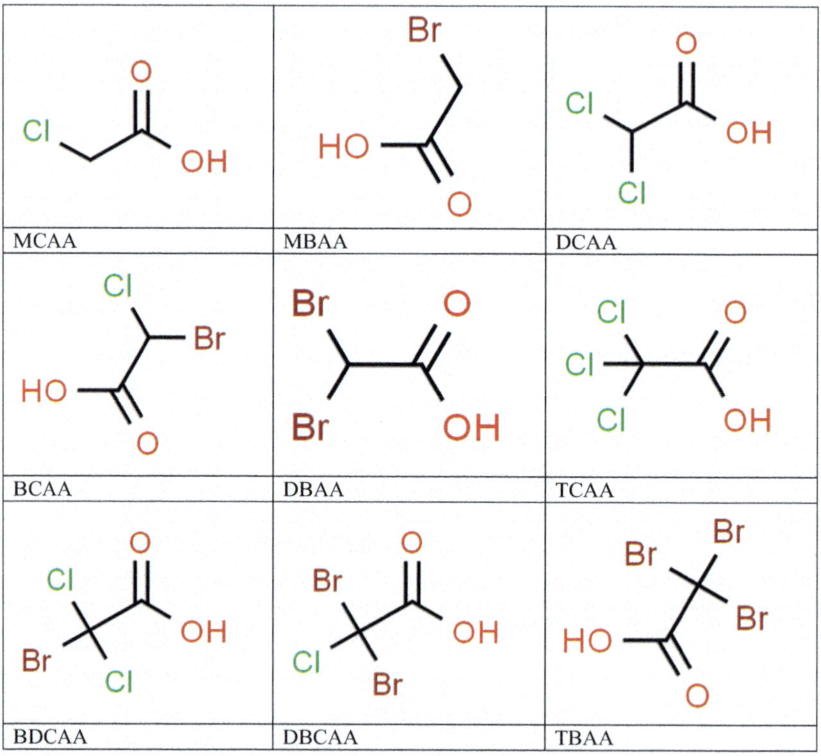

Fig. 2 Structures of HAAs

3 Precursors and Formation Mechanisms

DBPs are formed from the interaction of organic matter, bromide, and chlorine during disinfection. Organic matter in drinking water sources is complex but can comprise acidic, basic, and neutral fractions as well as aromatic and aliphatic components. In the aquatic environment, NOM is derived from both exterior (allochthonous) and interior (autochthonous) sources [7].

In the context of DBP reactivity, Kitis et al. [8] found the chemical properties (i.e., aromaticity) of organic matter were more important than the total organic content. They observed strong and simple linear trends between specific UV absorbance (UV absorbance at 254 nm divided by organic carbon content, SUVA) and THM formation potential. Studies using model compounds as organic matter surrogates support this finding, reporting that THMs form readily at high levels from activated aromatic compounds such as resorcinol [9–11], whereas HAAs have greater formation from aliphatic compounds such as carbonyls [12–14]. Different kinetics and precursors for tri-halogenated and di-halogenated haloacetic acids (TXAAs and DXAAs) have been elucidated [14–17].

The organic matter can be fractionated into hydrophobic and hydrophilic components to allow the reactivity of each to be determined. These can be further split into acid, base, and neutral fractions. After treatment by coagulation, the remaining organic matter is expected to be mainly hydrophilic in character and low in concentration [18]. However, this hydrophilic NOM can be as reactive as the hydrophobic NOM with respect to DBP formation.

Different water sources have fractions that have different reactivities when producing certain DBPs. For example, Kanokkantapong et al. [19] found the hydrophilic neutral fraction was the most reactive with respect to HAA formation, while Marhaba and Van [20] reported the hydrophobic neutral fraction to be the most reactive fractions for formation for both HAAs and THMs. Additionally, the humic (hydrophobic) fraction has been found to be very reactive in forming THMs [18, 21]. Kim and Yu [22] found the hydrophobic and the hydrophilic fractions had similar THM formation while the hydrophilic fraction formed more HAAs. It has also been reported that both hydrophobic and hydrophilic fractions produced similar DBP concentrations [23].

It has been hypothesized that THMs and tri-halogenated haloacetic acid (TXAAs) have common precursors, mainly hydrophobic organic matter while di-halogenated haloacetic acids (DXAAs) are linked to hydrophilic organic matter [24, 25]. Similarly, chlorination parameters such as pH correlate well with THM and TXAA formation potential but not with DXAA formation [26] although the mechanisms are difficult to elucidate.

There are different pathways by which chlorine reacts with organic matter precursors to form DBPs. These are oxidation, addition, and substitution. Chlorine is believed to undergo an addition reaction with saturated compounds that have a higher UV absorbance at 254 nm. It is reported that most chlorinated DBPs are the result of oxidation and substitution reactions [27]. Phenolic hydroxyl groups are reported to form more trichloromethane and TCAA during chlorination while

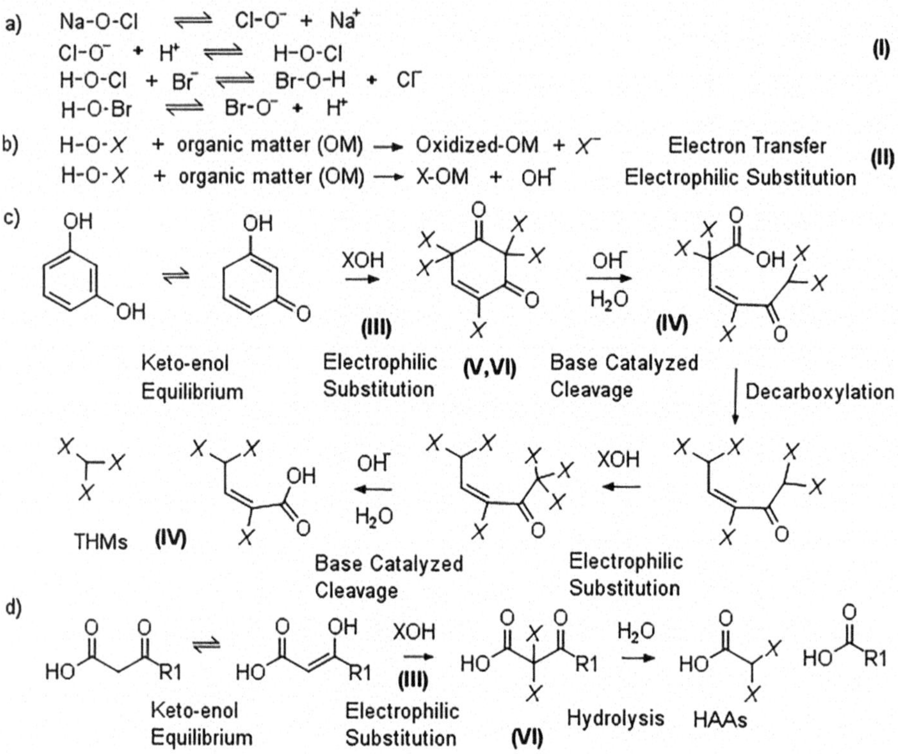

Fig. 3 Proposed key steps and mechanisms of formation for (**a**) hypohalous acids, (**b**) halogenated organic matter, (**c**) THMs from resorcinol, and (**d**) HAAs from β-dicarbonyls, where X is a halogen (Cl or Br) (from Grundy et al., submitted (with permission from the authors)) [28]

carboxyl and alcoholic hydroxyl groups were reported to form DCAA and brominated DBPs [25].

The key mechanistic steps in the formation of THMs and HAAs have been identified ([15], Fig. 3) as:

- The formation of hypohalous acid
- Activation of α carbon by keto-enol tautomerism
- Halogenation by electrophilic substitution
- Formation of target DBPs by hydrolysis and/or cleavage

4 Occurrence

4.1 THMs

The most comprehensive recent study of THM occurrence was carried out by Villanueva et al. [29] in 2023. This included 120 countries representing 94% of

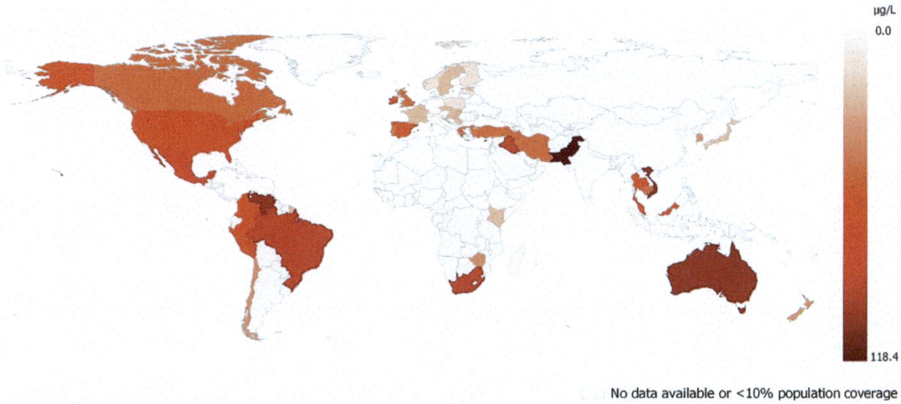

Fig. 4 Average trihalomethane levels in countries with >10% population coverage, 1997–2016 (from [29])

the global population and reported data from 116 of these from 1997 to 2016. The average levels present in each region are shown (Fig. 4).

4.2 HAAs

A recent global study of HAA occurrence was carried out in 2021 [30]. They reported THM and HAA levels. The HAA levels are reproduced here (Table 3). It should be noted that the data available for HAAs was limited compared to THMs due to the later regulation of HAAs in some locations and the lack of regulation in others. Additional data for HAA9 was less available than that for HAA5 and individual HAAs.

5 Regulation

The discovery of chloroform (trichloromethane) in drinking water in 1974 [2] coupled with advances in headspace gas chromatography which was used for the analysis of trihalomethanes (THMs) enabled a link to be established between the amount of aquatic organic matter present and levels of chloroform formed when disinfecting with chlorine. This allowed the US EPA to carry out a survey across the United States with 99% of sites containing THMs where chlorine or chloramine was the disinfectant [65]. Following this, the National Cancer Institute (NCI) recognized chloroform as a possible carcinogenic compound in small animals [66].

Table 3 Occurrence of THMs and HAAs in different countries or regions (adapted from [30])

Location	HAA levels (μg/L)	Reference
Americas		
Amherst, USA	34–161 (TCAA) 63–133 (DCAA)	[31]
North Carolina, USA	4–54 (TCAA)	[32]
USA	13–21 (HAA5)	[33]
USA	8–79 (DCAA) 15–64 (TCAA)	[34]
USA	Max: 89.8	[35]
North Carolina, USA	36–106 (HAA5) Mean: 77 Median: 81	[36]
Canada	<1–120 (DCAA) <1–473 (TCAA)	[37]
Canada	Max: 118 (HAA2)	[38]
USA	Min: 5 Max: 130 (HAA9)	[39]
Canada	25–115 (HAA9)	[40]
Canada	0–603 (HAA9)	[41]
Canada	Mean: 75 Max: 125 (HAA5)	[42]
Asia		
China	40 (HAA5)	[43]
Taiwan, China	<1–47 (HAA5)	[44]
Korea	50 (HAA5)	[45]
China	<1–81 (HAA9)	[46]
China	<1–60 (HAA6)	[47]
China	BDL-53	[48]
China	<1–11 (HAA3)	[49]
China	3–82 (HAA5)	[50]
China	6–39 (HAA5)	[51]
Europe		
Netherlands	0–15 (HAA9)	[52]
Spain	11–32 (HAA9)	[53]
Poland	Max: 120 (HAA5)	[54]
Finland	Mean: 26 Max: 108 (HAA6)	[55]
Spain	3–50 (HAA9) Min: 2 Max: 87	[56]
Cherepovets, Russia	Mean: 150 (HAA5)	[57]
UK	35–95 (HAA6) Max: 244	[58]
Scotland, UK	Min: 11 Max: 41 (HAA9)	[59]

(continued)

Table 3 (continued)

Location	HAA levels (μg/L)	Reference
England, UK	Min: 11 Max: 41 (HAA9)	[60]
Spain	<1–87 (HAA9)	[61]
Australia		
Australia-wide	10–244 (MCAA)	[62]
Africa		
Liberia	Min:9 Max: 92 Mean: 42 (HAA5)	[63]
Egypt	Max: 195 (HAA5), 219 (HAA9)	[64]

Haloacetic acid (HAA); Dichloroacetic acid (DCAA); Trichloroacetic acid (TCAA); Below Detection Limit (BDL)

Table 4 Key dates in DBP regulation and understanding (adapted from [67])

Year	Event	Reference
1974	Chlorination of organic matter in drinking water linked to chloroform formation	[2]
1975	USEPA survey of THMs in drinking water across the USA	[65]
1976	National Cancer Institute classifies chloroform as a suspected human carcinogen	[66]
1977	THMFP test developed	[68]
1979	HAAs identified in drinking water at levels similar to THMs	[69]
1979	THMs regulated at 100 μg/L by USEPA	
1989	UK regulations: THMs 100 μg/L	
1990	196 chlorination products identified in treated waters	[70]
1993	WHO Guidelines: DCAA 50 μg/L, TCAA 100 μg/L	
1998	Stage one USEPA D/DBP rule: THMs 80 μg/L, HAA_5 60 μg/L, based on running annual average	
2002	NDMA reported as DBP resulting from monochloramination	[71]
2006	600–700 DBPs reported for chlorine, chloramines, ozone, and chlorine dioxide	[39]
2006	Stage two of USEPA D/DBP rule: THMs 80 μg/L, HAA_5 60 μg/L, based on locational running annual average (LRAA)	
2006	WHO Guidelines: MCAA 20 μg/L, DCAA 50 μg/L, TCAA 200 μg/L	
2020	Recast EU Directive includes HAA_5 at 60 μg/L	

There have since been numerous studies and events involving regulation and recommendations to maintain the wholesomeness of drinking water and minimize any health effects that could arise from exposure to DBPs (Table 4).

While Scotland has adopted the Recast EU Directive into its legislation, England and Wales are yet to commit to any regulatory values concerning HAAs. Current research in the United States is discussing the inclusion of brominated HAAs in their D/DBP rule with no decisions made as yet [72].

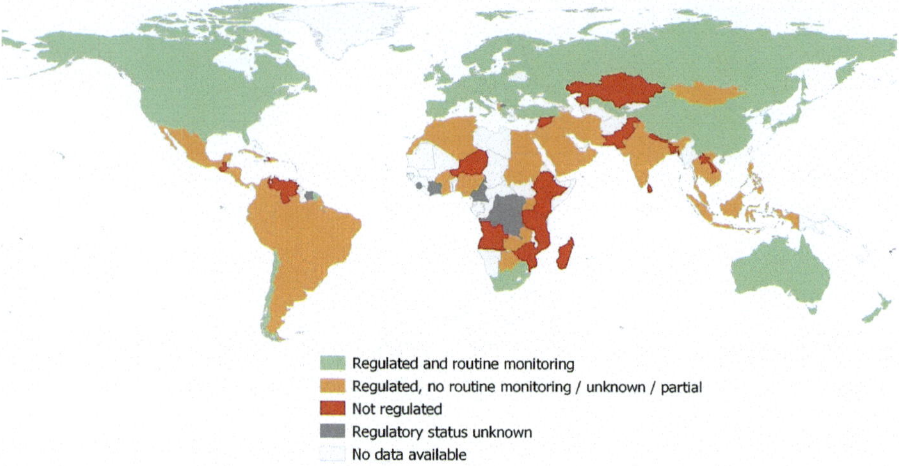

Fig. 5 Regulatory and routine monitoring status of trihalomethanes in the world in 2016 (from [29])

The most comprehensive recent study of THM regulation was included in the study on occurrence carried out by Villanueva et al. [29] in 2023. The extent of regulation is shown (Fig. 5) and it can be observed that regulation has become widespread across the world.

Villanueva et al. [29] reported that globally the MCL of total THMs ranged from 25 μg/L in Denmark and the Netherlands to 1,000 μg/L in Botswana. Low MCLs were reported for Austria, Zambia, and Italy (30 μg/L), as well as Hungary, Luxembourg, and Germany (50 μg/L). Countries reported to have high MCLs in addition to Botswana were Ecuador (500 μg/L), Australia (250 μg/L), Mexico and Colombia (both at 200 μg/L). MCL ranged from 50 to 150 μg/L in 48 countries with the majority (40) at 100 μg/L. The total THMs MCL was set as 1 for the sum of the ratios of the concentration of each individual trihalomethane (chloroform, bromodichloromethane, dibromochloromethane, and bromoform) to its respective maximum prescribed quantity as specified by WHO in 15 countries. Of these countries, five used WHO standards for individual THMs. Individual THMs were regulated in 16 countries, and three of these used WHO standards [29].

6 HALs

Haloacetaldehydes (HALs) are considered the third largest group of identified disinfection by-products (DBPs) by weight in drinking water [39]. HALs are not regulated in any country with the exception of chloral hydrate. This was regulated in the Chinese standard for drinking water (GB5749-2006) to a maximum of 10 μg/L but was not included in the revised standard in 2022. They have been reported as

Table 5 Haloacetaldehyde properties

Haloacetaldehyde	Molecular formula	Molecular weight (g/mol)	Boiling point (°C)	CAS number
Chloroacetaldehyde (CAL)	C_2H_3ClO	78.5	85	107-20-0
Bromochloroacetaldehyde (BAL)	C_2H_3BrO	122.95	93.7	17157-48-1
Dichloroacetaldehyde (DCAL)	$C_2H_2Cl_2O$	112.94	90.5	79-02-7
Bromochloroacetaldehyde (BCAL)	C_2H_2BrClO	157.39	109.2	98136-99-3
Dibromoacetaldehyde (DBAL)	$C_2H_2Br_2O$	201.85	136.3	3039-13-2
Trichloroacetaldehyde (TCAL) (also called chloral hydrate)	C_2HCl_3O	147.39	97.8	75-87-6
Bromodichloroacetaldehyde (BDCAL)	C_2HBrCl_2O	191.84	104.7	34619-29-9
Chlorodibromoacetaldehyde (DBCAL)	C_2HBr_2ClO	236.29	129.8	64316-11-6
Tribromoacetaldehyde (TBAL)	C_2HBr_3O	280.74	174	115-17-3
Iodoacetaldehyde (IAL)	C_2H_3IO	169.95	144.2	57782-51-9

being more cytotoxic compared to other DBPs. A recent paper [73] published in 2019 reported on the occurrence, formation, and control of HALs in drinking water and encompassed the previous 20 years of peer-reviewed research.

HALs are semi-volatile DBPs. Their basic properties are shown (Table 5).

6.1 *HAL Precursors and Formation Mechanisms*

Studies using model compounds have identified that direct precursors of haloacetaldehydes are aldehydes, which can derive from chlorination of amino acids or NOM oxidation by ozone [74]. Chlorination of amino acids can form aldehydes and nitriles with further chlorine substitution giving rise to trichloroacetaldehyde and dichloroacetonitrile, respectively [75, 76]. According to Gao et al. [73], studies reported tryptophan as highly reactive with chlorine in the formation of trichloroacetaldehyde and dichloroacetaldehyde [15, 77, 78].

Early studies investigating HAL formation mechanisms looked at aldehydes containing the acetyl functional group (e.g., acetaldehyde, methyl glyoxal) [76] but direct formation of these did not yield significant trichloroacetaldehyde levels [79]. More recent mechanistic research suggested that phenolic model compounds may be significant HAL precursors [80]. They reported that chlorination of lignin phenols (e.g., 4-hydroxybenzoic acid) formed 2,4,6-trichlorophenol as one

Table 6 HAL occurrence in Asia (adapted from [73])

Reference	Region, country	HAL concentrations (μg/L)		
		TCAL(CH)	BDCAL	DBCAL
[49]	Shenzhen, China	(2.8)	(0.2)	(0.1)
[81]	Beijing, China	N.D.-10.44 (0.50)		
[46]	Guangdong, China	N.D.-12.2 (2.1)		
[82]	Korea	N.D.-33.20 (0.98)		
[83]	Japan	N.D.-5.35		

intermediate. A key intermediate (2,3,5,5,6-pentachloro-6-hydroxy-cyclohexa-2-ene-1,4-dione) was identified at a branching point and was formed after additional Cl [+1] transfers. Chlorination of this intermediate formed HALs [73].

6.2 HAL Occurrence

Gao et al. [73] summarized concentrations of HALs from 20 years of studies from 1999 to 2019 These are separated by location and reported in the following (Tables 6, 7, and 8). Many studies only report trichloroacetaldehyde (chloral hydrate). In addition to the tables, Simpson and Hayes [62] reported TCAL values of 0.2–19 (μg/L) in Australia.

7 HKs

Haloketones (HKs) are detected in drinking water samples at levels an order of magnitude lower than THMs and HAAs. They are reported as unstable at high pH, to decrease with increasing contact time and are biodegraded in distribution networks [102]. Additionally, 1,1,1-trichloropropanone decomposes to form trichloromethane [103].

HKs are semi-volatile DBPs. Their basic properties are shown (Table 9).

7.1 HK Precursors and Formation Mechanisms

Ozonation of carbonaceous DBP precursors (such as phenol-type entities) is reported to occur via a Criegee-type reaction where aromatic rings are cleaved forming muconic-type and aliphatic products [104] including precursors of HKs [105]. This is further evidenced by the increased levels of HKs after chlorination of ozonated water which was shown to have increased methyl ketone species [105].

Table 7 HAL occurrence in North America (adapted from [73])

Reference	Region, country	HAL concentrations (μg/L)							
		CAL	DCAL	DBAL	BCAL	TCAL (CH)	TBAL	BDCAL	DBCAL
[84]	MI & GA, USA	N.Q.-0.3	N.Q.-0.5			N.Q.-23.0		N.Q.-0.5	
[85]	USA					N.D.-0.44/N.D.-0.79			
[86]	USA	0.2–0.3	0.3–2.0	N.D.-3.1	N.D.-2.2	N.D.-4.1	N.D.-12.56	1.1–2.2	N.D.-2.9
[87]	USA.								
[88]	USA	N.D.-2.4/N.D.-2.1	N.D.-14/N.D.-15		N.D.-4/N.D.-6	N.D.-16/N.D.-13	N.D.-3/N.D.-0.9		
[89]	USA					2.1–7.7/N.Q.-12			
[90]	USA					N.D.-46 (1.7)/N.D.-92 (4.0)			
[33]	USA					1.7–3.0			
[91]	Quebec, Canada		N.R.	N.D.	N.D.-0.44	N.D.-24.65	N.D.	N.R.	N.D.-0.66
[79]	Canada		1.0–3.3/1.1–3.6	N.D.-1.1/N.D.-1.0	0.6–2.7/0.4–3.1	0.7–7.8/N.D.-263	N.D.-2.0/N.D.-1.6	0.8–6.2/1.0–8.3	0.6–9.0/0.7–11.6
[92]	Canada		0.4–3.6	N.D.	.	0.2–12.2	N.D.	N.D.-5.4	N.D.-2.1
[93]	Canada		1.3–3.5	N.D.-1.7	0.1–2.7	1.8–12.0	N.D.-1.5	0.3–7.2	N.D.-11.5
[37]	Canada					N.Q.-15.1/N.Q.-22.5			
[94]	Canada					N.D.-23.4			

Note that [86] also reported BAL at N.D.-1.3 μg/L and IAL at N.D.-4.6 μg/L

Table 8 HAL occurrence in Europe (adapted from [73])

Reference	Region, country	HAL concentrations (μg/L)	
		DCAL	TCAL(CH)
[95]	Spain	N.D.-4.0	1.2–38
[61]	Spain		N.Q.-12.1 (N.D.)
[96]	Poland		N.D.-8.55
[97]	Poland		0.4–17
[98]	Greece		0.3–11.0/0.1–12.5
[99]	Greece		0.4–1.2
[100]	Greece		0.32–0.69/0.23–0.72
[101]	Europe		N.D.-23.2

Table 9 Haloacetone properties

Haloketone	Molecular formula	Molecular weight (g/mol)	Boiling point (°C)	CAS number
Chloroacetone (CA)	C_3H_5	92.52	120.0	78-95-5
1,1-dichloroacetone (1,1-DCA)	$C_3H_4Cl_2$	126.97	117.3	513-88-2
1,3-dichloroacetone (1,3-DCA)	$C_3H_4Cl_2$	126.97	173.0	534-07-6
1,1,1-trichloroacetone (1,1,1-TCA)	$C_3H_3Cl_3$	161.41	134.5	918-00-3
1,1,3-trichloroacetone (1,1,3-TCA)	$C_3H_3Cl_3$	161.41	170.7	921-03-9
1,1-Dibromoacetone (1,1-DBA)	$C_3H_4Br_2$	215.87	186.1	867-54-9
1,1,3,3-tetrachloroacetone (1,1,3,3-TeCA)	$C_3H_2Cl_4$	195.86	191.0	632-21-3
1,1,1,3-tetrachloroacetone (1,1,1,3-TeCA)	$C_3H_2Cl_4$	195.86	171.4	16995-35-0
1,1,1-tribromoacetone (1,1,1-TBA)	$C_3H_3Br_3$	219.50	294.767	3770-98-7
1,1-dibromo-3-chloroacetone (1,1-DB-3-CA)	$C_3H_3Br_2$	250.32	236.0	1578-18-3
1,1-dibromo-3,3-dichloroacetone (1,1-DB-3,3-DCA)	$C_3H_2Br_2$	284.76	251.9	62874-83-3
1,1,3-tribromoacetone (1,1,3-TBA)	$C_3H_3Br_3$	219.50	294.767	3475-39-6
1,3-dibromo-1,3-dichloroacetone (1,3-DB-1,3-DCA)	$C_3H_2Br_2$	284.76	249.4	62874-84-4
1,1,3-tribromo-3-chloroacetone (1,1,3-TB-3-CA)	$C_3H_2Br_3$	329.21	280.5	55716-01-3

It is known that THM formation increases with increasing pH, time, and chlorine dose. In contrast to this, HKs show opposing behavior with formation decreasing with increasing pH and chlorine dose due to hydrolysis reactions and reactivity with residual chlorine [106].

Formation is further complicated when considering chloramination [107]. 1,1-DCP formation was higher (and THM formation lower) in chloraminated water compared to chlorinated water while 1,1,1-TCP was detected in chlorinated water, but not in chloraminated water.

This points to 1,1-DCP being an intermediate product which can form 1,1,1-TCP when free chlorine is available but remains stable when chloramination is applied [107] because monochloramine cannot provide further chlorine substitution.

7.2 *HK Occurrence*

There have been relatively few studies on HK formation in drinking water. Of the studies that report HKs, these tend to report 1,1-DCA and 1,1,1-TCA due to the availability of standards. Where further standards have been available, 14 HKs have been reported in drinking water [39] with 1,1-DCA, 1,3-DCA, and 1,1,1-TCA occurring most often [39, 98, 100, 108].

Some potential health effects of HKs were identified [109] in 2003 thus 1,1-DCA and 1,1,1-TCA were included in the Information Collection Rule of the United States Environmental Protection Agency (U.S. EPA) with the sum of these reported as not detected to 9 μg/L with a median value of 2 μg/L [39]. Prior to this a study on 5 HKs (including three brominated ones) [53] in Spain, reported that tap water HKs were present at levels less than 0.5 μg/L with a U.S. Nationwide Occurrence Study also reporting sub μg/L levels in drinking waters [110].

References

1. WHO/SDE/WSH/03.04/64 Trihalomethanes in drinking-water background document for development of WHO Guidelines for drinking-water quality "Brominated trihalomethanes" and "Total trihalomethanes" originally published in Guidelines for drinking-water quality, 2nd edn., vol 2. Health criteria and other supporting information. Geneva, World Health Organization, 1996; "Chloroform" originally published in Guidelines for drinking-water quality, 2nd edn. Addendum to vol 2. Health criteria and other supporting information. Geneva, World Health Organization, 1998 © World Health Organization 2004
2. Rook JJ (1974) Formation of haloforms during chlorination of natural water. Water Treat Exam 23:234–243
3. Budavari S, O'Neill M, Smith A (eds) (1989) The Merck index: an encyclopedia of chemicals, drugs and biologicals.11th edn. Merck and Co, Rahway
4. ATSDR (1993) Toxicological profile for chloroform (update). US Department of Health and Human Services, Public Health Service, Agency for Toxic Substances and Disease Registry, Atlanta
5. Hansch C, Leo AJ (1985) Medchem project issue no. 26. Pomona College, Claremont
6. Singer PC (1999) Singer PC (ed) Formation and control of disinfection by-products in drinking water. American Water Works Association, Denver
7. Sillanpaa M (2014) Natural organic matter in water.1st edn. Elsevier

8. Kitis M, Karanfil T, Kilduff JE, Wigton A (2001) The reactivity of natural organic matter to disinfection by-products formation and its relation to specific ultraviolet absorbance. Water Sci Technol 43(2):9–16. https://doi.org/10.2166/wst.2001.0067
9. Kumar K, Margerum DW (1987) Kinetics and mechanism of general-acid-assisted oxidation of bromide by hypochlorite and hypochlorous acid. Inorg Chem 26(16):2706–2711. https://doi.org/10.1021/ic00263a030
10. Heeb MB, Criquet J, Zimmermann-Steffens SG, von Gunten U (2014) Oxidative treatment of bromide-containing waters: formation of bromine and its reactions with inorganic and organic compounds – a critical review. Water Res 48:15–42. https://doi.org/10.1016/j.watres.2013.08.030
11. Boyce SD, Hornig JF (1983) Reaction pathways of trihalomethane formation from the halogenation of dihydroxyaromatic model compounds for humic acid. Environ Sci Technol 17(4):202–211. https://doi.org/10.1021/es00110a005
12. Dickenson ERV, Summers RS, Croué J-P, Gallard H (2008) Haloacetic acid and trihalomethane formation from the chlorination and bromination of aliphatic β-dicarbonyl acid model compounds. Environ Sci Technol 42(9):3226–3233. https://doi.org/10.1021/es0711866
13. Psoras AW, McCoy SW, Reber KP, McCurry DL, Sivey JD (2023) Ipso substitution of aromatic bromine in chlorinated waters: impacts on trihalomethane formation. Environ Sci Technol. https://doi.org/10.1021/acs.est.3c00852
14. Li J, Jiang J, Manasfi T, von Gunten U (2020) Chlorination and bromination of olefins: kinetic and mechanistic aspects. Water Res 187:116424. https://doi.org/10.1016/j.watres.2020.116424
15. Bond T, Henriet O, Goslan EH, Parsons SA, Jefferson B (2009) Disinfection byproduct formation and fractionation behavior of natural organic matter surrogates. Environ Sci Technol 43(15):5982–5989. https://doi.org/10.1021/es900686p
16. Hong HC, Wong MH, Liang Y (2009) Amino acids as precursors of trihalomethane and haloacetic acid formation during chlorination. Arch Environ Contam Toxicol 56(4):638–645. https://doi.org/10.1007/s00244-008-9216-4
17. Hua G, Kim J, Reckhow DA (2014) Disinfection byproduct formation from Lignin precursors. Water Res 63:285–295. https://doi.org/10.1016/j.watres.2014.06.029
18. Goslan EH, Fearing DA, Banks J, Wilson D, Hills P, Campbell AT, Parsons SA (2002) Seasonal variations in the disinfection by-product precursor profile of a reservoir water. J Water Supply Res Technol Aqua 51(8):475–482. https://doi.org/10.2166/aqua.2002.0041
19. Kanokkantapong V, Marhaba TF, Pavasant P, Panyapinyophol B (2006) Characterization of haloacetic acid precursors in source water. J Environ Manage 80(3):214–221. https://doi.org/10.1016/j.jenvman.2005.09.006
20. Marhaba TF, Van D (2000) The variation of mass and disinfection by-product formation potential of dissolved organic matter fractions along a conventional surface water treatment plant. J Hazard Mater 74(3):133–147. https://doi.org/10.1016/S0304-3894(99)00190-9
21. Sinha S, Sohn J, Amy G. Reactivity of NOM in forming chlorinated DBPs. In: Proceedings of the American Water Works Association Annual Conference, June 15–19 (1997), Atlanta, GA, US
22. Kim HC, Yu MJ (2005) Characterization of natural organic matter in conventional water treatment processes for selection of treatment processes focused on DBPs control. Water Res 39(19):4779–4789. https://doi.org/10.1016/j.watres.2005.09.021
23. Croué J-P, Korshin GV, Benjamin MM (2000) Characterization of natural organic matter in drinking water, American Water Works Association Research Foundation report no. 90780, Denver, CO, US
24. Liang L, Singer PC (2003) Factors influencing the formation and relative distribution of haloacetic acids and trihalomethanes in drinking water. Environ Sci Technol 37(13):2920–2928. https://doi.org/10.1021/es026230q
25. Li A, Zhao X, Mao R, Liu H, Qu J (2014) Characterization of dissolved organic matter from surface waters with low to high dissolved organic carbon and the related disinfection

byproduct formation potential. J Hazard Mater 271:228–235. https://doi.org/10.1016/j.jhazmat.2014.02.009
26. Peng F, Lu Y, Wang Y, Yang L, Yang Z, Li H (2023) Predicting the formation of disinfection by-products using multiple linear and machine learning regression. J Environ Chem Eng 11(5):110612. https://doi.org/10.1016/j.jece.2023.110612
27. Thurman EM, Malcolm RL (1981) Preparative isolation of aquatic humic substances. Environ Sci Technol 15(4):463–466. https://doi.org/10.1021/es00086a012
28. Grundy PL, Jarvis PR, Jefferson B, Fawell J, Haley JA, Goslan EH Brominated disinfection by-products: using model compounds to reveal the chemistry of their formation. Environ Sci Technol. submitted
29. Villanueva CM, Evlampidou I, Ibrahim F, Donat-Vargas C, Valentin A, Tugulea A-M, Echigo S, Jovanovic D, Lebedev AT, Lemus-Pérez M, Rodriguez-Susa M, Luzati A, de Cássia dos Santos Nery T, Pastén PA, Quiñones M, Regli S, Weisman R, Dong S, Ha M, Phattarapattamawong S, Manasfi T, Egbenya Musah S-I, Eng A, Janák K, Rush SC, Reckhow D, Krasner SW, Vineis P, Richardson SD, Kogevinas M (2023) Global assessment of chemical quality of drinking water: the case of trihalomethanes. Water Res 230:119568. https://doi.org/10.1016/j.watres.2023.119568
30. Sinha R, Gupta AK, Ghosal PS (2021) A review on trihalomethanes and haloacetic acids in drinking water: global status, health impact, insights of control and removal technologies. J Environ Chem Eng 9(6):106511. https://doi.org/10.1016/j.jece.2021.106511
31. Uden PC, Miller JW (1983) Chlorinated acids and chloral in drinking water. J Am Water Works Ass 75:524–527. https://doi.org/10.1002/j.1551-8833.1983.tb05213.x
32. Norwood DL, Christman RF, Johnson JD, Hass JR (1986) Using isotope dilution mass spectrometry to determine aqueous trichloroacetic acid. J Am Water Works Ass 78:175–180. https://doi.org/10.1002/j.1551-8833.1986.tb05729.x
33. Krasner SW, Mcguire M, Jacangelo JG, Patania NL, Reagan K, Aieta EM (1989) The occurrence of disinfection by-products in US drinking water. J Am Water Works Ass 81: 41–53. https://doi.org/10.1002/j.1551-8833.1989.tb03258.x
34. Reckhow DA, Singer PC (1990) Chlorination by-products in drinking waters: from formation potentials to finished water concentrations. J Am Water Works Ass 82:173–180. https://doi.org/10.1002/j.1551-8833.1990.tb06949.x
35. Nieminski EC, Lumoreaux T (1993) The occurrence of DBPs in Utah drinking waters. J Am Water Works Ass 85:98–105. https://doi.org/10.1002/j.1551-8833.1993.tb06067.x
36. Singer PC, Obolensky A, Greiner A, Salem W, Salem W (1995) DBPs in chlorinated North Carolina drinking waters. J Am Water Works Ass 87:83–92. https://doi.org/10.1002/j.1551-8833.1995.tb06437.x
37. Williams DT, LeBel GL, Benoit FM (1997) Disinfection by-products in Canadian drinking water. Chemosphere 34:299–316. https://doi.org/10.1016/S0045-6535(96)00378-5
38. Rodriguez MJ, Sérodes JB, Levallois P (2004) Behavior of trihalomethanes and haloacetic acids in a drinking water distribution system. Water Res 38:4367–4382. https://doi.org/10.1016/j.watres.2004.08.018
39. Krasner SW, Weinberg HS, Richardson SD, Pastor SJ, Chinn R, Sclimenti MJ, Onstad GD, Thruston AD (2006) Occurrence of a new generation of disinfection byproducts. Environ Sci Technol 40:7175–7185. https://doi.org/10.1021/es060353j
40. Legay C, Rodriguez MJ, Sérodes JB, Levallois P (2010) The assessment of population exposure to chlorination by-products: a study on the influence of the water distribution system. Environ Health Glob Access Sci Source 9:1–14. https://doi.org/10.1186/1476-069X-9-59
41. Chowdhury S, Rodriguez MJ, Sadiq R (2011) Disinfection byproducts in Canadian provinces: Associated cancer risks and medical expenses. J Hazard Mater 187:574–584. https://doi.org/10.1016/j.jhazmat.2011.01.085
42. Guilherme S, Rodriguez MJ (2014) Occurrence of regulated and non-regulated disinfection by-products in small drinking water systems. Chemosphere 117:425–432. https://doi.org/10.1016/j.chemosphere.2014.08.002

43. Ye B, Wang W, Yang L, Wei J (2009) Factors influencing disinfection by-products formation in drinking water of six cities in China. J Hazard Mater 171:147–152. https://doi.org/10.1016/j.jhazmat.2009.05.117
44. Chang HH, Tung HH, Chao CC, Wang GS (2010) Occurrence of haloacetic acids (HAAs) and trihalomethanes (THMs) in drinking water of Taiwan. Environ Monit Assess 162:237–250. https://doi.org/10.1007/s10661-009-0792-1
45. Lee J, Kim ES, Roh BS, Eom SW, Zoh KD (2013) Occurrence of disinfection by-products in tap water distribution systems and their associated health risk. Environ Monit Assess 185: 7675–7691. https://doi.org/10.1007/s10661-013-3127-1
46. Gan W, Guo W, Mo J, He Y, Liu Y, Liu W, Liang Y, Yang X (2013) The occurrence of disinfection by-products in municipal drinking water in China's Pearl River Delta and a multipathway cancer risk assessment. Sci Total Environ 447:108–115. https://doi.org/10.1016/j.scitotenv.2012.12.091
47. Ding H, Meng L, Zhang H, Yu J, An W, Hu J, Yang M (2013) Occurrence, profiling and prioritization of halogenated disinfection by-products in drinking water of China. Environ Sci Process Impacts 15:1424–1429. https://doi.org/10.1039/c3em00110e
48. Pan S, An W, Li H, Su M, Zhang J, Yang M (2014) Cancer risk assessment on trihalomethanes and haloacetic acids in drinking water of China using disability-adjusted life years. J Hazard Mater 280:288–294. https://doi.org/10.1016/j.jhazmat.2014.07.080
49. Huang H, Zhu H, Gan W, Chen X, Yang X (2017) Occurrence of nitrogenous and carbonaceous disinfection byproducts in drinking water distributed in Shenzhen, China. Chemosphere 188:257–264. https://doi.org/10.1016/j.chemosphere.2017.08.172
50. Yu Y, Ma X, Chen R, Li G, Tao H, Shi B (2019) The occurrence and transformation behaviors of disinfection byproducts in drinking water distribution systems in rural areas of eastern China. Chemosphere 228:101–109. https://doi.org/10.1016/j.chemosphere.2019.04.095
51. Zhou X, Zheng L, Chen S, Du H, Gakoko Raphael BM, Song Q, Wu F, Chen J, Lin H, Hong H (2019) Factors influencing DBPs occurrence in tap water of Jinhua Region in Zhejiang Province, China. Ecotoxicol Environ Saf 171:813–822. https://doi.org/10.1016/j.ecoenv.2018.12.106
52. Peters RJB, Erkelens C, de Leer EWB, de Galan L (1991) The analysis of halogenated acetic acids in Dutch drinking water. Water Res 25:473–477. https://doi.org/10.1016/0043-1354(91)90084-4
53. Cancho B, Ventura F, Galceran MT (1999) Behavior of halogenated disinfection by-products in the water treatment plant of Barcelona, Spain. Environ Contam Toxicol 63:610–617. https://doi.org/10.1007/s001289901024
54. Dojlido J, Zbieć E, Ŝwietlik R (1999) Formation of the haloacetic acids during ozonation and chlorination of water in warsaw waterworks (Poland). Water Res 33(14):3111–3118. https://doi.org/10.1016/S0043-1354(99)00030-5
55. Nissinen TK, Miettinen IT, Martikainen PJ, Vartiainen T (2002) Disinfection by-products in Finnish drinking waters. Chemosphere 48:9–20. https://doi.org/10.1016/S0045-6535(02)00034-6
56. Villanueva CM, Kogevinas M, Grimalt JO (2003) Haloacetic acids and trihalomethanes in finished drinking waters from heterogeneous sources. Water Res 37:953–958. https://doi.org/10.1016/S0043-1354(02)00411-6
57. Egorov AI, Tereschenko AA, Altshu LM, Vartiainen T, Samsonov D, LaBrecque B, Mäki-Paakkanen J, Drizhd NL, Ford TE (2003) Exposures to drinking water chlorination by-products in a Russian city. Int J Hyg Environ Health 206:539–551. https://doi.org/10.1078/1438-4639-00244
58. Malliarou E, Collins C, Graham N, Nieuwenhuijsen MJ (2005) Haloacetic acids in drinking water in the United Kingdom. Water Res 39:2722–2730. https://doi.org/10.1016/j.watres.2005.04.052

59. Goslan EH, Krasner SW, Bower M, Rocks SA, Holmes P, Levy LS, Parsons SA (2009) A comparison of disinfection by-products found in chlorinated and chloraminated drinking waters in Scotland. Water Res 43:4698–4706. https://doi.org/10.1016/j.watres.2009.07.029
60. Zhang Y, Collins C, Graham N, Templeton MR, Huang J, Nieuwenhuijsen M (2010) Speciation and variation in the occurrence of haloacetic acids in three water supply systems in England. Water Environ J 24:237–245. https://doi.org/10.1111/j.1747-6593.2009.00200.x
61. Villanueva CM, Castaño-Vinyals G, Moreno V, Carrasco-Turigas G, Aragonés N, Boldo E, Ardanaz E, Toledo E, Altzibar JM, Zaldua I, Azpiroz L, Goñi F, Tardón A, Molina AJ, Martín V, López-Rojo C, Jiménez-Moleón JJ, Capelo R, Gómez-Acebo I, Peiró R, Ripoll M, Gracia-Lavedan E, Nieuwenhujsen MJ, Rantakokko P, Goslan EH, Pollán M, Kogevinas M (2012) Concentrations and correlations of disinfection by-products in municipal drinking water from an exposure assessment perspective. Environ Res 114:1–11. https://doi.org/10.1016/j.envres.2012.02.002
62. Simpson KL, Hayes KP (1998) Drinking water disinfection by-products: an Australian perspective. Water Res 32:1522–1528. https://doi.org/10.1016/S0043-1354(97)00341-2
63. Porter CK, Putnam SD, Hunting KL, Riddle MR (2005) The effect of trihalomethane and haloacetic acid exposure on fetal growth in a Maryland county. Am J Epidemiol 162:334–344. https://doi.org/10.1093/aje/kwi211
64. Smith EH, El-Deen KE (2009) Presence and formation of disinfection by-products in Cairo residential water supply. Water Sci Technol Water Supply 9:113–120. https://doi.org/10.2166/ws.2009.043
65. Symons JM, Bellar TA, Carswell JK, DeMarco J, Kropp KL, Robeck GG, Seeger DR, Slocum CJ, Smith BL, Stevens AA (1975) National organics reconnaissance survey for halogenated organics. J Am Water Works Ass 67:634–647. https://doi.org/10.1002/j.1551-8833.1975.tb02318.x
66. National Cancer Institute (NCI) (1976) Report on carciongenesis bioassay of chloroform, NTIS PB-264018, NCI, Washington, DC
67. Bond T, Goslan EH, Parsons SA, Jefferson B (2012) A critical review of trihalomethane and haloacetic acid formation from natural organic matter surrogates. Environ Technol Rev 1(1): 93–113. https://doi.org/10.1080/09593330.2012.705895
68. Stevens AA, Symons JM (1977) Measurement of trihalomethane and precursor concentration changes occurring during water treatment and distribution. J Am Water Works Ass 69:546–554. https://doi.org/10.1002/j.1551-8833.1977.tb06815.x
69. Christman RF, Johnson JD, Norwood DL, Liao WT, Hass JR, Pfaender FK, Webb MR, Bobenreith MJ (1979) Chlorination of aquatic humic substances: final report of USEPA project R-804430. US Environmental Protection Agency, Cincinnati
70. Stevens AA, Moore LA, Slocum CJ, Smith BK, Seeger DR, Ireland JC (1990) By-products of chlorination at 10 operating utilities. In: Jolley RL, Condie LW, Johnson JD, Katz S, Minear RA, Mattice JS, Jacobs VA (eds) Water chlorination – chemistry, environmental impact and health effects, vol 6. Lewis Publishers, Chelsea, pp 579–604
71. Choi J, Valentine RL (2002) Formation of N-nitrosodimethylamine (NDMA) from reaction of monochloramine: a new disinfection by-product. Water Res 36:817–824. https://doi.org/10.1016/S0043-1354(01)00303-7
72. Peterson ES, Raseman WJ, Stanford BD, Bruce GM, Klintworth H, Reckhow D (2023) Evaluating regulatory scenarios to limit U.S. nationwide exposure to cytotoxic haloacetic acids. AWWA Water Science, p e1351. https://doi.org/10.1002/aws2.1351
73. Gao J, Proulx F, Rodriguez MJ (2019) Halogenated acetaldehydes in water: a review of their occurrence, formation, precursors and control strategies. Crit Rev Environ Sci Technol 49(15): 1331–1385. https://doi.org/10.1080/10643389.2019.1571353
74. Krasner SW (2009) The formation and control of emerging disinfection by-products of health concern. Philos Trans R Soc A Math Phys Eng Sci 367:4077–4095. https://doi.org/10.1098/rsta.2009.0108

75. Trehy ML, Yost RA, Miles CJ (1986) Chlorination byproducts of amino acids in natural waters. Environ Sci Technol 20:1117–1122. https://doi.org/10.1021/es00153a006
76. McKnight A, Reckhow DA. Reactions of ozonation byproducts with chlorine and chloramines. In: Proceedings of the conference proceedings, AWWA annual conference, Vancouver, BC, Canada, 18–22 June 1992
77. Qian-Yuan W, Chao L, Ye D, Wen-Long W, Huang H, Hong-Ying H (2016) Elimination of disinfection byproduct formation potential in reclaimed water during solar light irradiation. Water Res 95:260–267. https://doi.org/10.1016/j.watres.2016.02.023
78. Wang JJ, Ng TW, Zhang Q, Yang XB, Dahlgren RA, Chow AT, Wong PK (2012) Technical note: reactivity of C1 and C2 organohalogens formation – from plant litter to bacteria. Biogeosciences 9:3721–3727. https://doi.org/10.5194/bg-9-3721-2012
79. Koudjonou B, Lebel GL, Dabeka L (2008) Formation of halogenated acetaldehydes, and occurrence in Canadian drinking water. Chemosphere 72:875–881. https://doi.org/10.1016/j.chemosphere.2008.03.062
80. Chuang YH, McCurry DL, Tung HH, Mitch WA (2015) Formation pathways and trade-offs between haloacetamides and haloacetaldehydes during combined chlorination and chloramination of lignin phenols and natural waters. Environ Sci Technol 49:14432–14440. https://doi.org/10.1021/acs.est.5b04783
81. Wei J, Ye B, Wang W, Yang L, Tao J, Hang Z (2010) Spatial and temporal evaluations of disinfection by-products in drinking water distribution systems in Beijing, China. Sci Total Environ 408:4600–4606. https://doi.org/10.1016/j.scitotenv.2010.06.053
82. Lee KJ, Kim BH, Hong JE, Pyo HS, Park SJ, Lee DW (2001) A study on the distribution of chlorination by-products (CBPs) in treated water in Korea. Water Res 35:2861–2872. https://doi.org/10.1016/S0043-1354(00)00583-2
83. Kawamoto T, Makihata N (2004) Distribution of bromine/chlorine-containing disinfection by-products in tap water from different water sources in the hyogo prefecture. J Health Sci 50: 235–247. https://doi.org/10.1248/jhs.50.235
84. Allen JM, Cuthbertson AA, Liberatore HK, Kimura SY, Mantha A, Edwards MA, Richardson SD (2017) Showering in flint, MI: is there a DBP problem? J Environ Sci (China) 58:271–284. https://doi.org/10.1016/j.jes.2017.06.009
85. Zeng T, Plewa MJ, Mitch WA (2016) N-Nitrosamines and halogenated disinfection byproducts in U.S. full advanced treatment trains for potable reuse. Water Res 101:176–186. https://doi.org/10.1016/j.watres.2016.03.062
86. Jeong CH, Postigo C, Richardson SD, Simmons JE, Kimura SY, Marinas BJ, Barcelo D, Liang P, Wagner ED, Plewa MJ (2015) Occurrence and comparative toxicity of haloacetaldehyde disinfection byproducts in drinking water. Environ Sci Technol 49:13749–13759. https://doi.org/10.1021/es506358x
87. Mitch WA, Krasner SW, Westerhoff P, Dotson A (2009) Occurrence and formation of nitrogenous disinfection by-products. Project #3014. Water Research Foundation and U.-S. Environmental Protection Agency, Denver and Washington
88. H.S. Weinberg, S.W. Krasner, S.D. Richardson, A.D. Thruston Jr., The occurrence of disinfection by-products of health concern in drinking water: results of a nationwide DBP occurrence study. U.S. Environmental Protection Agency, Washington., EPA/600/R-02/068 (NTIS PB2003-106823), 2002
89. Daiber EJ, DeMarini DM, Ravuri SA, Liberatore HK, Cuthbertson AA, Thompson-Klemish A, Byer JD, Schmid JE, Afifi MZ, Blatchley III ER, Richardson SD (2016) Progressive increase in disinfection byproducts and mutagenicity from source to tap to swimming pool and spa water: Impact of human inputs. Environ Sci Technol 50:6652–6662. https://doi.org/10.1021/acs.est.6b00808
90. McGuire MJ, McLain JL, Obolensky A (2003) Information collection rule data analysis. American Water Works Association, Denver
91. Legay C, Levallois P, Aranda-Rodriguez R, Dabeka TL, Hnatiw J, Rodriguez MJ (2015) Variability of non-regulated disinfection by-products in distribution systems: impact of the

storage tank, Chapter 18 of Recent advances in disinfection by-products, pp 341–362. https://pubs.acs.org. https://doi.org/10.1021/bk-2015-1190.ch018
92. Aranda-Rodriguez R, Koudjonou B, Jay B, Lebel GL, Benoit FM (2008) Disinfection by-products (DBPs) in drinking water from eight systems using chlorine dioxide. Water Qual Res J 43(1):11–22. https://doi.org/10.2166/wqrj.2008.003
93. Koudjonou BK, LeBel GL (2006) Halogenated acetaldehydes: analysis, stability and fate in drinking water. Chemosphere 64:795–802. https://doi.org/10.1016/j.chemosphere.2005.10.063
94. LeBel GL, Benoit FM, Williams DT (1997) A one-year survey of halogenated disinfection by-products in the distribution system of treatment plants using three different disinfection processes. Chemosphere 34:2301–2317. https://doi.org/10.1016/S0045-6535(97)00042-8
95. Serrano M, Silva M, Gallego M (2011) Micro liquid-liquid extraction combined with large-volume injection gas chromatography-mass spectrometry for the determination of haloacetaldehydes in treated water. J Chromatogr A 1218:8295–8302. https://doi.org/10.1016/j.chroma.2011.09.048
96. Włodyka-Bergier A, Bergier T, Kot M (2014) Occurrence of volatile organic chlorination by-products in water distribution system in Krakow (Poland). Desalin Water Treat 52:3898–3907. https://doi.org/10.1080/19443994.2014.887490
97. Dąbrowska A, Nawrocki J (2009) Controversies about the occurrence of chloral hydrate in drinking water. Water Res 43(8):2201–2208. https://doi.org/10.1016/j.watres.2009.02.022
98. Golfinopoulos SK, Nikolaou AD (2005) Survey of disinfection by-products in drinking water in Athens, Greece. Desalination 176(1–3):3–24. https://doi.org/10.1016/j.desal.2004.10.029
99. Nikolaou AD, Golfinopoulos SK, Lekkas TD, Kostopoulou MN (2004) DBP levels in chlorinated drinking water: effect of humic substances. Environ Monit Assess 93:301–319. https://doi.org/10.1023/B:EMAS.0000016798.53163.43
100. Golfinopoulos SK, Nikolaou AD, Lekkas TD (2003) The occurrence of disinfection by-products in the drinking water of Athens, Greece. Environ Sci Pollut Res 10:368–372. https://doi.org/10.1065/espr2002.06.124
101. Jeong CH, Wagner ED, Siebert VR, Anduri S, Richardson SD, Daiber EJ, McKague AB, Kogevinas M, Villanueva CM, Goslan EH, Luo W, Isabelle LM, Pankow JF, Grazuleviciene R, Cordier S, Edwards SC, Righi E, Nieuwenhuijsen MJ, Plewa MJ (2012) Occurrence and toxicity of disinfection byproducts in European drinking waters in relation with the HIWATE epidemiology study. Environ Sci Technol 46(21):12120–12128. https://doi.org/10.1021/es3024226
102. Arora H, LeChevallier MW, Dixon KL (1997) DBP occurrence survey. J Am Water Works Ass 89(6):60–68. https://doi.org/10.1002/j.1551-8833.1997.tb08242.x
103. Serrano M, Silva M, Gallego M (2014) Fast and green method for the analytical monitoring of haloketones in treated water. J Chromatogr A 1358:232–239. https://doi.org/10.1016/j.chroma.2014.06.103
104. Wenk J, Aeschbacher M, Salhi E, Canonica S, von Gunten U, Sander M (2013) Chemical oxidation of dissolved organic matter by chlorine dioxide, chlorine, and ozone: effects on its optical and antioxidant properties. Environ Sci Technol 47(19):11147–11156. https://doi.org/10.1021/es402516b
105. De Vera GA, Stalter D, Gernjak W, Weinberg HS, Keller J, José Farré M (2015) Towards reducing DBP formation potential of drinking water by favouring direct ozone over hydroxyl radical reactions during ozonation. Water Res 87:49–58. https://doi.org/10.1016/j.watres.2015.09.007
106. Singer PC (1994) Control of disinfection by-products in drinking water. J Environ Eng 120:727–744. https://doi.org/10.1061/(ASCE)0733-9372(1994)120:4(727)
107. Yang X, Shang C, Westerhoff P (2007) Factors affecting formation of haloacetonitriles, haloketones, chloropicrin and cyanogen halides during chloramination. Water Res 41(6):1193–1200. https://doi.org/10.1016/j.watres.2006.12.004

108. Golfinopoulos SK, Nikolaou AD (2001) Disinfection by-products and volatile organic compounds in the water supply system in Athens, Greece. J Environ Sci Health A 36:483–499. https://doi.org/10.1081/ESE-100103478
109. World Health Organisation (2003) Chloroacetones in drinking-water. Background document for preparation of WHO guidelines for drinking-water quality. http://www.who.int/water_sanitation_health/dwq/chemicals/chloroacetones.pdf
110. Serrano M, Silva M, Gallego M (2015) Determination of 14 haloketones in treated water using solid–phase microextraction and gas chromatography–mass spectrometry. J Chromatogr A 1407:208–215. https://doi.org/10.1016/j.chroma.2015.06.060

Iodinated DBPs

Tian-Yang Zhang and Bin Xu

Contents

1 Iodine Sources in Water Environment 114
2 Types and Characteristics of I-DBPs 115
2.1 Iodinated Trihalomethanes and Iodinated Acetic Acids 115
2.2 Iodinated Haloacetamides and Iodinated Haloacetonitriles 119
2.3 Aromatic I-DBPs 120
3 Water Quality Standards for I-DBPs 121
4 Detection Methods for I-DBPs 121
4.1 Pre-treatment Methods 121
4.2 Instrumental Quantitative Analysis Methods 122
5 The Generation Mechanism of I-DBPs 122
5.1 Pathways and Mechanisms of Iodide-Induced I-DBPs 123
5.2 Pathways and Mechanisms of Organic Iodine-Induced I-DBPs 124
5.3 Pathways and Mechanisms of Iodine-Based Disinfectants-Induced I-DBPs 124
5.4 Pathways and Mechanisms of Iodate-Induced I-DBPs 125
6 Control Strategies for I-DBPs in Drinking Water Treatment 126
6.1 Removal of Iodine Sources 126
6.2 Reduction of Organic Precursors 126
6.3 Formation Control of I-DBPs 127
6.4 Removal of Formed I-DBPs 127
References 128

Abstract Iodine is present in water environment mainly in the species of iodide, iodate, and organic iodine. During oxidation or disinfection processes, iodide or other iodine species in water can react with oxidants/disinfectants and organic precursors, generating a class of DBPs. Some of them have been proved to have "carcinogenic, teratogenic and mutagenic" properties, and were known as iodinated DBPs (I-DBPs). I-DBP includes many species. Several major species have been identified, such as iodinated trihalomethanes, iodinated acetic acids, iodinated haloacetamides, iodinated haloacetonitriles, and aromatic I-DBPs. Generally,

T.-Y. Zhang and B. Xu (✉)
Tongji University, Shanghai, People's Republic of China

Chao Chen, Susan Andrews, and Yuefeng Xie (eds.), *Disinfection By-Products in Water*, Hdb Env Chem (2024) 135: 113–130, DOI 10.1007/698_2024_1159,

Published online: 16 October 2024

I-DBPs are more toxic than the brominated and chlorinated analogs. Currently, there is still a lack of drinking water quality standards specific for I-DBPs in USEPA or WHO. But in China, iodinated trihalomethanes and iodinated acetic acids have been introduced into the national and local (Shanghai and Shenzhen City) drinking water standards. Overall, the formation mechanism of I-DBPs involves oxidizing iodine sources to reactive iodine species, which then undergo iodine substitution reactions with organic precursors in water to produce I-DBPs. However, factors affecting I-DBPs generation in water treatment processes are complex. The species of iodine source, the types of disinfection/oxidation method, and the structure of organic precursors all have important effects on I-DBP formation. From the perspective of iodine sources, iodide, organic iodine, and various iodine-based disinfectants can all produce I-DBPs. Iodate is chemically stable and does not directly produce I-DBPs during oxidation/disinfection processes. However, it can be converted to iodide through UV radiation or zero-valent iron reduction and indirectly produce I-DBPs during the oxidation/disinfection. The fundamental method to control the risk of I-DBPs is to remove iodine from the water source. However, there are technical difficulties in completely separating different species of iodine from water. The relevant study has become one of the hot issues in recent years. In addition, reducing organic precursors and optimizing or changing oxidation/disinfection processes are also effective in decreasing I-DBP formation. As for the formed I-DBPs, UV photolysis, advanced oxidation processes, and activated carbon adsorption have been reported as effective methods to remove I-DBPs directly.

Keywords Active iodine species, Control strategies, Detection methods, Generation mechanisms, Iodate, Iodide, Iodinated disinfection by-products, Organic iodine

1 Iodine Sources in Water Environment

Iodine is an essential trace element for the human body and is crucial for metabolism and growth and development. Iodine deficiency can lead to goiter, intellectual disability, stunted growth, or deformities. On the other hand, excessive iodine intake can cause hyperthyroidism, excessive growth, and even cancerous changes in thyroid gland. Iodine is present in the water environment, but not uniformly. There are various species of iodine in the water environment, with iodide, iodate, and organic iodine being the main species. There are very small amounts of active iodine species (HOI, I_2, I_3^-) and periodate (IO_4^-) present in water. Active iodine species are often present as the intermediate products in the oxidation process of water treatment and are easily consumed by the organic matter in water. In natural water bodies, seawater has the highest iodine content (the average concentration of iodide ions was 50 μg/L), including inorganic and organic iodine. The iodine content in freshwater is relatively low, but there are significant variations in the types and concentrations of iodine in different regions. Researchers conducted surveys on nearly 40 rivers in the United States, Canada, Western Europe, and found that the average concentration of iodide ions was 19.9 μg/L, with the range of 0.5–212 μg/L [1]. The difference in concentration is mainly influenced by factors such as geological conditions, climatic conditions, and irrigation methods. Another study reported the distribution of iodide concentration in raw water from 23 water treatment plants in the United States and Canada, which ranged from 0.13 to 104 μg/L [2]. In areas with seawater intrusion or geological changes, the iodide concentration in surface water and groundwater is generally higher, reaching above 50 μg/L, and this phenomenon is common in coastal areas [3].

Chinese researchers conducted surveys on water sources, water inlet of water treatment plants, treated water, and tap water in several cities along the Yangtze River [4]. They found that the total organic iodine concentration was the highest in water sources and water inlet, followed by iodide. However, in treated water and tap water, the proportion of iodate significantly increased, whereas iodide decreased significantly, and the total organic iodine concentration slightly decreased with little variation. Compared to the total iodine concentration in raw water, the total iodine concentration in finished water did not change significantly [4]. The selected water treatment processes have little removals of total iodine and organic iodine but they can transform iodide into iodate. The organic iodine species in the water environment are fairly complex. They include not only natural and artificial organic iodine compounds but also some I-DBPs. One typical and significant source of organic iodine is the iodinated contrast media (ICM) used in medical imaging of soft tissues [5]. The global consumption of ICM is currently around 3,500 tons per year. ICM has stable chemical properties. After being intravenously injected into human body, 95% of the ICM is excreted through urine and feces within 24 h and then discharged to the water environment. However, current sewage treatment processes cannot effectively remove ICM, which leaves residual ICM in effluent ranging from 0.1 to 10 μg/L [6]. Once ICM enters drinking water treatment plants, conventional drinking water treatment processes have limited removal efficiency too. Even ozone oxidation can only remove about 35–55% of non-ionized ICM and less than 20% of ionized ICM can be removed. These ICM compounds only represent a small fraction of the total organic iodine. Long-term consumption of water with high iodine levels not only poses direct health risks to human body but also results in formation of a class of highly toxic iodinated disinfection byproducts (I-DBPs) during water treatment processes. These I-DBPs are not only more toxic than their chlorinated and brominated analogs but also prone to causing foul odors (such as aromatic or medicinal smells), posing a threat to drinking water safety.

2 Types and Characteristics of I-DBPs

2.1 Iodinated Trihalomethanes and Iodinated Acetic Acids

I-DBPs are a group of emerging DBPs with strong “carcinogenic, teratogenic, mutagenic” properties that are formed by the combined action of organic matter,

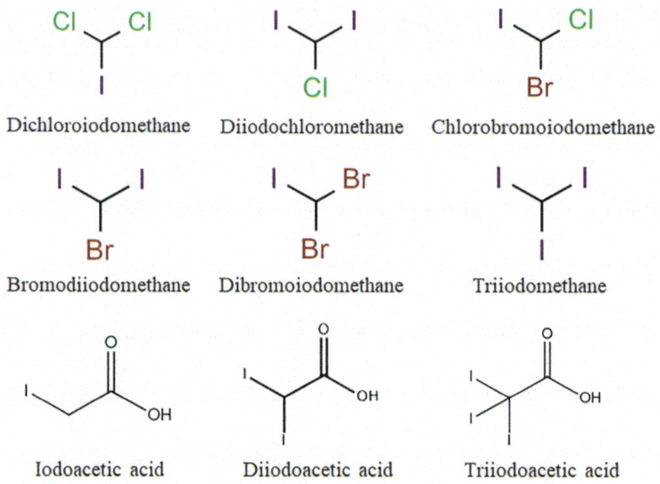

Fig. 1 Molecular formulas of I-THMs and I-HAAs

oxidants/disinfectants, and different iodine species in water. The two common types of I-DBPs are iodinated trihalomethanes (I-THMs) and iodinated haloacetic acids (I-HAAs). There are six main I-THMs, including dichloroiodomethane, chlorodiiodomethane, chlorobromoiodomethane, bromodiiodomethane, dibromoiodomethane, and triiodomethane. The three major I-HAAs are monoiodoacetic acid, diiodoacetic acid, and triiodoacetic acid. The chemical structures of each compound are shown in Fig. 1.

As early as in 1977, the first I-DBP, dichloroiodomethane, was identified [7]. So far, the presence of the aforementioned six iodinated trihalomethanes has been detected in the effluent of many water treatment plants around the world [8]. According to the existing reports [8], iodinated trihalomethanes are produced in chlorinated and chloraminated drinking water after treatment. The concentration of iodinated trihalomethanes in the effluent water, with low threshold levels of 0.02–0.5 μg L^{-1} [9], is lower than that of the conventional four trihalomethanes (such as chloroform, with average concentration of 23 μg L^{-1} [10]). Among them, dichloroiodomethane is the most commonly detected iodinated trihalomethane. For example, in a monitoring study of 111 water treatment plants in the United States, dichloroiodomethane was detected in 85 plants, with a higher detection frequency than bromoform [7].

There have been some studies on the cytotoxicity and genotoxicity of iodinated trihalomethanes in mammalian cells. Preliminary data indicate that iodoform has mutagenic effects on bacteria and higher cytotoxicity in mammalian cells, but no obvious genotoxicity. On the other hand, dichloroiodomethane shows significant cellular genotoxicity. Compared to the chronic cytotoxicity of brominated trihalomethanes on CHO cells, the cytotoxicity of iodinated trihalomethanes is stronger. The specific toxicity ranking is: trichloromethane > dibromoiodomethane >

Table 1 Comparison of cytotoxicity and genotoxicity between I-THMs and other THMs [11]

Disinfection by-product	Lowest cytotox. conc. (mol/L)	LC_{50} (mol/L)	Lowest genotox. conc. (mol/L)	50% TDNA or midpoint of Tail moment (mol/L)
Dichloroiodomethane	2.0×10^{-3}	4.13×10^{-3}	NS	NA
Chlorodiiodomethane	1.0×10^{-4}	2.41×10^{-3}	2.0×10^{-3}	2.95×10^{-3}
Bromochloroiodomethane	2.2×10^{-3}	2.42×10^{-3}	NS	NA
Bromodiiodomethane	1.5×10^{-3}	1.4×10^{-3}	NS	NA
Dibromoiodomethane	1.5×10^{-3}	1.91×10^{-3}	NS	NA
Triiodomethane	1.0×10^{-5}	6.6×10^{-5}	NS	NA
Trichloromethane	6.0×10^{-3}	9.62×10^{-3}	NS	NA
Bromodichloromethane	4.0×10^{-3}	1.15×10^{-2}	NS	NA
Chlorodibromomethane	7.5×10^{-4}	5.36×10^{-3}	NS	NA
Tribromomethane	1.0×10^{-4}	3.96×10^{-3}	NS	NA

NS Not statistically significant against the negative control
NA Not applicable or data not available

Table 2 Odors and threshold values of I-THMs

I-THMs	Odors	Odor threshold concentration (μg/L)
Dichloroiodomethane	Sweet, Syrupy	8.9
Bromochloroiodomethane	Sweet, Grassy, Aromatic, Winey	8.4
Dibromoiodomethane	Sweet, Aromatic, Asphalt	6.4
Chlorodiiodomethane	Medicinal, Sweet, Candy	0.2
Bromodiiodomethane	Sweet, Aromatic, Candy	0.2
Triiodomethane	Medicinal, Sweet, Aromatic, Chewing Gum Flavor	0.1

dichlorobromoiodomethane > dichloroiodomethane > bromochloroiodomethane > tribromomethane > dichloroiodomethane > dibromodichloromethane > trichloromethane > bromodichloromethane (as shown in Table 1).

Furthermore, studies have also shown that iodinated trihalomethanes have distinct odors with low odor thresholds [9]. For example, triiodomethane has a strong medicinal odor with an odor threshold concentration as low as 0.1 μg/L (as shown in Table 2).

I-HAAs were initially detected and identified in the finished water of water treatment plants after chloramine disinfection [12]. Generally, the concentrations

Table 3 Comparison of cytotoxicity and genotoxicity between I-HAAs and other HAAs [11]

Disinfection by-product	Lowest cytotox. conc. (mol/L)	LC50 (mol/L)	Lowest genotox. conc. (mol/L)	50% TDNA or midpoint of tail moment (mol/L)
Bromoacetic acid	2.0×10^{-6}	9.6×10^{-6}	1.3×10^{-5}	1.7×10^{-5}
Chloroacetic acid	2.5×10^{-4}	8.1×10^{-4}	3.0×10^{-4}	4.11×10^{-4}
Iodoacetic acid	5.0×10^{-7}	2.95×10^{-6}	5.0×10^{-6}	8.7×10^{-6}
Dibromoacetic acid	2.0×10^{-4}	5.9×10^{-4}	7.5×10^{-4}	1.76×10^{-3}
Dichloroacetic acid	2.0×10^{-3}	7.3×10^{-3}	NS	NA
Diiodoacetic acid	1.0×10^{-4}	3.32×10^{-4}	1.0×10^{-3}	1.98×10^{-3}
Bromochloroacetic acid	3.0×10^{-4}	7.78×10^{-4}	3.0×10^{-3}	3.64×10^{-3}
Bromoiodoacetic acid	2.5×10^{-4}	8.97×10^{-4}	2.5×10^{-3}	3.16×10^{-3}
Chloroiodoacetic acid	1.7×10^{-4}	3.04×10^{-4}	NA	NA
Tribromoacetic acid	5.0×10^{-6}	8.5×10^{-5}	3.0×10^{-3}	2.46×10^{-3}
Trichloroacetic acid	4.0×10^{-4}	2.4×10^{-3}	NS	NA
Chlorodibromoacetic acid	1.0×10^{-4}	2.02×10^{-4}	1.3×10^{-2}	1.36×10^{-2}
Bromodichloroacetic acid	5.0×10^{-4}	5.85×10^{-4}	NS	NA

NS Not statistically significant against the negative control
NA Not applicable or data not available

of I-HAAs after disinfection in water treatment plants are slightly lower than those of I-THMs. For example, the highest concentrations of iodoacetic acid and bromoiodoacetic acid of 1.7 and 1.4 μg/L, respectively, were observed in the finished water of 13 water treatment plants in Shanghai. The highest concentrations of tribromomethane and dichloroiodomethane were found to be 1.25 and 7.9 μg/L, respectively [13]. Among I-HAAs, iodoacetic acid has a strong mutagenic effect on *Salmonella* [14], with a mutagenic capability stronger than that of bromoacetic acid and chloroacetic acid. Studies on mammalian cell toxicity and genetic toxicity also indicate that iodoacetic acid has mutagenic effects on CHO cells that are 2 times and 47.2 times stronger than bromoacetic acid and chloroacetic acid, respectively. Additionally, the cytotoxicity of iodoacetic acid is 3.3 times and 274.6 times stronger than bromoacetic acid and chloroacetic acid, respectively (as shown in Table 3). Furthermore, research has confirmed that iodoacetic acid can affect mouse embryonic development at lower dosage levels [14].

2.2 *Iodinated Haloacetamides and Iodinated Haloacetonitriles*

Iodinated haloacetamides (I-HAMs) and iodinated haloacetonitriles (I-HANs) are nitrogenous I-DBPs, which are currently reported as the most cytotoxic and genotoxic disinfection byproducts. Among them, I-HANs are similar to conventional HANs and are chemically unstable, easily undergoing hydrolysis in water and forming I-HAMs. In recent years, there are some research reporting the occurrence of I-HAMs. Diiodoacetamide was first identified in drinking water treated with chlorination in the United States in 2008. Subsequently, Chinese researchers detected chloroiodoacetamide in water treated with chlorine, with the maximum concentration as 0.17 μg/L. Although the detected concentrations of I-HAMs are lower than those of I-THMs and I-HAAs, their cytotoxic and genotoxic effects are two orders of magnitude higher than those of I-THMs and I-HAAs (as shown in Table 4). Subcutaneous tests in mice have shown that iodinated acetamides have carcinogenic effects and can promote tumor formation in mice [7]. In addition, mammalian cell toxicity tests have shown that iodinated acetamides have strong cytotoxicity and DNA damage ability at the chromosomal level. The order of their chronic cell toxicity potential is as follows: diiodoacetamide > iodoacetamide > bromoacetamide > tribromoacetamide > bromoiodoacetamide > dibromochloroacetamide > chloroiodoacetamide > bromodichloroacetamide > dibromoacetamide > bromochloroacetamide > chloroacetamide > dichloroacetamide > trichloroacetamide. Regarding their cellular genotoxic potential, the order is as follows: tribromoacetamide > diiodoacetamide ≈ iodoacetamide > bromoacetamide > dibromo-chloroacetamide > bromoiodoacetamide > monobromo-dichloroacetamide > chloroiodoacetamide > bromochloroacetamide > dibromoacetamide > chloroacetamide > trichloroacetamide.

I-HANs mainly include iodochloroacetonitrile, diiodoacetonitrile, bromoiodoacetonitrile, and chloroiodoacetonitrile. However, only iodochloroacetonitrile has been identified before in the effluent of water treatment

Table 4 Cytotoxicity and genotoxicity of I-HAMs and I-HANs [11]

Disinfection by-product	Lowest cytotox. conc. (mol/L)	LC50 (mol/L)	Lowest genotox. conc. (mol/L)	50% TDNA or midpoint of tail moment (mol/L)
Iodoacetamide	5.0×10^{-7}	1.42×10^{-6}	3.0×10^{-5}	3.41×10^{-5}
Diiodoacetamide	2.5×10^{-8}	5.78×10^{-7}	2.5×10^{-5}	3.39×10^{-5}
Bromoiodoacetamide	2.0×10^{-6}	3.81×10^{-6}	2.5×10^{-5}	7.21×10^{-5}
Chloroiodoacetamide	2.0×10^{-6}	5.97×10^{-6}	2.0×10^{-4}	3.02×10^{-4}
Iodoacetonitrile	1.0×10^{-7}	3.3×10^{-6}	3.0×10^{-5}	3.71×10^{-5}

plants in previous reports. The concentration of iodochloroacetonitrile in the treated water after disinfection is usually higher than those of I-HAMs. For example, Chinese researchers found the highest concentration of iodochloroacetonitrile to be 4.2 μg/L in the effluent of one water treatment plant. Research data indicate that iodochloroacetonitrile can induce DNA damage in mammalian cells. Its cytotoxic potential is comparable to bromoacetonitrile and only lower than that of diiodoacetonitrile, but higher than the toxicity potentials of all other chloroacetonitriles. Additionally, iodochloroacetonitrile exhibits the strongest DNA damage ability at the chromosomal level among all haloacetonitriles.

2.3 *Aromatic I-DBPs*

Aromatic I-DBPs mainly include iodinated hydroxybenzaldehydes, iodinated hydroxybenzoic acids, and iodinated nitrophenols. Current research reports have identified 56 phenolic I-DBPs [15]. The chemical structures of some typical aromatic I-DBPs are shown in Fig. 2.

Generally, when water with high aromaticity (such as humic substances) is chlorinated, higher concentrations of aromatic I-DBPs can be produced. Due to the lack of standard substances for iodinated aromatic compounds for quantitative analysis, there are relatively few quantitative assessment studies on iodinated aromatic DBPs in drinking water. The detected concentrations of iodinated phenols in the treated water are only at the level of several ng/L. The cytotoxicity of iodinated aromatic byproducts is comparable to or higher than that of I-THMs and I-HAAs. Researchers have studied the cytotoxicity of aromatic DBPs using P. dumerilii cells and found that aromatic I-DBPs exhibit higher cytotoxicity than brominated and chlorinated aromatic DBPs, with the order being: 2,4,6-triiodophenol >2,4,6-tribromophenol >2,4,6-trichlorophenol; 2,6-diiodo-4-nitrophenol >2,6-dibromo-4-nitrophenol >2,6-dichloro-4-nitrophenol [16]. Moreover, the cytotoxicity of aromatic DBPs is generally higher than that of aliphatic analogs. For example, in the case of I-DBPs, the LC_{50} concentrations of 2,6-dichloro-4-nitrophenol and 2,4,6-triiodophenol on *P. dumerilii* cell are 2.02 × 10 μmol/L and 2.04 × 10 μmol/L, respectively, while the concentration of IAA required to reach LC_{50} is 3.67×10^3 μmol/L [16].

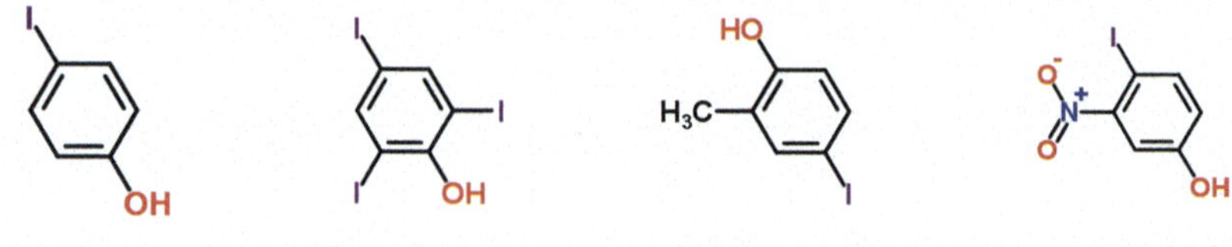

4-Iodophenol 2,4,6-Triiodophenol 4-Iodo-2-methylphenol 4-Iodo-3-nitrophenol

Fig. 2 Molecular formulas of some typical aromatic I-DBPs

3 Water Quality Standards for I-DBPs

I-DBPs, as emerging disinfection byproducts, are currently in the research stage, and there is a lack of water quality standard specifications internationally. In 2020, the city of Shenzhen, China, included iodide in the non-routine monitoring index in its local standard document "Standards for Drinking Water Quality" (DB4403/T60-2020) and set its limit value at 0.1 mg/L. Additionally, for the first time, dichloroiodomethane and iodoacetic acid were included in Appendix A as non-mandatory parameter with limiting values set at 0.01 and 0.02 mg/L, respectively. In 2022, in the new version of China's national standard document "Sanitary Standard for Drinking Water" (GB5749-2022), iodide and iodoacetic acid were also included in Appendix A as non-mandatory water quality parameters with threshold values set at 0.1 and 0.02 mg/L, respectively. In the future, research on the control of I-DBPs may be one of the key focuses and hotspots in the global water supply field.

4 Detection Methods for I-DBPs

The concentration of I-DBPs in drinking water is usually low, and the types and properties are diverse. Therefore, the detection of I-DBPs requires pre-treatment and quantitative analysis through relevant chromatographic instruments. The common pre-treatment methods for I-DBPs include liquid-liquid extraction, headspace technique, solid-phase extraction, and solid-phase microextraction. Instrumental quantitative analytical methods include gas chromatography with electron capture detection (GC-ECD), gas chromatography-mass spectrometry (GC-MS), ion chromatography, and ultra-high performance liquid chromatography triple quadrupole mass spectrometry (UPLC/ESI-tqMS) [17].

4.1 Pre-treatment Methods

Liquid-liquid extraction is the most commonly used sample pre-treatment method in DBPs detection, with the advantages of simple operation and equipment. However, it is only suitable for the pre-treatment of μg/L level I-DBPs, and it is difficult to accurately extract and quantify I-DBPs at lower concentrations. Moreover, the emulsion phenomenon during the extraction process is a common problem. The headspace technique can effectively avoid the excessive consumption of volatile organic solvents in liquid-liquid extraction and is commonly used for pre-treatment of volatile DBPs' detection. However, it has limitations such as low concentration factor, low sensitivity, and heat-induced conversion of some I-DBPs, and it is generally only suitable for pre-treatment of high concentration levels of I-THMs. Solid-phase extraction (SPE) has a high concentration factor and is a commonly used pre-treatment method for trace organic compounds in water. However, due to issues such as cumbersome operation and poor reproducibility, it is less commonly used in I-DBPs pre-treatment. Solid-phase microextraction (SPME) is a fairly new concentration and enrichment technique developed based on solid-phase extraction. Compared with SPE, it has simple operation, requires less sample volume, and has high sensitivity, making it suitable for pre-treatment of ng/L level I-DBPs. Currently, liquid-liquid extraction (LLE) and SPME are the two most commonly used pre-treatment methods for I-DBPs quantitative analysis of I-DBPs at μg/L and ng/L concentration levels, respectively.

4.2 Instrumental Quantitative Analysis Methods

GC-ECD is a commonly used quantitative analysis instrument for I-DBPs, and it can achieve a minimum detection concentration of 0.01–0.03 μg/L for I-THMs and other I-DBPs. It is an optimal instrument for quantitative analysis of I-DBPs and is often used in combination with liquid-liquid extraction. However, this method also has the disadvantage of poor qualitative structural capability. In complex water backgrounds, it needs to be combined with mass spectrometry for qualitative identification. GC-MS not only has the quantitative function of traditional gas chromatography analysis, but also has the separation and qualitative function for various complex components. It is also a commonly used quantitative analysis instrument for I-DBPs, and can be combined with liquid-liquid extraction and SPME pre-treatment methods. Ion chromatography is usually used for quantitative analysis of ionic DBPs and is only used for detection of I-HAAs. UPLC/ESI-tqMS, as a high-resolution mass spectrometry analysis method, has significant advantages in both qualitative and quantitative aspects. It is suitable for identification and quantification of various structurally complex I-DBPs, such as aromatic I-DBPs at low content. However, this method has disadvantages in expensive instrument equipment and complex operational procedures, making it unsuitable for daily monitoring of water treatment plants.

5 The Generation Mechanism of I-DBPs

The generation mechanism of I-DBPs in water treatment processes is complex and involves three key elements: iodine source, disinfection/oxidation process, and organic precursors. These three elements exist in different forms. The iodine source includes organic and inorganic iodine, the disinfection/oxidation process includes chemicals such as free chlorine, chloramines, chlorine dioxide, ozone, potassium permanganate, and ferrate, as well as advanced oxidation processes such as UV/H_2O_2, UV/O_3, UV/chlorine, and UV/persulfate. The structure of organic precursors is more complex, and different types of organic compounds have different reaction characteristics, such as humic substances, algae-derived organic matter, and exogenous chemicals. This chapter will focus on the pathways and mechanisms of I-DBP generation from different iodine sources.

5.1 Pathways and Mechanisms of Iodide-Induced I-DBPs

The main mechanism of iodide-induced I-DBPs is that iodide is oxidized to iodine species, including HOI, I_2, I_3^-, under the influences of disinfectants or oxidants [18]. These iodine species further react with dissolved organic matter in water to produce I-DBPs. In strong oxidizing processes such as free chlorine disinfection, ozone oxidation, ferrate oxidation, and advanced oxidation processes that can produce strongly oxidative free radicals, iodide is easily oxidized into iodate [18]. Iodate is chemically stable and does not react with organic matter to produce I-DBPs. Therefore, rapidly oxidizing iodide to stable iodate through strong oxidation processes is an effective method to control I-DBP generation in water treatment processes. In weak oxidation processes such as chloramine disinfection, chlorine dioxide disinfection, and potassium permanganate oxidation, iodide is oxidized but difficult to be converted into iodate [19]. Therefore, the generated active iodine species can react with dissolved organic matter in water to produce I-DBPs. Of course, the reactivity of the three common active iodine species differs during oxidation processes, with HOI being the most reactive with organic matter, while I_2 and I_3^- are relatively weaker. Due to the differences in the oxidation capability of oxidants, certain oxidants (such as chlorine dioxide and potassium permanganate) can only oxidize iodide to less reactive I_2 and I_3^-, while chloramines (including monochloramine and dichloramine) can oxidize iodide to more reactive HOI [20]. Therefore, among many disinfection and oxidation processes, chloramination poses a relatively higher risk for I-DBP generation (as shown in Fig. 3).

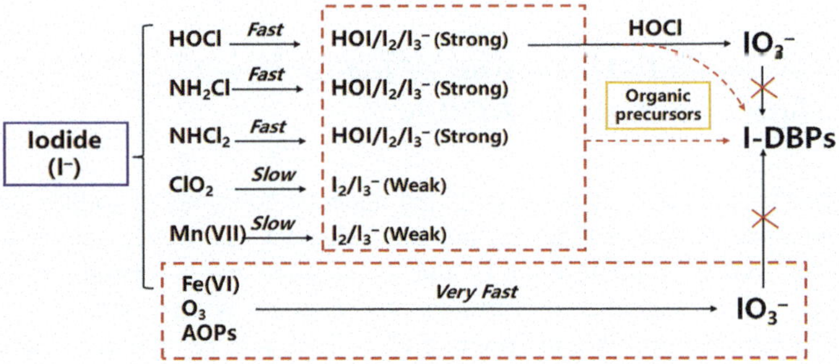

Fig. 3 Conversion pathways of iodine ions in different oxidation processes

5.2 Pathways and Mechanisms of Organic Iodine-Induced I-DBPs

Organic iodine is another important form of iodine in water environments, and its proportion in total iodine can even exceed that of inorganic iodide ions and iodate salts. It is also an important source of iodine for the formation of I-DBPs. In terms of the formation mechanism, the principle of I-DBPs formation by organic iodine is similar to that of inorganic iodine, which involves the generation of reactive iodine species first during disinfection or oxidation processes, which then react with organic matter in water to produce I-DBPs. However, there are some differences in this process. Organic iodine can provide a certain amount of organic precursors for the generation of I-DBPs. Therefore, even in the absence of organic precursors, direct chlorination or chloramination of organic iodine can produce I-DBPs, albeit usually in trace amounts due to its organic structure and low organic content [21]. However, when other soluble organic compounds are present, the concentration of I-DBPs produced significantly increases [21]. ICM as a typical organic iodine substance is widely detected in raw water, which has been the focus of current research on the transformation of organic iodine and formation of I-DBPs. Compared to iodide, ICM has more stable chemical properties and produces fewer reactive iodine species during disinfection or oxidation processes, resulting in relatively lower I-DBP formation. Additionally, the risks of I-DBP formation vary among different types of ICM. For example, ionized iodinated contrast agents are less oxidized, while non-ionized iodinated parabens are relatively more likely to generate I-DBPs [21]. Furthermore, the risk of organic iodine-induced I-DBPs can increase during ultraviolet treatment. This is due to the phenomenon of deiodination of organic iodine, where iodine is cleaved from the organic structure and becomes free iodide. In the subsequent chlorination or chloramination processes, iodide is easily oxidized to reactive iodine species, thereby increasing the risks of I-DBP formation [22].

5.3 Pathways and Mechanisms of Iodine-Based Disinfectants-Induced I-DBPs

Iodine-based disinfectants include saturated aqueous solutions of iodine and high iodine-containing amino acid tablets that can release I_2. However, due to the low solubility of iodine, a mixture of I_2 and I^- is often used, forming I_3^- in the formulation. I_3^- is included in formulations such as water solutions (Lugol's solution), water/ethanol mixtures (tincture of iodine), and I_3^- dissolved in coordination with weaker cationic polymers (polyvinylpyrrolidone or iodine carriers). Iodine-based disinfectants have some advantages in disinfection, such as equal effectiveness to chlorine in inactivating bacteria and viruses, and stronger ability than chlorine in killing protozoan cysts (such as amoeba). This is because nonpolar

I_2 is capable of penetrating cyst walls. Currently, iodine-based disinfectants are commonly used for special disinfection purposes, such as swimming pools, outdoor and military activity, glassware, and emergency water treatment, but have not been widely used in large-scale water treatment disinfection. Given the inherent characteristics of iodine-based disinfectants, which is a form of active iodine, its risk of generating I-DBPs during disinfection can hinder their large-scale application in water treatment.

5.4 *Pathways and Mechanisms of Iodate-Induced I-DBPs*

Generally speaking, iodate is considered non-toxic and non-reactive form of iodine and does not generate I-DBPs. Therefore, oxidizing iodide ions into iodates as much as possible is an effective way to control the risk of I-DBP formation. However, in recent years, studies have found that iodates can become sources of I-DBPs under specific conditions. For example, it has been found that iodates can be photo-reduced by UV radiation to form iodide, which can then react with monochloramine disinfectant to produce I-THMs (Fig. 4) [23]. The degradation and transformation of iodates by UV follow the first-order reaction kinetics. At the same time, when both iodates and iodide ions are present in water and exposed to UV radiation, iodide can generate reducing hydrated electrons, which can accelerate the reduction of iodates, resulting in production of iodide and greater risk of I-THM formation during subsequent disinfection processes [24]. Additionally, researchers have also found that applying zero-valent iron (ZVI) materials in water distribution systems can reduce iodates to iodide ions, leading to generation of I-THMs in the presence of monochloramine. This partially explains why iodide ions are almost absent in the finished water, but the concentration of I-DBPs gradually increases with the distance of water transmission [25].

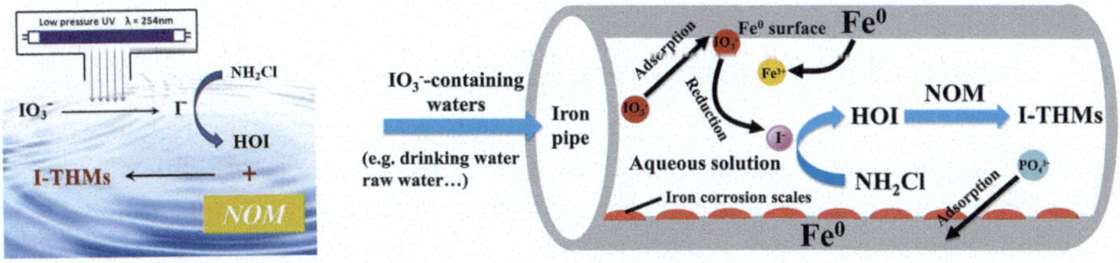

Fig. 4 A new pathway for the generation of I-DBPs from iodate

6 Control Strategies for I-DBPs in Drinking Water Treatment

6.1 Removal of Iodine Sources

The fundamental method to control I-DBPs in the water treatment process is to remove iodine from the source. However, efficient removal technique for various types of iodine is currently lacking, and that for directly separating iodine from aqueous phase is challenging. Currently, research on removal of iodine in water is mainly focused on the selective adsorption of new materials. For example, researchers in the United States have proposed the use of layered double hydroxide (LDH) ion exchange materials based on CoAl and NiAl to remove iodide and iodate, achieving the removal over 70% in real water samples [26]. Some scholars have also modified activated carbon with silver oxide which can adsorb iodide and organic iodine except iodate [27]. Japanese researchers have proposed using barite ($BaSO_4$) co-precipitation to remove radioactive iodate. The principle of this method is that iodate can be strongly bound to the crystal lattice of barite by substituting the SO_4^{2-} site in the structure, and the charge compensation problem from IO_3^- substitution in the SO_4^{2-} site is achieved by the substitution of Na^+-IO_3^- pairs at the nearest Ba^{2+} site [28]. Chinese scholars have recently proposed a method using bismuth-modified manganese dioxide materials for iodine removal. In this process, Bi_2O_3 plays a major role in the adsorption and enrichment of iodide and iodate, while MnO_2 can directly adsorb a certain concentration of iodide and also oxidize iodide to I_2 for further adsorption. This method achieves a total removal rate of 70–85% for iodide and iodate within 60 min [29]. Conventional pre-oxidants such as sodium hypochlorite can also oxidize iodide into active iodine, which can then be adsorbed by powdered activated carbon, resulting in an iodide removal of around 80% [30]. These methods can reduce the concentration of iodine in water from the source, thereby reducing the risk of I-DBPs formation during subsequent disinfection or oxidation processes.

6.2 Reduction of Organic Precursors

Dissolved organic matter has been well recognized for its important influence on the formation of carbonaceous and nitrogenous DBPs, but also it can impact the I-DBP formation. Researchers have found that when inorganic iodine ions are the iodine source in water, the concentration of I-DBPs produced during chloramine disinfection is significantly higher with the matrix of humic substances compared to that of algal organic matter. However, when organic iodine (such as iopamidol, an ICM) is the main iodine source in water, chloramination of algal organic matter produces far higher I-DBPs than that of humic and fulvic acid (as shown in Fig. 5) [31]. Therefore, tailored treatment measures should be taken for raw water containing different iodine sources and different organic precursors. Generally, advanced treatment processes

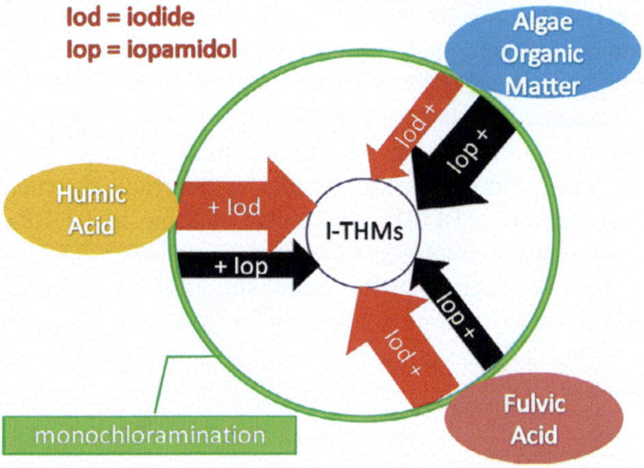

Fig. 5 The influence of organic precursor types on the formation of I-DBPs

such as ozone-biological activated carbon and nanofiltration have good removal efficiency for organic matter and can play a positive role in reducing various DBP precursors, including I-DBPs. These are currently mature means of controlling DBPs.

6.3 Formation Control of I-DBPs

Choosing appropriate disinfectants or oxidants and optimizing their dosage methods are key means to suppress the formation of reactive iodine species (particularly HOI) in the reaction process, which can control the generation of I-DBPs. In Sect. 5.1, we have mentioned the differences in iodine conversion characteristics and the risk of I-DBPs generation associated with different oxidants and disinfectants. For example, using chlorine dioxide or potassium permanganate instead of conventional sodium hypochlorite for pre-oxidation of iodide can reduce the formation of reactive iodine species, especially highly reactive HOI [19], leading to low yield of I-DBPs. Alternatively, using stronger oxidants like high ferrate pre-oxidants can rapidly oxidize iodide or organic iodine in raw water into iodate, thus reducing the generation of I-DBPs [32, 33]. When using chloramines for disinfection, employing short-term free chlorine contact followed by ammonia supplementation can replace direct chloramine disinfection [20].

6.4 Removal of Formed I-DBPs

Targeted treatment for removal of I-DBPs produced already during water treatment process is also an important supplementary measure to control the risk of I-DBPs.

Such removal methods for I-DBPs include adsorption and degradation. Studies have found that activated carbon has good adsorption for I-THMs and its adsorption capacity on I-THMs is significantly higher than those on chlorine and bromine-substituted THMs [34]. I-DBPs have certain UV photosensitivity and can undergo photo-degradation [35, 36]. In advanced oxidation processes such as UV/chlorine, the generation of strongly oxidative free radicals, such as ·OH, can accelerate the degradation of I-DBPs and convert iodine in I-DBPs into stable iodate, thereby significantly reducing the cytotoxicity of water [35, 36].

In conclusion, I-DBPs control should be systematically designed throughout the entire water treatment train. By a combination of precursor reduction, process control, and end-point removal measures, efficient reduction of I-DBPs can be achieved.

References

1. Moran JE, Oktay SD, Santschi PH (2002) Sources of iodine and iodine 129 in rivers. Water Res 38(8):1149–1158
2. Richardson SD, Fasano F, Ellington JJ et al (2008) Occurrence and mammalian cell toxicity of iodinated disinfection byproducts in drinking water. Environ Sci Technol 42(22):8330–8338
3. Muramatsu Y, Wedepohi KH (1998) The distribution of iodine in the earth's crust. Chem Geol 147(3–4):201–216
4. Yang X, Zheng Q, He M et al (2021) Bromine and iodine species in drinking water supply system along the Changjiang River in China: occurrence and transformation. Water Res 202: 117401
5. Duirk SE, Lindell C, Cornelison CC et al (2011) Formation of toxic iodinated disinfection by-products from compounds used in medical imaging. Environ Sci Technol 45:6845–6854
6. Kormos JL, Schuiz M, Ternes TA (2011) Occurrence of iodinated X-ray contrast media and their biotransformation products in the urban water cycle. Environ Sci Technol 45(20): 8723–8732
7. Dong H, Qiang Z, Richardson SD (2019) Formation of iodinated disinfection byproducts (I-DBPs) in drinking water: emerging concerns and current issues. Acc Chem Res 52:896–905
8. Hansson RC, Henderson MJ, Jack P et al (1987) Iodoform taste complaints in chloramination. Water Res 21:1265–1271
9. Richardson SD, Plewa MJ, Wagner ED et al (2007) Occurrence, genotoxicity, and carcinogenicity of regulated and emerging disinfection by-products in drinking water: a review and roadmap for research. Mutat Res 636:178–242
10. Brass HJ, Feige MA, Halloran T et al (1977) The national organic monitoring survey: sampling and analyses for purgeable organic compounds. In: Pojasek RB (ed) Drinking water quality enhancement through source protection. Ann Arbor Science, Ann Arbor, pp 393–416
11. Wagner ED, Plewa MJ (2017) CHO cell cytotoxicity and genotoxicity analyses of disinfection byu-products: an updated review. J Environ Sci 58:64–76
12. Plewa MJ, Wagner ED, Richardson SD et al (2004) Chemical and biological characterization of newly discovered iodoacid drinking water disinfection byproducts. Environ Sci Technol 38: 4713–4722
13. Wei X, Chen X, Wang X et al (2013) Occurrence of regulated and emerging iodinated DBPs in the Shanghai drinking water. PLoS One 8(3):e59677

14. Attene-Ramos MS, Wagner ED, Plewa MJ (2010) Comparative human cell toxicogenomic analysis of monohaloacetic acid drinking water disinfection byproducts. Environ Sci Technol 44:7206–7212
15. Liberatore HK, Plewa MJ, Wagner ED et al (2017) Identification and comparative mammalian cell cytotoxicity of new iodo-phenolic disinfection byproducts in chloraminated oil and gas wastewaters. Environ Sci Technol Lett 4:475–480
16. Yang MT, Zhang XR (2013) Comparative developmental toxicity of new aromatic halogenated DBPs in a chlorinated saline sewage effluent to the marine polychaete platynereis dummerilii. Environ Sci Technol 47(19):10868–10876
17. Wu T, Karimi-Maleh H, Dragoi EN et al (2023) Traditional methods and biosensors for detecting disinfection by-products in water: a review. Environ Res 237:116935
18. Ye T, Zhang TY, Tian FX et al (2021) The fate and transformation of iodine species in UV irradiation and UV-based advanced oxidation processes. Water Res 206:117755
19. Zhang TY, Xu B, Hu CY et al (2015) A comparison of iodinated trihalomethane formation from chlorine, chlorine dioxide and potassium permanganate oxidation processes. Water Res 68: 394–403
20. Zhang S, Lin YL, Zhang TY et al (2022) Insight into the formation of iodinated tr ihalomethanes dur ing chlor ination, monochloramination, and dichloramination of iodide-containing water. J Environ Sci 117:285–294
21. Ye T, Xu B, Wang Z et al (2014) Comparison of iodinated trihalomethanes formation during aqueous chlor(am)ination of different iodinated X-ray contrast media compounds in the presence of natural organic matter. Water Res 66:390–398
22. Tian FX, Xu B, Lin YL et al (2014) Photodegradation kinetics of iopamidol by UV irradiation and enhanced formation of iodinated disinfection by-products in sequential oxidation processes. Water Res 58:198–208
23. Zhang TY, Lin YL, Wang AQ et al (2016) Formation of iodinated trihalomethanes during UV/chloramination with iodate as the iodine source. Water Res 98:199–205
24. Tang LZ, Lin YL, Xu B et al (2021) Photodegradation pathway of iodate and formation of I-THMs during subsequent chloramination in iodate-iodide-containing water. Water Res 193: 116851
25. Xia Y, Lin YL, Xu B et al (2017) Iodinated trihalomethane formation during chloramination of iodate-containing waters in the presence of zero valent iron. Water Res 124:219–226
26. Kang J, Cintron-Colon F, Kim H et al (2022) Removal of iodine (I^- and IO_3^-) from aqueous solutions using CoAl and NiAl layered double hydroxides. Chem Eng J 430:132788
27. Li D, Kaplan DI, Price KA et al (2019) Iodine immobilization by silver-impregnated granular activated carbon in cementitious systems. J Environ Radioact 208:106017
28. Tokunaga K, Takahashi Y, Tanaka K et al (2021) Effective removal of iodate by coprecipitation with barite: behavior and mechanism. Chemosphere 266:129104
29. Wang N, Xiong RX, Zhang G et al (2022) Species transformation and removal mechanism of various iodine species at the Bi_2O_3@MnO_2 interface. Water Res 223:118965
30. Pan R, Lin YL, Zhang TY et al (2023) Sequential combination of pre-chlorination and powdered activated carbon adsorption on iodine removal and I-THMs control in drinking water. Chemosphere 313:137529
31. Wang Z, Xu B, Lin YL et al (2014) A comparison of iodinated trihalomethane formation from iodide and iopamidol in the presence of organic precursors during monochloramination. Chem Eng J 257:292–298
32. Zhang MS, Xu B, Wang Z et al (2016) Formation of iodinated trihalomethanes after ferrate pre-oxidation during chlorination and chloramination of iodide-containing water. J Taiwan Inst Chem Eng 60:453–459
33. Li M, Zhang TY, Xu B et al (2021) Iodinated trihalomethanes formation in iopamidol-contained water during ferrate/chlor(am)ination treatment. Chemosphere 272:129568

34. Qian H, Lin YL, Xu B et al (2018) Adsorption of haloforms onto GACs: effects of adsorbent properties and adsorption mechanisms. Chem Eng J 349:849–859
35. Wang AQ, Lin YL, Xu B et al (2017) Kinetics and modeling of iodoform degradation during UV/chlorine advanced oxidation process. Chem Eng J 323:312–319
36. Liu Z, Lin YL, Xu B et al (2020) Degradation of diiodoacetamide in water by UV/chlorination: kinetics, efficiency, influence factors and toxicity evaluation. Chemosphere 240:124761

Formation, Analysis, and Removal of Halogenated Cyclic DBPs

Ruixue Meng, Meijiao Zhou, Yuefeng F. Xie, and Wei Wang

Contents

1 Halobenzoquinones 132
 1.1 Analytical Characterization 133
 1.2 Precursor and Formation 146
 1.3 Transformation and Removal 150
2 Mutagen X (3-Chloro-4-(Dichloromethyl)-5-Hydroxy-2(5*H*)-Furanone, MX) 152
 2.1 MX Occurrence in Drinking Water 153
 2.2 MX Precursors 154
 2.3 MX Analysis 155
 2.4 Removal and Control 156
 2.5 Health Risks and Regulatory Considerations 157
3 Other Cyclic DBPs 157
References 161

Abstract In recent years, halogenated cyclic disinfection byproducts (DBPs) have received widespread attention due to their greater developmental toxicity compared to aliphatic ones, and were recognized to be the primary contributors to the overall developmental toxicity of chlorinated drinking water. This chapter provided a brief summary for the recent research progress on the analysis, occurrence, formation, and elimination strategies of halogenated cyclic DBPs, primarily including halobenzoquinones (HBQs), mutagen X (MX), trihalo-hydroxy-cyclopentene

R. Meng, M. Zhou, and W. Wang (✉)
Department of Environmental Science, Zhejiang University, Hangzhou, China

Zhejiang Provincial Key Laboratory of Organic Pollution Process and Control, Hangzhou, China
e-mail: 12214038@zju.edu.cn; mjzhou@zju.edu.cn; ww1@zju.edu.cn

Y. F. Xie (✉)
Department of Civil, Construction and Environmental Engineering, The Pennsylvania State University, Middletown, PA, USA
e-mail: yxx4@psu.edu

Chao Chen, Susan Andrews, and Yuefeng Xie (eds.), *Disinfection By-Products in Water*, Hdb Env Chem (2024) 135: 131–168, DOI 10.1007/698_2024_1158,

Published online: 19 November 2024

diones (trihalo-HCDs), halonaphthoquinones (HNQs), chlorophenylacetonitriles (CPANs), halogenated pyridinols, haloanilines, haloquinone chloroimides (HQCs), halophenylacetamides (HPAcAms), halobenzamides (HBZAMs), and halogenated nucleic acids. The authors aim to provide some basic information for understanding the health risk associated with halogenated cyclic DBPs and potential future research for scholars working in the field of drinking water.

Keywords Formation, Halobenzoquinones, Halogenated cyclic DBPs, Mutagen X, Precursor, Removal, Transformation

Over 50% of the halogenated substances produced in chlorinated drinking water remained unknown, particularly the one- and two-carbon-atom DBPs of current interest accounted for only ~16% of the cytotoxicity of chlorinated drinking water, there is a need to identify toxicity drivers within the poorly characterized higher-molecular-weight (more than two carbon atoms) DBP fraction [1, 2]. Halogenated cyclic DBPs have been intensively attracting concerns due to their significantly greater toxicity and the possibility of degradation into the regulated DBPs. Therefore, their impact on drinking water safety cannot be ignored.

1 Halobenzoquinones

Halobenzoquinones (HBQs) were first identified and confirmed as emerging DBPs in 2010 [3]. Four HBQs, 2,6-dichloro-1,4-benzoquinone (2,6-DCBQ), 2,6-dichloro-3-methyl-1,4-benzoquinone (DCMBQ), 2,3,5-trichloro-1,4-benzoquinone (TriCBQ), and 2,6-dibromo-1,4-benzoquinone (2,6-DBBQ) were detected in treated tap water at ng/L levels. In 2020, three new iodinated HBQs, including 2-chloro-6-iodo-1,4-benzoquinone (2,6-CIBQ), 2-bromo-6-iodo-1,4-benzoquinone (2,6-BIBQ), and 2,6-diiodo-1,4-benzoquinone (2,6-DIBQ), were detected in drinking water for the first time. Using the newly developed pseudo-targeted LC-MS/MS (liquid chromatography-mass spectrometry/mass spectrometry) method, 46 HBQs could be quantified and 30 of which were lack of chemical standards. It is noted that 29 of the 30 newly identified HBQs were halo-methyl-benzoquinones (HMBQs) [4]. The iodinated HBQs were generally more cytotoxic compared with the chlorinated or the brominated HBQs [5]. Although the concentrations of HBQs in water were low to ng/L levels, they exhibited higher toxicity compared to the regulated DBPs such as trihalomethanes (THMs) and haloacetic acids (HAAs) [6, 7].

In view of the potential health risks of long-term exposure to HBQs, this section summarized the analytical characterization, precursor and formation pathways, transformation and removal mechanism of HBQs, aiming to provide a comprehensive summary of the current progress and future direction of research related to HBQ DBPs, and provide a scientific basis for the assessment of potential health risks and the development of related regulation standards of drinking water.

1.1 Analytical Characterization

1.1.1 Chemical Properties of HBQs

It is essential to comprehend the physical and chemical properties of HBQs for developing qualifying and quantifying methods. The fundamental composition of HBQs includes halogen, alkyl, and hydroxyl groups attached to a benzoquinone (BQ) ring. Exploring the distinctive chemical characteristics of BQ, notably its redox chemistry, electrophilic reactivity, and substitution reactions, is imperative for gaining insights into the chemical properties of HBQs [8]. On one hand, BQ can undergo reversible two-electron reductions to form hydroquinone (HQ) and be oxidized back to BQ as shown in Fig. 1. This redox activity is essential in biological processes like cellular respiration and photosynthesis [10]. On the other hand, BQ is electrophilic due to the presence of electron-withdrawing carbonyl groups, which can readily react with nucleophiles (substances with electron-rich centers) in addition reactions [9]. The substituents on the BQ ring of HBQs, including halogens and additional methyl groups, affected the chemical properties of HBQs, such as pK_a values and reduction potential (E^0) [11]. For example, the pK_a of 2,6-DCBQ, DCMBQ, TriCBQ, and 2,6-DBBQ were 6.3, 5.9, 5.7, and 6.0, respectively. In vitro experiments show that the number of halogens and methyl groups significantly affects the cytotoxicity and genotoxicity of HBQs [12, 13].

The identification of HBQs was restricted using targeted LC-MS/MS analysis in previous studies. Therefore, only 16 HBQs were identified in drinking water due to the lack of chemical standards [5]. In 2022, Lou [4] developed a pseudo-targeted LC-MS/MS method for detecting and quantifying HBQs. Distinct fragment characteristics of HBQs were observed according to the halogen substituent effects, which was extended to develop the multiple-reaction-monitoring (MRM) method for

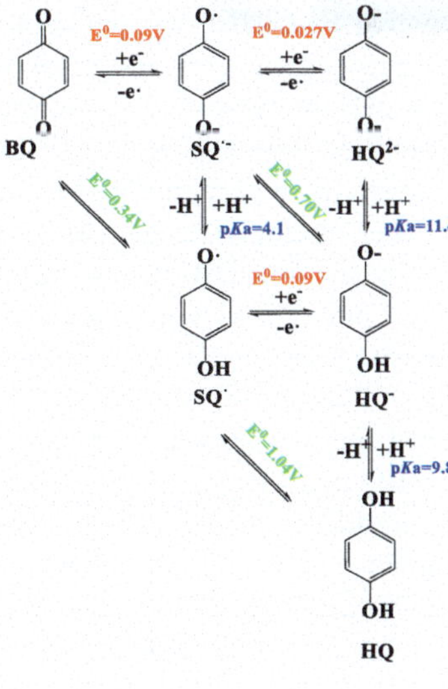

Fig. 1 The reversible redox reactions and potentials between varied redox statuses of benzoquinone (BQ), semiquinone (SQ), and hydroquinone (HQ) [8, 9]

quantifying HBQs, and 46 HBQs were finally observed in simulated drinking water using the pseudo-targeted analysis. 29 of the 30 newly identified HBQs were HMBQs, which were predicted to be carcinogens related to drinking-water bladder cancer risk and be more toxic than non-methyl HBQs. Structures and properties of 30 common HBQs are shown in Table 1.

1.1.2 Analytical Techniques for HBQs Determination

High-sensitive analytical methods were the prerequisite to detect HBQs in drinking water at trace levels. In the early stages, ultraviolet-visible (UV-vis) spectroscopy and electron spin resonance (ESR) were successively employed for HBQs detection [15]. Both of these methods were only applicable for the detection of HBQs at μg/L ~ mg/L levels, not for the analysis of HBQ DBPs at ng/L levels. The state-of-the-art mass spectrometry instruments, including gas chromatography-mass spectrometry (GC-MS) and ultrahigh-performance liquid chromatography-tandem mass spectrometry (UPLC-MS/MS), facilitated the accurate analysis of HBQs [4, 16]. HBQs are non-volatile and thermally unstable, thus cannot be directly detected by GC-MS without derivatization. Derivatization easily introduced unavoidable matrix interference, impacting the limits of detection and precision of the analytical methods, thus GC-MS was unsuitable for high-sensitive detection of trace HBQs in drinking water. UPLC-MS/MS, which employed a soft ionization ion source (electrospray ionization, ESI) to connect liquid chromatography and mass spectrometry, exhibited significant superiority in the analysis of HBQs [3, 16].

A major challenge in the UPLC-MS/MS analysis of HBQs is achieving effective ionization. It was difficult to generate stable regular ionization of $[M + 1]^+$ or $[M-1]^-$ ions of HBQs in neither positive nor negative ESI mode, respectively. Instead, $[M + H]^-$ ions were consistently observed under optimized negative mode, which can be explained based on the electrochemistry of HBQs on the ESI tip. During negative ESI, HBQs underwent an electrochemical reduction at the spraying tip, resulting in the conversion of both C=O groups to C-OH, thus generating an [M + 2H] intermediate. The intermediate further went through a rapid deprotonation to produce $[M + H]^-$ (Fig. 2). This process was consistent with the electrochemistry of benzoquinone, namely, quinone+$2H^+$+2e → dihydroquinone [3].

After obtaining stable $[M + H]^-$ parent ion signals, the characteristic fragment ions for the four HBQs were identified to include $[M + H - HX]^-$, $[M + H - HX - CO]^-$, and X^- (where X represents Cl or Br) [17]. Two stable ion pairs of high intensity were selected for optimizing MRM conditions, including collision energy (CE), delustering voltage (DP), exit voltage (CXP), and entrance voltage (EP). MRM method significantly enhanced the sensitivity, selectivity, and repeatability of HBQ analysis.

Liquid chromatography is a crucial component of UPLC-MS/MS. C18 chromatographic column was applied for the separation of HBQs [11, 18–21]. The mobile phase is composed of organic and inorganic phases. According to the measured pK_a of HBQs and MHBQs [19], the mobile phase for the analysis of HBQs was methanol

Table 1 Structures and properties of 30 common HBQs [4, 14]

Name	Abbreviation	Molecular formula	Structural formula	Molecular weight	MRM (m/z) (Q1 → Q3)
2,6-dichloro-1,4-benzoquinone	2,6-DCBQ	$C_6H_2Cl_2O_2$	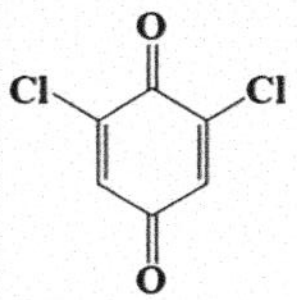	176.98	177 → 141
2,5-dichloro-1,4-benzoquinone	2,5-DCBQ	$C_6H_2Cl_2O_2$	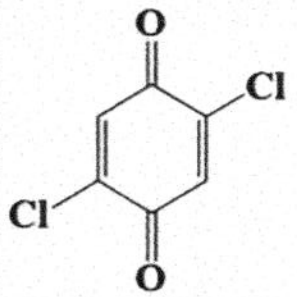	176.98	177 → 113
2,3,6-trichloro-1,4-benzoquinone	TriCBQ	$C_6HCl_3O_2$	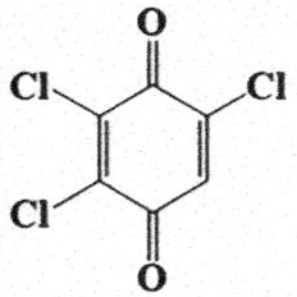	211.43	211 → 175 211 → 147
2,3,5,6-tetrachloro-1,4-benzoquinone	TetraC-1,4-BQ	$C_6Cl_4O_2$	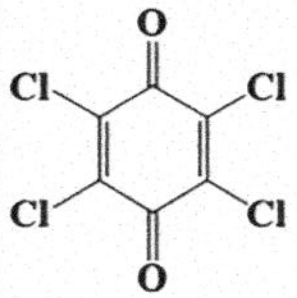	245.88	245 → 35 247 → 35
3,4,5,6-tetrachloro-1,2-benzoquinone	TetraC-1,2-BQ	$C_6Cl_4O_2$	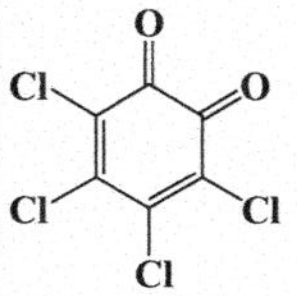	245.88	247 → 211 247 → 209

(continued)

Table 1 (continued)

Name	Abbreviation	Molecular formula	Structural formula	Molecular weight	MRM (m/z) (Q1 → Q3)
2,6-dibromo-1,4-benzoquinone	2,6-DBBQ	$C_6H_2O_2Br_2$	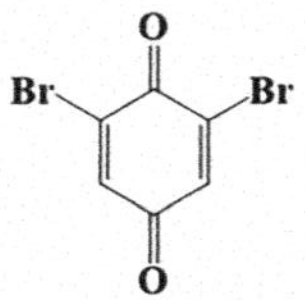	265.89	267 → 79 267 → 81
2,5-dibromo-1,4-benzoquinone	2,5-DBBQ	$C_6H_2O_2Br_2$	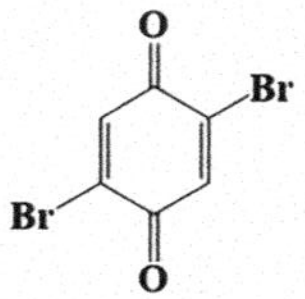	265.89	267 → 79 267 → 81
2,3,5,6-tetrabromo-1,4-benzoquinone	TetraB-1,4-BQ	$C_6O_2Br_4$	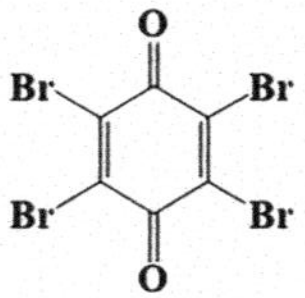	423.68	425 → 79 425 → 81
3,4,5,6-tetrabromo-1,2-benzoquinone	TetraB-1,2-BQ	$C_6O_2Br_4$	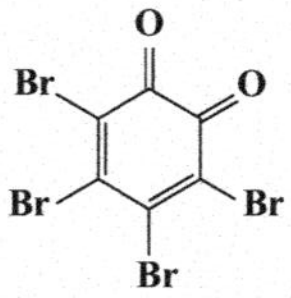	423.68	423 → 79 423 → 81
2-chloro-6-iodo-1,4-benzoquinone	2,6-CIBQ	$C_6H_2ClIO_2$	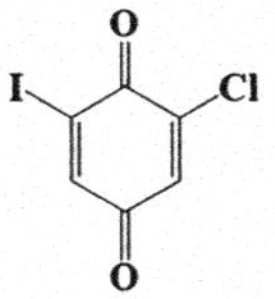	268.436	269 → 127 271 → 127

2,6-diiodo-1,4-benzoquinone	2,6-DIBQ	$C_6H_2I_2O_2$	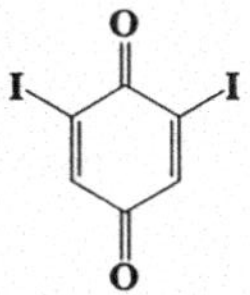	359.89	361 → 127 361 → 205
2-chloro-1,4-benzoquinone	2-MCBQ	$C_6H_3ClO_2$	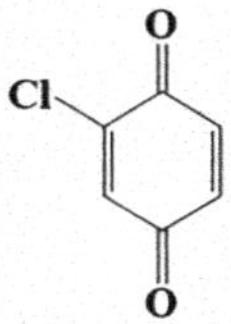	142.54	143 → 79 143 → 51
2-bromo-1,4-benzoquinone	2-MBBQ	$C_6H_3BrO_2$	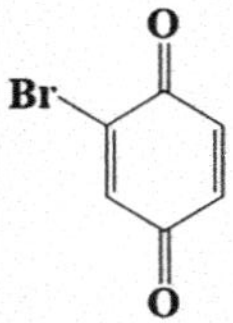	186.99	187 → 119 187 → 79
2-iodo-1,4-benzoquinone	2-MIBQ	$C_6H_3IO_2$	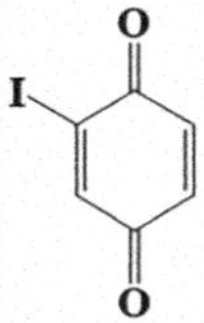	233.99	235 → 127 235 → 205
2,6-dichloro-3-methyl-1,4-benzoquinone	2,6-DC-3-MBQ	$C_7H_4Cl_2O_2$	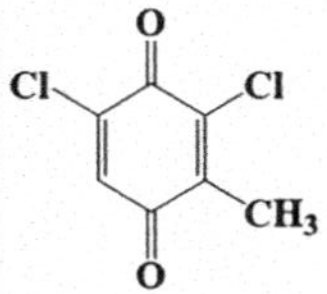	191.01	191 → 127 191 → 155

(continued)

Table 1 (continued)

Name	Abbreviation	Molecular formula	Structural formula	Molecular weight	MRM (m/z) (Q1 → Q3)
2-bromo-6-chloro-3,5-dimethyl-1,4-benzoquinone	2-Br-5-Cl-3,6-DMBQ	$C_8H_6BrClO_2$	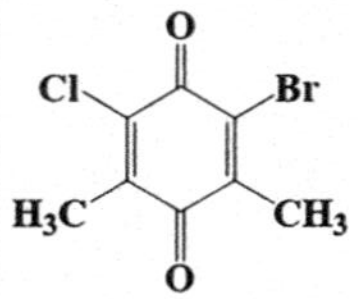	249.49	249 → 81 249 → 79
2-chloro-6-methyl-1,4-benzoquinone	2-Cl-6-MBQ	$C_7H_5ClO_2$	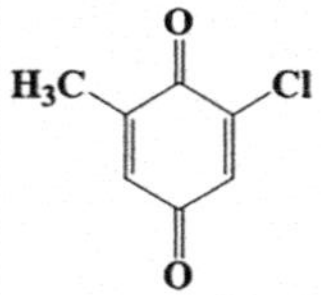	156.57	157 → 93 157 → 35
5-chloro-2,3-dimethyl-1,4-benzoquinone	5-Cl-2,3-DMBQ	$C_8H_7ClO_2$	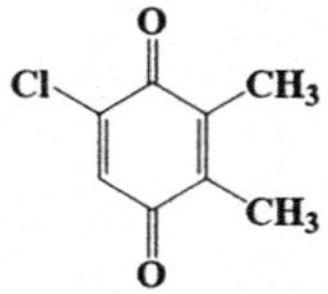	170.59	171 → 143 171 → 35
2,3-dichloro-5,6-dimethyl-1,4-benzoquinone	2,3-DC-5,6-DMBQ	$C_8H_6Cl_2O_2$	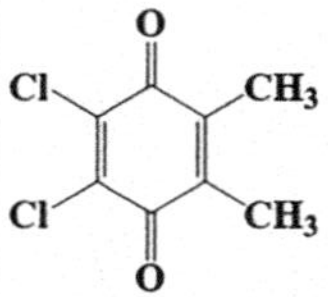	205.03	204 → 35 206 → 35
2,6-dibromo-3-chloro-5-methyl-1,4-benzoquinone	2,6-DB-3-Cl-5-MBQ	$C_7H_3Br_2ClO_2$	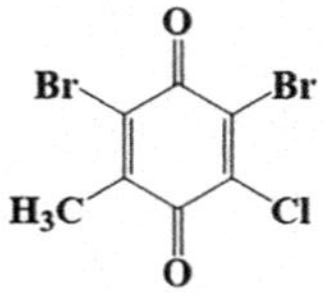	314.36	312 → 79 314 → 81

2,3-dibromo-5,6-dimethyl-1,4-benzoquinone	2,3-DB-5,6-DMBQ	$C_8H_6Br_2O_2$	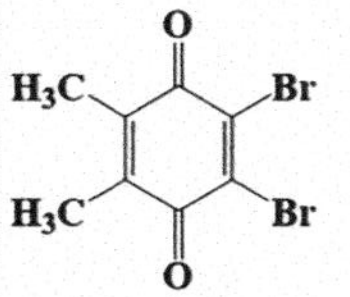	293.94	293 → 81 293 → 249
2-bromo-6-methyl-1,4-benzoquinone	MBMBQ	$C_7H_5BrO_2$	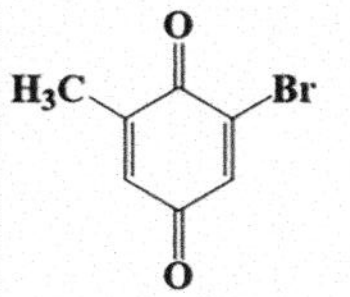	201.02	201 → 157 201 → 142 201 → 81
3-bromo-5-chloro-2-methyl-1,4-benzoquinone	MCMBMBQ	$C_7H_4BrClO_2$	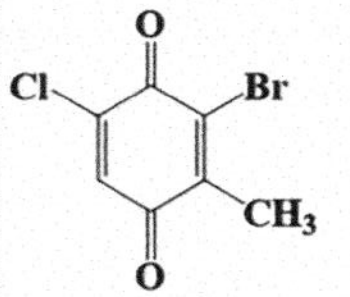	235.46	235 → 81 235 → 147
2-bromo-3,5,6-trimethyl-1,4-benzoquinone	MBTriMBQ	$C_9H_9BrO_2$	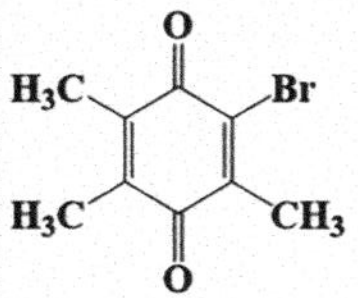	229.07	229 → 214 229 → 79
2-iodo-6-methyl-1,4-benzoquinone	MIMBQ	$C_7H_5IO_2$	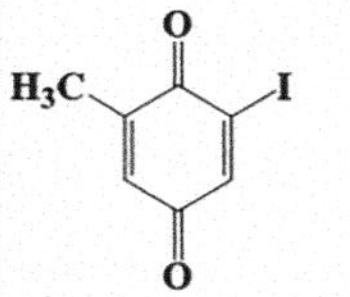	248.02	249 → 234 249 → 127

(continued)

Table 1 (continued)

Name	Abbreviation	Molecular formula	Structural formula	Molecular weight	MRM (m/z) (Q1 → Q3)
3,5-diiodo-2-methyl-1,4-benzoquinone	DIMBQ	$C_7H_4I_2O_2$	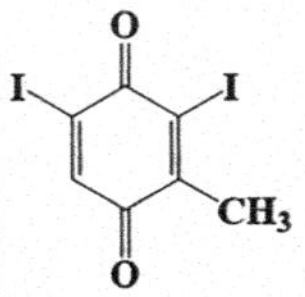	373.92	375 → 360 375 → 127
5-chloro-3-iodo-2-methyl-1,4-benzoquinone	MIMCMBQ	$C_7H_4ClIO_2$	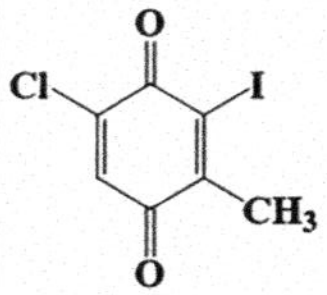	282.46	283 → 251 283 → 127
2,6-diiodo-3,5-dimethyl-1,4-benzoquinone	DIDMBQ	$C_8H_6I_2O_2$	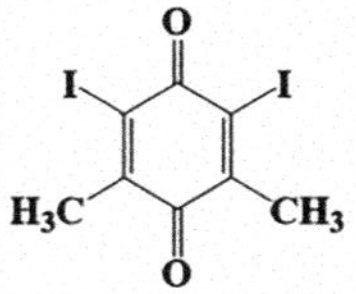	387.94	389 → 345 389 → 127
2-iodo-3,5-dimethyl-1,4-benzoquinone	MIDMBQ	$C_8H_7IO_2$	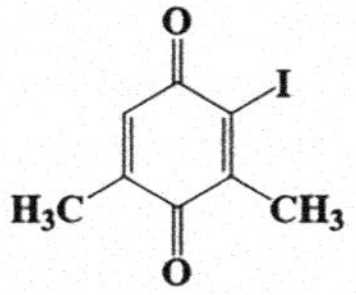	262.05	263 → 219 263 → 127

2-chloro-6-iodo-3,5-dimethyl-1,4-benzoquinone	MCMIDMBQ	$C_8H_6ClIO_2$	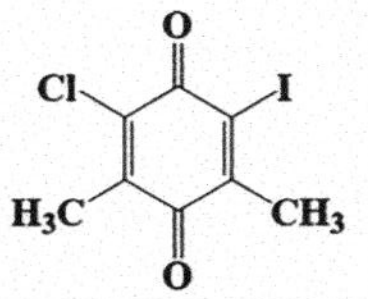	296.49	297 → 127 297 → 237

Fig. 2 Proposed ESI pathway of DCBQ

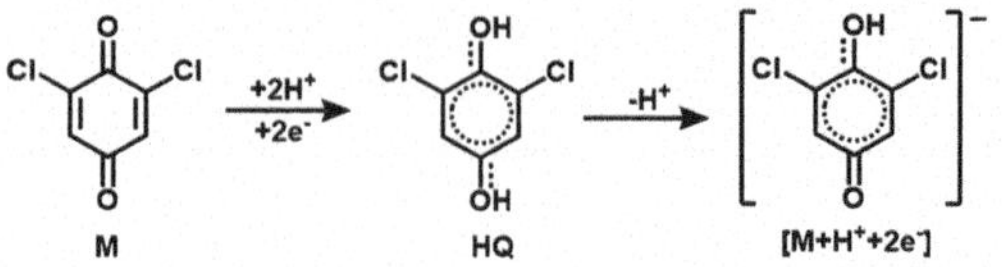

(0.1% formic acid) as the organic phase and water (0.1% formic acid) as the inorganic phase [3]. Whereas the signal was inhibited by formic acid in the detection of HMBQs, and then the one for the analysis of MHBQs was methanol (no formic acid) as the organic phase and water (no formic acid) as the inorganic phase [4].

1.1.3 Occurrence of HBQs in Disinfected Water

Studies on the occurrence of HBQs in water showed that the concentration levels of HBQs varied significantly among disparate water media and disinfection methods.

HBQs in Drinking Water

In 2010, Qin [3] utilized LC-MS/MS technology to analyze drinking water samples collected from two water treatment plants (WTPs) and water distribution systems after chloramination for disinfection. 2,6-DCBQ was detected at concentrations in the range of 5.3–54.6 ng/L, which was decreased with the increased distance to the WTP.

In the same year 2010, Zhao et al. tested chlorinated drinking water samples collected in Canada [22]. They found that the concentration of 2,6-DCBQ reached as high as 165.1 ng/L. Additionally, three other HBQs were detected for the first time, including DCMBQ, TriCBQ, and 2,6-DBBQ, with concentrations ranging from 0.5 to 9.1 ng/L.

In 2012, Zhao [11] tested HBQs in 16 post-disinfection water samples collected from nine WTPs in Canada and the United States. The involved disinfection methods included chlorination, chloramination, chlorination with chloramination, and ozonation with chloramination. 2,6-DCBQ is the most abundant HBQ in all samples, with its concentration reaching up to 274.5 ng/L. Simulated chlorination experiments showed the highest production rate of 2,6-DCBQ among the four tested HBQs, and pre-ozonation increased 2,6-DBBQ formation in the presence of bromide. 2,6-DCBQ was stable at pH < 6.8, but was easily hydrolyzed to form mostly 3-hydroxyl-2,6-DCBQ at pH > 7.6 in drinking water.

Huang [23] developed an analytical method for 12 types of HBQs using UPLC-MS/MS with MRM mode combined with standards. By optimizing the liquid chromatographic elution gradient, this method can efficiently separate 12 types of HBQs within 18 min, including four classes of isomers such as dichlorobenzoquinone, dibromobenzoquinone, tetrachlorobenzoquinone, and tetrabromobenzoquinone. Using SPE as a pretreatment technique, the SPE-UPLC-

MS/MS detection method achieved HBQ recovery rates ranging from 51 to 95%, with quantification limits ranging from 0.6 to 22.1 ng/L.

Nakai [24] surveyed 12 WTPs in Japan in the summer and winter of 2013. 2,6-DCBQ was detected in 21 out of 24 water samples, with concentrations ranging from 8.0 to 51.0 ng/L. Besides, a significant relationship was observed between 2,6-DCBQ concentration and chloroform concentration in drinking water according to Spearman's rank-order correlation coefficient.

Li [25] analyzed HBQs in water samples collected from a WTP and its water distribution network in Tianjin, China during winter and summer of 2021. Consistent with previous surveys, 2,6-DCBQ was identified as the predominant HBQ, and the concentration was decreased with the increase of the transportation distance. Furthermore, they observed that the concentration of detected 2,6-DCBQ in summer ranged from 0 to 72.6 ng/L, with an average concentration of 23.7 ng/L, significantly higher than that in winter. This difference may be related to the high organic matter content in source water in summer.

In recent years, there have been continuous reports of new HBQs discovered in drinking water. In addition to 2,6-DCBQ, DCMBQ, TriCBQ, and 2,6-DBBQ, Hu [5] observed the presence of 2,6-ICBQ, 2,6-DIBQ, and 2,6-IBBQ in drinking water samples collected from Nanjing and Shanghai in 2020, with concentration ranges of 0.7–1.3, 1.8–8.0, and 0.4–15.9 ng/L, respectively. Compared with 2,6-DBBQ, the iodinated HBQs were generally more cytotoxic in HepG2 cells. The formation of 2,6-DIBQ during chloramination was significantly affected by the iodide concentration, pH, and the characteristics of natural organic matters (NOMs).

In 2023, Gao [26] studied the life cycles of TetraC-1,4-BQ and 2,6-DBBQ in drinking water. As the water–transmission distance (water age) increased, the levels of bromo-benzoquinones (Br-BQs) decreased and those of chloro-benzoquinones (Cl-BQs) remained relatively stable in drinking water distribution system in the presence of residual chlorine. TetraC-1,4-BQ was the most abundant Cl-BQ while 2,6-DBBQ was the most abundant Br-BQ in drinking water, with maximum concentrations of 60.2 and 181.4 ng/L, respectively.

HBQs in Swimming Pool Water

Swimming is one of the most popular recreational activities worldwide, and the input or recirculating water used in swimming pools must be disinfected to inactivate pathogens in order to prevent waterborne diseases [27]. Common disinfection methods used in public swimming pools include chlorination, ozonation, ultraviolet irradiation, and combinations of these treatments [28, 29].

The formation of HBQs in swimming pools is complex. Besides the input tap water containing NOM, the sharing of swimming pool by a large number of swimmers introduced additional carbonaceous contaminants, such as body fluids (urine and sweat), skin particles, hair, and personal care products (cosmetics, lotions, and sunscreen). Compared to drinking water, swimming pool water often exhibited higher levels of total organic carbon (TOC) and applied higher chlorine dosages, which led to higher concentrations of HBQs in swimming pool water. In 2013, Wang [14] collected water samples from 10 indoor swimming pools in Canada, in which municipal tap water was further treated with either chlorine alone or a combination of chlorine and ultraviolet treatment. 2,6-DCBQ was detected in all tested swimming pool water samples at concentrations ranging from 19 to 299 ng/L, which was 100 times higher than the concentrations in the source tap water (1–6 ng/L). Additionally, TriCBQ, 2,6-DBBQ, and DBDMBQ were detected in four of the pool waters, which were not found in the input tap water. DBDMBQ was reported for the first time as a DBP.

In 2019, Wu [30] investigated the types and concentrations of HBQs in seven public swimming pools in Nanning, China. In their study, 2,6-DCBQ was found in all seven swimming pool water samples, with concentrations ranging from 4.56 to 45.30 ng/L. The levels of 2,6-DCBQ (median 14.15 ng/L) were significantly higher than those of 2,6-DBBQ (median < 0.38 ng/L) and TetraC-1,2-BQ (median < 0.54 ng/L). Further analysis revealed that the concentrations of HBQs in the indoor swimming pools were positively correlated with chloride ions, humic acid, COD, and UV_{254} but showed no correlation with bromide, free chlorine, pH, and water temperature. However, no correlation was observed between any water quality parameters of outdoor pools and the HBQ concentrations [28], which is probably due to the degradation of HBQs under natural UV light [31].

HBQs in Tea

In 2019, Lou [32] revealed that various HBQs occurred in tea due to the HBQs in tap water, the leachate from tea leaves, and the chlorination of tea polyphenols. 2,6-DCBQ was identified in drinking water at the concentration of 2.32 ± 0.12 ng/L, which was significantly higher than in the four teas (with an average concentration of 0.24 ± 0.01 ng/L). The concentrations of TetraC-1,2-BQ (with an average concentration of 1.40 ± 0.37 ng/L) were all significantly higher in the four teas than in drinking water (0.59 ± 0.01 ng/L) and followed the order black tea (1.76 ± 0.10 ng/L) $>$ pu-erh tea (1.31 ± 0.14 ng/L) $>$ oolong tea (1.17 ± 0.02 ng/L) $>$ green tea (1.03 ± 0.09 ng/L). HBQ concentrations in tea infusions increased as a function of fermentation level. The fermentation-related enzymes, such as peroxidase and polyphenol oxidase, were found to promote HBQ formation. The HBQs in tea infusions were confirmed to form from the reaction between inorganic chloride and phenols in the presence of chloroperoxidase. Additionally, the application of chlorine-free drinking water to prepare tea may effectively reduce HBQs in tea water.

Commonly detected concentrations of HBQs in different environmental media were shown in Table 2. In summary, HBQs were detected in WTPs, swimming pools, and also tea water. Among the common HBQs detected, 2,6-DCBQ has the highest detection frequency and concentration. The types and concentrations of HBQs formed in the disinfected water were related to various factors such as the carbonous precursors, the type and dosage of disinfectants, and disinfection

Table 2 Commonly detected concentrations of HBQs in different environmental media

Environmental media	Water sampling location or type	Country or region	HBQs and their concentration, ng/L	Disinfection methods	Reference
In drinking water	WTPs and water distribution systems	Canada	5.3–54.6 for 2,6-DCBQ	Chloramination	[3]
	WTPs	Japan	8.0–51.0 for 2,6-DCBQ	Chloramination	[24]
	WTPs	Canada	165.12 for 2,6-DCBQ; 0.5–9.1 for DCMBQ, TriCBQ, and 2,6-DBBQ	Chloramination	[22]
	WTPs	Canada and the USA	4.5–274.5 for 2,6-DCBQ, below DL–37.9 for 2,6-DBBQ, below DL–6.5 for 2,6-DC-3-MBQ, and below DL–9.1 for 2,3,6-TriCBQ	Chlorination, chloramination, chlorination with chloramination, and ozonation with chloramination	[11]
	WTPs and water distribution systems	China (Tianjin)	0–72.6 for 2,6-DCBQ	Chloramination	[25]
	WTPs	China (Nanjing and Shanghai)	2,6-DCBQ, DCMBQ, TriCBQ, 2,6-DBBQ, 0.7–1.3 for 2,6-ICBQ, 1.8–8.0 for 2,6-DIBQ, and 0.4–15.9 for 2,6-IBBQ	Chloramination	[5]
	Finished water and tap water	China (Nanning)	<2.6–19.7 for 2,6-DCBQ and <0.38–1.8 for 2,6-DBBQ	Chlorine dioxide and chlorine	[33]
	Plant effluent	China (Hangzhou)	2.1–42.7 for 2,6-DCBQ, 1.1–2.5 for 2,6-DBBQ, and 1.6–2.4 for TetraC-1,2-BQ	Chlorination	[34]

(continued)

Table 2 (continued)

Environmental media	Water sampling location or type	Country or region	HBQs and their concentration, ng/L	Disinfection methods	Reference
	Tap water	China (Hangzhou)	1.2–12.7 for 2,6-DCBQ, 0.7 for 2,6-DBBQ, and 1.9 for TetraC-1,2-BQ	Chlorination	[34]
In swimming pool water	Indoor swimming pools	Canada	19.0–299.0 for 2,6-DCBQ, TriCBQ, 2,6-DBBQ, and DBDMBQ	Chlorine alone or a combination of chlorine and ultraviolet	[14]
	Public swimming pools	China (Nanning)	14.15 for 2,6-DCBQ, 0.38 for 2,6-DBBQ and 0.54 for TetraC-1,2-BQ	Chlorination	[30]
In tea	Black tea, pu-erh tea, oolong tea, and green tea	China	An average concentration of 0.24 ± 0.01 for 2,6-DCBQ, and an average concentration of 1.40 ± 0.37 for TetraC-1,2-BQ	Chlorination	[32]

conditions (pH value, temperature, etc.). Furthermore, HBQs might also exist in other media besides aquatic environment, such as various biological organisms.

1.2 *Precursor and Formation*

1.2.1 Natural Organic Matter

NOM present in the source water exhibited complicated chemical composition and structure. NOM was typically divided into humic and non-humic substances. It is generally believed that the chlorination of humic substances led to the formation of primary DBPs [3]. The humic substances in NOM included humin, fulvic acid, and humic acid, among which fulvic acid and humin collectively referred to as humic acid. Humic acids typically contained an aromatic core skeleton and were formed from the conjugation of numerous aliphatic compounds, fatty acids, carbohydrates, and nitrogen-containing compounds through the formation of ether, ester, and other bonds. Several research utilized characteristic techniques like three-dimensional fluorescence spectroscopy and Fourier transform infrared (FTIR) spectroscopy to study the morphology and structure of humic acids during chlorination, and the result showed that chlorination disrupted the aromatic structure of humic acids. Subsequently, subsequent chlorine substitution reactions led to the formation of C-Cl bonds and the transformation of carbon-functional groups such as carboxyl, carbonyl, and methoxy into C-Cl bonds [35]. It was observed that chlorination of humic acids generated various small-molecule aromatic molecules, such as phenol, 4-hydroxyphthalic acid, chlorophenols, and others. These compounds underwent substitution and oxidation reactions to form HBQs [36].

NOM in natural water was confirmed to be an important precursor for the formation of HBQs [37]. In 2013, Diemert et al. [38] investigated the correlation between NOM characteristics and the formation of 2,6-DCBQ during chlorination of three representative source waters in Canada (Lake Ontario, Otonabee River, and Grand River). NOM fractions were separated and analyzed using liquid chromatography-organic carbon detection (LC-OCD), and HBQs were quantified using liquid chromatography-triple quadrupole mass spectrometry. 2,6-DCBQ was identified in all waters after chlorination, and significantly correlated with UV absorbance at 254 nm. Pooled data from the three waters revealed that the 2,6-DCBQ formation strongly correlated with biopolymer and humic substance fractions of NOM ($R^2 = 0.78$, $P < 0.001$). In 2022, Lou [16] separated river NOM according to their molecule weight using molecule cutoff, and studied the composition of the size-fractionated river NOM using Orbitrap-MS. Pearson correlation analysis of HBQ formation and source water NOM components indicated that highly oxygen compounds and lignin/carboxyl-rich alicyclic molecules-like components in large molecular weight fractions (>50 kDa) were the primary precursors of HBQs. These results provided comprehensive insights into the molecular mechanism of HBQ formation by NOM, which will be helpful for the prediction of HBQ generation of different source water and the employment of more effective treatment technologies towards HBQ control at drinking water treatment plants. These researches indicated that aromatic components of NOM, from macromolecules like lignin and tannins to smaller molecules like catechol and HQ, serve as potential precursors for HBQ formation [39].

Phenolic moieties contributed to the formation of HBQs [8]. In 2012, Zhao [11] investigated the formation of HBQs in chlorination of phenol. When phenol and sodium hypochlorite were mixed in a molar ratio of 1:10 under pH 6.0 conditions, TriCBQ, 2,6-DCBQ, and the intermediate 2,4,6-trichlorophenol (TCP) were detected in the mixture using UPLC-MS/MS. With the addition of bromide ions, noticeable signals of 2,6-DBBQ and 2,4,6-tribromophenol (TBP) were observed. Owing to a higher dissociation energy O-Cl bond than O-Br bond, bromine tended to react faster with phenol than chlorine. During chlorine or chloramine disinfection, phenol underwent electrophilic reactions with bromine or chlorine to form TBP or TCP, which further underwent ion free radical reaction forming 2,6-DBBQ or 2,6-DCBQ. 2,6-DCBQ continued to undergo chlorination reactions to form TriCBQ. When ozone and chloramine were used in combination for disinfection, phenol was first rapidly oxidized by ozone to form hydroxylated 1,4-dihydroxybenzene and 1,2-dihydroxybenzene, which were further oxidized to the corresponding benzoquinones. Benzoquinones then reacted with bromide ions to form 2,6-DBBQ or reacted with chloramine to form 2,6-DCBQ.

A recent study in 2023 studied the formation and decomposition of 2,6-DCBQ and the chlorination of lignin phenol. 2,6-DCBQ and the corresponding hydroxyl DCBQ (OH-DCBQ) were identified as the intermediate products in the chlorination of the three typical lignin phenol precursors (p-hydroxybenzoic acid, protocatechuic acid, and gallic acid). 2,6-DCBQ and OH-DCBQ were generated within 2–6 h, and the relative yields of mol carbon atoms in 2,6-DCBQ from the lignin phenol precursors were about 0.01%–14.37% under varied pH. 2,6-DCBQ and OH-DCBQ were further decomposed into HAAs and other unknown DBPs [40].

Wang et al [41] evaluated the formation of HBQs from chlorination of precursors such as bisphenol A (BPA), TCP, 2-chlorophenol, N-methylaniline, p-hydroxybenzoic acid, and tyrosine. It was found that the formation potential of 2,6-DCBQ from BPA was the highest, reaching 14.86 μg/L after 1 h, followed by tyrosine, 2-chlorophenol, P-hydroxybenzoic acid, trichlorophenol, and N-methylaniline.

Algal-derived organic matter (AOM), including both extracellular organic matter (EOM) and intracellular organic matter (IOM), was verified to be HBQ precursors. Ge [42] fractionated *C. vulgaris* into protein-rich and IOM and polysaccharide-rich EOM. 2,6-DCBQ formation potential of IOM (1.4 μg/mg C) was two-fold higher (0.7 μg/mg C) than that of EOM, indicating proteins as potent 2,6-DCBQ precursors. (The yields of HBQs were expressed as the ratio of the generated HBQs to the initial TOC concentration and expressed in μg/mg C.) A linear correlation between 2,6-DCBQ production and the concentration of total algal protein was developed ($R^2 = 0.98$). *Microcystis aeruginosa* and *Chlorella* algae cells were also precursors of 2,6-DCBQ. The production of 2,6-DCBQ increased first and then decreased, indicating that HBQs were unstable in water. THMs were detected during chlorination, and the concentration increased with reaction time. 2,6-DCBQ production decreased and 2,6-DBBQ production increased with increment of bromide concentration and the bromide promoted the formation of tribromomethane. The production of 2,6-DCBQ decreased with increase of pH, and the maximum production was 141.38 μg/L at a pH of 5 [41].

1.2.2 Biological Organic Matter

Free Aromatic Amino Acids

Chlorination of free aromatic amino acids, such as tryptophan (Trp), tyrosine (Tyr), and phenylalanine (Phe), generated 2,6-DCBQ, TriCBQ, TetraCBQ, and DCMBQ [43]. Under the same conditions, the concentrations of HBQs in chlorinated Trp or Tyr samples were significantly higher than in chlorinated Phe samples, possibly because Trp and Tyr had electron-rich double bonds which are conducive to singlet oxygen oxidation [44]. Furthermore, the yield of chlorinated HBQs during the chlorination of these free aromatic amino acids is significantly influenced by the dose of free chlorine, pH, and temperature.

Other Aromatic Compounds

Para-substituted phenolic compounds and para-substituted aromatic amines also functioned as 2,6-DCBQ precursors. In 2017, Kosaka [45] investigated 2,6-DCBQ formation from 31 aromatic compounds via chlorination to understand the structural characteristics of 2,6-DCBQ precursors, including HQ, BQ, *p*-quinone-4-chloroimide (QC), 2,6-dichloroquinone-4-chloroimide, cresols, anisole and its derivative 4-methoxyphenol (MP), 4-hydroxybenzoic acid (HBA), aromatic amines, and the ingredients in PCPs. 2,6-DCBQ was formed from 21 of 31 target compounds after 60 min of chlorination with the molar formation yields ranging from 0.0008% to 4.9%. The proposed formation mechanism of 2,6-DCBQ was in agreement with the previous study, as it was reported that 2,6-DCBQ was formed from phenol through TCP as an intermediate [11]. In addition, several industrial products that are commonly present in environmental water (e.g., 4-hydroxybenzoic acid methyl ester, BPA, and 4-nonylphenol) were shown to serve as 2,6-DCBQ precursors, suggesting that discharge control of such compounds in water sources and their efficient treatment during the water purification process may be important to reduce HBQ concentrations in drinking water [46, 47].

1.2.3 Anthropogenic Organic Matter

Artificial Sweeteners

Acesulfame and sucralose are the most commonly used artificial sweeteners. They are synthetic or semi-synthetic organic compounds, and are widely used in food, drugs, beverages, and personal care products [48]. They are recalcitrant to metabolism within the human body. In 2017, Li [49] studied the removal of acesulfame through chlorination and found that various brominated DBPs were produced during chlorination, with higher toxicity compared to chlorinated disinfection byproducts.

Pharmaceuticals and Personal Care Products

Wang [14] assessed the formation of HBQs from four body lotions and four sunscreens commonly available in markets. All chlorinated samples of body lotions and sunscreens produced 2,6-DCBQ, with the highest concentration of 5,420 ± 373 ng/L. In addition, DCMBQ, TriCBQ and TetraB-1,4-BQ were also detected in some body lotion or sunscreen samples. Some common ingredients in lotions, such as benzyl alcohol, lecithin, para-hydroxybenzoic acid esters, and certain plant extracts containing vitamins, amino acids, peptides, and proteins, can act as precursors for HBQs. Chemical substances in sunscreens that provide UV blocking and absorption, such as avobenzone, octocrylene, and benzophenone-4, might also be precursors for HBQs.

Residual antibiotics in water were found to be precursors for various DBPs. Tetracyclines, quinolones, and other antibiotics consist of quinone-like skeletons, which may be potential precursors for HBQs [50].

1.3 Transformation and Removal

Control methods for DBPs in drinking water can be broadly categorized into four main classes: source control, process control, end-of-pipe control, and coordinated control. Source control refers to reduce the concentration of DBP precursors in water through source protection or treatment processes at the drinking water plant before the addition of disinfectants. Process control aims to minimize the generation of DBPs or alter the types of DBPs produced by changing disinfection methods and optimizing disinfection processes. End-of-pipe control involves using physical or chemical methods to remove a range of already generated DBPs from the water. Coordinated control combines source, process, and end-of-pipe control methods to achieve efficient control of various types of DBPs. These control methods are implemented to ensure that the levels of potentially harmful DBPs in drinking water are minimized to protect public health.

1.3.1 Transformation

HBQs underwent spontaneous hydrolysis and hydrolysis in neutral and alkaline water, forming hydroxyl HBQs (OH-HBQs) [51], and the hydrolysis rate of 2,6-DCBQ was proportionally correlated to pH [25]. No THMs and HAAs were generated during the spontaneous hydrolysis of 2,6-DCBQ, while 2,6-DCBQ underwent further transformation in the presence of chlorine, leading to the formation of OH-DCBQ, THMs (2.9% to 10.0%) and HAAs (0.1% to 2.2%) [34]. A negative correlation between the hydrolysis rate of HBQs and residual chlorine was observed [25], and the decay of HBQs was fit in a pseudo-first-order mathematical model [6]. Chloramines did not significantly affect the stability of HBQs [52]. The transformation rates of HBQs and the final products were affected by pH, temperature, water age, inorganic ions, and some other organic compounds [26, 53].

Amino acids (AAs), such as histidine (His), cysteine (Cys), threonine (Thr), and phenylalanine (Phe), were capable of influencing the transformation of HBQs [54]. The transformation rates of HBQs in the presence of AAs were up to 95.3%, with the highest rate induced by Cys. The reactivity of three HBQs with AAs followed the order 2-chloro-1,4-benzoquinone (MCBQ) $<$ 2,6-DCBQ $<$ TetraCBQ, consistent with their halogenation degree and calculated electron affinity (EA). In the presence of AAs, initial spontaneous hydrolysis of HBQs led to OH-HBQs formation, followed by nucleophilic addition/substitution of AAs, resulting in AA-HBQs/AA-HBQs-OH adducts. These adducts were oxidized into corresponding decarboxylated forms. Higher AA dosages and elevated pH levels favored HBQ transformation. Free aromatic AAs, such as tryptophan (Trp), tyrosine (Tyr), and phenylalanine (Phe), also acted as precursors of chlorinated HBQs during chlorination, forming 2,6-DCBQ, TriCBQ, TetraCBQ, and DCMBQ [43]. Tyr and Trp exhibited relatively higher yields of chlorinated HBQs than Phe. Electrophilic addition, electrophilic substitution, oxidation, deacidification, and dehydration reactions were involved in the interactions between AAs and HBQs.

1.3.2 Control and Removal

The strategies for the control and removal of HBQs primarily involved in the removing of HBQ precursors before disinfection, optimizing disinfection parameters to reduce HBQs formation, or removing generated HBQs in post-disinfection processes.

The effects of pre-disinfection treatments on the removal of HBQ precursors were studied. Alum was found to be effective in partially removing, but was unable to completely remove the precursors of HBQs, such as humic substances and high molecular weight organic matter [38]. Alum coagulation in WTPs reduced dissolved organic carbon (DOC) by up to 56% and HBQ precursors by up to 39% for 2,6-DCBQ [55]. WTP survey found that granular activated carbon (GAC) removed 10–20% of HBQ FPs (formation potential) with GAC, and 0.2–4.7% of DOC removal was minimal, indicating that GAC primarily functioned in the mode of biodegradation other that adsorption. Ozonation reduced 10–30% of HBQ FPs, while anthracite/sand filtration and UV irradiation showed little impact on HBQ FPs. Pre-ozonation significantly enhances the removal of HBQ precursors in the joint application with GAC [50]. Overall, combined water treatments can partially remove HBQ precursors, but cannot substantially eliminate HBQ precursors.

Appropriate disinfection was capable of minimizing the formation of HBQs. High residual chlorine significantly increased the formation of HBQs [33]. Substitution reactions were dominant at lower chlorine dosages, while oxidation reactions became more significant at higher dosages. However, the generation of HBQs tends to decrease at excessively high dosages, possibly due to the further reaction of HBQs to form OH-HBQs and aliphatic DBPs. Chlorine dioxide disinfection resulted in much lower HBQs generation compared to chlorine disinfection, possibly attributed to the effective removal of phenolic compounds in water by chlorine dioxide oxidation [11]. The ionization of chloramines in water built a weak alkaline environment, leading to instability and decomposition of HBQs [22]. Ozone/chlorine combined disinfection generated less HBQs compared to chlorine or chloramine disinfection possibly due to the reduced chlorine dosage [56].

UV disinfection facilitated the photodegradation of HBQs. The molar absorptivity of 2,6-DCBQ, 2,6-DBBQ, DCMBQ, and TriCBQ at 254 nm UV light are 4,258, 2,828, 6,882, and 4,449 $L \cdot mol^{-1} \cdot cm^{-1}$, providing a theoretical basis for the UV degradation of HBQs in water [8, 18]. A large fraction of 2,6-DCBQ, DCMBQ, TriCBQ, and 2,6-DBBQ were degraded in water after UV treatment, and found that DCMBQ is the most easily removed HBQ by UV_{254} [18]. Factors such as UV intensity, pH, and initial concentration can influence the degradation efficiency of HBQs. The direct UV irradiation can degrade DCBQ to some extent but the UV dose required can be much higher than what is practical in a WTP. The UV advanced oxidation processes (UV/AOPs) provided a higher removal rate of HBQs. For example, UV/H_2O_2, UV/O_3, and UV/H_2O_2/O_3 removed 88.5%, 92.4%, and 96.7% of 2,6-DCBQ, respectively [57].

Activated carbon is a highly porous adsorbent material with a high specific surface area that can effectively remove pollutants from water. Combining ozone with GAC can significantly enhance the removal efficiency of HBQs [58]. Activated carbon, ozone, and ozone-activated carbon demonstrated removal rates of 29.64%, 32.81%, and 84.93% for 2,6-DCBQ, respectively [59]. The ozone-activated carbon combined treatment showed a much significant removal effect compared to activated carbon or ozone alone, since the continuous replenishment of effective absorption sites on activated carbon after ozonation.

Non-thermal plasma is an efficient and environmentally friendly technology that offers potential for removing DBP precursors [60]. In 2023, Gao [61] evaluated the performance of bubbling corona discharge on the degradation of 24 halogenated alicyclic and aliphatic DBPs present in drinking water at concentrations ranging from ng/L to μg/L. THMs, HNMs, and HBQs were effectively removed within 5 min under a discharge power of 28 W. Degradation rates surpassed other conventional treatment technologies. The required energy consumption was in the range of 5–30 kW h/m^3/order. The breakdown of HBQs did not produce secondary contamination such as aliphatic DBPs. The carbon in DBPs was primarily converted to formic acid, acetic acid, and oxalic acid, and the halogens were mainly converted to halogen ions. Additionally, luminescent bacteria toxicity testing confirmed that plasma treatment reduced the acute toxicity of water samples. These findings demonstrated the great application prospect of plasma techniques as post-treatments at the household level.

In summary, the effects of conventional water treatment on the removal of HBQs were limited, and may facilitate the transformation of HBQs to small-molecule DBPs. Future studies may direct to clean techniques, such as chlorine-free disinfections, to eliminate the formation of DBPs including HBQs, in order to effectively reduce the health risks related to drinking water.

2 Mutagen X (3-Chloro-4-(Dichloromethyl)-5-Hydroxy-2(5*H*)-Furanone, MX)

MX, or Mutagen X, was the code name for a compound in chlorinated pulp bleaching waters with high mutagenicity but at that time not chemically identified. This compound was later identified as 3-chloro-4-(dichloromethyl)-5-hydroxy-2

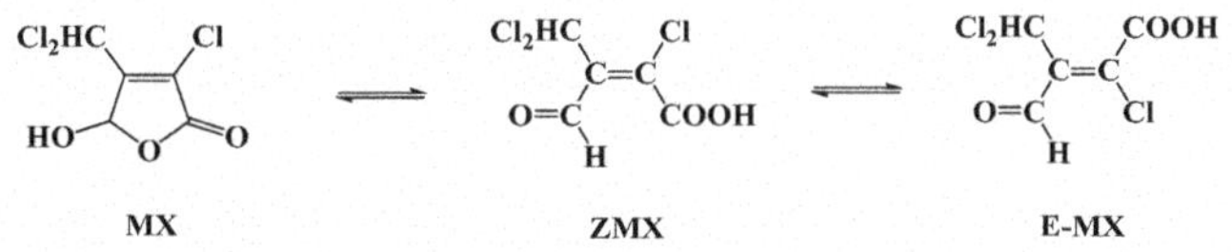

Fig. 3 Structures of MX, ZMX, and E-MX

(5H)-furanone. MX also has one open-ring form, (Z)-2-chloro-3-(dichloromethyl)-4-oxobutenoic acid (ZMX). The ring form of MX occurs in water at pH less than 5.5 and open ring form, ZMX, occurs in the typical pH range of drinking water. There is also a geometric isomer, (E)-2-chloro-3-(dichloromethyl)-4-oxobutenoic acid (E-MX or EMX). The chemical structures of MX, ZMX, and E-MX are shown in Fig. 3.

2.1 *MX Occurrence in Drinking Water*

MX was first reported at 20–60 ng/L in two Finnish surface water treatment plant chlorinated effluents in 1986 [62]. E-MX was first reported at 8–41 ng/L in chlorinated tap water from three Finnish water plants, where MX was detected at 15–67 ng/L [63]. MX in chlorinated drinking water was also reported at 2–33, 5–67, 3–9, 27, and 11 ng/L in the USA, Finland, Japan, Canada, and China, respectively.[64–68] The occurrence of E-MX was also reported in these studies.

As a strong mutagen, MX is a main contributor to the mutagenic activity in chlorinated drinking water. The percentages of MX contribution were reported as high as 46, 34, 23, and 35% in Finland, USA, Japan, and China, respectively [63, 64, 66, 68]. E-MX also contributes to the observed mutagenic activity in water at percentages which are much lower than MX. The study conducted in Finland reported the E-MX contribution at 2% or lower to the observed mutagenic activities and this contribution was more than 10 times lower than that of MX [69].

Like many DBPs, MX and E-MX also have analogues, including brominated MXs in water high in bromide and oxidized MX in water under oxidized conditions. In water high in bromide, sequential substitution of three chlorine atoms in MX with bromine atoms could result in three brominated MX analogues, BMX-1 (3-chloro-4 (bromochloromethyl)-5-hydroxy-2(5H)-furanone), BMX-2 (3-chloro-4-(dibromomethyl)-5-hydroxy-2(5H)-furanone), and BMX-3 (3-bromo-4-(dibromomethyl)-5-hydroxy-2(5H)-furanone). The chemical structures of BMX-1, 2, and 3 are shown in Fig. 4. These three brominated MXs are also mutagenic but slightly lower than that of MX, with specific mutagenic activities at 55, 35, 38, and 27 net revertants/ng for MX, BMX-1, BMX-2, and BMX-3 respectively [70]. In water low in bromide, MX should be the predominated MX species. Relatively high levels of brominated MXs are expected in water high in bromide. The concentrations of brominated MX analogues (2.3, 5.7, and 11.1 ng/L, respectively) were reported much higher than that of MX (0.9 ng/L) in one drinking water sample likely high in

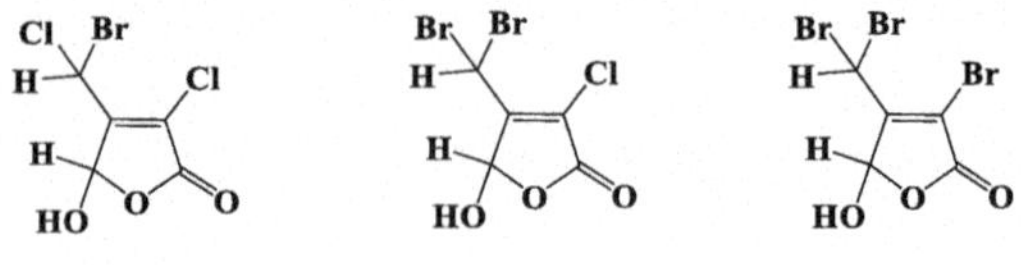

Fig. 4 Structures of BMX-1, 2, and 3

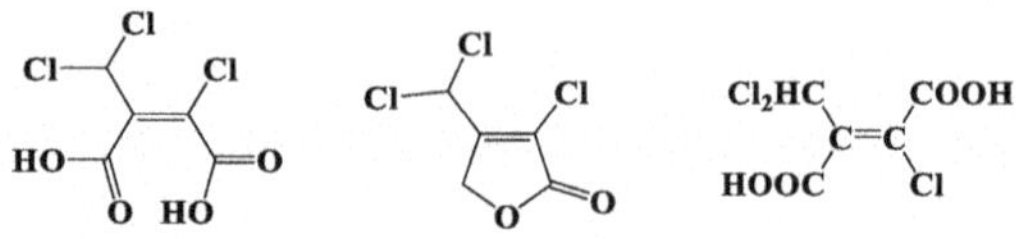

Fig. 5 Structure of ox-MX, red-MX, and ox-EMX

bromide [70]. The contribution of MX (2%) to the observed mutagenic activity in this water sample was much lower than the contribution of three brominated analogues (3.4, 9.0, and 12.6%, respectively). Although the bromide level was not reported for this water, the strong correlations between the normalized MX, BMX-1, BMX-2, and BMX-3 levels (to total MXs) and the normalized chloroform, bromodichloromethane, chlorodibromomethane and bromoform (to total THMs) indicate bromide was the main factor which enhanced the formation of brominated MXs.

Oxidized and reduced forms of MX (ox-MX and red-MX) and oxidized E-MX (ox-E-MX) are another group of MX and E-MX analogues. Their structures are shown in Fig. 5. Ox-MX, red-MX, and ox-E-MX were reported in chlorinated drinking water at 53, 58, and 251 ng/L in water containing 13 ng/L of MX and 20 ng/L of E-MX [71]. These analogues occurred in drinking water at levels several times higher than MX and E-MX; however, these analogues did not contribute significantly to mutagenic activity observed. Of these analogues, only ox-MX is mutagenic and with a potency 70 times lower than that of MX [71].

A comprehensive study was conducted to survey the occurrence of 12 halogenated furanones (MX-analogues) in 12 drinking water plants in the USA [72]. MXsum (MX, ZMX, and E-MX) was detected in 37 out of 67 samples analyzed, with a median concentration of 50 ng/L and maximum of 850 ng/L. BMXsum, six brominated MXs, including brominated MXs and brominated E-MX, was detected in 29 out of 47 samples, with a median concentration of 120 ng/L and maximum 2100 ng/L. The highest BMXsum, 2100 ng/L, was detected in water with the highest bromide level, 0.25 mg/L. Reduced MX, Red-MX, was also detected at 580 ng/L in one plant effluent and at 280 ng/L in its distribution system.

2.2 MX Precursors

Formation of MX in drinking water is generally associated with chlorinated surface water. Natural organic matter is the main precursor of MX and its analogues. This is supported by the fact that much higher levels of MX were detected in chlorinated fulvic and humic acids solutions than in chlorinated drinking water. MX and E-MX were reported at 190–380 and 390–700 ng/L, respectively, in chlorinated humic solutions (25 mg/L TOC), while the typical MX and E-MX levels were at 36 and 29 ng/L, respectively, in chlorinated drinking water [69]. In another study, MX and E-MX were reported at 675 and 1200 ng/L and 260 and 526 ng/L in chlorinated fulvic and humic acid solutions, respectively [71]. Ox-MX, Red-MX, and ox-E-MX were also detected at similar or much higher levels in chlorinated fulvic and humic acid solutions in this study.

Three compounds, 3,4,5-trimethoxybenzaldehyde, 1-dichloromethyl-3,4,5-trimethoxy-benzene and 3,4,5-trimethoxyphenylacetic acid, were identified as MX precursors [73]. Syringaldehyde, acetosyringone, ferulic acid, 3-ethoxy-4-hydroxy-benzaldehyde, 3,4,5-trimethoxy-benzaldehyde, vanillin, tryptophan, and tyrosine were also reported as MX precursors [74]. Of these precursors, syringaldehyde gave the highest MX yield after chlorination. The MX formation pathways from these precursors were investigated as well. These studies suggested that aldehyde groups in the aromatic rings of natural humic material play an important role in the formation of MX.

2.3 MX Analysis

For MX determination, the original method includes XAD resin concentration, acidic methanol methylation, and gas chromatography and mass spectrometry (GC-MS) analysis [62, 69]. The identification of MX and E-MX was done using mass spectra and quantification was conducted using selected ion monitoring. This procedure was also used to analyze other MX analogues, including brominated MXs, brominated E-MXs, ox-MX, red-MX, and ox-E-MX. This procedure is relatively complicated and time-consuming. It requires 2L water sample and a solvent extract concentration step with a rotary evaporator.

Another improved method involves solid phase extraction, acidic 2-propanol derivatization, and GC-MS analysis [75]. The derivatization with 2-propanol enhanced the sensitivity of MS detection and reduced water sample volume from 2L to 500 mL. This method can be used for the quantification of MX and brominated MXs. However, the type and operating conditions of the GC injector and column need to be optimized because of instability of propylated derivatives of BMXs.

The third method consists of liquid–liquid extraction, boron trifluoride methanol derivatization, and GC with microelectron capture detection (μECD) [76]. The boron trifluoride methanol derivatization and the use of μECD greatly enhanced the instrument sensitivity for the detection of MX and its analogues and reduced the water sample volume requirement to 250 mL. However, the method has a higher detection limit, 20 ng/L and cannot simultaneously quantify red-MX, MX-, and BMX-analogues.

Recently, a liquid–liquid extraction and liquid chromatography electrospray ionization tandem mass spectrometry (LC-ESI-MS/MS) detection method were also developed for the determination of MX, mucobromic acid (MBA), BMX-1 and BMX-3 [77]. The practical limits of detection for MBA, MX, BMX-1, and BMX-3 were reported at 7, 10, 6, and 2 ng/L, respectively. The method does not require sample derivatization; however, it does require a sophisticated analytical instrument. Addition work is needed to optimize this method.

2.4 *Removal and Control*

As DBPs, MX and its analogues in drinking water can be controlled by natural organic matter removal and modification, chlorination optimization, and removal after their formation. However, due to sophistication of the MX analysis, limited studies were conducted on MX control and removal.

Chlorination is a critical factor for MX and E-MX formation. As reported that increasing chlorine dose from 0.5 to 1 and 2 mg/L increased the formation of MX from 1.3 to 4 and 5.3 ng/L, respectively [78]. Alternative disinfectants, chloramine (10 mg/L) and chlorine dioxide (20 mg/L), reduced the formation of MX and E-MX from 194 and 243 ng/L with chlorine (10 mg/L) to 43 and 59 ng/L and 20 and 2 ng/L, respectively, in humic acid solutions [79]. The study also indicated that preozonation at 10 mg/L reduced the formation of MX and E-MX from 194 and 393 ng/L to 128 and 219 ng/L, respectively, in chlorinated (20 mg/L) humic acid solution. The effects of alternative disinfectants on MX and E-MX were similar to those on chloroform [79]. Therefore, reducing chlorine dose and using of alternative disinfectants generally reduce the formation of MX and its analogues.

Unlike trihalomethanes (THMs) and haloacetic acids (HAAs), the effect of coagulation on the formation of MX and E-MX is inconclusive. A study in Finland reported slightly higher MX and E-MX levels when alum coagulation was included in the treatment process [63]. A substantial increase in MX and E-MX levels, from 8 and 3 ng/L to 24 and 104 ng/L, respectively, after ferric coagulation was also reported in China [68].

Granular activated carbon (GAC) is effective in removing MX and E-MX and their precursors. One Canadian study reported that GAC process completely removed the MX (83 ng/L) formed during prechlorination and prevented the formation of MX in postchlorinated water [67]. A study in China reported that GAC process, or biologically active carbon because of TOC saturation, completely removed the MX and E-MX (20 and 57 ng/L) formed during prechlorination, coagulation, and filtration and reduced their formation to 11 and 57 ng/L after postchlorination [68]. The reduction of MX and E-MX from 24 and 104 ng/L to 20 and 57 ng/L was also observed during rapid sand filtration in this study. Both GAC adsorption and biological degradation played a role in removing MX, E-MX, and their precursors.

The effective removal of MX precursors by biofiltration was also reported in a pilot study [80]. The study indicated that the empty bed contact time did not impact the performance of biofiltration. However, cyclically operated filters performed better than continuously operated filters indicating the beneficial impact of anoxic conditions.

2.5 Health Risks and Regulatory Considerations

As a potent mutagen, MX is the major contributor to the mutagenic activity in chlorinated water. MX also induced cancer at multiple sites in rats, enhanced tumor yields in genetically modified rodents, and induced genotoxic effects in numerous in vitro and in vivo test systems [81]. However, the risks associated with MX are also reported low in comparison with other DBPs. Bromoacetic acid was 31 times more cytotoxic and 14 times more genotoxic in CHO cells than MX [82]. Many other nitrogenous or iodinated DBPs also showed much higher cytotoxicity and/or genotoxicity than MX.

California Environmental Protection Agency (EPA) determined MX's No Significant Risk Level (NSRL), a daily intake level posing a 10^{-5} lifetime risk of cancer, as 0.11 μg/day (California Environmental Protection Agency 2001). Assuming an adult drinking 2 liters of water per day, the concentration would be 55 ng/L, the typical level in chlorinated drinking water. World Health Organization (WHO) determined the MX concentration associated with a 10^{-5} risk as 1.8 μg/L (World Health Organization 2004). Because of the health risks and occurrence levels associated with MX and halogenated furanones, WHO decided not to propose a formal guideline value for MX and other halogenated furanones in drinking-water. MX and several halogenated furanones were also included in the USEPA Six-Year Review 3. MX and other halogenated furanones were listed as unregulated DBPs. No regulatory revision was recommended for MX and other halogenated furanones based on available information and data [83].

3 Other Cyclic DBPs

In addition to the aforementioned HBQs and MXs, many other halogenated cyclic DBPs have been discovered in recent years. Typical representatives included trihalo-hydroxy-cyclopentene-diones (trihalo-HCDs), halonaphthoquinones (HNQs), chlorophenylacetonitriles (CPANs), halogenated pyridinols, haloanilines, haloquinone chloroimides (HQCs), halophenylacetamides (HPAcAms), halohydroxybenzonitriles (HHBNs), halobenzamides (HBZAMs), halogenated nucleobases, and chlorinated nucleotides. We roughly classified them into DBPs formed from exogenous organic matter (Fig. 6) and those formed from endogenous organic matter (Fig. 7).

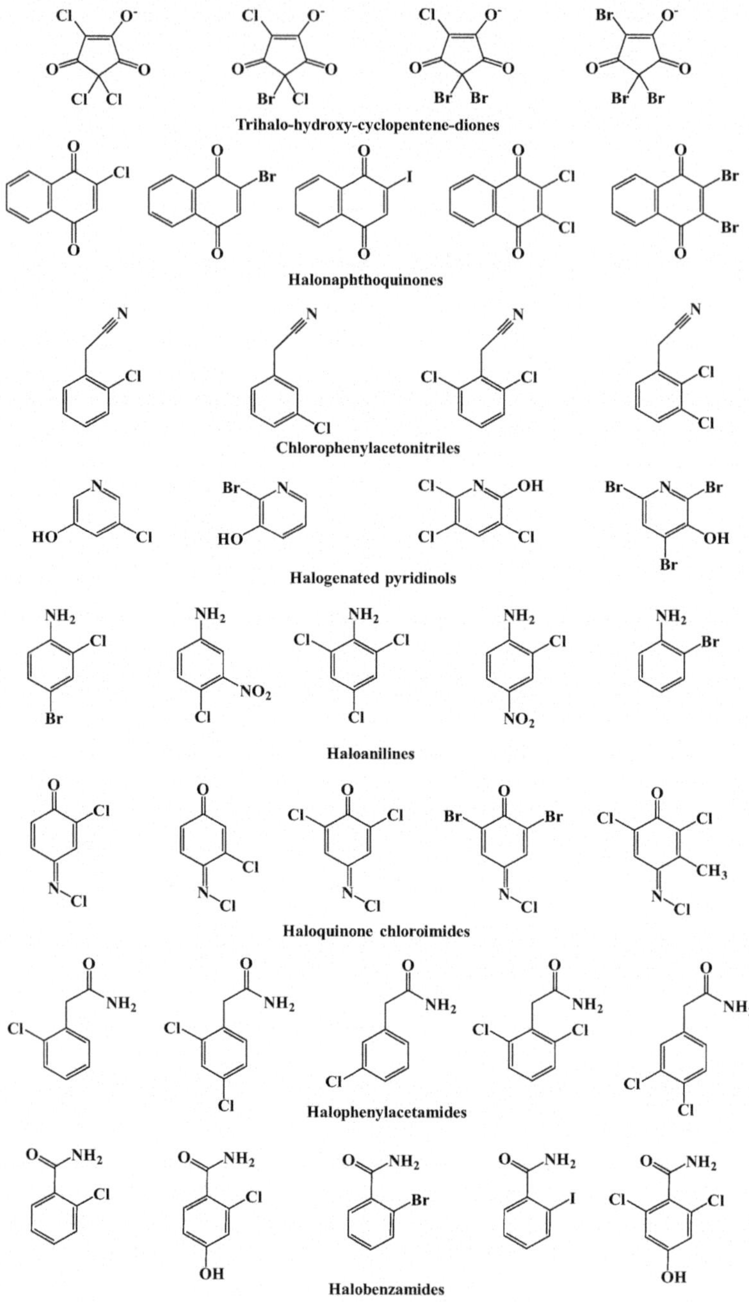

Fig. 6 Other cyclic DBPs formed from exogenous organic matter

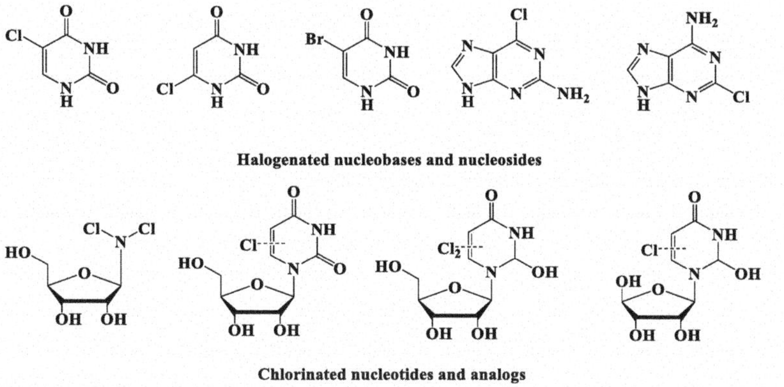

Fig. 7 Other heterocyclic DBPs formed from endogenous organic matter

The successful synthesis and purification paved a path for the identification of trihalo-HCDs [84]. The drinking-water survey suggested that chloramination of source water containing higher levels of bromide under less alkaline condition may enhance the formation of trihalo-HCDs (especially the more brominated ones).

Aromatic halogenated DBPs garnered significant attention in recent years owing to their elevated toxicity. However, most of the relevant researches at present focuses merely on halo-monocyclic DBPs. Wang proposed that halo-polycyclic DBPs may be the potential DBPs of high occurrence and toxicity in drinking water [85]. A highly sensitive analytical method was developed for five representative HNQs, which were identified using this method with 100% detection frequency at concentrations up to 136.7 ng/L in drinking water originated from seven WTPs. The cytotoxicity of the five tested HNQs in CHO-K1 cells was comparable to the most toxic carbonaceous DBP in drinking water, i.e., iodoacetic acid, which is hundreds to thousands of times more toxic than regulated DBPs.

Several groups of nitrogenous aromatic halogenated DBPs were proposed to be toxicity drivers and were identified in drinking water. A new SPE-GC/MS method was developed to determine eight CPANs [86]. The concentrations of the eight CPANs in nine finished drinking waters were determined to be at concentrations ranging from 0.5 to 155 ng/L. Seven CPANs were detectable in all samples. CPANs were detected at concentrations between 0.8 and 155 ng/L in chlorinated waters, and from 0.5 to 15 ng/L in chloraminated waters. Across all waters, the sum of all CPANs in chloraminated waters was 13% of that in chlorinated systems, and 2-CPAN was the most abundant CPAN species in chlorinated and chloraminated waters.

To enhance extraction efficiency, the pH of water samples was typically adjusted to an acidic level (e.g., below 0.5) to maintain the target DBPs in the neutral states. However, the approach may overlook certain amphoteric DBPs, particularly nitrogenous DBPs with diverse functional groups, which can acquire protons at low pH and release them at high pH. Tang initially reported the detection and identification of a novel class of heterocyclic nitrogenous DBPs, halogenated pyridinols, through meticulous selection and optimization of extraction pH in simulated drinking water [87]. The optimized method accurately detected eight novel DBPs in two actual drinking water samples. Additionally, the developmental and acute toxicity in zebrafish embryos of these new DBPs were discovered to be relatively lower than HBQs but significantly higher, up to dozens of times, compared to commonly known DBPs like tribromomethane and iodoacetic acid.

The toxicity of aromatic N-DBPs is frequently higher than their nitrogenous or aromatic analogs [88], and the metabolic rate of aromatic nitrogenous groups is significantly lower than non-nitrogenous aromatic groups [89]. However, since only about 50% of total organic halogen has been identified, greater attention is required to identify unknown DBPs that could be driving DBP-induced toxicity [90]. Zhang reported the identification of a new category of aromatic N-DBPs, haloanilines, in actual chloraminated drinking waters [90]. Eight haloanilines were detected in chloraminated drinking water, with total concentrations up to 443 ng/L. The most prevalent was 2-bromoaniline. The EC_{50} values of the eight haloanilines were 1–2 orders of magnitude lower than the regulated DBPs in HepG2 cell. The lowest toxic concentration of 2-chloro-4-nitroaniline was 1 μM, 500 times lower than dichloroacetic acid (regulated DBPs).

In recent years, HQCs, with double-bound oxygen of HBQs replaced with a chloramine group, were proposed to be toxicity-significant DBPs in drinking water [91]. Wang developed a detection method for five HQCs, and the five HQCs were identified as new DBPs in drinking water samples from nine WTPs and water distribution systems with concentrations of up to 23.1 ng/L [7]. The nitrogenous aromatic substructures present in NOM, such as aminophenols and quinone imines, led to the formation of HQCs after chlorination [92]. The cytotoxicity of the five tested HQCs in HepG2 cells was found to be higher than or comparable to that of 2,6-DCBQ.

After HQCs, HPAcAms were successfully identified for the first time in both the laboratory experiments and realistic drinking water [93]. The majority of the identified HPAcAms exhibited considerably greater cytotoxicity compared to dichloroacetamide, the most prevalent aliphatic haloacetamide DBP. The QSAR (Quantitative Structure-Activity Relationship) model highlighted the significance of cellular uptake efficiency and polarized distributions of electrons in the cytotoxicity mechanisms of HPAcAms.

More recently, HHBNs were identified in drinking water with a maximum concentration of 36 ng/L and a detection frequency of 100% [94]. Previous studies showed that halo-nitriles could undergo hydrolysis to form halo-amides, representing the generation of haloacetamide from the hydrolysis of haloacetronitrile [95]. Additionally, halo-amides can be formed through pathways independent of halo-nitrile hydrolysis, as indicated by N^{15} labeling experiments [96]. Consequently, HBZAMs were considered a potential new category of nitrogenous DBPs. Eight HBZAMs were quantified in 10 drinking water samples, with concentrations ranging from 2.4 to 7.2 ng/L and a detection frequency of 100% [97]. The cytotoxicity of the eight quantified HBZAMs in CHO-K1 cells exhibited variations, with the cytotoxicity of 3,5-DBBZAM being over 10-fold higher than that of the aliphatic dichloroacetamide.

In addition to NOM, DBPs from endogenous organic matter may be bio-homologous. Nucleic acids and their constituents, including DNA, RNA, epigenetically modified nucleic acids, nucleosides, nucleotides, and nucleobases, stand out as potential precursors of nucleotide DBPs. As these nucleic substances are ubiquitous components found in all organisms, their concentrations have been reported to range from tens to hundreds of μg/L in the environment [98]. The endogenous hypochlorous acid reacted with cellular nucleic acids, leading to the formation of chlorinated nucleobases and nucleosides [99]. These chlorinated derivatives were incorporated into nucleobases, initiating mutations, strand breaks, and cross-links in nucleic acids during cell/tissue proliferation, representing an initial molecular event associated with chronic inflammation and carcinogenesis [99, 100]. Wang for the first time identified chlorinated nucleobases and nucleosides that could emerge as mutagenic DBPs in drinking water, forming from the reaction between chlorine disinfectants and nucleic acids [101]. The detection frequency of the five halogenated nucleobases ranged from 73 to 100% with a maximum concentration of up to 65.3 ng/L in the representative drinking water samples. Additionally, mutagenic chlorinated nucleotides have been suggested as potential DBPs in drinking water [98].

References

1. Mitch WA, Richardson SD, Zhang X, Gonsior M (2023) High-molecular-weight by-products of chlorine disinfection. Nat Water 1:336–347. https://doi.org/10.1038/s44221-023-00064-x
2. Krasner SW, Weinberg HS, Richardson SD, Pastor SJ, Chinn R, Sclimenti MJ, Onstad GD, Thruston AD (2006) Occurrence of a New Generation of Disinfection Byproducts. Environ Sci Technol 40:7175–7185. https://doi.org/10.1021/es060353j
3. Qin F, Zhao Y-Y, Zhao Y, Boyd JM, Zhou W, Li X-F (2010) A toxic disinfection by-product, 2,6-dichloro-1,4-benzoquinone, identified in drinking water. Angew Chem 122:802–804. https://doi.org/10.1002/ange.200904934
4. Lou J, Lu H, Wang W, He S, Zhu L (2022) Quantitative identification of halo-methylbenzoquinones as disinfection byproducts in drinking water using a pseudo-targeted LC-MS/MS method. Water Res 218:118466. https://doi.org/10.1016/j.watres.2022.118466
5. Hu S, Gong T, Zhu H, Wang J, Li Z, Chen H, Huang Z, Zhang M, Xian Q (2020) Formation and decomposition of new iodinated halobenzoquinones during chloramination in drinking water. Environ Sci Technol 54:5237–5248. https://doi.org/10.1021/acs.est.9b07564
6. Hu S, Chen X, Zhang B, Liu L, Gong T, Xian Q (2022) Occurrence and transformation of newly discovered 2-bromo-6-chloro-1,4-benzoquinone in chlorinated drinking water. J Hazard Mater 436:129189. https://doi.org/10.1016/j.jhazmat.2022.129189
7. Xu S, Hu S, Zhu L, Wang W (2021) Haloquinone chloroimides as toxic disinfection byproducts identified in drinking water. Environ Sci Technol 55:16347–16357. https://doi.org/10.1021/acs.est.1c01690
8. Wang W, Moe B, Li J, Qian Y, Zheng Q, Li X-F (2016) Analytical characterization, occurrence, transformation, and removal of the emerging disinfection byproducts halobenzoquinones in water. TrAC Trends Anal Chem 85:97–110. https://doi.org/10.1016/j.trac.2016.03.004

9. Song N, Gagliardi CJ, Binstead RA, Zhang M-T, Thorp H, Meyer TJ (2012) Role of proton-coupled electron transfer in the redox interconversion between benzoquinone and hydroquinone. J Am Chem Soc 134:18538–18541. https://doi.org/10.1021/ja308700t
10. Bolton JL, Trush MA, Penning TM, Dryhurst G, Monks TJ (2000) Role of quinones in toxicology. Chem Res Toxicol 13:135–160. https://doi.org/10.1021/tx9902082
11. Zhao Y, Anichina J, Lu X, Bull RJ, Krasner SW, Hrudey SE, Li X-F (2012) Occurrence and formation of chloro- and bromo-benzoquinones during drinking water disinfection. Water Res 46:4351–4360. https://doi.org/10.1016/j.watres.2012.05.032
12. Uchimiya M, Stone AT (2009) Reversible redox chemistry of quinones: impact on biogeochemical cycles. Chemosphere 77:451–458. https://doi.org/10.1016/j.chemosphere.2009.07.025
13. Du H, Li J, Moe B, McGuigan CF, Shen S, Li X-F (2013) Cytotoxicity and oxidative damage induced by halobenzoquinones to T24 bladder cancer cells. Environ Sci Technol 47:2823–2830. https://doi.org/10.1021/es303762p
14. Wang W, Qian Y, Boyd JM, Wu M, Hrudey SE, Li X-F (2013) Halobenzoquinones in swimming pool waters and their formation from personal care products. Environ Sci Technol 47:3275–3282. https://doi.org/10.1021/es304938x
15. Gan D, Jia M, Vaughan PP, Falvey DE, Blough NV (2008) Aqueous photochemistry of methyl-benzoquinone. J Phys Chem A 112:2803–2812. https://doi.org/10.1021/jp710724e
16. Lou J, Lu H, Wang W, Zhu L (2022) Molecular composition of halobenzoquinone precursors in natural organic matter in source water. Water Res 209:117901. https://doi.org/10.1016/j.watres.2021.117901
17. Yang M, Zhang X, Liang Q, Yang B (2019) Application of (LC/)MS/MS precursor ion scan for evaluating the occurrence, formation and control of polar halogenated DBPs in disinfected waters: a review. Water Res 158:322–337. https://doi.org/10.1016/j.watres.2019.04.033
18. Qian Y, Wang W, Boyd JM, Wu M, Hrudey SE, Li X-F (2013) UV-induced transformation of four halobenzoquinones in drinking water. Environ Sci Technol 47:4426–4433. https://doi.org/10.1021/es305044k
19. Pan Y, Zhang X (2013) Four groups of new aromatic halogenated disinfection byproducts: effect of bromide concentration on their formation and speciation in chlorinated drinking water. Environ Sci Technol 47:1265–1273. https://doi.org/10.1021/es303729n
20. Zhai H, Zhang X, Zhu X, Liu J, Ji M (2014) Formation of brominated disinfection byproducts during chloramination of drinking water: new polar species and overall kinetics. Environ Sci Technol 48:2579–2588. https://doi.org/10.1021/es4034765
21. Zhai H, Zhang X (2011) Formation and decomposition of new and unknown polar brominated disinfection byproducts during chlorination. Environ Sci Technol 45:2194–2201. https://doi.org/10.1021/es1034427
22. Zhao Y, Qin F, Boyd JM, Anichina J, Li X-F (2010) Characterization and determination of chloro- and bromo-benzoquinones as new chlorination disinfection byproducts in drinking water. Anal Chem 82:4599–4605. https://doi.org/10.1021/ac100708u
23. Huang R, Wang W, Qian Y, Boyd JM, Zhao Y, Li X-F (2013) Ultra pressure liquid chromatography–negative electrospray ionization mass spectrometry determination of twelve halobenzoquinones at ng/L levels in drinking water. Anal Chem 85:4520–4529. https://doi.org/10.1021/ac400160r
24. Nakai T, Kosaka K, Asami M, Akiba M (2015) Analysis and occurrence of 2,6-dichloro-1,4-benzoquinone in drinking water by liquid chromatography-tandem mass spectrometry. J Jpn Soc Water Environ 38:67–73. https://doi.org/10.2965/jswe.38.67
25. Li Y, Zhang L, Yang L, Zhang Y, Niu Z (2021) Hydrolysis characteristics and risk assessment of a widely detected emerging drinking water disinfection-by-product – 2,6-dichloro-1,4-benzoquinone – in the water environment of Tianjin (China). Sci Total Environ 765:144394. https://doi.org/10.1016/j.scitotenv.2020.144394

26. Gao X, Wang X, Jiang Y, Kong D, Pan X, Ma J, Liu Y (2024) Chloro- and bromo-benzoquinone formation and transformation mechanisms in a drinking water-distribution system. J Hazard Mater 461:132692. https://doi.org/10.1016/j.jhazmat.2023.132692
27. Zwiener C, Richardson SD, De Marini DM, Grummt T, Glauner T, Frimmel FH (2007) Drowning in disinfection byproducts? Assessing swimming pool water. Environ Sci Technol 41:363–372. https://doi.org/10.1021/es062367v
28. Richardson SD (2004) Environmental mass spectrometry: emerging contaminants and current issues. Anal Chem 76:3337–3364. https://doi.org/10.1021/ac040060d
29. El-Athman F, Zehlike L, Kämpfe A, Junek R, Selinka H-C, Mahringer D, Grunert A (2021) Pool water disinfection by ozone-bromine treatment: assessing the disinfectant efficacy and the occurrence and *in vitro* toxicity of brominated disinfection by-products. Water Res 204: 117648. https://doi.org/10.1016/j.watres.2021.117648
30. Wu H, Long K, Lu D, Mo Y, Yang Q, Wei X (2019) Occurrence and formation of halobenzoquinones in indoor and outdoor swimming pool waters of Nanning City, Southwest China. Environ Sci Pollut Res 26:31537–31545. https://doi.org/10.1007/s11356-019-06341-1
31. Chen B, Lee W, Westerhoff PK, Krasner SW, Herckes P (2010) Solar photolysis kinetics of disinfection byproducts. Water Res 44:3401–3409. https://doi.org/10.1016/j.watres.2010.03.014
32. Lou J, Wang W, Zhu L (2019) Occurrence, formation, and oxidative stress of emerging disinfection byproducts, halobenzoquinones, in tea. Environ Sci Technol 53:11860–11868. https://doi.org/10.1021/acs.est.9b03163
33. Wu H, Long K, Sha Y, Lu D, Xia Y, Mo Y, Yang Q, Zheng W, Yang M, Wei X (2021) Occurrence and toxicity of halobenzoquinones as drinking water disinfection byproducts. Sci Total Environ 770:145277. https://doi.org/10.1016/j.scitotenv.2021.145277
34. Lou J, Wang W, Zhu L (2021) Transformation of emerging disinfection byproducts Halobenzoquinones to haloacetic acids during chlorination of drinking water. Chem Eng J 418:129326. https://doi.org/10.1016/j.cej.2021.129326
35. Kanokkantapong V, Marhaba T, Panyapinyophol B, Pavasant P (2006) FTIR evaluation of functional groups involved in the formation of haloacetic acids during the chlorination of raw water. J Hazard Mater 136:188–196. https://doi.org/10.1016/j.jhazmat.2005.06.031
36. Kosaka K, Ohkubo K, Akiba M (2016) Occurrence and formation of haloacetamides from chlorination at water purification plants across Japan. Water Res 106:470–476. https://doi.org/10.1016/j.watres.2016.10.029
37. Bond T, Henriet O, Goslan EH, Parsons SA, Jefferson B (2009) Disinfection byproduct formation and fractionation behavior of natural organic matter surrogates. Environ Sci Technol 43:5982–5989. https://doi.org/10.1021/es900686p
38. Diemert S, Wang W, Andrews RC, Li X-F (2013) Removal of halo-benzoquinone (emerging disinfection by-product) precursor material from three surface waters using coagulation. Water Res 47:1773–1782. https://doi.org/10.1016/j.watres.2012.12.035
39. Bull RJ, Reckhow DA, Li X, Humpage AR, Joll C, Hrudey SE (2011) Potential carcinogenic hazards of non-regulated disinfection by-products: Haloquinones, halo-cyclopentene and cyclohexene derivatives, N-halamines, halonitriles, and heterocyclic amines. Toxicology 286:1–19. https://doi.org/10.1016/j.tox.2011.05.004
40. Li Y, Niu Z, Wang Y, Zhang L, Zhang Y (2023) The convergence of 2,6-dichloro-1,4-benzoquinone in the whole process of lignin phenol precursor chlorination. Chemosphere 312:137290. https://doi.org/10.1016/j.chemosphere.2022.137290
41. Wang Y, Gao S, Yan X, Wang S, Zhang R, Zhou Y, Ren L, Li C (2024) The impact of water quality on the formation of halogenated benzoquinones and the adsorption efficiency by activated carbon. J Environ Sci 135:693–702. https://doi.org/10.1016/j.jes.2023.02.002
42. Ge F, Xiao Y, Yang Y, Wang W, Moe B, Li X-F (2018) Formation of water disinfection byproduct 2,6-dichloro-1,4-benzoquinone from chlorination of green algae. J Environ Sci 63: 1–8. https://doi.org/10.1016/j.jes.2017.10.001

43. Zhao J, Hu S, Zhu L, Wang W (2022) Formation of chlorinated halobenzoquinones during chlorination of free aromatic amino acids. Sci Total Environ 825:153904. https://doi.org/10.1016/j.scitotenv.2022.153904
44. Yettella RR, Min DB (2010) Effects of Trolox and ascorbic acid on the riboflavin photosensitised oxidation of aromatic amino acids. Food Chem 118:35–41. https://doi.org/10.1016/j.foodchem.2009.04.022
45. Kosaka K, Nakai T, Hishida Y, Asami M, Ohkubo K, Akiba M (2017) Formation of 2,6-dichloro-1,4-benzoquinone from aromatic compounds after chlorination. Water Res 110:48–55. https://doi.org/10.1016/j.watres.2016.12.005
46. Haman C, Dauchy X, Rosin C, Munoz J-F (2015) Occurrence, fate and behavior of parabens in aquatic environments: a review. Water Res 68:1–11. https://doi.org/10.1016/j.watres.2014.09.030
47. Benotti MJ, Trenholm RA, Vanderford BJ, Holady JC, Stanford BD, Snyder SA (2009) Pharmaceuticals and endocrine disrupting compounds in U.S. Drinking Water. Environ Sci Technol 43:597–603. https://doi.org/10.1021/es801845a
48. Chattopadhyay S, Raychaudhuri U, Chakraborty R (2014) Artificial sweeteners – a review. J Food Sci Technol 51:611–621. https://doi.org/10.1007/s13197-011-0571-1
49. Li AJ, Wu P, Law JC-F, Chow C-H, Postigo C, Guo Y, Leung KS-Y (2017) Transformation of acesulfame in chlorination: kinetics study, identification of byproducts, and toxicity assessment. Water Res 117:157–166. https://doi.org/10.1016/j.watres.2017.03.053
50. Qiao R, Liang S, Chen C, Xiong L, Guan Q, Wang L, Fu Z, Pan Y, Liu H, Zhu J, Hu Y, Li L, Huang G (2022) 2,6-Dichloro-1,4-benzoquinone formation from chlorination of substituted aromatic antioxidants and its control by pre-ozonation in drinking water treatment plant. Chemosphere 299:134498. https://doi.org/10.1016/j.chemosphere.2022.134498
51. Wang W, Qian Y, Li J, Moe B, Huang R, Zhang H, Hrudey SE, Li X-F (2014) Analytical and toxicity characterization of halo-hydroxyl-benzoquinones as stable halobenzoquinone disinfection byproducts in treated water. Anal Chem 86:4982–4988. https://doi.org/10.1021/ac5007238
52. Mohan A, Reckhow DA (2021) Hydrolysis and Chlorination of 2,6-Dichloro-1,4-benzoquinone under conditions typical of drinking water distribution systems. Water Res 200:117219. https://doi.org/10.1016/j.watres.2021.117219
53. Zhai H, Zhao J, Wang R, Yan Y, Yu S, Zhao Y (2022) Formation of trihalomethanes and haloacetic acids from 2,6-dichloro-1,4-benzoquinone during chlorination: decomposition kinetics, conversion rates, and pathways. Chemosphere 291:132729. https://doi.org/10.1016/j.chemosphere.2021.132729
54. Li J, Wang W, Zhang H, Le XC, Li X-F (2014) Glutathione-mediated detoxification of halobenzoquinone drinking water disinfection byproducts in T24 Cells. Toxicol Sci 141:335–343. https://doi.org/10.1093/toxsci/kfu088
55. Wang W, Qian Y, Jmaiff LK, Krasner SW, Hrudey SE, Li X-F (2015) Precursors of halobenzoquinones and their removal during drinking water treatment processes. Environ Sci Technol 49:9898–9904. https://doi.org/10.1021/acs.est.5b02308
56. Kim J, Chung Y, Shin D, Kim M, Lee Y, Lim Y, Lee D (2003) Chlorination by-products in surface water treatment process. Desalination 151:1–9. https://doi.org/10.1016/S0011-9164(02)00967-0
57. Pan Z, Du Z, Jia J, Lin A, Wang Y, Song W, Sun S, Wang H, Jia R, Hou L (2022) Degradation of 2,6-dichloro-1,4-benzoquinone by UV/H2O2/O3 treatment: effectiveness, water matrix effects, and degradation mechanism. Chemosphere 296:134014. https://doi.org/10.1016/j.chemosphere.2022.134014
58. Chu W, Gao N, Yin D, Deng Y, Templeton MR (2012) Ozone–biological activated carbon integrated treatment for removal of precursors of halogenated nitrogenous disinfection by-products. Chemosphere 86:1087–1091. https://doi.org/10.1016/j.chemosphere.2011.11.070

59. Ding C, Miao S, Zhao S, Xiao M, Zou S (2016) Removal of halogenated carbon disinfection by-product 2,6-dichloro-1,4-benzoquinone by ozone and granular activated carbon. China Environ Sci 37:2173–2178
60. Hama Aziz KH, Miessner H, Mahyar A, Mueller S, Kalass D, Moeller D, Omer KM (2019) Removal of dichloroacetic acid from aqueous solution using non-thermal plasma generated by dielectric barrier discharge and nano-pulse corona discharge. Sep Purif Technol 216:51–57. https://doi.org/10.1016/j.seppur.2019.01.074
61. Gao X, Wang X, Ma J, Liu Y (2023) Potential and mechanism of disinfection by-products removal in drinking water by bubbling corona discharge. Water Res 245:120624. https://doi.org/10.1016/j.watres.2023.120624
62. Hemming J, Holmbom B, Reunanen M, Kronberg L (1986) Determination of the strong mutagen 3-chloro-4-(dichloromethyl)-5-hydroxy-2(5H)-furanone in chlorinated drinking and humic waters. Chemosphere 15:549–556. https://doi.org/10.1016/0045-6535(86)90003-2
63. Kronberg L, Vartiainen T (1988) Ames mutagenicity and concentration of the strong mutagen 3-chloro-4-(dichloromethyl)-5-hydroxy-2(5H)-furanone and of its geometric isomer E-2-chloro-3-(dichloromethyl)-4-oxo-butenoic acid in chlorine-treated tap waters. Mutat Res 206:177–182. https://doi.org/10.1016/0165-1218(88)90158-9
64. Meier JR, Knohl RB, Coleman WE, Ringhand HP, Munch JW, Kaylor WH, Streicher RP, Kopfler FC (1987) Studies on the potent bacterial mutagen, 3-chloro-4-(dichloromethyl)-5-hydroxy-2(5H)-furanone: aqueous stability, XAD recovery and analytical determination in drinking water and in chlorinated humic acid solutions. Mutat Res Toxicol 189:363–373. https://doi.org/10.1016/0165-1218(87)90044-9
65. Kronberg L, Christman RF (1989) Chemistry of mutagenic by-products of water chlorination. Sci Total Environ 81–82:219–230. https://doi.org/10.1016/0048-9697(89)90128-9
66. Suzuki N, Nakanishi J (1990) The determination of strong mutagen, 3-chloro-4-(dichloromethyl)-5-hydroxy-2(5H)-furanone in drinking water in Japan. Chemosphere 21: 387–392. https://doi.org/10.1016/0045-6535(90)90008-H
67. Andrews RC, Daignault SA, Laverdure C, William DT, Huck PM (1990) Occurrence of the mutagenic compound 'MX' in drinking water and its removal by activated carbon. Environ Technol 11:685–694. https://doi.org/10.1080/09593339009384913
68. Xie Y, Wang Z (1990) Effect of water treatment on mutagenic activity and concentration of MX in water. In: Proceedings of American Water Works Association Water Quality Technology Conference. American Water Works Association, pp 1423–1434
69. Leif K, Bjarne H, Markku R, Leena T (1988) Identification and quantification of the Ames mutagenic compound 3-chloro-4-(dichloromethyl)-5-hydroxy-2(5H)-furanone and of its geometric isomer (E)-2-chloro-3-(dichloromethyl)-4-oxobutenoic acid in chlorine-treated humic water and drinking water extracts. Environ Sci Technol 22:1097–1103. https://doi.org/10.1021/es00174a016
70. Suzuki N, Nakanishi J (1995) Brominated analogues of MX (3-chloro-4-(dichloromethyl)-5-hydroxy-2(5H)-furanone) in chlorinated drinking water. Chemosphere 30:1557–1564. https://doi.org/10.1016/0045-6535(95)00049-E
71. Kronberg L, Christman RF, Singh R, Ball LM (1991) Identification of oxidized and reduced forms of the strong bacterial mutagen (Z)-2-chloro-3-(dichloromethyl)-4-oxobutenoic acid (MX) in extracts of chlorine-treated water. Environ Sci Technol 25:99–104. https://doi.org/10.1021/es00013a009
72. Onstad GD, Weinberg HS, Krasner SW (2008) Occurrence of halogenated furanones in U.S. drinking waters. Environ Sci Technol 42:3341–3348. https://doi.org/10.1021/es071374w
73. Långvik V-A, Hormi O (1994) Possible reaction pathways for the formation of 3-chloro-4-(dichloromethyl)-5-hydroxy-2(5H)-furanone (MX). Chemosphere 28:1111–1117. https://doi.org/10.1016/0045-6535(94)90329-8
74. Huixian Z, Junhe L, Zhuo C, Chengyong Y, Jinqi Z, Wen Z (2000) Screening the precursors of strong mutagen [3-chloro-4-(dichloromethyl)-5-hydroxy-2(5H)-furanone] MX from chlorinated water. Water Res 34:225–229. https://doi.org/10.1016/S0043-1354(99)00106-2

75. Rantakokko P, Yritys M, Vartiainen T (2004) Matrix effects in the gas chromatographic–mass spectrometric determination of brominated analogues of 3-chloro-4-(dichloromethyl)-5-hydroxy-2(5H)-furanone. J Chromatogr A 1028:179–188. https://doi.org/10.1016/j.chroma.2003.12.008
76. Onstad GD, Weinberg HS (2005) Evaluation of the stability and analysis of halogenated furanones in disinfected drinking waters. Anal Chim Acta 534:281–292. https://doi.org/10.1016/j.aca.2004.11.046
77. Padilla LPR (2015) Development of an LC-MS/MS method for the characterization of halogenated furanones in drinking water. Master's Thesis. University of Washington
78. Backlund P, Wondergem E, Voogd K, de Jong A (1989) Mutagenic activity and presence of the strong mutagen 3-chloro-4-(dichloromethyl)-5-hydroxy-2(5H)-furanone (MX) in chlorinated raw and drinking waters in The Netherlands. Sci Total Environ 84:273–282. https://doi.org/10.1016/0048-9697(89)90389-6
79. Backlund P, Kronberg L, Tikkanen L (1988) Formation of AMES mutagenicity and of the strong bacterial mutagen 3-chloro-4-(dichloromethyl)-5-hydroxy-2(5H)-furanone and other halogenated compounds during disinfection of humic water. Chemosphere 17:1329–1336. https://doi.org/10.1016/0045-6535(88)90085-9
80. McKie MJ, Taylor-Edmonds L, Andrews SA, Andrews RC (2015) Engineered biofiltration for the removal of disinfection by-product precursors and genotoxicity. Water Res 81:196–207. https://doi.org/10.1016/j.watres.2015.05.034
81. Mcdonald TA, Komulainen H (2005) Carcinogenicity of the chlorination disinfection by-product MX. J Environ Sci Health Part C 23:163–214. https://doi.org/10.1080/10590500500234988
82. Plewa MJ, Kargalioglu Y, Vankerk D, Minear RA, Wagner ED (2002) Mammalian cell cytotoxicity and genotoxicity analysis of drinking water disinfection by-products. Environ Mol Mutagen 40:134–142. https://doi.org/10.1002/em.10092
83. US Environmental Protection Agency (2016) Six-year review 3 technical support document for disinfectants/disinfection byproducts rules. US Environmental Protection Agency, Office of Water
84. Pan Y, Li W, Li A, Zhou Q, Shi P, Wang Y (2016) A new group of disinfection byproducts in drinking water: trihalo-hydroxy-cyclopentene-diones. Environ Sci Technol 50:7344–7352. https://doi.org/10.1021/acs.est.6b00798
85. Jiang H, Kaw HY, Zhu L, Wang W (2022) Halonaphthoquinones: a group of emerging disinfection byproducts of high toxicity in drinking water. Water Res 217:118421. https://doi.org/10.1016/j.watres.2022.118421
86. Zhang D, Bond T, Krasner SW, Chu W, Pan Y, Xu B, Yin D (2019) Trace determination and occurrence of eight chlorophenylacetonitriles: an emerging class of aromatic nitrogenous disinfection byproducts in drinking water. Chemosphere 220:858–865. https://doi.org/10.1016/j.chemosphere.2018.12.127
87. Tang H, Zhong H, Pan Y, Zhou Q, Huo Z, Chu W, Xu B (2021) A new group of heterocyclic nitrogenous disinfection byproducts (DBPs) in drinking water: role of extraction pH in unknown DBP exploration. Environ Sci Technol 55:6764–6772. https://doi.org/10.1021/acs.est.1c00078
88. Zhang Z, Zhu Q, Huang C, Yang M, Li J, Chen Y, Yang B, Zhao X (2020) Comparative cytotoxicity of halogenated aromatic DBPs and implications of the corresponding developed QSAR model to toxicity mechanisms of those DBPs: binding interactions between aromatic DBPs and catalase play an important role. Water Res 170:115283. https://doi.org/10.1016/j.watres.2019.115283
89. Liu J, Zhang X (2014) Comparative toxicity of new halophenolic DBPs in chlorinated saline wastewater effluents against a marine alga: halophenolic DBPs are generally more toxic than haloaliphatic ones. Water Res 65:64–72. https://doi.org/10.1016/j.watres.2014.07.024
90. Zhang D, Bond T, Pan Y, Li M, Luo J, Xiao R, Chu W (2022) Identification, occurrence, and cytotoxicity of haloanilines: a new class of aromatic nitrogenous disinfection byproducts in

chloraminated and chlorinated drinking water. Environ Sci Technol 56:4132–4141. https://doi.org/10.1021/acs.est.1c07375
91. Liu X, Chen L, Yang M, Tan C, Chu W (2020) The occurrence, characteristics, transformation and control of aromatic disinfection by-products: a review. Water Res 184:116076. https://doi.org/10.1016/j.watres.2020.116076
92. Huang H, Zheng H, Jiao J, Lei Y, Zhou Y, Qiu J, Yang X (2022) Trichloramine and hydroxyl radical contributions to dichloroacetonitrile formation following breakpoint chlorination. Environ Sci Technol 56:12592–12601. https://doi.org/10.1021/acs.est.2c03701
93. Hu S, Kaw HY, Zhu L, Wang W (2022) Formation and cytotoxicity of halophenylacetamides: a new group of nitrogenous aromatic halogenated disinfection byproducts in drinking water. Environ Sci Technol 56:3181–3192. https://doi.org/10.1021/acs.est.1c08419
94. Hu S, Kaw HY, Zhu L, Wang W (2022) Halohydroxybenzonitriles as a new group of halogenated aromatic DBPs in drinking water: are they of comparable risk to halonitrophenols? Water Res 219:118547. https://doi.org/10.1016/j.watres.2022.118547
95. Vu TN, Kimura SY, Plewa MJ, Richardson SD, Mariñas BJ (2019) Predominant N-haloacetamide and haloacetonitrile formation in drinking water via the aldehyde reaction pathway. Environ Sci Technol 53:850–859. https://doi.org/10.1021/acs.est.8b02862
96. Huang H, Wu Q-Y, Hu H-Y, Mitch WA (2012) Dichloroacetonitrile and dichloroacetamide can form independently during chlorination and chloramination of drinking waters, model organic matters, and wastewater effluents. Environ Sci Technol 46:10624–10631. https://doi.org/10.1021/es3025808
97. Lou J, Yin L, Meng Z, Fang S, Pan X (2023) Occurrence, stability and cytotoxicity of halobenzamides: a new group of nitrogenous disinfection byproducts in drinking water. Water Res 245:120670. https://doi.org/10.1016/j.watres.2023.120670
98. Sun G, Kaw HY, Zhou M, Guo P, Zhu L, Wang W (2023) Chlorinated nucleotides and analogs as potential disinfection byproducts in drinking water. J Hazard Mater 452:131242. https://doi.org/10.1016/j.jhazmat.2023.131242
99. Macer-Wright JL, Stanley NR, Portman N, Tan JT, Bursill C, Rayner BS, Hawkins CL (2019) A role for chlorinated nucleosides in the perturbation of macrophage function and promotion of inflammation. Chem Res Toxicol 32:1223–1234. https://doi.org/10.1021/acs.chemrestox.9b00044
100. Macer-Wright JL, Sileikaite I, Rayner BS, Hawkins CL (2020) 8-Chloroadenosine alters the metabolic profile and downregulates antioxidant and DNA damage repair pathways in macrophages. Chem Res Toxicol 33:402–413. https://doi.org/10.1021/acs.chemrestox.9b00334
101. Sun G, Guo P, Kaw HY, Zhou M, Wang W (2023) Uncovering an emerging group of halogenated nucleobase-derived disinfection byproducts in drinking water: prioritization of the highly cytotoxic 2-chloroadenine. Environ Sci Technol 57:8768–8775. https://doi.org/10.1021/acs.est.3c01484

Nitrogenous DBPs

Wenhai Chu, Di Zhang, and Zhenqi Du

Contents

1 Haloacetonitriles (HANs) 170
1.1 Stability 170
1.2 Precursor 171
1.3 Formation Mechanism 172
1.4 Occurrence 173
1.5 Regulation 174
1.6 Health Implications 175
2 Halonitromethanes (HNMs) 176
2.1 Basic Properties 176
2.2 Precursor 176
2.3 Formation Mechanism 177
2.4 Occurrence and Regulation 179
2.5 Health Implications 179
3 Haloacetamides (HAMs) 180
3.1 Basic Properties 180
3.2 Precursor 181
3.3 Formation Mechanism 182
3.4 Occurrence 182
3.5 Health Implications 183
4 Cyanogen Halides (CNX) 183
4.1 Basic Properties 183
4.2 Precursors and Formation of CNX 184
4.3 Occurrence and Regulation 186
4.4 Health Implications 187
5 Summary 187
References 187

Abstract Nitrogenous disinfection byproducts (N-DBPs) have gained increased attention because toxicological research suggests that they may pose greater health

W. Chu (✉), D. Zhang, and Z. Du
State Key Laboratory of Pollution Control and Resources Reuse, College of Environmental Science and Engineering, Tongji University, Shanghai, China

Chao Chen, Susan Andrews, and Yuefeng Xie (eds.), *Disinfection By-Products in Water*, Hdb Env Chem (2024) 135: 169–194, DOI 10.1007/698_2024_1163,

Published online: 23 October 2024

risk compared to carbonaceous disinfection byproducts (C-DBPs). The increased interest is especially noteworthy since utilities are required to use source water that might be affected by wastewater effluents or algal blooms, both commonly containing high levels of organic nitrogen, acting as precursors of N-DBPs. Four classes of typical N-DBPs, haloacetonitriles (HANs), halonitromethanes (HNMs), haloacetmides (HAMs), and cyanogen halides (CNX), have been identified and are the focus of critical review in this chapter. The following key aspects are typically addressed: basic properties, essential precursors, formation mechanisms, occurrences, regulations, and health implications. The review of these themes aims to provide a comprehensive understanding of N-DBPs, enabling researchers, policymakers, and water utility professionals to make informed decisions regarding water treatment processes and the safeguarding of public health in the face of population growth and the increasing utilization of impaired water sources.

Keywords DBP precursors, Formation mechanism, Nitrogenous disinfection byproducts, Regulation

The research on disinfection byproducts (DBPs) has historically focused on carbonaceous DBPs (C-DBPs), particularly trihalomethanes (THMs), and haloacetic acids (HAAs), since their discovery in the 1970s. However, there has been a growing interest in nitrogenous DBPs (N-DBPs) due to their relatively high toxicity. This shift in focus is especially significant as water treatment utilities explore alternatives to chlorine disinfection, such as chloramine disinfection, in efforts to reduce the formation of regulated THMs and HAAs. However, the use of chloramine may inadvertently promote the formation of N-DBPs, adding complexity to water treatment strategies. The toxicity of N-DBPs has been reported to be higher than that of C-DBPs. Among the families of N-DBPs that are currently of interest are haloacetonitriles (HANs), halonitromethanes (HNMs), haloacetamides (HAMs), and cyanogen halides (CNX). In this chapter, the discussion has likely encompassed the following aspects for each of these four classes of N-DBPs: basic properties, essential precursors, formation mechanisms, occurrences, regulations, and health implications.

1 Haloacetonitriles (HANs)

1.1 Stability

HANs constitute a significant class of N-DBPs found in almost all drinking waters, with the total mass typically representing approximately 10% of the THMs [1, 2]. HANs are reactive and can undergo considerable transformation on time scales relevant to system distribution (i.e., from a few hours to a week or more).

Yu et al. examined the stability of seven mono-, di-, and trihaloacetonitriles under a variety of conditions including different pH levels (6, 7, 8, 8.5, and 9) and disinfectant doses that are typical of drinking water distribution systems [3]. Results showed that all seven HANs are most stable at pH 6, and the rate of loss increases with both increasing pH and the number of halogens. It was consistent with the results reported previously that HANs can undergo considerable degradation through neutral and base-catalyzed hydrolysis on time scales relevant to distribution system residence times [4], and their hydrolysis rates increase with increasing pH [5, 6]. As HANs decreased, there was a rapid increase in the corresponding haloacetamide (HAM) concentrations, followed by a slower formation of HAAs. An activating effect of the electron withdrawing halogens on nitrile reactivity was also demonstrated, leading to decreasing HAN stability with increasing degree of halogenation of the substituents. Furthermore, with the same number of halogens in the substituents, HAN hydrolysis rate decreased as the halogens shifted from chlorine to bromine, resulting in the following hierarchy of HAN hydrolytic stability: monobromoacetonitrile (MBAN) > monochloroacetonitrile (MCAN) > dibromoacetonitrile (DBAN) > bromochloroacetonitrile (BCAN) > dichloroacetonitrile (DCAN) > bromodichloroacetonitrile (BDCAN) > trichloroacetonitrile (TCAN) [3].

1.2 Precursor

DBP precursors encompass various organic and inorganic substances in water that can react with disinfectants to generate DBPs. It is commonly acknowledged that organic nitrogenous compounds in water, particularly dissolved organic nitrogen (DON), serve as the primary precursors of N-DBPs. DON is widely present in aquatic environments, and the conventional water treatment process has low removal efficiency in DON. These compounds can react with free chlorine or chloramines during disinfection, generating both halogenated N-DBPs (e.g., HANs) and non-halogenated N-DBPs (e.g., *N*-dimethylnitrosamine).

DON is composed of a variety of compounds containing different functional groups. These compounds encompass proteins, amino acids, amides, nitriles, pyrrole, purines, and pyrimidines [7]. Given the highly complex composition and structure of DON, studying its reaction mechanism with disinfectants directly posed challenges. Therefore, research on N-DBP precursors mainly employs specific nitrogen-containing organic compounds (e.g., free amino acids). By using these model compounds, the formation pathways of N-DBPs were well studied.

The concentration of free amino acids in water is very low, and at such levels, it is insufficient to generate HANs at the detected concentration level. This indicates that other unknown precursors play an important role in the formation of HANs. Bond et al. highlighted that algae organic matter (AOM) and sewage secondary effluent organic matter (EfOM) are important components of DON [8]. AOM, primarily consisting of proteins and carbohydrates, serves as a principal source of organic nitrogen in surface water. The organic nitrogen in AOM is characterized by two structures: the amine structure ($N(R)_3$, exemplified by amino acids and fatty amines) and the amide structure ($N(R)_2$-(C=O)R, represented by peptide bonds) [9].

Research has identified a range of nitrogenous organic compounds in the effluent DON of urban sewage treatment plants, including urea, amino acids, low molecular weight amines, and chelating agents [10]. Conventional sewage treatment processes encounter challenges in effectively removing DON from water. If the secondary effluent from urban sewage treatment plants undergoes disinfection and is discharged into rivers or lakes, it may potentially contribute to a downstream drinking water source, either directly or indirectly. In a study by Zhang et al., chlorine disinfection of soluble microbial products (SMPs) in activated sludge revealed the production of DCAN and TCAN during disinfection, indicating that SMPs can also serve as precursors of HANs [11].

Chuang et al. and Yang et al. used ^{15}N isotope chloramine for disinfecting organic compounds and observed the generation of both ^{14}N-DCAN and ^{15}N-DCAN [12, 13]. This suggests that chloramine itself serves as nitrogen sources for DCAN formation. Yang et al. highlighted those compounds containing the R1-CH_2-CH (NH_2)R_2 structure or intermediate products with the [-$CH_2CH(NH_2)$-] structure, formed during chlorination, can generate DCAN during the disinfection process [13].

1.3 Formation Mechanism

HANs contain up to three halogens on the α-carbon bound to nitrile group. Two pathways have been characterized for HANs formation. In the first pathway, known as the "decarboxylation pathway", free chlorine or chloramines react with the α-amine group, leading to halogenation reactions that result in the generation of HANs. The nitrile nitrogen in this process originates from the primary amine constituents of DON in water. The second major pathway characterized for nitrile formation is the "aldehyde pathway", involving reactions between inorganic chloramine and aldehydes. In this pathway, inorganic chloramines serve as the source of the nitrile nitrogen [14]. The process starts with the nucleophilic attack of inorganic monochloramine on formaldehyde, resulting in the formation of chloroaminomethanol. Subsequent dehydration eliminates HCl, leading to the formation of HANs. It is noteworthy that the formation of HANs through this pathway increases with increasing pH. Yang et al. conducted disinfection experiments using both ^{14}N-NH_2Cl and ^{15}N-NH_2Cl on glutamic acid. They proposed potential DCAN formation pathways (Fig. 1). It could be seen that the nitrogen of N-DBPs generated through pathway 1 originates from the pre-added ^{15}N-NH_2Cl. The nitrogen in N-DBPs generated through pathway 2 is provided by nitrogen-containing organic compounds [13].

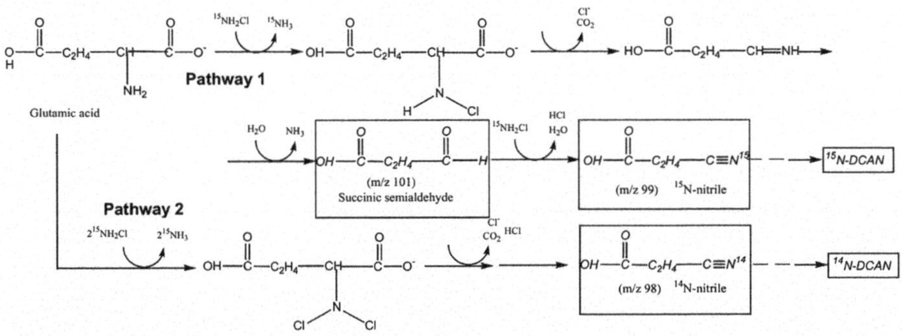

Fig. 1 Reaction pathways during chloramination of glutamic acid [13] (Note: compounds in the frames were identified during intermediate studies)

Table 1 Levels of HANs reported in drinking water of various countries

Country name	Source water	HANs (μg/L)	References
China	DWDS	N.D.-39.2	[19, 20]
Thailand	DWTPs	30[a]	[21]
Hungary	DWTPs	<1.0–17	[22]
UK	Finished water	2.8[b]	[23]
	DWDS	2.6[b]	
USA	DWTPs	3–14	[24]

DWTPs Drinking water treatment plants, *DWDS* Drinking water distribution system, *N.D.* (not detected)

[a] Maximum value

[b] Average value

1.4 Occurrence

Worldwide reported concentrations of HANs in drinking water have been presented in Table 1. The concentrations of HANs detected in the drinking water distribution system of some Chinese cities are in the range of N.D.-39.2 μg/L. In Shanghai, Xiamen, Chengdu, and Zhengzhou, chlorinated HANs are the primary HANs, constituting 53% to 76% of the total HANs, whereas in Guangzhou and Shenzhen, brominated HANs predominate [15–17]. The observed phenomenon may be attributed to a reduction in surface runoff in coastal areas, leading to an increased presence of bromide ions (Br^-). TCAN is typically not detected in most cities due to its susceptibility to alkaline-catalyzed hydrolysis, especially when pH exceeds 5 [18]. In addition, the highest concentration of HANs in drinking water treatment plants in Thailand was detected as 30 μg/L, which is higher than that in Hungary (<1.0–17 μg/L), UK (average concentration = 2.6 μg/L), and the USA (3–14 μg/L).

1.5 Regulation

Only a few countries or organizations have included HANs into the drinking water standards.

World Health Organization (WHO)
In the first edition of the "Guidelines" by the WHO published from 1983 to 1984, no provision was made for any HANs. In the second edition of the "Guidelines" published in three volumes from 1993 to 1997, there was some evidence at the time suggesting potential toxic effects of DCAN, DBAN, and TCAN, but limited data on health effects were available. Therefore, only provisional guidance values of 90, 100, and 1 μg/L were set, respectively. The third edition of the "Guidelines" published in 2004 provided guidance values of 20 and 70 μg/L for DCAN and DBAN, respectively. The fourth edition of the "Guidelines" has the same limit values for relevant HANs as the third edition [25]. Looking back at the development process of WHO drinking water standards, the limits for HANs have gradually become strict and rigorous, aiming to better lay the foundation for countries to establish drinking water quality standards.

Japan
In the water quality standards implemented in Japan in 1993, DCAN is a monitored parameter with a limit of 80 μg/L. The current water quality standards in Japan have undergone seven revisions. The latest update was issued according to the Ministry of Health and Welfare Notification No. 101 in 2003. In the current drinking water quality standards, DCAN is listed as a "water quality target management project" with a limit of 10 μg/L. TCAN, BCAN, and DBAN are listed as "projects to be reviewed", and only a DBAN limit of 60 μg/L is set [26].

New Zealand
The earliest drinking water standards in New Zealand were promulgated in 1984 and have undergone multiple revisions since then. In the revised water quality standards in 2000, high-risk indicators were monitored as priority levels 1 and 2, while low-risk indicators were monitored as levels 3 and 4. DBPs are classified as level 2 water quality indicators for monitoring, with limits of 100, 200, and 1 μg/L for DCAN, DBAN, and TCAN, respectively. The subsequent revised drinking water quality standards have always followed this monitoring rule. The current drinking water quality standards have been revised twice in 2008 and 2018 based on the water quality standards published in 2005 [27]. Strict regulations have been implemented for DCAN and DBAN, with limits set at 20 and 80 μg/L, respectively.

China
The latest "Standards for Drinking Water Quality (GB5749-2022)" was approved by the National Standardization Administration in 2022 and none of HANs was included. The group standard "Standards for N-Nitrosodimethylamine, Dichloroacetonitrile, Dibromoacetonitrile in Drinking Water (T/SAWP0001-2020)" released by Shanghai Water Purification Technology Institute, China in

2020, establishes limits of 35 and 10 μg/L for DCAN and DBAN in drinking water, respectively.

Other Countries
The control of HANs in Singapore and the Philippines is consistent with the current WHO drinking standards [28, 29]. South Korea and Malaysia have established limits for DCAN, DBAN, and TCAN in accordance with the second edition of the WHO guidelines [30, 31]. Egypt and Sudan established limits for DCAN and DBAN in their drinking water standards issued in 1995 and 2002 [32].

1.6 Health Implications

Muellner et al. determined the toxicity of seven kinds of HANs using the assay with Chinese hamster ovary (CHO) cells. The results showed that DBAN had the highest cytotoxicity, and the LC_{50} of DBAN is 2.85×10^{-6} (M) [33]. The cytotoxicity of iodoacetonitrile (IAN) and bromoacetonitrile (BAN) is higher than that of chloroacetonitrile (CAN). In addition, the toxicity of most HANs is higher than that of C-DBPs, such as THMs and HAAs [34]. Chen et al. used HepG2 cells to verify and compare the cytotoxicity of DCAN and HAAs. The results showed that the cytotoxicity ranking of the four DBPs was DCAN > dibromoacetic acid (DBAA) > dichloroacetic acid (DCAA) > trichloroacetic acid (TCAA) [35].

Studies showed that HANs can cause various genetic damages such as gene mutation, DNA damage, and chromosomal aberration [36–39]. The toxicity of HANs is related to the halogenated groups and cyano group (-CN) [40]. Halogenated groups on α-carbon can undergo nucleophilic reactions with DNA, leading to DNA damage. The strength of nucleophilic reactions is also related to the degree of halogenation of halogen atoms. The higher the degree of halogenation, the smaller the discrete energy, and the nucleophilic reactivity of iodine atoms is stronger than that of bromine and chlorine atoms [41]. In addition, the more halogen groups substituted on α-carbon, the easier it is for the cyano group of HANs to lose electrons and become more electrophilic. The dispersion energy of halogen group decreases after obtaining electrons. For example, the genotoxicity of TCAN is higher than that of DCAN, which may be related to the strong nucleophilic reaction of cyano groups.

Given that HANs have stronger genetic toxicity than the regulated THMs and HAAs, their potential carcinogenicity has been of concern. As early as 1985, Bull et al. conducted an experiment with 2 weeks skin exposure of CAN, DCAN, TCAN, BCAN, and DBAN as initiators to induce skin cancer in SENCAR mice, and with a 20-week promotion schedule of 12-O tetradecanoylphorbol-13-acetate applications. The results showed that, except for DCAN and TCAN, the other 3 HANs (CAN, BCAN, and DBAN) can all induce the occurrence of skin tumors in mice, and Br-HANs can generally induce more tumors than Cl-HANs (DBAN > BCAN > CAN) [42].

Although some HANs have potential carcinogenicity in short-term carcinogenic tests, there is still a lack of long-term animal carcinogenic test data. The International Agency for Research on Cancer classified HANs as category three in the chemical carcinogenicity classification, indicating that they are not yet classified as carcinogenic to humans and are suspected to be carcinogenic to humans [43, 44]. However, among the 10 detected HANs in drinking water, there is still a lack of carcinogenicity data for the other 5 HANs (BAN, TBAN, BDCAN, DBCAN, IAN). Actually, the IAN showed stronger genotoxicity than other HANs in the single-cell gel electrophoresis test of CHO AS52 cells [33]. Therefore, further research is needed to determine whether these highly genotoxic HANs have potential carcinogenicity.

2 Halonitromethanes (HNMs)

2.1 Basic Properties

Generally, HNMs consist of nine chlorine- and bromine-substituted nitromethanes, namely monochloro-(CNM), dichloro-(DCNM), trichloro-(TCNM), monobromo-(BNM), dibromo-(DBNM), tribromo-(TBNM), bromochloro-(BCNM), bromodichloro-(BDCNM), and dibromochloro-(DBCNM). Halonitromethanes (HNMs) could be formed during the chlorine, chloramine, ozone-chlorine, or ozone-chloramine disinfection of drinking water [45], wastewater [46], and swimming pool water [47]. At present, HNMs are not regulated in drinking water in any country. Even though the concentration of HNMs is lower than the regulated trihalomethanes (THMs) and haloacetic acids (HAAs), their cytotoxicity and genotoxicity are several orders of magnitude higher [48, 49]. Therefore, HNMs are selected and monitored by the USEPA as high-priority non-regulated DBPs [45].

2.2 Precursor

Free or bound amino acids, integral components of hydrophilic NOM, are widespread in various water bodies and play a crucial role as significant precursors for HNMs [50]. The formation potential of HNMs is intricately linked to the specific type of amino acids present and the method used for disinfection. Amino acids demonstrated a significantly more efficient production of HNMs during ozonation-chlorination compared to chlorination alone. Among amino acids, glycine, lysine, and tryptophan generally exhibited a higher formation potential for HNMs during chlorine disinfection [51].

Algal organic matter (AOM) contains a higher concentration of amino acids compared to NOM and plays a crucial role as a significant source of HNMs in eutrophic lakes. AOM produced by different algae species also has different formation potential of HNMs. Even at low concentrations of amines (μg/L) in water, the corresponding functional groups can be converted to nitro groups, which subsequently serve as crucial precursors for HNMs [52]. Dissolved organic matter (DOM) introduced by various atmospheric deposition pathways, such as dry deposition, wet deposition, snow deposition, and dust storm deposition, has been shown to be an important source of precursors for high-risk DBPs [51, 53–55]. Specifically, unsaturated non-aromatic organic nitrates, which is widely available in the atmosphere, have been confirmed as important precursors of HNMs [56]. The formation of HNMs in surface water is also associated with the nitrification of alkanes and aromatics [57]. Other related studies have also mentioned that amino sugars, azo compounds, pyrimidines, and purines also produce HNMs during chlorination or chloramination [58]. In the presence of chlorine, HNMs are generated through reactions with organic matter, nitrites and nitrates.

Prerequisite substances for the formation of HNMs from anthropogenic sources encompass organic nitrogen species in wastewater. Additionally, various anthropogenic discharges such as pharmaceuticals, personal care products, estrogens, sunscreens, and other trace pollutants, as well as organics in wastewater treatment plant effluents, and human sweat or urine from swimming pools, contribute to the precursors of HNMs [51, 52, 58–61].

2.3 Formation Mechanism

For chlorination, the formation of HNMs typically increases with increasing chlorine dose and NOM levels. pH affects both the form of chlorine species and the distribution of different forms of organic matter as well as the stability of HNMs, which in turn affects the formation of HNMs [62].

For the commonly used disinfectants and disinfection processes, the order of HNMs generation potential is as follows: ozone pre-oxidation + chlorine disinfection $>>$ chlorine disinfection $>$ ozone pre-oxidation + chloramine disinfection $>$ chloramine disinfection $>$ ozone disinfection [58]. Many studies have found that ozone pre-oxidation can increase the generation potential of HNMs in the subsequent chlorine disinfection process [45, 63].

The formation of HNMs mainly involves the gradual replacement of nitroalkanes by chlorine, catalysis of HNM formation by NO_2^-, promotion of HNM formation by ultraviolet (UV) irradiation, and HNM formation through the oxidation of O_3 and addition of chlorine.

1. Nitroalkanes are gradually replaced by chlorine to form HNMs

For nitrogenous compounds such as primary amines, amino acids, dipeptides, nucleic acids, etc., the chlorine disinfectant oxidizes the amino group into the nitro group to form a nitroalkane compound [64]. The nitroalkane is further deprotonated by several substrates (e.g., carbonates, hydroxides, hypochlorite, etc.), and the chlorine or chloramine further chlorinates the remaining deprotonated nitroalkane [65]. Nitro functional groups will increase the acidity of C-H bond on α-C, which leads to halogenation reaction on α-C, while the electron withdrawal characteristics of chlorine atoms will increase the acidity of HNMs. When chlorine atoms substitute on nitromethane, the deprotonation reaction will accelerate, resulting in the accumulation of HNMs.

2. NO_2^- catalyzes the generation of HNMs

In addition, NO_2^- significantly increased the HNM formation potential of organic matter from different sources [66, 67]. NO_2^- promoted the formation of HNMs from eight amino acids as well as four aminoglycans, in both individual chlorination and re-chlorination after O_3 oxidation [51]. Despite the low concentrations of nitrophenols in water, the chlorination of phenolics generates nitrophenols in the presence of NO_2^- [56, 65], making nitrophenols important precursors of HNMs. The reaction mechanism mainly involves the formation of electrophilic $ClNO_2$ intermediates (nitration reagent) with reaction of free chlorine and NO_2^- [68]. $ClNO_2$, in turn, further produces another nitration reagent, N_2O_4, through interaction with NO_2^- [69]. Aromatic compounds can be nitrated by these nitration reagents [65], and the subsequent chlorination leads to aromatic ring breaking. Although the concentration of NO_2^- in natural waters is quite low and less stable than ammonia and nitrate, it can exceed 2 mg/L in anoxic environments or during the cold season when nitrifying bacteria are restrained to oxidize nitrite. Therefore, the catalytic role of NO_2^- should not be ignored under these specific conditions.

3. UV reacts with inorganic nitrogen to form HNMs

During medium-pressure UV photolysis, NO_2^- will be transformed into NO· and ·OH [70]. Then, ·OH reacts with NO_2^- to produce the nitration reagent NO_2·, which further produces other nitration reagents: N_2O_4 and ONOOH [70, 71]. NO_3^- will be transformed into NO_2· and ONOOH during photolysis, which are nitration reagents that nitrate aromatic structural compounds [65]. Although NO_2^- is less abundant in natural waters, which can be generated from NO_3^- under UV irradiation. The NO_2^- produced from the photolysis of NO_3^- can participate in various reactions with organic matter in the water, including hydroxylation, nitrification, nitrosation, and others. These processes ultimately result in the formation of HNMs [72].

4. O_3 oxidation to form HNM precursors

TCNM formation potential was elevated in both natural and sewage waters after O_3 oxidation [45, 49, 53, 73]. O_3 can oxidize amines in the water into nitro-compounds, which exhibit a higher capability for producing HNMs [74]. However, the effect of O_3 on HNM formation potential varies with different pollutants. Some compounds have low HNM yields under direct chlorination, but their ozonation products present significantly higher HNM yields under the subsequent chlorination [75, 76]. For example, the TCNM formation potential of aspartic acid and histidine significantly increased after O_3 oxidation, and the magnitude of elevation was positively correlated with O_3 concentration [75]. Still, it has also been shown that O_3 could reduce HNM formation potential for some specific pollutants. Studies have shown that both ranitidine and nizatidine exhibit a low potential for TCNM formation, even as the concentration of O_3 increases [77]. O_3 oxidizes primary and secondary amines into nitro compounds, which are more prone to forming HNMs [74].

2.4 Occurrence and Regulation

HNMs have been frequently detected and reported in drinking water since 2000, leading to an increased focus on their distribution pattern. However, the pattern is affected by water treatment processes, disinfection methods, and operating conditions. The circulation flow of water bodies and climatic conditions can also affect the distribution of HNMs. It was reported that the average concentrations of the two most detected HNMs, i.e., TCNM and DBNM, ranged from undetectable to 3.4 μg/L in drinking water in England treated with chlorine or monochloramine [78]. TCNM levels in wastewater treatment plant effluents in United States ranged from 0.9 to 1.5 μg/L [62]. The persistence of high concentrations of free chlorine residue as well as human excreta in public swimming pool makes it an important source of HNMs. TCNM concentrations ranged from 0.7 to 1.7 μg/L in indoor swimming pool water [79], and from 0.4 to 2.3 μg/L in both indoor and outdoor swimming pools [80].

The median and maximum concentrations of total HNMs detected in drinking water supply systems in the UK were 0.2 and 7.0 μg/L, respectively. This data indicates lower levels in comparison to findings from US drinking water treatment plants, where the reported equivalents were 3.0 and 10.0 μg/L, for median and maximum concentrations, respectively [23]. The lower concentration of bromine in drinking water treatment plants in UK contributes to lower TBNM concentrations compared to those in the United States [45]. The median and maximum TCNM concentrations in water works in UK were 0.2 and 0.5 μg/L, respectively, which are lower than the values of 0.4 and 2.0 μg/L in the US survey of 2000–2002 [45].

2.5 Health Implications

There has been a shift in disinfectants from chlorine to chloramines to control the production of regulated DBPs such as THMs and HAAs. However, this shift in turn has increased the production of nitrogenous DBPs such as HNMs, which may become an even higher health risk.

All nine HNMs are mutagenic, genotoxic, and cytotoxic [81–83], with brominated HNMs being more toxic than chlorinated HNMs. HNMs can cause DNA damage in Chinese hamster ovary (CHO) cells by disrupting the cholesterol synthesis process [49, 84]. The LC_{50} values for CHO cells in BNM, CNM, DBNM,

DCNM, BCNM, TBNM, TCNM, BDCNM, and DBCNM were 7.06×10^{-6} M, 5.29×10^{-4} M, 6.09×10^{-6} M, 3.73×10^{-4} M, 4.05×10^{-5} M, 8.57×10^{-6} M, 5.36×10^{-4} M, 1.32×10^{-5} M, and 6.88×10^{-6} M, respectively [49]. However, there is no report to confirm the effects of HNMs on human health. At present, there is still controversy about the carcinogenicity of HNMs, but it has been shown that HNMs have hepatotoxicity [83].

The high cytotoxicity and genotoxicity of HNMs are related to their molecular structure. The higher cytotoxicity and genotoxicity of HNMs compared to halomethanes (HMs) have been attributed to the greater electron adsorption capacity of the nitro group in comparison to hydrogen. This leads to a more unstable dissociated halogen and higher reactivity of the resulting carbon-positive ion [84]. Furthermore, in the study of quantitative constitutive relationship (QSAR) method, the quantum chemical descriptor of BNM was found to be strongly correlated with the molecular toxicity endpoint and phenotypic toxicity endpoint. The genotoxicity of BNM is mainly affected by quantum chemical parameters. Additionally, the ability to provide electron donors and acceptors may also be related to the genotoxicity of BNM [65]. Meanwhile, oxidative stress is also an important toxicity mechanism of HNMs [85]. It has been demonstrated that BNM, BCNM, and TCNM can cause a significant reduction in glutathione peroxidase and catalase activities [83]. This allows the production of excessive reactive oxygen radicals (ROS) beyond the elimination capacity of the antioxidant system, leading to disruption of amino acid, lipid, and carbohydrate metabolism in mice [86, 87]. Thus, oxidative stress was inferred to be the mechanism of toxicity of HNMs.

Additionally, due to their low concentrations in water, HNMs are challenging to enrich and detect, which often results in research focusing primarily on TCNMs. This means that the risk of HNMs is only partially recognized and the impact on human health is not yet known. The ecological risks posed to aquatic plants and animals when HNMs enter the water cycle remain largely unknown. Understanding the health effect, formation mechanism of HNMs and exploring the control technologies of HNMs are of high demand.

3 Haloacetamides (HAMs)

3.1 Basic Properties

Haloacetamides (HAMs) were first identified as an emerging class of nitrogenous DBPs in the 2000–2002 national survey in the United States. HAMs are produced when halogen atoms are substituted on acetamide molecules, such as dichloroacetamide (DCAM), bromochloroacetamide (BCAM), monobromoacetamide (MBAM), dibromoacetamide (DBAM), chloroiodoacetamide (CIAM), bromoiodoacetamide (BIAM), diiodoacetamide (DIAM), trichloroacetamide (TCAM), dibromochloroacetamide (BDCAM), bromodichloroacetamide (DBCAM), and tribromoacetamide (TBAM). The molecules of HAMs are characterized by their small size, high water solubility, strong polarity, and hydrophilicity. The presence of hydrogen and the amide group in the molecular structure contributes to a high degree of stability, rendering them resistant to volatilization.

Ding et al. studied the stability of 10 HAMs including DCAM, BCAM, DBAM, CIAM, BIAM, DIAM, TCAM, BDCAM, DBCAM, and TBAM in the presence and absence of chlorine and chloramines [88]. The findings indicate that HAMs undergo an alkaline hydrolysis reaction, which is influenced by the concentration of hydroxide ions and HAMs. In general, both the hydrolysis rate and the chlorination rate of HAM increase with the increase of pH and the number of halogens substituted on methyl groups. The hydrolysis rate and chlorination rate of HAMs decreased successively: TCAM > BDCAM > DBCAM > TBAM > DCAM > BCAM > DBAM > BIAM > DIAM > MBAM > MIAM. In addition, due to the weakened electron absorption effect of chlorine atoms when compared with bromine or iodine atoms, the reactivity of chlorinated HAMs is stronger than that of brominated HAMs, followed by iodinated ones.

3.2 Precursor

HAMs are closely associated with the content of DON in water. Generally, DON in water comprises amino compounds, nitrile, purine, pyrimidine, and other compounds. Among these, bound amino acids and free amino acids are the most important components of DON. Chu et al. [89, 90] found that the low molecular weight amino acids are the important precursors of HAMs. Research has shown that HANs in water are also an important source of HAMs. Yu et al. [3] found that hydroxides, hypochlorous acid, and their ionized products could react with HANs to form HAMs and HAAs by breaking the carbon-nitrogen triple bond ($C \equiv N$) in HANs through nucleophilic attack.

Studies have shown that a low Cl/N value is conducive to the formation of DCAM, and a high dose of chlorine will inhibit the formation of DCAM [90]. In addition, extending the residence time of drinking water after disinfection can reduce the production of DCAM and TCAM, but increase the DCAA concentration. The yield of DCAM, DCAN, and DCAA changed significantly with the change of DOC concentration. Since DON is the important precursor of HAMs, removal of DON before disinfection is an effective way to prevent HAM formation. The process of coagulation, precipitation, and filtration can remove some precursors and reduce the formation of HAMs. However, conventional water treatment to reduce turbidity only removes some of the high molecular weight and hydrophobic organic compounds such as humus. The low molecular weight and strong hydrophilicity of many precursors therefore present a challenge to the treatment process because of high solubility in water.

3.3 Formation Mechanism

In the case of formation mechanism, most studies are mainly focused on DCAM. Currently, the exploration of formation mechanisms and the source of nitrogen elements involves the application of experiments and the utilization of density functional theory (DFT). The nitrogen in the amide is mainly derived from the amino acid in the water body. The formation of HAMs mainly comes from the hydrolysis of the nitrile (Fig. 2), so the reaction of organics and chlorine/chloramines to form HAMs is like HANs. In other words, it includes two typical pathways, namely "decarboxylated pathway" and "aldehyde pathway" [91]. The study of Chu et al. demonstrated that aspartic acid reacts with the disinfectant sodium hypochlorite to form DCAN via a series of substitution and decarboxylation reactions [92]. The DCAN then undergoes hydrolysis, resulting in the production of DCAM.

3.4 Occurrence

In the United States and Canada, HAMs have been widely detected in drinking water treatment plants and networks, especially in the water treatment plants that use chloramine disinfection. According to Krasner and Richardson et al. [93, 94], the main HAMs detected in United States were dihaloacetamide (DHAMs) and TCAM, among which DCAM was the most dominant species with the maximum concentration of 5.6 μg/L. Other researchers reported the total HAM concentrations in drinking water treatment plants in the United States was as high as 14 μg/L, with the main constituents including monochloroacetamide, monobromoacetamide, dichloroacetamide, dibromoacetamide, and trichloroacetamide. Chlorinated acetamides are mainly formed during chlorination or chloramination. If bromide or iodide ions are present in water, an amount of brominated or iodinated acetamides will be generated. Studies have shown that the chloramination can effectively reduce the production of regulated DBPs such as trihalomethane and haloacetic acid, but increase the production of N-DBPs such as HAMs.

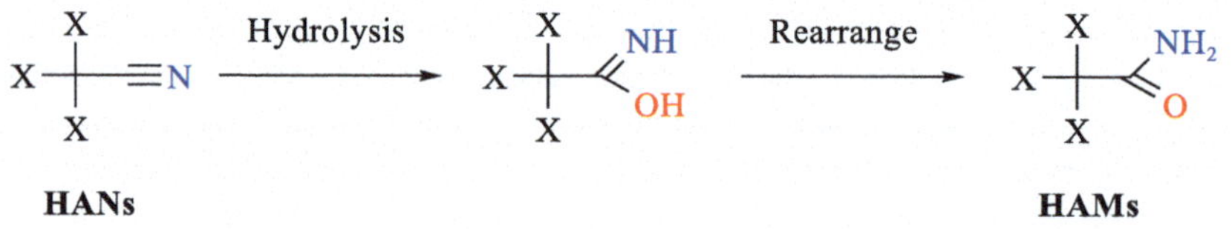

Fig. 2 Formation of HAMs from the hydrolysis of HAMs [91]

3.5 Health Implications

Although HAMs are generally much lower in concentration than regulated DBPs (i.e., THMs and HAAs), cytotoxicity test based on mammalian cell assays showed that HAMs (the values of LC_{50} are in the level of 10^{-6} M and diiodoacetamide is the most toxic with a LC_{50} of 6.78×10^{-7} M) are two orders of magnitude more cytotoxic and genotoxic than THMs (the values of LC_{50} are in the level of 10^{-4} M and triiodomethane is the most toxic with a LC_{50} of 6.60×10^{-5} M) and HAAs (the values of LC_{50} are in the level of 10^{-4} M and iodoacetic acid is the most toxic with a LC_{50} of 2.95×10^{-6} M) [30]. Jeongden et al. found that HAMs were slightly more genotoxic than HANs and HNMs and much more genotoxic than THMs and halogenated acetaldehydes (HALs) [85]. Considering the universality of chloramine disinfection and the high toxicity of HAMs, the health impact of halogenated amides in drinking water cannot be neglected.

4 Cyanogen Halides (CNX)

4.1 Basic Properties

Cyanogen halides (CNX) encompass cyanogen chloride (CNCl) and cyanogen bromide (CNBr). CNCl and CNBr have been identified as prominent byproducts generated from the process of chloramination [95].

CNCl is a DBP commonly detected in chlorinated or chloraminated drinking water as well as wastewater [96]. This triatomic pseudo halogen is an easily condensed colorless gas. CNCl is recognized for its gradual hydrolysis in aqueous environments, and this hydrolysis process can be facilitated by the presence of hydroxide ions (OH^-) [96].

$$CNCl + H_2O \xrightarrow{k_{H_2O}} CNO^- + Cl^- + 2H^+ \quad (1)$$

$$CNCl + OH^- \xrightarrow{k_{OH}} CNO^- + Cl^- + H^+ \quad (2)$$

where k_{H2O} is the first-order rate constant of water hydrolysis and k_{OH} is the second-order rate constant of hydroxide-assisted hydrolysis. Reported value of k_{H2O} is 5.48×10^{-7} s^{-1} under 25°C. [96]. By reaction 1, CNCl has a half-life of hundreds of hours [96]. By reaction 2, values of k_{OH} are somewhat more uncertain, studies suggest that k_{OH} is less than 10 M^{-1} s^{-1} over the temperature range of 5 to 25°C [96]. Based solely on the reactions 1 and 2, it can be inferred that CNCl would exhibit relatively higher stability in water under typical drinking water conditions. However, CNCl is unstable in alkaline solutions and chlorinated water. Research showed that CNCl may continuously undergo hydrolysis to form cyanate (CNO^-) in alkaline solutions with pH 8.5, and a slight excess of chlorine may accelerate the rate of CNCl hydrolysis [97].

CNBr is polar but does not spontaneously ionize in water. It dissolves in both water and polar organic solvents. CNBr is formed during the chlorination or chloramination of bromide-rich water, and higher concentrations of bromide ions produce more CNBr [98]. CNBr is hydrolyzed to release hydrogen cyanide and hypobromous acid. Hydrogen cyanide is very toxic. The hydrolysis is base-catalyzed and enhanced with the increase in pH [98].

$$CNBr + H_2O \rightarrow HCN + HOBr \tag{3}$$

Like other cyanogen compounds, CNBr undergoes an exothermic trimerization process, resulting in the formation of cyanuric bromide ($(BrCN)_3$). This reaction is catalyzed by minute amounts of bromine, metal salts, acids, or bases.

(4)

4.2 Precursors and Formation of CNX

The formation of CNX is highly sensitive to the structure of organic nitrogen precursors, the ratio of free chlorine to organic nitrogen and the presence/absence of ammonium ions.

CNCl is formed during chlorination or chloramination of amino acids, peptides, proteins, purines, pyrimidines, formaldehyde, or humic acid. Amino acids predominate among these precursors, and glycine shows the greatest potential for production among them, with a molecular yield up to 72% [99]. Two linked mechanistic pathways appear to be significant in the formation of CNCl.

The first pathway begins with pre-oxidation which forms formaldehyde, and then followed by a step called post-chloramination. The reaction between monochloramine and formaldehyde is rapid and generates CNCl (Fig. 3) with molecular yield nearly 100% [100]. Formaldehyde is known to be produced during drinking water treatment at levels of 4–10 μg/ L, and these concentrations were

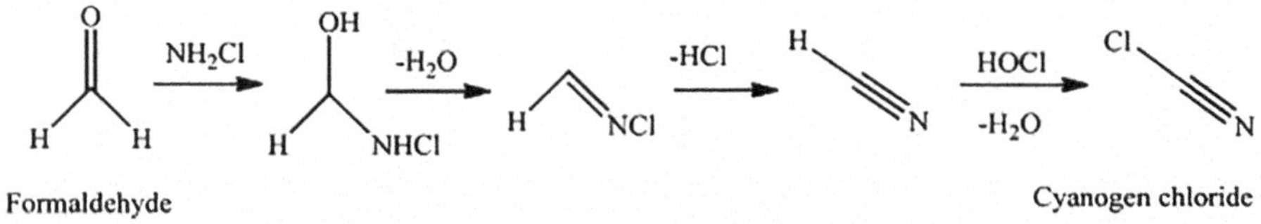

Fig. 3 Formation of CNX from formaldehyde [100]

calculated as being capable of generating 8.6 and 20.3 μg/L CNCl (assuming 100% conversion) under typical chlorination and ozonation conditions, respectively. It should be noted that CNCl can be produced directly under high chlorine dose and low pH [101].

The second pathway is the monochloramination of amino acids to generate CNCl directly. Through this pathway, CNCl was formed by the reaction of chloramine with humic acid, peptides, and purines [98]. This mechanism is like the chlorination of amino acid. The nitrogen in CNCl can be provided by either chloramines or nitrogenous organic compounds in water. Research showed that these two formation pathways were of roughly equal importance and operated concurrently. Nevertheless, the possibility that chloramine or amino acids serve as the primary nitrogen source for CNCl also exists. This phenomenon may be attributed to the variations in organic matter characteristics present in water samples.

In a previous study, it was demonstrated that CNCl concentration increased with rising glycine concentration. This was observed when 8 mg/L (as Cl_2) free chlorine was added to solutions containing 5 mg/L humic acid, 2 mg/L N ammonium ions and varied amounts (0.067–2 mg/L N) of glycine [102]. CNCl formation had a maximum yield when chlorine was added to water containing humic acid and ammonium ions at Cl:N weight ratios of 8–9:1, and further increasing free chlorine dosages reduced CNCl formation [103]. CNCl was stable at pH 7 in monochloramine solutions [98].

In bromide-rich water, chlorine oxidizes bromide into HOBr which favors the formation of CNBr. In chloramination of fulvic acid with different bromide concentrations, cyanogen bromide (CNBr) became the predominant form, while the total molar concentration of the two CNX species was only slightly affected [98]. A direct reaction of HOBr with the organic nitrogenous precursors that provides the nitrogen for CNBr is suggested. CNBr concentration in water is determined by the formation and decomposition reactions, as shown in the equations below:

$$\mathrm{HOCl} + \mathrm{Br}^{-} \xrightarrow{k_1} \mathrm{HOBr} + \mathrm{Cl}^{-} \tag{5}$$

$$\text{organic N precursors} + \mathrm{HOBr} \xrightarrow{k_2} \mathrm{CNBr} + \mathrm{H_2O} \tag{6}$$

$$\mathrm{CNBr} + \mathrm{OH}^{-} \xrightarrow{k_3,\,\text{chlorine catalysis}} \mathrm{CNO}^{-} + \mathrm{Br}^{-} + \mathrm{H}^{+} \tag{7}$$

The oxidation of bromide by free chlorine, denoted as Reaction 5, occurs rapidly. Due to the higher rate constant ($k_1 > k_2$), this reaction leads to the production of brominated DBPs. CNCl undergoes decomposition through basic hydrolysis, while CNBr decomposes according to Reaction 7. During the initial stages, the reactant concentrations are high in Reaction 6, and $k_2 > k_3$. As time goes by, the residual halogen and nitrogen precursors become depleted, causing the rate of Reaction 6 to become lower than that of Reaction 7. Studies revealed that bromine and chlorine can catalyze CNBr hydrolysis. Thus, the higher the oxidant concentration is, the faster the decomposition of CNBr is. On a molar basis, CNBr was the major volatile product formed at 1.5 mg/L chlorine. pH has a significant influence on the formation of CNX, which means CNX concentration decreases notably as pH increases. CNX are also subject to pH-dependent decomposition; hydrolysis is alkaline-catalyzed and accelerates with higher pH levels.

The concentration of CNBr formed during combined disinfection with chlorine dioxide and chloramine is higher than CNCl. These phenomena may attribute to the ability of ClO_2 to oxidize some organic molecules to form new precursors to CNBr. As mentioned earlier, enhancement of CNCl production was reported when ozone was applied prior to chloramination. And ClO_2 is a strong oxidizing agent to form aldehydes when being added in natural water. It is possible that the formaldehyde produced by ClO_2 reacts with the mixed halamine to form CNBr [98].

4.3 Occurrence and Regulation

CNCl has been widely detected in drinking water treated with both free chlorine and chloramine. The United States Environmental Protection Agency (USEPA) and the California Department of Health Services (CDHS) undertook a comprehensive survey encompassing 35 treatment facilities throughout the United States between 1987 and 1989. According to this survey, the concentrations of CNCl varied from 0 to 7.5 μg/L, depending on the disinfectants employed [95]. Later, according to the EPA's Information Collection Rule (ICR) requirement, large-scale water treatment facilities in the United States reported high levels of CNCl, reaching 20.6 μg/L which using chloramine [104]. A survey of 11 US water treatment plants with sources impacted by algae or wastewater was conducted between 2006 and 2007. The median concentration of CNX (including CNCl and CNBr) in the chloraminated and chlorinated waters was determined to be 2.6 μg/L and not detected, respectively [100]. This difference can be attributed to the rapid decomposition of CNCl in chlorinated water, while CNCl exhibits relatively higher stability in monochloraminated water.

CNCl has been listed on the US Environmental Protection Agency (USEPA) First Acute Exposure Guideline Level (AEGL) chemical priority list since 1997 [105], while CNBr and cyanogen iodide (CNI) have been included on the Second AEGL chemical priority list since 2002. Notably, both the World Health Organization (WHO) and the Chinese Standards for Drinking Water Quality (Ministry of Health of the People's Republic of China, 2022) have established a guideline value of 70 μg/L (as cyanide) for CNCl. Similarly, the Agricultural Research Management Council of Australia and New Zealand has established a guideline value of 80 μg/L based on cyanide for CNCl [106]. The USEPA has set a guideline value of 200 μg/L for CNCl (as cyanide). It is crucial to bear in mind the significant toxicity of CNBr and CNI toward aquatic organisms, as they can potentially cause long-term adverse effects within the aquatic ecosystem.

4.4 Health Implications

CNCl undergoes rapid metabolism in the body, converting into cyanide, a highly toxic compound known for its acute toxicity [107]. High doses of CNX can induce thyroid toxicity by inhibiting the uptake of iodine from detoxified thiocyanate. The 24-h and 48-h median lethal concentrations (LC_{50}) of CNCl were found to be 1.40×10^{-6} M and 1.06×10^{-6}65 M, respectively, for Daphnia magna. The median lethal concentration for aquatic organisms over a 48-h period does not exceed 150 μg/L [108], while high concentrations of CNCl can cause death in humans within minutes. Based on the WHO report from 2009, cyanogen chloride is deemed unlikely to occur in drinking water at concentrations posing toxicological risks. Consequently, the development of a formal guideline for cyanogen chloride is deemed unnecessary.

CNBr exhibits volatility and can be readily absorbed through the skin or gastrointestinal tract, making inhalation, physical contact, or ingestion potential routes of toxic exposure. It possesses high acute toxicity, manifesting in various nonspecific symptoms. Even minimal exposure can lead to convulsions or fatalities. For mice, the median lethal dose of CNBr falls within the range of 25 to 50 mg/kg.

5 Summary

Toxicological researches indicate that N-DBPs may pose greater health risks compared to C-DBPs. N-DBPs are frequently detected, particularly in the drinking water sourced from the water sources rich in dissolved organic nitrogen materials. By integrating advanced analysis and detection technologies, along with concentration investigation results, formation and transformation mechanisms, and assessments of health risks, some categories of N-DBPs including HANs should be paid more attention due to their health risk in drinking water. Moreover, more investigations are proposed for possible inclusion in drinking water quality standards to enhance effective supervision and control.

References

1. Krasner SW, McGuire MJ, Jacangelo JG, Patania NL, Reagan KM, Aieta EM (1989) The occurrence of disinfection by-products in United-States drinking water. J Am Water Works Ass 81(8):41–53
2. Oliver BG (1983) Dihaloacetonitriles in drinking water: algae and fulvic acid as precursors. Environ Sci Technol 17(2):80–83
3. Yu Y, Reckhow DA (2015) Kinetic analysis of haloacetonitrile stability in drinking waters. Environ Sci Technol 49(18):11028–11036
4. Reckhow D, Platt T, MacNeill A, McClellan J (2001) Formation and degradation of dichloroacetonitrile in drinking waters. J Water Supply Res Technol AQUA 50(1):1–13

5. Bieber TI, Trehy ML (1981) Dihaloacetonitriles in chlorinated natural water. Water Chlorination Environmental Impact & Health Effects
6. Glezer V, Harris B, Tal N, Iosefzon B, Lev O (1999) Hydrolysis of haloacetonitriles: linear free energy relationship, kinetics and products. Water Res 33(8):1938–1948
7. Ding SK, Chu WH, Tom B, Wang Q, Gao NY, Xu B, Du ED (2018) Formation and estimated toxicity of trihalomethanes, haloacetonitriles, and haloacetamides from the chlor(am)ination of acetaminophen. J Hazard Mater 341:112–119
8. Bond T, Templeton MR, Graham N et al (2012) Precursors of nitrogenous disinfection by-products in drinking water - a critical review and analysis. J Hazard Mater 235:1–16
9. Fang JY, Yang X, Ma J, Shang C, Zhao Q (2010) Characterization of algal organic matter and formation of DBPs from chlor(am)ination. Water Res 44(20):5897–5906
10. Chu WH, Yao D, Deng Y, Sui MH, Gao NY (2017) Production of trihalomethanes, haloacetaldehydes and haloacetonitriles during chlorination of microcystin-LR and impacts of pre-oxidation on their formation. J Hazard Mater 237:153–160
11. Zhang BB, Xian QM, Zhu JP et al (2015) Characterization, DBPs formation, and mutagenicity of soluble microbial products (SMPs) in wastewater under simulated stressful conditions. Chem Eng J 279:258–263
12. Chuang YH, Tung HH (2015) Formation of trichloronitromethane and dichloroacetonitrile in natural waters: precursor characterization, kinetics and interpretation. J Hazard Mater 283: 218–226
13. Yang X, Fan C, Shang C, Zhao Q (2010) Nitrogenous disinfection byproducts formation and nitrogen origin exploration during chloramination of nitrogenous organic compounds. Water Res 44(9):2691–2702
14. Pedersen EJ, Urbansky ET, Marinas BJ, Margerum DW (1999) Formation of cyanogen chloride from the reaction of monochloramine with formaldehyde. Environ Sci Technol 33(23):4239–4249
15. Hou YK, Chu WH et al (2012) Carbonaceous and nitrogenous disinfection by-product formation in the surface and ground water treatment plants using Yellow River as water source. J Environ Sci 24(7):1204–1209
16. Dong L, Wang HY, Cai HQ et al (2016) Investigation on nitrogenous disinfection by-products in drinking water in six cities, China. J Environ Health 33(3):232–235
17. Li YZ, Ma X, Chen R et al (2021) Levels and health risk assessments of halomethanes, haloacetic acids, and haloacetonitriles disinfection by-products in source and finished water in Shanghai. J Environ Occup Med 38(5):460–466
18. Bougeard CMM, Goslan EH, Jefferson B et al (2010) Comparison of the disinfection by-product formation potential of treated waters exposed to chlorine and monochloramine. Water Res 44(3):729–740
19. Huang H, Zhu HH, Gan WH, Chen X, Yang X (2017) Occurrence of nitrogenous and carbonaceous disinfection byproducts in drinking water distributed in Shenzhen, China. Chemosphere 188:257–264
20. Ding H, Meng L, Zhang H et al (2013) Occurrence, profiling and prioritization of halogenated disinfection by-products in drinking water of China. Environ Sci: Processes Impacts 15(7): 1424–1429
21. Ratpukdi T, Sinorak S, Kiattisaksiri P, Punyapalakul P, Siripattanakul-Ratpukdi S (2019) Occurrence of trihalomethanes and haloacetonitriles in water distribution networks of Khon Kaen Municipality, Thailand. Water Supply 19(6):1748–1757
22. Stefán D, Erdélyi N, Izsák B, Záray G, Vargha M (2019) Formation of chlorination by-products in drinking water treatment plants using breakpoint chlorination. Microchem J 149:104008
23. Bond T, Templeton MR, Kamal NHM, Graham N, Kanda R (2015) Nitrogenous disinfection byproducts in English drinking water supply systems: occurrence, bromine substitution and correlation analysis. Water Res 85:85–94

24. Krasner SW, Weinberg HS, Richardson SD, Pastor SJ, Chinn R, Sclimenti MJ, Onstad GD, Thruston AD (2006) Occurrence of a new generation of disinfection byproducts. Environ Sci Technol 40(23):7175–7185
25. WHO (2017) Guidelines for drinking-water quality: fourth edition incorporating the first addendum. WHO, Geneva
26. Ministry of Health, Labour and Welfare (2015) Drinking water quality standards in Japan. Ministry of Health, Labour and Welfare, Tokyo
27. Ministry of Health (2018) Drinking-water standards for New Zealand 2005 (revised 2018). Ministry of Health, Wellington
28. Ministry of the Environment and Water Resources (2019) Environmental public health (water suitable for drinking) (No.2) regulations 2019. Ministry of the Environment and Water Resources, Singapore
29. Department of Health (2017) Philippine nation standards for drinking water of 2017. Department of Health, Manila
30. Ministry of the Environment (2020) Drinking water management ACT. Ministry of the Environment, Seoul
31. Ministry of Health Malaysia (2004) National standard for drinking water quality. Engineering Services Division, Kuala Lumpur
32. WHO (2006) A compendium of drinking-water quality Standards in the eastern Mediterranean Region. WHO, Geneva
33. Muellner MG, Wagner ED, McCalla K et al (2007) Haloacetonitriles vs. regulated haloacetic acids: are nitrogen-containing DBPs more toxic. Environ Sci Technol 41:645–651
34. Plewa MJ, Wagner ED, Jazwierska P et al (2004) Halonitromethane drinking water disinfection by-products: chemical characterization and mammalian cell cytotoxicity and genotoxicity. Environ Sci Technol 38:62–68
35. Chen Z (2009) Cytotoxicity and genotoxicity analysis for chlorinated disinfection by-products of drinking water in HepG2 cells. Huazhong University of Science and Technology, Wuhan
36. Bull RJ, Meier JR, Robinson M et al (1985) Evaluation of mutagenic and carcinogenic properties of brominated and chlorinated acetonitriles: by -products of chlorination. Fundam Appl Toxicol 5:1065–1074
37. Osgood C, Sterling D (1991) Dichloroacetonitrile, a by-product of water chlorination, induces aneuploidy in Drosophila. Mutat Res 261:85–91
38. Le Curieux F, Giller S, Gauthier L et al (1995) Study of the genotoxic activity of six halogenated acetonitriles, using the SOS chromotest, the Ames-fluctuation test and the newt micronucleus test. Mutat Res 341:289–302
39. Muller-Pillet V, Joyeux M, Ambroise D et al (2000) Genotoxic activity of five haloacetonitriles: comparative investigations in the single cell gel electrophoresis (comet) assay and the Ames-fluctuation test. Environ Mol Mutagen 36:52–58
40. Lin EL, Guion CW (1989) Interaction of haloacetonitriles with glutathione and glutathione-S-transferase. Biochem Pharmacol 38:685–688
41. Loudon GM (2009) Organic chemistry. Roberts & Company Publishers, Greenwood Village, pp 398–399
42. Bull RJ, Robinson M (1983) Carcinogenic activity of haloacetonitrile and haloacetone derivatives in the mouse skin and lung. Water Chlorination Chem Environ Impact Health Effects 5
43. International Agency for Research on Cancer (1991) Chlorinated drinking water, chlorination by-products, some other halogenated compounds, cobalt and cobalt compounds. IARC, Lyon, pp 269–296
44. International Agency for Research on Cancer (1999) Re-evaluation of some organic chemicals, hydrazine and hydrogen peroxide. IARC, Lyon, pp 1291–1537
45. Zhang RH, Wang FF, Fang C, Luan XM, Yang X, Chu WH (2020) Occurrence of CX_3R-type disinfection byproducts in drinking water treatment plants using DON-rich source water. ACS ES&T Water 1(3):553–561

46. Peng LQ, Wang FF, Zhang D, Fang C, Jan PH, Chu WH (2021) Effect of oxidation ditch and anaerobic-anoxic-oxic processes on CX3R-type disinfection by-product formation during wastewater treatment. Sci Total Environ 770:145344
47. Liviac D, Wagner ED, Mitch WA, Altonji MJ, Plewa MJ (2010) Genotoxicity of water concentrates from recreational pools after various disinfection methods. Environ Sci Technol 44(9):3527–3532
48. Plewa MJ, Wagner ED, Jazwierska P, Richardson SD, Chen PH, McKague AB (2004) Halonitromethane drinking water disinfection byproducts: chemical characterization and mammalian cell cytotoxicity and genotoxicity. Environ Sci Technol 38(1):62–68
49. Plewa MJ, Wagner ED, Muellner MG, Hsu KM, Richardson SD (2007) Comparative mammalian cell toxicity of N-DBPs and C-DBPs. Abstr Papers Ame Chem Soc 233:372–372
50. Xiao R, Deng Y, Xu ZX, Chu WH (2023) Disinfection byproducts and their precursors in drinking water sources: origins, influencing factors, and environmental insights. Engineering. In press
51. He JJ, Wang FF, Zhao TT, Liu SG, Chu WH (2020) Characterization of dissolved organic matter derived from atmospheric dry deposition and its DBP formation. Water Res 171: 115368
52. Chu WH, Gao NY, Deng Y, Michael RT, Yin DQ (2011) Formation of nitrogenous disinfection by-products from pre-chloramination. Chemosphere 85(7):1187–1191
53. Hou MT, Chu WH, Wang FF, Deng Y, Gao NY, Zhang D (2018) The contribution of atmospheric particulate matter to the formation of CXR-type disinfection by-products in rainwater during chlorination. Water Res 145:531–540
54. He JJ, Shi MH, Wang FF, Duan YL, Zhao TT, Shu SH, Chu WH (2020) Removal of CX3R-type disinfection by-product precursors from rainwater with conventional drinking water treatment processes. Water Res 185:116099
55. Du ZQ, Ding SK, Xiao R, Fang C, Jia RB, Chu WH (2023) Disinfection by-product precursors introduced by sandstorm events: composition, formation characteristics and potential risks. Water Res 171:115368
56. Du ZQ, Ding SK, Xiao R, Fang C, Song WC, Jia RB, Chu WH (2022) Does snowfall introduce disinfection by-product precursors to surface water? Environ Sci Technol 56(20): 14487–14497
57. Fu J, Wang X, Bai W, Yang H, Xie YF (2017) Azo compound degradation kinetics and halonitromethane formation kinetics during chlorination. Chemosphere 174:110–116
58. Fang C, Yang X, Ding SK, Luan XM, Xiao R, Du ZQ, Wang P, An W, Chu WH (2021) Characterization of dissolved organic matter and its derived disinfection byproduct formation along the Yangtze River. Environ Sci Technol 55(18):12326–12336
59. Merlet N, Thibaud H, Dore M (1985) Chloropicrin formation during oxidative treatments in the preparation of drinking-water. Sci Total Environ 47:223–228
60. Wei XY, Gao NY, Li CJ (2017) Evaluation and reduction of THMs, DCAN and TCNM formation potential from chlorination and chloramination of verapamil. J Environ Chem Eng 5(5):4397–4405
61. Yang X, Shang C, Shen Q, Chen B, Westerhoff P, Peng J, Guo W (2012) Nitrogen origins and the role of ozonation in the formation of haloacetonitriles and halonitromethanes in chlorine water treatment. Environ Sci Technol 46(23):12832–12838
62. Song H, Addison JW, Hu J, Karanfil T (2010) Halonitromethanes formation in wastewater treatment plant effluents. Chemosphere 79(2):174–179
63. Hoigne J, Bader H (1988) The Formation of trichloronitromethane (chloropicrin) and chloroform in a combined ozonation chlorination treatment of drinking-water. Water Res 22(3): 313–319
64. Orvik JA (1980) Kinetics and mechanism of nitromethane chlorination – new rate expression. J Am Chem Soc 102(2):740–743

65. Thibaud H, Delaat J, Merlet N, Dore M (1987) Chloropicrin formation in aqueous-solution – effect of nitrites on precursors formation during the oxidation of organic-compounds. Water Res 21(7):813–821
66. Duguet JP, Tsutsumi Y, Bruchet A, Mallevialle J (1988) Chloropicrin in potable water – conditions of formation and production during treatment processes. Environ Technol Lett 9(4): 299–310
67. Hong HC, Qian LY, Xiao ZQ, Zhang JQ, Chen JR, Lin HJ, Yu HY, Shen LG, Liang Y (2015) Effect of nitrite on the formation of halonitromethanes during chlorination of organic matter from different origin. J Hydrol 531:802–809
68. Gan GJ, Qiu L, Wu H, Hong HC, Mazumder A, Pan XL, Liang Y (2017) Effect of nitrite on the formation of trichloronitromethane (TCNM) during chlorination of polyhydroxy-phenols and sugars. Water Air Soil Pollut 228(6)
69. Johnson DW, Margerum DW (1991) Nonmetal redox kinetics – a reexamination of the mechanism of the reaction between hypochlorite and nitrite ions. Inorg Chem 30(25): 4845–4851
70. Mack J, Bolton JR (1999) Photochemistry of nitrite and nitrate in aqueous solution: a review. Journal of Photochemistry and Photobiology A-Chemistry 128(1–3):1–13
71. Shah AD, Dotson AD, Linden KG, Mitch WA (2011) Impact of UV disinfection combined with chlorination/chloramination on the formation of halonitromethanes and haloacetonitriles in drinking water. Environ Sci Technol 45(8):3657–3664
72. Martijn AJ, Boersma MG, Vervoort JM, Rietjens IMCM, Kruithof JC (2014) Formation of genotoxic compounds by medium pressure ultraviolet treatment of nitrate-rich water. Desalin Water Treat 52(34–36):6275–6281
73. Chu WH, Gao NY, Yin DQ, Deng Y, Templeton MR (2012) Ozone-biological activated carbon integrated treatment for removal of precursors of halogenated nitrogenous disinfection by-products. Chemosphere 86(11):1087–1091
74. McCurry DL, Quay AN, Mitch WA (2016) Ozone promotes chloropicrin formation by oxidizing amines to nitro compounds. Environ Sci Technol 50(3):1209–1217
75. Liu ZG, Chen W, Yu H, Tao H, Xu H, Yu JJ, Gu YM, Wan Z (2017) Effects of pre-oxidation and adsorption on haloacetonitrile and trichloronitromethane formation during subsequent chlorination. Environ Sci Pollut Res 24(27):21836–21845
76. Shan J, Hu J, Kaplan-Bekaroglu SS, Song H, Karanfil T (2012) The effects of pH, bromide and nitrite on halonitromethane and trihalomethane formation from amino acids and amino sugars. Chemosphere 86(4):323–328
77. Wang J, Rong LJ, Li DZ, Lu SP (2016) Effect of welding thermal cycles on the oxidation resistance of 9 wt.% Cr heat resistant steels in 550 °C lead-bismuth eutectic. Appl Surf Sci 389: 930–941
78. Bougeard CMM, Goslan EH, Jefferson B, Parsons SA (2010) Comparison of the disinfection by-product formation potential of treated waters exposed to chlorine and monochloramine. Water Res 44(3):729–740
79. Kanan A, Karanfil T (2011) Formation of disinfection by-products in indoor swimming pool water: the contribution from filling water natural organic matter and swimmer body fluids. Water Res 45(2):926–932
80. Montesinos I, Cardador MJ, Gallego M (2011) Determination of halonitromethanes in treated water. J Chromatogr A 1218(18):2497–2504
81. Giller S, LeCurieux F, Gauthier L, Erb F, Marzin D (1995) Genotoxicity assay of chloral hydrate and chloropicrine. Mutat Res Lett 348(4):147–152
82. Kristiana I, Liew D, Henderson RK, Joll CA, Linge KL (2017) Formation and control of nitrogenous DBPs from Western Australian source waters: Investigating the impacts of high nitrogen and bromide concentrations. J Environ Sci 58:102–115
83. Yin J, Wu B, Zhang XX, Xian Q (2017) Comparative toxicity of chloro- and bromo-nitromethanes in mice based on a metabolomic method. Chemosphere 185:20–28

84. Kundu B, Richardson SD, Swartz PD, Matthews PP, Richard AM, DeMarini DM (2004) Mutagenicity in *salmonella* of halonitromethanes: a recently recognized class of disinfection by-products in drinking water. Mutat Res Genet Toxicol Environ Mutag 562(1–2):39–65
85. Lan JQ, Rahman SM, Gou N, Jiang T, Plewa MJ, Alshawabkeh A, Gu AZ (2018) Genotoxicity assessment of drinking water disinfection byproducts by DNA damage and repair pathway profiling analysis. Environ Sci Technol 52(11):6565–6575
86. Paduraru M, Ciobica A, Dobrin I, Stefanescu C (2010) Evaluation of antioxidant enzymes activities and lipid peroxidation in schizophrenic patients treated with typical and atypical antipsychotics. Neurosci Lett 479(3):317–320
87. Pae HO, Kim EC, Chung HT (2008) Integrative survival response evoked by heme oxygenase-1 and heme metabolites. J Clin Biochem Nutr 42(3):197–203
88. Ding S, Chu WH, Krasner SW, Yu Y, Fang C, Xu B, Gao NY (2018) The stability of chlorinated, brominated, and iodinated haloacetamides in drinking water. Water Res 142: 490–500
89. Chu WH, Gao NY, Deng Y (2010) Formation of haloacetamides during chlorination of dissolved organic nitrogen aspartic acid. J Hazard Mater 173(1):82–86
90. Chu WH, Gao NY, Deng Y, Krasner SW (2010) Precursors of dichloroacetamide, an emerging nitrogenous DBP formed during chlorination or chloramination. Environ Sci Technol 44(10): 3908–3912
91. Chu WH, Hu JL, Bond T, Gao NY, Xu B, Yin DQ (2016) Water temperature significantly impacts the formation of iodinated haloacetamides during persulfate oxidation. Water Res 98: 47–55
92. Mack S, J. and Bolton, J.R. (1999) Photochemistry of nitrite and nitrate in aqueous solution: a review. J Photochem Photobiol A Chem 128(1–3):1–13
93. Chu WH, Krasner SW, Gao NY, Templeton MR, Yin DQ (2015) Contribution of the antibiotic chloramphenicol and its analogues as precursors of dichloroacetamide and other disinfection byproducts in drinking water. Environ Sci Technol 50(1):388–396
94. Chu WH, Gao NY, Yin DQ, Krasner SW, William AM (2014) Impact of UV/H_2O_2 pre-oxidation on the formation of haloacetamides and other nitrogenous disinfection byproducts during chlorination. Environ Sci Technol 48(20):12190–12198
95. Ding SK, Chu WH (2017) Recent advances in the analysis of nitrogenous disinfection by-products. Trends Environ Anal Chem 14:19–27
96. Na CZ, Olson TM (2004) Stability of cyanogen chloride in the presence of free chlorine and monochloramine. Environ Sci Technol 38(22):6037–6043
97. Hu JL, Chu WH, Sui MH, Xu B, Gao NY, Ding SK (2018) Comparison of drinking water treatment processes combinations for the minimization of subsequent disinfection by-products formation during chlorination and chloramination. Chem Eng J 335:352–361
98. Heller-Grossman L, Idin A, Limoni-Relis B, Rebhun M (1999) Formation of cyanogen bromide and other volatile DBPs in the disinfection of bromide-rich lake water. Environ Sci Technol 33(6):932–937
99. Zhu HH, Xue L, Liu JS, Chen YG, Jin X, Yuan M (2020) Types, hazards and pollution status of chlorinated disinfection by-products in surface water. Res Environ Sci 33:1640–1648
100. Ding SK, Deng Y, Bond T, Fang C, Cao ZQ, Chu WH (2019) Disinfection byproduct formation during drinking water treatment and distribution: a review of unintended effects of engineering agents and materials. Water Res 160:313–329
101. Alouini Z, Seux R (1987) Kinetics and mechanisms of hypochlorite oxidation of alpha-amino-acids at the time of water disinfection. Water Res 21(3):335–343
102. Diehl AC, Speitel GE, Symons JM, Krasner SW, Hwang SJ, Barrett SE (2000) DBP formation during chloramination. J Am Water Works Ass 92(6):76–90
103. Ding SK, Chu WH, Tom B, Cao ZQ, Xu B, Gao NY (2018) Contribution of amide-based coagulant polyacrylamide as precursors of haloacetamides and other disinfection by-products. Chem Eng J 350:356–363

104. Obolensky A, Singer PC, Shukairy HM (2007) Information collection rule data evaluation and analysis to support impacts on disinfection by-product formation. J Environ Eng 133(1):53–63
105. USEPA (1996) ICR Manual for Bench- and Pilot-Scale Treatment Studies. EPA 814/B-96-003
106. Simpson KL, Hayes KP (1998) Drinking water disinfection by-products: An Australian perspective. Water Res 32(5):1522–1528
107. Yang X, Shang C (2005) Quantification of aqueous cyanogen chloride and cyanogen bromide in environmental samples by MIMS. Water Res 39(9):1709–1718
108. Jiang FY, Xie YF, Dong K, Wang DQ, Li HX (2022) Detection and stability of cyanogen bromide and cyanogen iodide in drinking water. Water 14(10):14

Nitrosamines

Yu Wang, Chao Chen, and Susan Andrews

Contents

1 Basic Properties of Nitrosamines ... 196
  1.1 Physical-Chemical Properties ... 196
  1.2 Health Implications ... 197
  1.3 History of Nitrosamine Studies ... 201
2 Nitrosamine Occurrence ... 202
  2.1 Global Occurrence ... 202
  2.2 Factors Influencing Nitrosamine Formation ... 204
3 Formation Mechanism ... 208
  3.1 Nitrosation Pathway ... 208
  3.2 Formation During Chloramination ... 209
  3.3 Formation During Ozonation ... 211
  3.4 Formation by Other Oxidants ... 212
4 Nitrosamine Precursors ... 212
  4.1 NOM ... 212
  4.2 Artificial Sources ... 213
  4.3 Reagents and Materials Used in Water Industry ... 213
  4.4 Microbiological Metabolism of Amino Acids ... 214
  4.5 Characterization of Nitrosamine Precursors ... 215
5 Control of Nitrosamines in Drinking Water Treatment Plants ... 216
  5.1 Removal of Precursors before Municipal Disinfection ... 216
  5.2 Alternative Disinfection Processes ... 218
  5.3 Direct Removal of Nitrosamines ... 218

Y. Wang and C. Chen (✉)
State Key-Lab of Environmental Simulation and Pollution Control, School of Environment, Tsinghua University, Beijing, China
e-mail: chen_water@tsinghua.edu.cn

S. Andrews
University of Toronto, Toronto, ON, Canada
e-mail: susan.andrews@utoronto.ca

Chao Chen, Susan Andrews, and Yuefeng Xie (eds.), *Disinfection By-Products in Water*, Hdb Env Chem (2024) 135: 195–230, DOI 10.1007/698_2024_1146,

Published online: 12 September 2024

6 Nitrosamine Regulation 219
6.1 WHO Guidelines in Drinking Water 219
6.2 National Regulations in Drinking Water 220
6.3 Regulation of Nitrosamines in Wastewater/Reclaimed Water 222
7 Summary 222
References 223

Abstract Nitrosamines, a family of emerging contaminants with strong carcinogenicity, have been widely detected in water, wastewater, air, soil, food, beverage, tobacco, pharmaceuticals, and industrial reagents. Their occurrence as disinfection by-products (DBPs) has been widely reported in drinking water treatment plants in the U.S. and Canada since the late twentieth century, and later in Europe, Asia, Oceania, and South America. *N*-nitrosodimethylamine (NDMA) is the most frequently detected nitrosamine species in water. Chloramination, intended or de facto, is the key factor for nitrosamine occurrence during drinking water treatment, however, properties of their precursors, the presence of other disinfectants, and contact time are also strong factors associated with their occurrence. The unsymmetric dimethylhydrazine (UDMH) pathway is often considered to be the main mechanism for the formation of NDMA. Nitrosamine precursors are quite complex and from diverse sources, including basic amines, certain pesticides, pharmaceuticals, natural organic matter, industrial wastewater, and domestic wastewater, including the effluent from wastewater treatment plants. Amino acid metabolism has also been found to be a source of nitrosamine precursors. Nitrosamine control strategies include source control, process control, and end-of-pipe removal. Conventional processes, e.g., coagulation, sedimentation, and filtration, remove few nitrosamine precursors while advanced treatment processes, e.g., ozonation and (biological) activated carbon processes, can reduce nitrosamine precursor concentrations effectively. Modifications to disinfection processes can also help to minimize the formation of nitrosamines. High doses of UV irradiation are also able to degrade NDMA, and while this is not often used to minimize NDMA as a disinfection by-product it may be a method to remove NDMA from contaminated water. Due to its wide occurrence and health impacts, NDMA has been regulated in some countries including the United States, Canada, the Netherlands, Australia, South Korea, and China.

Keywords Chloramines, Nitrosamine, *N*-nitrosodimethylamine, Precursor, Regulation, Treatment process

1 Basic Properties of Nitrosamines

1.1 Physical-Chemical Properties

Nitrosamines, identifiable by their general structure $R_1N(R_2)$-N=O, can be found at trace levels in various media including surface water, drinking water, wastewater, air, soil, and diverse industrial applications including rocket fuel, cosmetics, pesticides, and polymers production [1]. This family has about 200 chemicals and showcases a vast spectrum of diversity, with classifications ranging from alkyl-substituted to aromatic, cycloalkyl, and polycyclic variants [2]. *N*-nitrosodimethylamine (NDMA), as a prime example, is characterized by its simple dialkylnitrosamine structure and is the most abundant species in various media and industrial applications.

Nitrosamines generally appear as colorless, oily liquids, soluble in water and most organic solvents due to their low molecular weight (MW) and polar nature. Their volatility varies with the type of alkyl substituents, making NDMA among the most volatile in this group. They exhibit stability at room temperature but can be decomposed by strong acids, bases, UV irradiation, or high temperatures [2, 3]. The presence of the nitroso group renders nitrosamines prone to reduction and other chemical reactions, facilitating the formation or release of nitrosation reagents, which contribute to their toxic and carcinogenic nature (Table 1).

1.2 Health Implications

Researches since the 1950s have shown that almost all of the over 300 nitrosamines and nitrosamides (another family of *N*-nitroso chemicals) exhibit some level of toxicity and carcinogenicity to animals or humans [4]. NDMA is of special concern owing to its high toxicity and carcinogenicity and widespread occurrence. NDMA acts as a hepatotoxin, inducing liver fibrosis and cancer in various animal species. British scientists John Barnes and Peter Magee first reported its toxic effects in 1956 while investigating chemicals used as solvents in the dry-cleaning industry [5]. Since their discovery, NDMA has been widely identified in food, drinking water, soil, air, etc. [1, 6–8].

The carcinogenic mechanism of nitrosamines is initiated by enzymatic α-hydroxylation mediated by cytochrome P450, followed by dealkylation to form dealkylated primary nitrosamines. These intermediates then decompose to form diazonium ions, which alkylate DNA and potentially lead to genetic alterations and cellular mutations, ultimately leading to tumorigenesis [9]. In contrast, nitrosamides are widely acknowledged to be directly carcinogenic, obviating the requirement for enzymatic activation via mixed-function oxidases for tumorigenesis.

Toxicological studies have shown their potential for mutagenicity, teratogenicity, and carcinogenicity. The carcinogenic potential of NDMA has been investigated most widely among the nitrosamines [10]. According to experiments with rats, mice, and dogs, NDMA can induce tumors in organs like the liver and kidneys through various exposure routes, including oral, inhalation, or intranasal administration. Pregnant rats and mice exposed to NDMA have their offspring with increased percentages suffering liver and kidney tumors. Long-term oral administration of low doses, less than 2 mg/kg/day, of NDMA in rats and mice showed increased incidences of liver tumors, but not kidney tumors. Studies have indicated that human S9 fraction exhibits a higher genotoxicity-promoting effect in the Ames test compared to rat S9 fraction, suggesting a particular sensitivity of humans to the

Table 1 Basic properties of NAs

Compounds	Abbr.	CAS	MW	$LogK_{ow}$	Boling point (°C)	Structure formula
N-Nitrosodimethylamine	NDMA	62-75-9	74.082	−0.57	152.0 ± 0.0	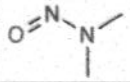
N-Nitrosodiethylamine	NDEA	55-18-5	102.135	0.48	173.9 ± 9.0	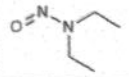
N-Nitrosodi-n-propylamine	NDPA	621-64-7	130.188	1.36	206.0 ± 0.0	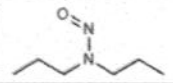
N-Nitrosodi-n-butylamine	NDBA	924-16-3	158.241	2.63	250.6 ± 9.0	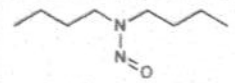
N-Nitrosomethylethylamine	NMEA	10595-95-6	88.108	0.04	154.4 ± 9.0	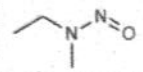
N-Nitrosopiperidine	NPIP	100-75-4	114.146	0.36	229.8 ± 9.0	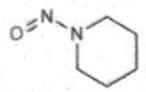
N-Nitrosopyrrolidine	NPYR	930-55-2	100.119	−0.19	214.0 ± 9.0	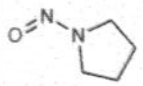
N-Nitrosomorpholine	NMOR	59-89-2	116.118	−0.44	226.1 ± 15.0	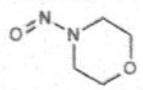
N-Nitrosodiethanolamine	NDELA	1116-54-7	134.134	−1.28	375.9 ± 27.0	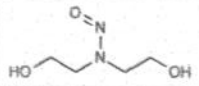
N-Nitrosodiphenylamine	NDPhA	86-30-6	198.221	3.13	346.5 ± 15.0	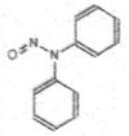

Note: $LogK_{ow}$ data from ChemSpider, Exper. Database match

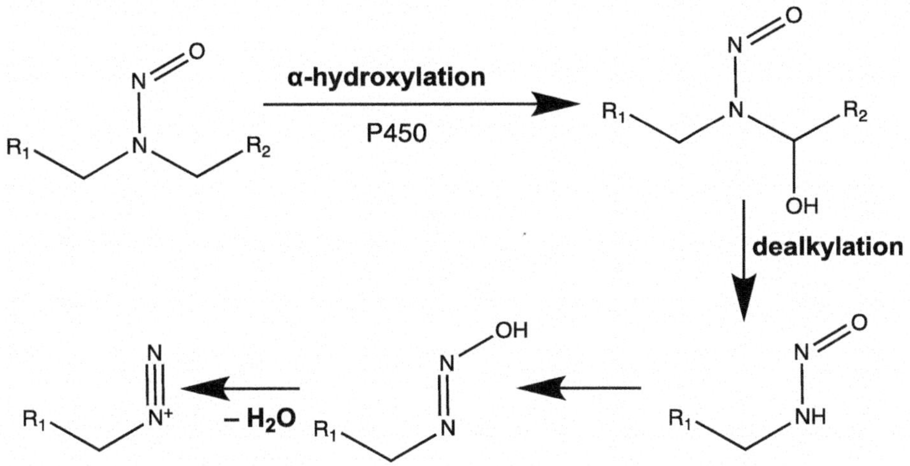

Fig. 1 Transformation of nitrosamine into diazonium via enzymatic α-hydroxylation (redrawn from Beard and Swager [9])

carcinogenic effects of NDMA. NDBA and NDPhA have been identified as bladder carcinogens amongst known disinfection by-products (DBPs) by toxicological studies [11]. These findings support the view that NDMA poses adverse health effects, particularly its strong carcinogenic potential to humans. Epidemiological studies also suggest the link between nitrosamine exposure via food or water and the increased risks of bladder, esophageal, colorectal, pancreatic, and liver cancers in human [12–15] (Fig. 1).

The International Agency for Research on Cancer (IARC) has classified several specific nitrosamines as Group 1 or Group 2A carcinogens, indicating they are known or probable human carcinogens. This classification is based on evidence from animal and human studies, as well as their ability to cause DNA damage and promote tumor development. The Environmental Protection Agency (EPA) Integrated Risk Information System (IRIS) also provides cancer slope factors for several nitrosamines which means the lifetime cancer risk per dose of contaminant, highlighting the risk assessment of these compounds [16]. Nitrosamine cancer slope factors published in IRIS range from 4.9×10^{-3} (mg/kg/day)$^{-1}$ for NDPhA to 150 (mg/kg/day)$^{-1}$ for NDEA (see Table 2).

According to the U.S. EPA carcinogen classification system, nitrosamines were categorized as B2 carcinogens, indicating sufficient evidence of carcinogenicity in animals but inadequate or no evidence in humans. In the U.S. EPA's Proposed Guidelines for Carcinogen Risk Assessment (1999), this classification corresponds to "likely to be carcinogenic to humans." The IRIS database from the U.S. EPA estimates a remarkably low concentration of NDMA (0.7 ng/L) in drinking water to be associated with an excess lifetime cancer risk of 10^{-6}, which is the second only to NDEA with the threshold of 0.2 ng/L for the same level of risk [17].

Table 2 Toxicity-related parameters of nitrosamines

Agent	Abbreviation	CAS no.	Group by IARC	Additional information	Oral slope factor (mg/kg/day)$^{-1}$	Skin penetration factor (cm·h^{-1})	Drinking water unit risk (μg·L^{-1})
N-Nitroso-dimethyl-amine	NDMA	62-75-9	2A	Overall evaluation upgraded to Group 2A with supporting evidence from other relevant data	51	2.57×10^{-4}	1.4×10^{-3}
N-Nitroso-diethyl-amine	NDEA	55-18-5	2A		150	8.82×10^{-4}	4.3×10^{-3}
N-Nitroso-di-n-propylamine	NDPA	621-64-7	2B	/	7	2.34×10^{-3}	2.0×10^{-4}
N-Nitroso-di-n-butylamine	NDBA	924-16-3	2B	/	5.4	1.13×10^{-4}	1.6×10^{-4}
N-Nitroso-methylethyl-amine	NMEA	10595-95-6	2B	/	22	5.41×10^{-4}	6.3×10^{-4}
N-Nitroso-piperidine	NPIP	100-75-4	2B	/	/	6.22×10^{-4}	/
N-Nitroso-pyrrolidine	NPYR	930-55-2	2B	/	2.1	3.27×10^{-4}	6.1×10^{-5}
N-Nitroso-morpholine	NMOR	59-89-2	2B	/	/	1.78×10^{-4}	/
N-Nitroso-diphenyl-amine	NDPhA	86-30-6	3	/	0.0049	/	1.4×10^{-7}

From: List of Classifications, Agents classified by the IARC Monographs, Volume 17, Sup 7, issued in 1987; U.S. EPA, IRIS Assessments

Beyond carcinogenicity, recent research has also paid attention to the reproductive and developmental toxicity of nitrosamines. For example, Luo et al. detected associations of maternal exposure to drinking water nitrosamines during pregnancy with decreased body weight and increased risks of small for gestational age and preterm delivery [18].

1.3 History of Nitrosamine Studies

Nitrosamines have received increasing attention since the 1950s as being potentially harmful to human health after the milestone research of Magee and Barnes [5]. They were found in cured meats like sausage and malt beverages like beer due to their formation from preservatives like nitrites in food reacting with amines during food processing [8]. Tobacco-specific nitrosamines have also been among the hot topics of carcinogens [19].

Subsequent detection of NDMA in the air raised epidemiological concerns about the relationship between airborne nitrosamines in some industries and increased cancer risks for workers [20]. Later, the presence of nitrosamines in water sources garnered more attention. The first recorded report of NDMA in water dates back to the 1970s when an industrial facility producing asymmetrical dimethylhydrazine in Baltimore, Maryland contaminated the nearby groundwater [21]. The occurrence of NDMA (150 ng/L) in drinking water wells near a rocket engine testing facility in Sacramento County, CA, indicated a link to UDMH-based rocket fuels [22]. The early results concerning NDMA occurrence in the drinking water supplies industries are shown in Table 3.

Studies in California, the United States, and Ontario, Canada since the late twentieth century revealed that NDMA presence was not limited to contamination from certain industries. Instead, NDMA detected in some locations appeared to be related to chloramine disinfection of water and wastewater [24, 25]. With increasingly stringent regulations on trihalomethanes (THMs) and haloacetic acids (HAAs) in drinking water, monochloramine has emerged as an alternative disinfectant for free chlorine. However, the likelihood of NDMA formation was more closely associated with chloramination rather than free chlorination [26]. Chloramine disinfection and industrial pollution were later considered as the primary contributors to NDMA occurrence in drinking water, and the discharge of wastewater or reclaimed wastewater also posed a potential risk of nitrosamines [6]. Early studies concentrated on the impacts of chlor(am)ination and industrial pollution on the occurrence of NDMA, revealing their critical roles in the drinking water treatment processes [27, 28].

In the early 2000s, most nitrosamine researchers were from North America while some were from Western Europe, including the U.K., Netherlands, and Germany. The low occurrence of nitrosamines in Western Europe was due to the non-disinfection approach in their water treatment processes [29]. In 2008, WHO provided a guideline value for NDMA of 100 ng/L and later included it formally in

Table 3 Occurrence of nitrosamines in the water industry [23]

Year	Location	Substance & concentration	Possible sources
1994	Ohsweken, Southern Ontario, Canada	NDMA 10–115 ng/L	Not specified, but related to chloramination disinfection
1998	Eastern Sacramento County, California, U.S.	NDMA ~150 ng/L	Concern about the presence of NDMA contamination at an aerospace facility
1998	San Gabriel Basin, California, U.S.	NDMA 79–3000 ng/L	The well with 3000 ng/L NDMA was previously taken out of service due to trichloroethylene and perchlorate contamination
1999	Not specified	NDMA <10 ng/L	Related to disinfection processes
1999	Not specified	NDMA in wastewater	Proposed recycled water projects involving wastewater discharges and groundwater recharge
2000	Orange County, California, U.S.	NDMA ~30–40 ng/L	Nearby groundwater recharge operation involving injection of treated wastewater
2000	Los Angeles County, California, U.S.	NDMA 32–76 ng/L	Past production of chemicals used in the aerospace industry
2000	Los Angeles County, California, U.S.	NDMA ~30 ng/L	Resins used in water treatment for nitrate removal
2000	Los Angeles County, California, U.S.	NDMA 49–91 ng/L	Treated wastewater blended for use as groundwater recharge
2005	Not specified	NDPA and NDMA	Disposable resins being evaluated for drinking water treatment

the Guidelines for Drinking Water Quality (third edition) in 2011 [30]. Afterwards, more and more countries started to investigate, regulate, and control nitrosamines in drinking water, including but not limited to Australia, Japan, South Korea, China, and Brazil.

Over the past fifty years, research on nitrosamines has primarily focused on their sources, occurrence, formation mechanisms, and control techniques [6, 26, 31]. Building on these findings, subsequent research has expanded the understanding of the public health impacts of nitrosamines [32]. These studies will inform stricter water quality standards and regulation strategies.

2 Nitrosamine Occurrence

2.1 Global Occurrence

The U.S. and Canada were the first two countries conducting large-scale surveys on nitrosamines in drinking water. According to the Unregulated Contaminant Monitoring Rule 2 (UCMR2) in the U.S [33] and the Ontario Drinking Water Surveillance

Program (DWSP) datasets [34], NDMA is the most frequently detected nitrosamine in drinking water. About 10% of UCMR2 samples had NDMA detected, with lower occurrence of other nitrosamines such as NDEA, NPYR, NDBA, and NMEA. In addition, 48% of chloraminated water samples contained detectable levels of NDMA, compared to only 3.8% of chlorinated samples [35]. Data from the Ontario DWSP from 1998 to 2002 indicated a similar trend, with higher NDMA concentrations in water treated with chloramine than those treated with free chlorine [36]. The median concentrations of NDMA were 1.3 ng/L in 277 chloramination samples and < 1 ng/L in 1429 chlorination samples, respectively. Moreover, a significant increase in NDMA average concentration during the distribution process was observed. Other nitrosamines than NDMA turned out to be of minor importance. The state of California in the U.S. has published monitoring data on all occurrences of NDMA in drinking water from January 2011 through October 2023 [23].

Other developed countries have also conducted investigations into the occurrence of N-nitrosamines in drinking water. A survey in Japan in 2009 revealed that NDMA average concentrations in finished water (0.9 ng/L) were generally lower than in raw water (1.9 ng/L) in East Japan area since chloramine disinfection is rarely used in Drinking Water Treatment Plants (DWTPs) to reduce THM formation. Notably, one DWTP employing ozonation exhibited significantly higher NDMA concentrations in finished water [37]. The South Korean government has conducted a long-term water quality monitoring program covering 33 DWTPs since 2013. Based on this monitoring data, the concentrations of total nitrosamines in finished waters from Korean DWTPs showed a notable decrease from 2013 to 2020. Maximum concentrations of NDMA decreased from 11.4 ng/L to 5.2 ng/L after 2018, which might be associated with limiting the application of chloramines [38]. One survey on nitrosamines in eleven DWTPs and distribution systems in Spain revealed the maximum NDMA concentration of 4.2 ± 0.2 ng/L was detected in a water distribution system without pre-ozonation, which was much lower than WHO guidelines as well as the public health goals in many regions and countries [39].

Some developing countries reported higher concentrations of nitrosamines in tap water, especially in China. In a three-year study covering 23 provinces and 44 cities across China from 2012–2015, Bei et al. reported a higher occurrence of nitrosamines [40]. NDMA was the most prevalent among the nine commonly detected nitrosamines, occurring in 33% of finished water samples and 41% of tap water samples, especially in the Yangtze River Delta and the Pearl River Delta. The average concentrations of NDMA were reported as 11 ng/L in finished water and 13 ng/L in tap water. Another investigation reported the occurrence of nitrosamines in the finished water of 54 DWTPs from 30 cities across China [41]. The results showed that NDMA, NDEA, and NDBA were the most abundant in DWTPs, with detection frequencies of 57.4%, 53.7%, and 37% in finished waters, respectively. One investigation of the Pearl River water samples showed occurrences of different N-nitrosamine species, including NDMA (5.7 ng/L), NDEA (1.7 ng/L), NPYR (2.2 ng/L), NMOR (2.2 ng/L) and NDBA (4.9 ng/L) [42]. Another survey in the lower drainage of the Yangtze River in China reported a mean NDMA concentration of 35.83 ng/L and a mean NMEA concentration of 26.73 ng/L in finished waters [32].

Turkey [43], Brazil [44], and South Africa [45] also have shown increasing concern about nitrosamines and their precursors in water. These countries commonly use chlorination or chloramination for drinking water disinfection but lack proper infrastructure. Table 4 summarizes publications on the occurrence of nitrosamines in raw water, finished water of full-scale facilities, surface water, and groundwater. For example, the first detection of nitrosamines in Brazilian drinking water was conducted in 2021, with eighteen samples from 13 cities in the Metropolitan Region of Campinas (São Paulo, Brazil). NDMA was the most frequently detected nitrosamine (89%) and of the highest concentration (up to 67 ng/L). This area has significantly expanded its industrial and agricultural activities over the last few decades, which is of considerable relevance to nitrosamine presence.

2.2 *Factors Influencing Nitrosamine Formation*

According to early investigations on nitrosamines, chloramination is the key factor for nitrosamine occurrence; precursors and dosage of chloramine are also strong factors associated with nitrosamine occurrence. Laboratory experiments have shown that NDMA formation is directly linked to the dosage of chloramines [26], and that the chlorine-to-ammonia nitrogen ratio during the application of chloramine also impacts NDMA formation [51]. The results are consistent with almost every survey of nitrosamines worldwide. A survey of nitrosamines in 37 water treatment plants across North America indicated that a plant with chloramination, when shifting to free chlorination temporarily, showed an associated decrease in NDMA concentrations, averaging a reduction from 34 ng/L to 4 ng/L [52]. Additionally, samples from distribution systems with long residence times generally had significantly higher levels of NDMA compared to the finished water at the treatment plants.

The source water quality and water treatment processes significantly impact nitrosamine formation in the finished water. For instance, DWTPs in the same city, using different water sources such as Taihu Lake and the Yangtze River, have been shown to have different nitrosamine concentrations. Disinfection methods and advanced treatment techniques also play a crucial role [53]. The formation of nitrosamines is also related to certain treatment processes, such as ozonation and pre-chlorination. For example, the highest NDMA levels in Japan were detected in finished water samples collected from the Yodo River basin DWTP with ozonation, indicating the presence of ozone-reactive nitrosamine precursors in the water source [37]. Another study reported that the duration of pre-chlorination at the plant inversely impacted NDMA levels, with longer pre-chlorination times associated with reduced NDMA [54]. The treatment processes predominantly affect precursor levels, such as the use of specific chemicals like polyDADMAC, which increases precursor concentration, and the application of ozone or granular activated carbon

Table 4 Global reports of nitrosamine occurrence

Countries or regions	Time	Sampling information	Results	Detected NA species	Reference
Finished water					
U.S.	1999	32 DWTPs in California	Chlorination: The maximum concentration of 9.4 ng/L in raw water, 3.3 ng/L in finished water, and 2.5 ng/L in distribution system Chloramination: The maximum concentration of 3.9 ng/L in raw water, 18.3 ng/L in finished water, and 15.8 ng/L in distribution system Ozone-Chlorine Disinfection: The maximum concentration of 1.3 ng/L in raw water, 3.9 ng/L in finished water, and 6.8 ng/L in distribution system	NDMA	[36]
Canada	1994–2002	179 DWTPs in Ontario	Chlorination: Up to 40 ng/L in finished water and 66 ng/L in the distribution system. Chloramination: Up to 65 ng/L in finished water and 18 ng/L in distribution system	NDMA	[36]
Canada	2004	20 DWTPs in Alberta	Distribution system: 30% detection, <5–100 ng/L	NDMA, NPyr, NMor	[36]
U.K.	2008	41 DWTPs in England and Welsh	Finished water: up to 5.8 ng/L was detected at		[46]

(continued)

Table 4 (continued)

Countries or regions	Time	Sampling information	Results	Detected NA species	Reference
			3 out of 41 DWTPs		
U.S.	2008–2010	1200 DWTPs	Chloramination: Detectable levels of NDMA were found in 17% of samples and 25% of treatment plants, NDEA, NPyr, NDBA, and NMEA in less than 1% of samples	NDMA, NDEA, NPyr, NDBA, NMEA	[35, 47]
Japan	2008	Finished water from DWTPs in six areas across Japan	Concentration ranged from N. D.-2.2 ng/L in Summer and N. D.-10.6 ng/L in Winter	NDMA	[37]
China	2012–2016	44 cities or towns, in 23 provinces	NDMA: N.D.-189 ng/L (11 ng/L average); NDPA: N.D.–79 ng/L; NMOR: N.D.–40 ng/L	NDMA: 33% of finished water and 41% of tap water NDPA: 16% of finished water and 31% of tap water NMOR: 81% of finished water and 74% of tap water	[40]
Spain	2018	11 DWTPs	Maximum NDMA concentration of 4.2 ng/L in finished water	NDMA, NDEA, NDBA	[39]
Brazil	2019	The 13 most representative cities throughout the Metropolitan Region of Campinas, in the State of São Paulo, Brazil	NDMA (8–67 ng/L), NPIP (12–20 ng/L), NPYR (7–33 ng/L), NDEA (5–20 ng/L), NDPA (7–17 ng/L), NDBA (4–18 ng/	NDMA: 89%; NPIP: 78%; NPYR: 72%; NDEA: 56%; NDPA: 44%; NDBA: 39% NMEA: 17%;	[44]

(continued)

Table 4 (continued)

Countries or regions	Time	Sampling information	Results	Detected NA species	Reference
			L), NMEA (7–14 ng/L)		
South Korea	2013–2020	Finished water of 33 DWTPs in South Korea	NDMA (0.3–11.4 ng/L), NMEA (0.1–3.6 ng/L, NDEA (0.1–8.0 ng/L), NMOR (0.1–9.46 ng/L), NDBA (0.15–15.3 ng/L)	NDMA: 44%; NMEA: 10%; NDEA: 27.8%; NMOR: 28.5%; NDPA: 0%; NDBA: 9.73	[38]
Raw water, surface water, and groundwater					
Japan	2008	Raw water of DWTPs in six area across Japan	NDMA in summer: ~50% detection, up to 2.6 ng/L NDMA in winter: ~30% detection, up to 4.3 ng/L	NDMA	[37]
Japan	2009–2010	23 Groundwater and 18 River water	NDMA in groundwater: <0.5–5.2 ng/L (Median: 0.9 ng/L) NDMA in river water: <0.5–3.4 ng/L (Median: 2.2 ng/L)	NDMA	[48]
Korea	2011	27 river samples in the Nakdong River	Nitrosamines in river water: 26–5,180 ng/L	NAs	[49]
China	2012	12 Groundwater on the Jialu River basin in Central China	Secondary amines in shallow groundwater: 360–4,380 ng/L NAs in shallow groundwater: ND–101.1 ng/L	NDMA NDEA NDBA	[50]

N.D. Not Detected

treatments, known for their effectiveness in precursor removal. A more comprehensive discussion on these aspects is provided in the precursor section.

In addition to the direct impact of precursors, property changes in precursors caused by pre-oxidation can also affect the formation of nitrosamines [55]. On one hand, the variation of water quality indicators such as TOC and UV_{254} after different pre-oxidation processes show differences that can be attributed to the differing oxidation ability of various oxidants. On the other hand, different reaction patterns and sites during oxidation can result from the different characteristics of precursors in terms of their MW, polarity, and electric charge. These changes subsequently influence the formation of DBPs. Furthermore, inorganic ions present in source water have been found to exert an influence [56]. The presence of bromide in source water has a dual effect on NDMA formation which may cause varying effects on NDMA formation [57]. On one hand, bromide reacts with chloramines to form unstable substances like NH_2Br and NHBrCl, which react more readily with precursors to form NDMA. On the other hand, bromide can transform into other halogenated by-products, competing with NDMA formation.

The formation of nitrosamines in distribution systems fluctuates to some extent due to the effects of pipe materials, orthophosphate, flow condition, and bioactivity [58]. Metallic ions in distribution systems such as copper can also influence NDMA formation in tap water as they act as catalysts [59].

3 Formation Mechanism

NDMA and other nitrosamines are carcinogenic DBPs formed during chloramination and chlorination processes in DWTPs. The formation mechanism is highly dependent on the amine precursors and disinfectants that may be present.

3.1 Nitrosation Pathway

Nitrosation predominantly occurs in acidic conditions, where a nitrogen atom in nitrogenous organic compounds is substituted with a nitroso (–N=O) group. This reaction is a major pathway for nitrosamine formation in foods, involving nitrite salts as nitrosating agents. Organic nitrogen groups like secondary and tertiary amines are activated in this process. For example, dimethylamine (DMA) upon nitrosation yields NDMA. This mechanism is significant in the formation of NDMA from nitrites in cured vegetables and fishes, as well as being a primary method for laboratory synthesis of NDMA. However, the formation of NDMA in water bodies cannot be solely attributed to the presence of nitrites. Experiments have shown that under natural water conditions (pH = 7), a reaction between 100 μM nitrite and 100 μM DMA for 10 h produces less than 10–12 ng/L of NDMA. This indicates that nitrosation at neutral pH is relatively slow, hence not the main pathway for NDMA

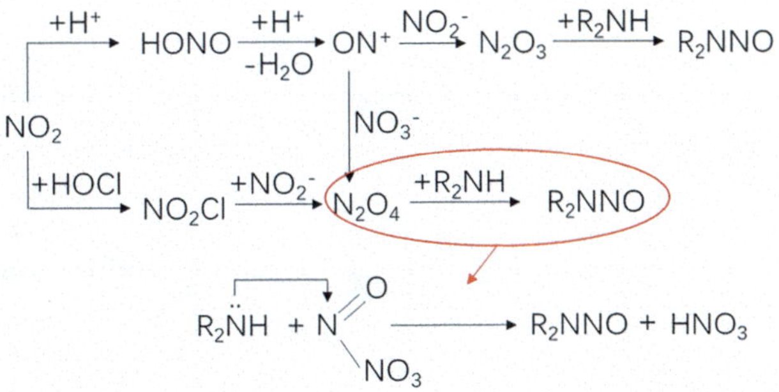

Fig. 2 NDMA formation in the presence of nitrite and free chlorine

formation in drinking water treatment processes. Nevertheless, in disinfection processes, free chlorine can catalyze and promote the nitrosation pathway, leading to nitrosamine formation, as illustrated in Fig. 2.

3.2 Formation During Chloramination

The mechanism of NDMA formation during chloramination is complex, involving multiple chemical reaction pathways. Early research proposed a potential mechanism for NDMA formation through a nucleophilic substitution reaction between unprotonated secondary amines (such as DMA) and monochloramine [27, 60]. As shown in Fig. 3a, this process initially forms an intermediate, unsymmetric dimethylhydrazine (UDMH), which is then oxidized by monochloramine to form NDMA. This process is known as the monochloramine hypothesis.

However, as research progressed, the monochloramine hypothesis encountered difficulties in explaining experimental data. For instance, at least 2 orders of magnitude higher NDMA concentrations formed following NH_2Cl application to DMA than to equivalent concentrations of UDMH. Additionally, the monochloramine pathway could not account for the oxygen source in NDMA [51].

Considering that the auto-decomposition of monochloramine can lead to the formation of dichloramine, both monochloramine and dichloramine often coexist in real chlorination conditions. Researchers proposed the dichloramine hypothesis [62]. This hypothesis suggests that NDMA formation is actually induced by a nucleophilic substitution reaction between DMA and dichloramine, forming Cl-UDMH as an intermediate (see Fig. 3b). Compared to the monochloramine hypothesis, the dichloramine hypothesis more adequately explains the experimental observations of NDMA formation, especially the phenomenon of increasing NDMA production with rising dissolved oxygen concentrations.

Liu et al. proposed a four-step formation pathway for NDMA involving nucleophilic substitution by chloramine, oxidation, and dehydration followed by nitrosation [63]. They investigated the formation mechanisms of NDMA from ranitidine and trimethylamine (TMA) during chloramination by using density functional theory (DFT), and proposed NDMA formation pathways from ranitidine

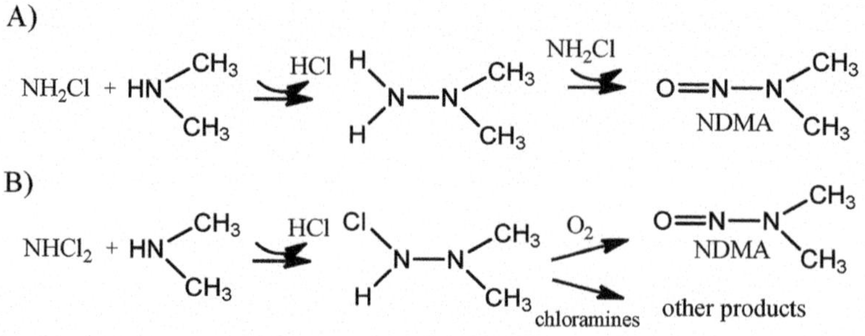

Fig. 3 (**a**) Initial UDMH pathway by monochloramine hypothesis; (**b**) revised Cl-UDMH pathway by dichloramine hypothesis [61]

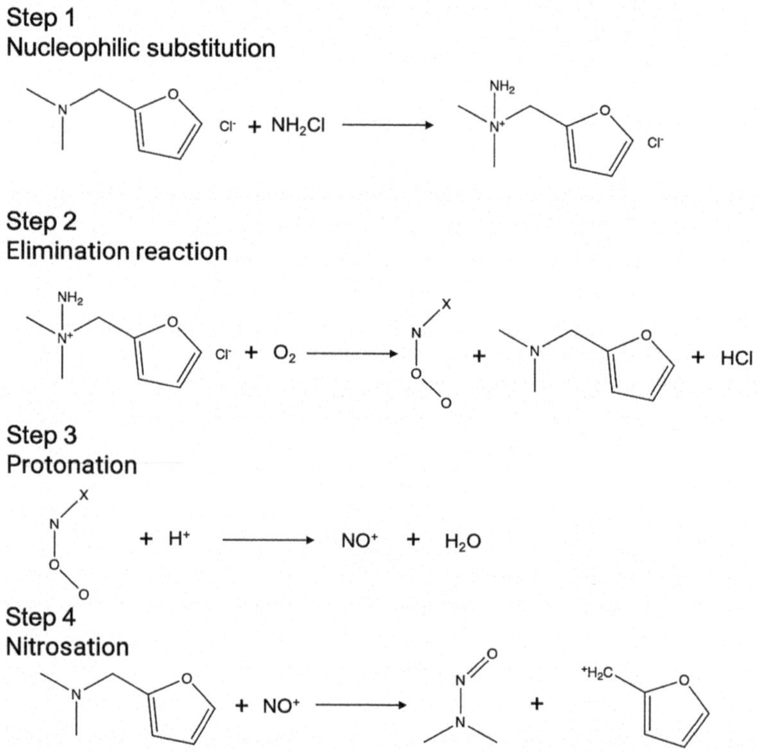

Fig. 4 Proposed NDMA formation pathways from the ranitidine model during chloramination (adapted from Liu et al. [63])

during chloramination (see Fig. 4). Additionally, the DFT calculation results suggested that the nitrosation reaction was the rate-limiting step and determined the NDMA yield for tertiary amines.

3.3 Formation During Ozonation

Ozonation can also contribute to NDMA formation in DWTPs, particularly from chloramine-reactive precursors that commonly possess DMA structures. Some organic compounds featuring DMA or -NH_2 groups separated by leaving groups like -SO_2 or -CO_2 can form NDMA during ozonation, even though pre-ozonation is considered as an effective method for controlling NDMA precursors in subsequent chlorination or chloramination processes [64].

Ozonation of DMA-containing compounds directly yields NDMA. The formation rate of NDMA increases with pH, suggesting a mechanism beyond the formaldehyde-catalyzed nitrosation pathway. Hydrazine, a significant nitrogen-containing inorganic precursor, plays a vital role in NDMA formation, although it may be rapidly consumed by ozone. The reaction of DMA with ozone primarily includes three stages: 1) Partial oxidation of DMA to form hydrazine; 2) Reaction of hydrazine with DMA to produce UDMH; 3) Further rapid reaction of UDMH with ozone to form NDMA, with the second step being rate-limiting [65, 66] (Fig. 5).

As research advances, an increasing variety of new ozone-reactive NDMA precursors have been identified [67], such as 3-(dimethylamino)-1-propylamine [68], and metformin [69]. Additionally, some organics do not directly form NDMA with ozone but do so in the presence of bromide ions or ammonia. For instance, the reaction of *N,N*-dimethylsulfamide (DMS) with ozone forms NDMA in the presence of bromide ions [70], while dimethylformamide (DMF) reacts with ozone to form NDMA only when ammonia is present [71]. The chemical yield of NDMA varies significantly among different ozone-reactive precursors. Some secondary amines, dyes, and their related compounds have very low molar conversion rates to NDMA, while certain hydrazines (like UDMH) and sulfonamides can produce more NDMA upon ozonation, with molar yields exceeding 50% [72].

It is noteworthy that precursors identified in the ozonation pathway do not necessarily exhibit high NDMA conversion rates during chloramination. However, the ozonation products of these precursors might continue to form NDMA through chloramination pathways due to changes in their polarity [73]. Therefore, DWTPs employing ozonation as a pre-oxidation process need to carefully assess the levels of NDMA formation in subsequent disinfection processes.

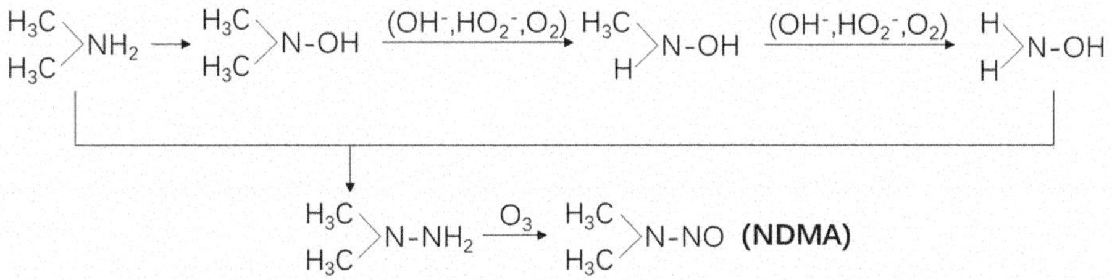

Fig. 5 The pathway of nitrosamine formation via ozonation (redrawn from Yang et al. [66])

3.4 Formation by Other Oxidants

Other (pre)oxidation processes involve the generation of NDMA. Gan et al. reported that the use of ClO_2 has the potential to form NDMA in waters containing precursors with hydrazine moieties [74]. The hydrolytic release of UDMH is a crucial step, supported by gas chromatography-mass spectrometry analysis.

The ability of potassium permanganate ($KMnO_4$) to form NDMA has been verified. Test results indicate that the reaction of KMnO4 with DMA in aqueous solutions produces NDMA, primarily through nitrosation. The experimental reaction mixture, comprising phosphate buffer, DMA, and permanganate without ammonium ions or chlorine, supports this pathway by excluding alternative NDMA formation routes via unsymmetrical dimethylhydrazine or chlorodimethylhydrazine. Consequently, NDMA likely forms from DMA oxidation by permanganate through a multistage process: initially generating nitrate, nitrite, and formic acid, followed by the formation of a nitrosating agent that reacts with DMA to produce NDMA, a reaction facilitated by MnO_2 suspension produced during DMA oxidation [75].

The investigation of NDMA formation during peroxymonosulfate (PMS) oxidation without activation reports a distinct mechanism from that observed with ozonation and chloramination. The degradation of TMDS by PMS predominantly involves side chain cleavage, dealkylation, and oxygen addition. NDMA generation pathways are primarily through UDMH formation and oxygen addition processes. The crucial steps for NDMA production include the cleavage of amide nitrogen in oxygen addition products and primary amine nitrogen in UDMH [76].

4 Nitrosamine Precursors

Characterizing nitrosamine precursors is fundamental for controlling their formation in water. The sources of these precursors include natural organic matter, industrial and domestic wastewater, chemicals used in water supply processes, and microbial metabolic products [1, 28, 77, 78]. A significant portion of these precursors remains unidentified, making the fractionation and characterization of precursors a promising tool to predict nitrosamine formation.

4.1 NOM

Water sources rich in NOM, including humic substances and algal matter, can lead to a high yield of NDMA during chloramination. Chen and Valentine reported an NDMA formation potential (NDMAFP) of 112 ng/L in Iowa River water with a dissolved organic carbon (DOC) content of 3.4 mg/L [79]. Li et al. reported that the NDMAFP from extracellular and intracellular algal organic matter of Microcystis in

Taihu Lake was 9.6 and 18.9 ng/mg C, respectively, contributing to NDMA concentration of 56.73 and 28.95 ng/L at typical DOC levels [80]. Algal-derived NDMA precursors likely lead to NDMA concentrations in the finished water of over 10 ng/L, exceeding the notification level in California. Even though intracellular organic matter (IOM) has more protein-like substances than extracellular organic matter (EOM), its NDMAFP is about half that of EOM, suggesting that NDMA precursors are more prevalent in EOM.

4.2 *Artificial Sources*

Domestic wastewater often contains a high NDMAFP, likely due to the nitrogenous organic compounds that are present. Besides the direct discharge of municipal wastewater, the effluent organic matter from wastewater treatment plants (WWTPs) is a significant source of NDMA in downstream surface waters [77, 81]. Zeng and Mitch found that many NDMA precursors in U.S. domestic wastewater originated from laundry wastewater, particularly from dyes [82]. Chen et al. reported an NDMAFP of about 1000 ng/L in secondary treatment effluent and about 300 ng/L in tertiary treatment effluent [83].

Some industries, such as chemical, pharmaceutical, pesticide, paper, leather, and dye industries, utilize amine-containing compounds as reagents. In turn, they discharge wastewater rich in nitrosamine precursors. Wang et al. found thousands of ng/L of NDMA precursors in the influent and effluent of wastewater treatment plants primarily treating industrial wastewater while hundreds of ng/L of NDMA precursors were detected in domestic wastewater treatment plant effluents [84].

Pharmaceuticals containing DMA or diethylamine structures also act as NDMA precursors. For example, Shen and Andrews measured the NDMA yield of 20 pharmaceuticals following chloramination [85]. The results showed ranitidine rendered the highest conversion (89.9–94.2%), followed by doxylamine (8.0–9.7%), sumatriptan (6.1%), chlorphenamine (5.2–5.5%), nizatidine (4.5–4.8%), diltiazem (2.1–2.6%), carbinoxamine (1.0–1.4%), and then tetracycline (0.8–1.2%). The concentrations of ranitidine and other pharmaceuticals in drinking water sources are low and so NDMA concentrations resulting from chloramination of it during treatment should also be low. However, taking ranitidine with glasses of chloraminated drinking water could form NDMA concentrations of potential human health concern. Therefore, further researches on the chloramination of ranitidine, sumatriptan, and other pharmaceuticals (e.g., methadone) continue to be conducted to better understand the reaction kinetics and mechanisms and any potential health concerns [86–89].

4.3 *Reagents and Materials Used in Water Industry*

Some reagents and materials used in water treatment plants have DMA functional groups and can be potent NDMA precursors when reacting with disinfectants. This includes the quaternary ammonium-based flocculants and anion exchange resins. The rubber pipes and sealing rings used in water systems also contain some nitrosamine impurities during production [90].

In the U.S., many water treatment facilities have traced their primary source of NDMA to the use of polyammonium salt flocculants. Najm and Trussell reported that ion-exchange resins containing trimethyl or dimethyl-ethanol quaternary amine functional groups can form NDMA upon reaction with disinfectants, and that resins with triethyl and tripropyl groups may be precursors to NDEA and NDPA [26]. Park et al. discovered that polychlorohydrin dimethylamine (PA) and polydiallyldimethylammonium chloride (PolyDADMAC) flocculants reacted with chloramine to produce significant levels of NDMA, particularly noting that PA demonstrated a higher NDMA formation potential than PolyDADMAC [91]. Krasner et al. conducted a source-to-tap evaluation of the origin and fate of NDMA precursors at 21 full-scale drinking water plants and concluded that wastewater in water sources, polymers (PA and polyDADMAC) in coagulation and biofilter are known sources of precursors [92]. This NDMA yield was found to be closely linked to the degree of polymer degradation and the amount of DMA liberated. The replacement of rubber gaskets in storage tanks led to the detection of not only NDMA but also NPIP in the water [93, 94].

In the last thirty years, researchers have identified dozens of chemicals as nitrosamine precursors, such as secondary amines, pesticides, pharmaceutical and personal care products, and certain chemical reagents. However, these specific compounds are only present at trace concentrations in source water. Many precursors have a low molar yield to nitrosamines less than 5% except for ranitidine. Therefore, these chemicals can only explain a small proportion of the whole formation potential of nitrosamines in water sources. Industrial and municipal wastewater are known to have a high potential to generate NDMA but the components as nitrosamine precursors have not been clearly identified.

4.4 Microbiological Metabolism of Amino Acids

Amino acids, the most abundant nitrogenous organic material deriving from wastewater and microbes, have been widely detected in source water at much higher concentrations than the aforementioned identified nitrosamine precursors. Despite this, amino acids usually fail to produce nitrosamines via nitrosation pathways and do not react with chloramine to form nitrosamines under gentle conditions. Actually, 19 amino acids have been tested as unable to contribute nitrosamine formation potential (NDMAFP) with chloramination [95]. A recent study introduced a new mechanism that amino acids might be transformed into nitrosamine precursors through microbial metabolism [78]. The aerobic metabolism of amino acids produced up to 0.12 nmol/L (peak) or 0.09 nmol/L (average) NDMAFP. Correlating with the comprehensive survey of NDMAFP in aquatic environments [40], this new mechanism accounts for approximately 38% (median) and 27% (average) of the observed NDMAFP. This mechanism has been further verified by a city-level field study that the amino acids metabolism in one lake contributed to 31% of NDMAFP and other sources are river transportation from upstream and the discharge of local wastewater treatment plant effluent [81].

4.5 Characterization of Nitrosamine Precursors

Formation Potential (FP) tests indirectly represent the total amount of nitrosamine precursors in water. This method involves the slow oxidation of precursors to nitrosamines by adding excess disinfectant until equilibrium is reached [28]. However, more specific characterization is needed to understand these complex precursors rather than just their potential total NDMA formation. Understanding the nature of these precursors enables the development of targeted treatment strategies to reduce or control NDMA formation.

The applications of resin fractionation, the Polar Rapid Assessment Method (PRAM) and molecular weight fractionation through ultrafiltration or gel permeation chromatography have also enhanced our understanding of nitrosamine precursors [83, 96]. As mentioned before, the characterization of specific individual precursor chemicals does not fully reflect the NDMA formation potential in water, however, these model precursors can be used to test the performance of various characterization methods.

Chen and Valentine discovered that the NDMA formation potential in Iowa River water was higher for hydrophilic organic matter than hydrophobic, with basic components being more significant than acidic ones [79]. Wang et al. found similar trends in the NDMA formation potential of different organic fractions in Chinese rivers [97]. However, resin enrichment methods have limitations, including complex operations, low recovery rates of organic matter, and potential alterations in organic matter properties due to pH adjustments during processing.

Developed by Rosario-Ortiz et al. [98], PRAM is a novel method for analyzing the polarity distribution of organic matter in water. It uses various solid-phase extraction cartridges to quantitatively describe the polarity distribution of organic matter in water. This method has also been applied to study nitrosamine precursors in wastewater treatment plants [83] and source waters [96]. PRAM is more efficient, less labor-intensive, and provides a more accurate representation of the natural state of organic matter with more reliable quality control when compared to the traditional resin fractionation methods.

Methods like ultrafiltration and gel permeation chromatography help fractionize organic substances based on molecular weight. Studies have shown that substances with molecular weights less than 1,000 Da have a significant NDMA formation potential, suggesting that small molecule substances are the main nitrosamine precursors in natural waters. Research by Mantas et al. [99] and Wang et al. [97] showed that small molecule precursors (molecular weight $<$ 1,000 Da) play a significant role in NDMA formation.

5 Control of Nitrosamines in Drinking Water Treatment Plants

According to the known formation mechanisms and characteristics of nitrosamines, there are two main sources of nitrosamines in drinking water: the nitrosamines present in the raw water and the transformation of precursors during water treatment processes. Nitrosamine control strategies include source control, process control, and end-of-pipe removal. Source control focuses on the removal of precursors; process control reduces nitrosamine formation by adjusting disinfection methods and process parameters; and end-of-pipe control applies physical and chemical means to remove the formed nitrosamines. Since nitrosamines are much more recalcitrant to normal treatment processes than their precursors, source control and process control are the more appropriate routes for controlling nitrosamines.

5.1 *Removal of Precursors before Municipal Disinfection*

Removal of DBP precursors is the most effective method to control the DBPs formation in the finished water. The nitrosamine precursors, including specific precursors such as DMA, certain pharmaceuticals and personal care products (PPCPs), certain pesticides, and nitrogen-containing organics with dialkylamine, tertiary amine or quaternary amine functional groups, can react with chloramine or chlorine to form nitrosamines. Removing these nitrosamine precursors before disinfection can reduce the formation of nitrosamines during the chloramination process.

Conventional processes are usually more effective at removing particles, organic components with large molecular weight (MW), and hydrophobic components from the source water. However, due to the hydrophilic and low molecular weight characteristics of nitrosamine precursors, their removal is fairly limited by conventional processes. Advanced treatment processes with O_3-BAC show better removal efficiency for nitrosamine precursors, being considered as an effective process. Moreover, cation exchange can also effectively remove nitrosamine precursors, but they are less commonly used in water treatment plants due to cost and feasibility. Below is the technique performance of the removal of NA precursors by conventional and advanced treatment processes in DWTP.

Conventional processes, i.e., coagulation, sedimentation, and filtration, have limited effectiveness in removing the soluble nitrosamine precursors. When ferric chloride or alum coagulation are applied, NDMAFP removal is less than 10% [96]. However, when quaternary amine-based coagulants such as polyDADMAC are applied [100], the concentration of nitrosamines in the finished water will increase significantly.

Regular oxidation treatments in DWTPs including chlorination, ozonation, permanganate oxidation [101], hydrogen peroxide oxidation [102], and chlorine dioxide oxidation [103] have varying performance in removing precursors. Particularly,

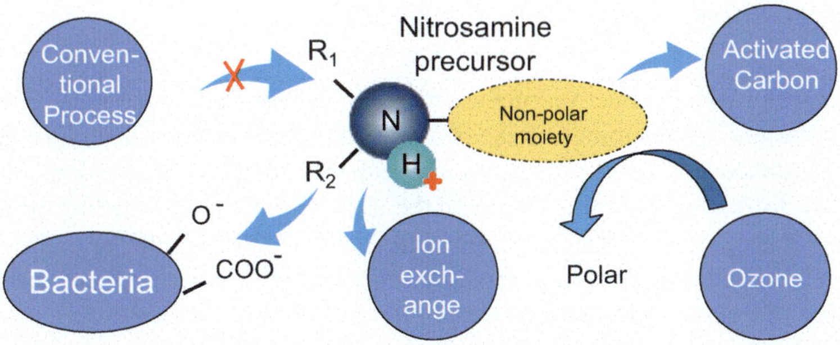

Fig. 6 Mechanism of nitrosamine precursor removal by different water treatment processes

ozonation exhibits a higher reaction rate constant with amines and other nitrosamine precursor compounds and is therefore perceived as a superior oxidant for their removal. It has been reported that pre-ozonation can decrease NDMA formation by 49.7% during subsequent chloramination. In contrast, pre-chlorination and permanganate pre-oxidation achieve lesser reductions of 33.9% and 24.2%, respectively [55]. Another research revealed that ozonation at 1 mg/L can eliminate NDMAFP, NDEAFP, and DOC by 26%, 42%, and 20%, respectively [96].

It is important to recognize that the formation of nitrosamines has been found during the pre-oxidation processes. The efficiency of this process is influenced by various source water characterization, including DON, ammonia nitrogen, the electrochemical characteristics of organic matter, and the specific oxidation conditions [104]. Consequently, it would be necessary to conduct a pilot test to ascertain the most effective parameters for the pre-oxidation process.

Activated carbon adsorption serves as a highly effective technology for eliminating precursors for many DBPs. Adsorption via GAC and PAC demonstrates a superior percentage on nitrosamine formation potential than that of the regulated DBP precursors such as THMs and HAAs [105]. Consequently, O_3-BAC is recognized as the best available technology for mitigating nitrosamine precursors or restraining their formation after disinfection. Using PRAM to determine the organic matter properties after each unit process, Liao et al. found that activated carbon preferentially adsorbs non-polar nitrosamine precursors, ozonation transforms some non-polar precursors into the polar ones, while the microbial population prefers to interact and degrade the positively charged precursors [96] (Fig. 6).

High-pressure membrane filtration systems, such as reverse osmosis (RO) and nanofiltration (NF), can remove nitrosamine precursors effectively [106]. RO has been widely applied in desalination plants all over the world and for direct potable reuse in the United States. NF has been put into practice in DWTPs in Europe and China, thanks to lower energy consumption and a higher rejection of calcium and magnesium when compared to RO.

5.2 Alternative Disinfection Processes

Optimization of the disinfection process and parameters also benefits the control of DBPs, including nitrosamines. Research indicates that nitrosamine formation is more pronounced during chloramination compared to free chlorination, resulting in a higher formation potential. Among the chloramines, dichloramine presents a higher nitrosamine yield than monochloramine. Thus, the restraint of dichloramine formation by prepared monochloramine and elevated pH control proves to be pivotal in reducing nitrosamine formation in chloramination [62]. Implementing a sequential dosing strategy, i.e., the sequential chlorination with short-term chlorination and subsequent chloramination was designed to restrain the formation of THMs and HAAs, while maintaining sufficient inactivation efficiency on pathogens [107]. It has been proven to be able to oxidize the nitrosamine precursors quickly by short-term free chlorine. Thus, the final NDMA formation is successfully restrained.

Actually, the nitrosamine formation varies with the characterization of precursors present in different water sources. This variability affects the removal of nitrosamine via sequential chlorination. Li et al. examined the reactions of free chlorine disinfection, chloramine disinfection, and sequential chlorination (initial chlorination followed by chloramination) with four NDMA precursors – dimethylamine, imidacloprid, glyphosate, and acetamiprid. They found that sequential chlorination reduced NDMA formation from dimethylamine and imidacloprid more effectively than chloramination alone, but increased NDMA formation from glyphosate and acetamiprid [108].

While removing nitrosamine precursors and optimizing disinfection processes is essential for nitrosamine formation control, it can lead to higher concentrations of other DBPs [109]. The trade-off of these outcomes is a critical issue. Considering practical water treatment scenarios, prioritizing precursor removal at the source may be more effective than optimizing the disinfection process solely for nitrosamine control. This approach addresses multiple requirements such as microbial inactivation, control of both regulated and unregulated DBPs, and maintaining residual disinfectant in the distribution system.

5.3 Direct Removal of Nitrosamines

The formed nitrosamines are quite recalcitrant to normal water treatment processes. Although activated carbon adsorption and ozonation are practical for nitrosamine precursors removal, they cannot eliminate nitrosamines effectively due to their strong stability, high hydrophilicity, and high solubility. Fortunately, ultraviolet technology can destroy nitrosamines under high dosages. It is reported that UV irradiation at 1000 mJ/cm^2 can decrease NDMA concentrations by an order of magnitude, yet it only removes 30% of the NDMA precursors. This dosage is much higher than that for pathogen inactivation (40 mJ/cm^2) which will hinder the practice of using UV to eliminate nitrosamines. Furthermore, the products of UV irradiation on nitrosamines in WTPs are amines which may lead to the partial rebound of nitrosamines in the chloramine-treated distribution system.

In terms of water reuse, RO is the primary method for removing trace organic pollutants in reclaimed water. The rejection of NDMA by RO has been reported to be low and highly variable, between 14% and 78% [110, 111]. However, the combination of UV after RO is successful in eliminating NDMA, achieving a significant low level to less than 2 ng/L. Additionally, it is important to consider the potential permeation of small molecule precursors like DMA through the RO membranes, as this may contribute to the reformation of NDMA [112, 113]. Low pH of RO effluent benefits dichloramine existence and the pH elevation after RO treatment to maintain chemical stability in the distribution system neutralizes the nitrosamine precursors. Either situation will favor the regrowth of NDMA in RO effluent.

Chuang et al. conducted a pilot test and compared several wastewater reuse treatment trains for their performance on NDMA control [114]. The results showed that the O_3/BAC/MF/RO train followed by UV/H_2O_2 treatment was the most effective for controlling N-nitrosamines and their precursors. The median NDMA concentrations measured after chloramination were lowest for the sequential operation of O_3/BAC/MF/RO (~14 ng/L), while the concentrations were similar (~30 ng/L) for the parallel-operated O_3/BAC and MF/RO effluents. Further AOP treatment of UV/H_2O_2 or UV/HOCl AOPs reduced NDMA and NMOR to below detection limits and reduced NDMA precursors from 30 ng/L to 10 ng/L.

6 Nitrosamine Regulation

The widespread occurrence of nitrosamines as DBPs in drinking water has prompted international and national attention to their regulation. Regulations typically include mandatory requirements and health guidelines to ensure compliance. Some countries and regions have established their own national regulations for nitrosamines, often referencing and adapting the WHO guidelines to fit their specific situations.

6.1 WHO Guidelines in Drinking Water

The Guidelines for Drinking-Water Quality (GDWQ) by the WHO are one of the organization's longstanding normative publications, first published in 1958. GDWQ serves as an international reference point for establishing national or regional regulations and water safety standards. They are primarily addressed to water and health regulatory bodies, policymakers, and their advisors, aiming to assist in developing national standards. GDWQ is also widely used as a source of information on water quality and health, as well as effective management practices [30].

Since 2008, NDMA has been included in the WHO Guidelines for Drinking-Water Quality, with a guideline value of 100 ng/L. In the second addendum to the third edition of the Guidelines, published in 2008, a guideline value of 0.0001 mg/L was established. This value was carried forward to the fourth edition of the Guidelines, published in 2011.

According to WHO statistics from 2018 [115], NDMA is among the least specified organic parameters, with only three countries and regions specifying a mandatory value. In the second edition of the overview in 2021, this number increased to five. Among these, three countries reference the WHO's GDWQ, while two have standards stricter than GDWQ. As of 2023, national and regional standards for nitrosamines in water have been further updated. Countries with standards stricter than the GDWQ include the Netherlands (12 ng/L) Canada (40 ng/L) and South Korea (70 ng/L), while China, Japan, Australia, Singapore, and the UAE have NDMA limits in their water quality standards that are the same as the WHO's GDWQ. It is noteworthy that considering monitoring capabilities and treatment technologies, some regulations are non-mandatory, but this does not diminish the profound recognition of the health threats posed by nitrosamines.

6.2 National Regulations in Drinking Water

In the U.S., water quality standards derive from the Safe Drinking Water Act (SDWA), enforced by the EPA. States have the flexibility to establish stricter criteria or include additional test items than the EPA mandates. The National Primary Drinking Water Regulations (NPDWR) comprise the baseline standards and treatment methods for public water systems, designed to safeguard public health by managing pollutants. Moreover, the EPA, under the SDWA, must evaluate unregulated contaminants and may later introduce regulations. It publishes the Contaminant Candidate List (CCL) every five years to deliberate on whether to regulate at least five substances listed. The CCL identifies contaminants not regulated by the NPDWR but foreseen in Public Water Systems (PWS), flagging them for potential regulation. Integral to the SDWA's regulatory framework are the CCL and the Unregulated Contaminant Monitoring Rule (UCMR), which generates supportive data for determining public health interventions by detecting contaminants in PWS. This monitoring, both assessment and screening, applies to PWS serving over 10,000 people and a representative subset of smaller systems, thereby guiding regulatory decisions and furnishing future CCL updates. Notably, the EPA broadcast CCL 3 on October 8, 2009, including NDMA, NDEA, NDPA, NDPhA, and NPYR, and CCL 4 on November 17, 2016, first listing NDBA. Most recently, the EPA released CCL 5 on November 14, 2022, maintaining its focus on nitrosamines without alteration (Table 5).

The European Union's current benchmark for drinking water quality is the Drinking Water Directive, updated in 2020. Some European countries, including

Table 5 Regulation process on nitrosamines in drinking water in the U.S.

Year	Regulatory action	Substance & concentration	Regulatory body or source
1998	Established an action level based on a de minimis cancer risk level	NDMA 2 ng/L	CDHS (now CPDH)
1999	Revised the action level to accommodate studies on NDMA's production in drinking water treatment	NDMA 20 ng/L	CDHS
2002	Requested a public health goal (PHG) for NDMA	NDMA	CDHS
2002	Posted results of studies on NDMA's production in drinking water treatment, and a revised notification level	NDMA 10 ng/L	CDHS
2004	Initiated a risk assessment for NDMA that will lead to a PHG	NDMA	OEHHA
2004	Added a notification level for another nitrosamine	NDEA 10 ng/L	CDHS
2005	Added a notification level for N-nitroso-n-propylamine (NDPA)	NDPA 10 ng/L	CDHS
2006	Released a draft public health goal (PHG) for NDMA	NDMA 3 ng/L draft PHG	OEHHA
2006	Established the PHG for NDMA	NDMA 3 ng/L	OEHHA
2007	The drinking water program became part of the new California Department of Public Health	Not specified	Transition to CDPH

CDHS California Department of Health Services, *CPDH* California Department of Public Health, *OEHHA* Office of Environmental Health Hazard Assessment

the Netherlands regulated NDMA in their national water standards and Germany gave a recommendation with the low criteria of around 10 ng/L [116].

The Australian Drinking Water Guidelines health limit for NDMA is 100 ng/L, which is based on health considerations. The guideline encourages adopting action to reduce NDMA but must not compromise disinfection, as non-disinfected water poses a significantly greater risk than NDMA.

The drinking water quality in China is regulated by the Sanitary Standards for Drinking Water (GB5749–2022). This new revision aligned with WHO standards, introducing an NDMA threshold of 100 ng/L and adding it to the appendix. Although not compulsory, these measures indicate a recognition of nitrosamine risks and guide future regulatory development. Moreover, some pioneering cities in China including Shanghai, Shenzhen, Suzhou, and Zhangjiakou have announced stricter local standards, setting the NDMA limit in drinking water at 100 ng/L, to address public demand for better water quality.

In other Asian countries, South Korea regulates NDMA in drinking water with criteria of 70 ng/L [38]. Japan's Ministry of Health, Labour and Welfare continuously evolves its drinking water standards based on Ordinance No. 101, including NDMA in the "need further study" category, also with a 100 ng/L limit (Table 6).

Table 6 Global regulation on nitrosamines in drinking water

Regions	Type	Years	Criteria (ng/L)
Ontario, Canada	MAC	2002	9
Massachusetts, U.S.	GV	2004	10
California, U.S.	PHG	2006	3
WHO	GV	2008	100
California, U.S.	NL	2009	10
Japan	GV	2010	100
Canada	MAC	2011	40
Australia	MAC	2011	100
Singapore	MAC	/	100
Hongkong, China	MAC	/	100
Shanghai, China	MAC	2018	100
The Netherlands	MAC	2018	12
UAE	MAC	2019	100
Zhangjiakou, China	MAC	2019	100
Shenzhen, China	MAC	2020	100
Suzhou, China	MAC	2021	100

MAC Maximum Allowable Concentration, *GV* Guideline Value, *PHG* Public Health Goal, *NL* Notification Level

6.3 Regulation of Nitrosamines in Wastewater/Reclaimed Water

In the United States, California requires that when reclaimed water is reused as drinking water, it must meet the U.S. EPA drinking water quality standards. In particular, the NDMA concentration should not exceed 10 ng/L.

In Australia, the data on chemicals detected in secondary treated sewage indicate exceedances for nitrosamines. For example, the maximum concentration of NDMA is 550 ng/L. These guidelines are more stringent for reclaimed water than for conventional drinking water. The Australian Guidelines for Water Recycling (Phase 2) have a guideline value of 10 ng/L for NDMA. NDEA is also regulated in Australia's recycled water, with a guideline value of 10 ng/L.

7 Summary

This chapter has traversed the complex landscape of nitrosamines, from their fundamental properties and health implications to the intricacies of their formation and the global regulatory response. We have shown how numerous researchers established that nitrosamines, particularly NDMA, pose a significant health risk due to their carcinogenic potential, necessitating their close monitoring and control in water systems.

Our exploration of the formation mechanisms highlighted the various pathways through which nitrosamines can arise, especially in the context of water treatment processes. We recognized the role of natural organic matter, artificial sources, water treatment chemicals, and microbiological metabolism as key precursors in nitrosamine formation.

The occurrence of nitrosamines is a global issue, influenced by regional industrial and human activities, and shaped by the specifics of local water treatment processes. To mitigate their presence, a range of specific treatment methods are employed, including disinfection optimization, enhancing conventional treatment process performance, pre-oxidation, bio-activated carbon, and membrane technology. Among all control and treatment strategies, source control is considered the best practice.

Lastly, this review discussed the regulatory landscape, noting that while guidelines exist, such as those provided by WHO and different national bodies, these are continually evolving with advancing scientific understanding.

In summary, nitrosamines represent a significant and ongoing challenge in water quality management, demanding a dynamic and evidence-based approach to ensure the safety of drinking water supplies worldwide.

Acknowledgement This research was supported by the National Natural Science Foundation of China (Grant No. 22376114, 22076091), and the State Key Joint Laboratory of Environment Simulation and Pollution Control, China (Grant No. 21L01ESPC).

References

1. Krasner SW, Mitch WA, McCurry DL et al (2013) Formation, precursors, control, and occurrence of nitrosamines in drinking water: a review. Water Res 47:4433–4450. https://doi.org/10.1016/j.watres.2013.04.050
2. Xu H (1988) N-nitroso compounds in the environment. Science Press, Beijing
3. Plumlee MH, Reinhard M (2007) Photochemical attenuation of N-Nitrosodimethylamine (NDMA) and other nitrosamines in surface water. Environ Sci Technol 41:6170–6176. https://doi.org/10.1021/es070818l
4. Green LC (2023) N-Nitroso compounds. In: Patty's toxicology. Wiley, pp 1–11
5. Magee PN, Barnes JM (1956) The production of malignant primary hepatic tumours in the rat by feeding Dimethylnitrosamine. Br J Cancer 10:114–122. https://doi.org/10.1038/bjc.1956.15
6. Mitch WA, Sharp JO, Trussell RR et al (2003) N-Nitrosodimethylamine (NDMA) as a drinking water contaminant: a review. Environ Eng Sci 20:389–404. https://doi.org/10.1089/109287503768335896
7. Ray SD, Brown D, Yang N (2014) Dimethylnitrosamine. In: Wexler P (ed) Encyclopedia of toxicology3rd edn. Academic Press, Oxford, pp 598–600
8. Li X, Bei E, Qiu Y et al (2021) Intake of volatile nitrosamines by Chinese residents in different provinces via food and drinking water. Sci Total Environ 754:142121. https://doi.org/10.1016/j.scitotenv.2020.142121
9. Beard JC, Swager TM (2021) An organic chemist's guide to N-nitrosamines: their structure, reactivity, and role as contaminants. J Org Chem 86:2037–2057. https://doi.org/10.1021/acs.joc.0c02774

10. IARC (1987) Overall evaluations of carcinogenicity: an updating of IARC monographs volumes 1 to 42. IARC, Lyon
11. Diana M, Felipe-Sotelo M, Bond T (2019) Disinfection byproducts potentially responsible for the association between chlorinated drinking water and bladder cancer: a review. Water Res 162:492–504. https://doi.org/10.1016/j.watres.2019.07.014
12. Cui J (2016) Relationship between N-nitrosodimethylamine and risk of digestive tract cancers: a meta analysis based on cohort studies. Chin J Epidemiol 37:725–729. https://doi.org/10.3760/cma.j.issn.0254-6450.2016.05.029
13. Zhao C, Lu Q, Gu Y et al (2019) Distribution of N-nitrosamines in drinking water and human urinary excretions in high incidence area of esophageal cancer in Huai'an, China. Chemosphere 235:288–296. https://doi.org/10.1016/j.chemosphere.2019.06.124
14. Hidajat M, McElvenny DM, Ritchie P et al (2019) Lifetime exposure to rubber dusts, fumes and N-nitrosamines and cancer mortality in a cohort of British rubber workers with 49 years follow-up. Occup Environ Med 76:250–258. https://doi.org/10.1136/oemed-2018-105181
15. McGwin G (2020) The association between ranitidine use and gastrointestinal cancers. Cancers (Basel) 13:24. https://doi.org/10.3390/cancers13010024
16. USEPA (1987) Integrated risk information system: N-Nitrosodimethylamine. https://cfpub.epa.gov/ncea/iris2/chemicalLanding.cfm?substance_nmbr=45
17. U.S. EPA N-Nitrosodimethylamine; CASRN 62-75-9. National Center for Environmental Assessment
18. Luo Q, Miao Y, Liu C et al (2023) Maternal exposure to nitrosamines in drinking water during pregnancy and birth outcomes in a Chinese cohort. Chemosphere 315:137776. https://doi.org/10.1016/j.chemosphere.2023.137776
19. Li X, Ye Z, Wang J et al (2024) Intake of tobacco nitrosamines of smokers in various provinces of China and their cancer risk: a meta-analysis. J Environ Sci 141:249–260. https://doi.org/10.1016/j.jes.2023.05.018
20. Shapley D (1976) Nitrosamines: scientists on the trail of prime suspect in urban cancer. Science 191:268–270. https://doi.org/10.1126/science.191.4224.268
21. Fine DH, Rounbehler DP, Rounbehler A et al (1977) Determination of dimethylnitrosamine in air, water, and soil by thermal energy analysis: measurements in Baltimore, Md. Environ Sci Technol 11:581–584. https://doi.org/10.1021/es60129a003
22. SWRCB CA, U.S. (2017) Groundwater fact sheet N-Nitrosodimethylamine (NDMA). https://www.waterboards.ca.gov/gama/docs/ndma.pdf
23. California Department of Health Services (2003) A brief history of NDMA findings in drinking water supplies. http://www.dhs.ca.gov/ps/ddwem/chemicals/ndma/history.htm
24. Graham JE (1996) Factors affecting NDMA formation during drinking water treatment. In: Proceedings of the 1995 Water Quality Technology Conference. pp 757–772
25. Jobb DB, Hunsinger RB, Meresz O (1994) Removal of N-nitrosodimethylamine from the ohsweken (six nations) water supply. Ontario Ministry of Environment and Energy, Science and Technology Branch, Toronto
26. Najm I, Trussell RR (2001) NDMA formation in water and wastewater. J AWWA 93:92–99. https://doi.org/10.1002/j.1551-8833.2001.tb09129.x
27. Choi J, Valentine RL (2002) Formation of N-nitrosodimethylamine (NDMA) from reaction of monochloramine: a new disinfection by-product. Water Res 36:817–824. https://doi.org/10.1016/S0043-1354(01)00303-7
28. Mitch WA, Gerecke AC, Sedlak DL (2003) A N-Nitrosodimethylamine (NDMA) precursor analysis for chlorination of water and wastewater. Water Res 37:3733–3741. https://doi.org/10.1016/S0043-1354(03)00289-6
29. Nawrocki J, Andrzejewski P (2011) Nitrosamines and water. J Hazard Mater 189:1–18. https://doi.org/10.1016/j.jhazmat.2011.02.005
30. WHO Chronicle (2011) Guidelines for drinking-water quality, vol 216. 4th edn. WHO Chronicle, pp 303–304

31. Vellingiri K, Kumar PG, Kumar PS et al (2023) Status of disinfection byproducts research in India. Chemosphere 330:138694. https://doi.org/10.1016/j.chemosphere.2023.138694
32. Chen Z, Yang L, Huang Y et al (2019) Carcinogenic risk of N-nitrosamines in Shanghai drinking water: indications for the use of ozone pretreatment. Environ Sci Technol 53:7007–7018. https://doi.org/10.1021/acs.est.8b07363
33. US EPA OW (2015) Second unregulated contaminant monitoring rule. https://www.epa.gov/dwucmr/second-unregulated-contaminant-monitoring-rule. Accessed 12 Feb 2024
34. Ontario Drinking Water Surveillance Program. https://data.ontario.ca/dataset/drinking-water-surveillance-program. Accessed 12 Feb 2024
35. Woods GC, Dickenson ERV (2015) Evaluation of the final UCMR2 database: nationwide trends in NDMA. J AWWA 107:E58–E68. https://doi.org/10.5942/jawwa.2015.107.0009
36. Charrois JWA, Boyd JM, Froese KL, Hrudey SE (2015) Occurrence of N-nitrosamines in Alberta public drinking-water distribution systems. J Environ Eng Sci 6(1):103–114. https://doi.org/10.1139/s06-031
37. Asami M, Oya M, Kosaka K (2009) A nationwide survey of NDMA in raw and drinking water in Japan. Sci Total Environ 407:3540–3545. https://doi.org/10.1016/j.scitotenv.2009.02.014
38. Hashemi S, Park J-H, Yang M et al (2022) Long-term monitoring and risk assessment of N-nitrosamines in the finished water of drinking water treatment plants in South Korea. Environ Sci Pollut Res 29:3930–3943. https://doi.org/10.1007/s11356-021-15814-1
39. Farré MJ, Insa S, Lamb A et al (2019) Occurrence of N-nitrosamines and their precursors in Spanish drinking water treatment plants and distribution systems. Environ Sci Water Res Technol 6:210–220. https://doi.org/10.1039/C9EW00912D
40. Bei E, Shu Y, Li S et al (2016) Occurrence of nitrosamines and their precursors in drinking water systems around mainland China. Water Res 98:168–175. https://doi.org/10.1016/j.watres.2016.04.013
41. Wang W, Yu J, An W, Yang M (2016) Occurrence and profiling of multiple nitrosamines in source water and drinking water of China. Sci Total Environ 551–552:489–495. https://doi.org/10.1016/j.scitotenv.2016.01.175
42. Chen W, Chen Y, Huang H et al (2019) Occurrence of N-nitrosamines in the Pearl River delta of China: characterization and evaluation of different sources. Water Res 164:114896. https://doi.org/10.1016/j.watres.2019.114896
43. Aydin E, Yaman FB, Ates Genceli E et al (2012) Occurrence of THM and NDMA precursors in a watershed: effect of seasons and anthropogenic pollution. J Hazard Mater 221–222:86–91. https://doi.org/10.1016/j.jhazmat.2012.04.012
44. Vizioli BDC, Hantao LW, Montagner CC (2021) Drinking water nitrosamines in a large metropolitan region in Brazil. Environ Sci Pollut Res Int 28:32823–32830. https://doi.org/10.1007/s11356-021-12998-4
45. Mhlongo SH, Mamba BB, Krause RW (2009) Monitoring the prevalence of nitrosamines in South African waters and their removal using cyclodextrin polyurethanes. Phys Chem Earth Parts A/B/C 34:819–824. https://doi.org/10.1016/j.pce.2009.07.008
46. Dillon G, Blake S, Rumsby P (2008) NDMA - concentrations in drinking water and factors affecting its formation. Department for Environment Food & Rural Affairs
47. Russell CG, Blute NK, Via S et al (2012) Nationwide assessment of nitrosamine occurrence and trends. J AWWA 104:E205–E217. https://doi.org/10.5942/jawwa.2012.104.0037
48. Van Huy N, Murakami M, Sakai H et al (2011) Occurrence and formation potential of N-nitrosodimethylamine in ground water and river water in Tokyo. Water Res 45:3369–3377. https://doi.org/10.1016/j.watres.2011.03.053
49. Lee J-H, Oh J-E (2016) A comprehensive survey on the occurrence and fate of nitrosamines in sewage treatment plants and water environment. Sci Total Environ 556:330–337. https://doi.org/10.1016/j.scitotenv.2016.02.090
50. Ma F, Wan Y, Yuan G et al (2012) Occurrence and source of nitrosamines and secondary amines in groundwater and its adjacent Jialu River basin, China. Environ Sci Technol 46:3236–3243. https://doi.org/10.1021/es204520b

51. Schreiber IM, Mitch WA (2006) Nitrosamine formation pathway revisited: the importance of chloramine speciation and dissolved oxygen. Environ Sci Technol 40:6007–6014. https://doi.org/10.1021/es060978h
52. Krasner SW, Roback S, Qian Y et al (2020) Occurrence of nitrosamines and their precursors in North American drinking waters. AWWA Water Sci 2:e1208. https://doi.org/10.1002/aws2.1208
53. Cai H, Shen C, Xu H et al (2023) Seasonal variability, predictive modeling and health risks of N-nitrosamines in drinking water of Shanghai. Sci Total Environ 857:159530. https://doi.org/10.1016/j.scitotenv.2022.159530
54. Leavey-Roback SL, Sugar CA, Krasner SW, Suffet IH(M) (2016) NDMA formation during drinking water treatment: a multivariate analysis of factors influencing formation. Water Res 95:300–309. https://doi.org/10.1016/j.watres.2016.02.060
55. He Z, Cheng Y, Liao X et al (2022) Which pre-oxidation methods to choose? From perspective of DBPs formation and toxicities reduction. Process Saf Environ Prot 161:118–125. https://doi.org/10.1016/j.psep.2022.02.072
56. Jasemizad T, Sun P, Padhye LP (2022) Aqueous N-nitrosamines: precursors, occurrence, oxidation processes, and role of inorganic ions. Crit Rev Environ Sci Technol 52:3604–3650. https://doi.org/10.1080/10643389.2021.1930439
57. Xia R, Liao X, Liu T et al (2023) Effects of Br− on NDMA formation during Chloramination: a review of influencing factors, mechanisms, and control. Chem Res Chin Univ 39:370–377. https://doi.org/10.1007/s40242-023-3048-7
58. Zhang H, Andrews SA (2013) Effects of pipe materials, orthophosphate, and flow conditions on chloramine decay and NDMA formation in modified pipe loops. J Water Supply Res Technol - AQUA 62:107–119. https://doi.org/10.2166/aqua.2013.077
59. Zhang H, Andrews SA (2013) Factors affecting catalysis of copper corrosion products in NDMA formation from DMA in simulated premise plumbing. Chemosphere 93:2683–2689. https://doi.org/10.1016/j.chemosphere.2013.08.067
60. Mitch WA, Sedlak DL (2002) Formation of N-Nitrosodimethylamine (NDMA) from dimethylamine during chlorination. Environ Sci Technol 36:588–595. https://doi.org/10.1021/es010684q
61. Shah AD, Mitch WA (2012) Halonitroalkanes, Halonitriles, Haloamides, and N-nitrosamines: a critical review of nitrogenous disinfection byproduct formation pathways. Environ Sci Technol 46:119–131. https://doi.org/10.1021/es203312s
62. Schreiber IM, Mitch WA (2005) Influence of the order of reagent addition on NDMA formation during Chloramination. Environ Sci Technol 39:3811–3818. https://doi.org/10.1021/es0483286
63. Liu YD, Selbes M, Zeng C et al (2014) Formation mechanism of NDMA from ranitidine, trimethylamine, and other tertiary amines during Chloramination: a computational study. Environ Sci Technol 48:8653–8663. https://doi.org/10.1021/es500997e
64. Liao X, Chen C, Xie S et al (2015) Nitrosamine precursor removal by BAC: a case study of adsorption versus biotreatment. J AWWA 107:E454–E463. https://doi.org/10.5942/jawwa.2015.107.0123
65. Padhye L, Luzinova Y, Cho M et al (2011) PolyDADMAC and dimethylamine as precursors of N-Nitrosodimethylamine during ozonation: reaction kinetics and mechanisms. Environ Sci Technol 45:4353–4359. https://doi.org/10.1021/es104255e
66. Yang L, Chen Z, Shen J et al (2009) Reinvestigation of the nitrosamine-formation mechanism during ozonation. Environ Sci Technol 43:5481–5487. https://doi.org/10.1021/es900319f
67. Schmidt CK, Brauch H-J (2008) N,N-Dimethylsulfamide as precursor for N-Nitrosodimethylamine (NDMA) formation upon ozonation and its fate during drinking water treatment. Environ Sci Technol 42:6340–6346. https://doi.org/10.1021/es7030467
68. Gao M, Jiang Z, Liao X et al (2022) NDMA formation during ozonation of DMAPA: influencing factors, mechanisms, and new pathway exploration. Sci Total Environ 825:153881. https://doi.org/10.1016/j.scitotenv.2022.153881

69. Liao X, Shen L, Jiang Z et al (2021) NDMA formation during ozonation of metformin: roles of ozone and hydroxyl radicals. Sci Total Environ 796:149010. https://doi.org/10.1016/j.scitotenv.2021.149010
70. von Gunten U, Salhi E, Schmidt CK, Arnold WA (2010) Kinetics and mechanisms of N-Nitrosodimethylamine formation upon ozonation of N,N-Dimethylsulfamide-containing waters: bromide catalysis. Environ Sci Technol 44:5762–5768. https://doi.org/10.1021/es1011862
71. Sgroi M, Roccaro P, Oelker GL, Snyder SA (2014) N-Nitrosodimethylamine formation upon ozonation and identification of precursors source in a municipal wastewater treatment plant. Environ Sci Technol 48:10308–10315. https://doi.org/10.1021/es5011658
72. Song Y, Feng S, Qin W et al (2022) Formation mechanism and control strategies of N-nitrosodimethylamine (NDMA) formation during ozonation. Sci Total Environ 823: 153679. https://doi.org/10.1016/j.scitotenv.2022.153679
73. Marti EJ, Pisarenko AN, Peller JR, Dickenson ERV (2015) N-nitrosodimethylamine (NDMA) formation from the ozonation of model compounds. Water Res 72:262–270. https://doi.org/10.1016/j.watres.2014.08.047
74. Gan W, Bond T, Yang X, Westerhoff P (2015) Role of chlorine dioxide in N-Nitrosodimethylamine formation from oxidation of model amines. Environ Sci Technol 49:11429–11437. https://doi.org/10.1021/acs.est.5b01729
75. Andrzejewski P, Nawrocki J (2009) N-nitrosodimethylamine (NDMA) as a product of potassium permanganate reaction with aqueous solutions of dimethylamine (DMA). Water Res 43: 1219–1228. https://doi.org/10.1016/j.watres.2008.12.005
76. Shen L, Chen Z, Kang J et al (2022) N-nitrosodimethylamine formation during oxidation of N, N-dimethylhydrazine compounds by peroxymonosulfate: kinetics, reactive species, mechanism and influencing factors. J Hazard Mater 428:128191. https://doi.org/10.1016/j.jhazmat.2021.128191
77. Sgroi M, Vagliasindi FGA, Snyder SA, Roccaro P (2018) N-Nitrosodimethylamine (NDMA) and its precursors in water and wastewater: a review on formation and removal. Chemosphere 191:685–703. https://doi.org/10.1016/j.chemosphere.2017.10.089
78. Bei E, Li X, Wu F et al (2020) Formation of N-nitrosodimethylamine precursors through the microbiological metabolism of nitrogenous substrates in water. Water Res 183:116055. https://doi.org/10.1016/j.watres.2020.116055
79. Chen Z, Valentine RL (2007) Formation of N-Nitrosodimethylamine (NDMA) from humic substances in natural water. Environ Sci Technol 41:6059–6065. https://doi.org/10.1021/es0705386
80. Li L, Gao N, Deng Y et al (2012) Characterization of intracellular & extracellular algae organic matters (AOM) of microcystic aeruginosa and formation of AOM-associated disinfection byproducts and odor & taste compounds. Water Res 46:1233–1240. https://doi.org/10.1016/j.watres.2011.12.026
81. Qiu Y, Bei E, Li X et al (2021) Quantitative analysis of source and fate of N-nitrosamines and their precursors in an urban water system in East China. J Hazard Mater 415:125700. https://doi.org/10.1016/j.jhazmat.2021.125700
82. Zeng T, Mitch WA (2015) Contribution of *N* -nitrosamines and their precursors to domestic sewage by greywaters and Blackwaters. Environ Sci Technol 49:13158–13167. https://doi.org/10.1021/acs.est.5b04254
83. Chen C, Leavey S, Krasner SW, Suffet IH (2014) Applying polarity rapid assessment method and ultrafiltration to characterize NDMA precursors in wastewater effluents. Water Res 57: 115–126. https://doi.org/10.1016/j.watres.2014.02.052
84. Wang L, Li Y, Shang X, Shen J (2014) Occurrence and removal of N-nitrosodimethylamine and its precursors in wastewater treatment plants in and around Shanghai. Front Environ Sci Eng 8:519–530. https://doi.org/10.1007/s11783-013-0610-4

85. Shen R, Andrews SA (2011) Demonstration of 20 pharmaceuticals and personal care products (PPCPs) as nitrosamine precursors during chloramine disinfection. Water Res 45:944–952. https://doi.org/10.1016/j.watres.2010.09.036
86. Shen R, Andrews SA (2011) NDMA formation kinetics from three pharmaceuticals in four water matrices. Water Res 45:5687–5694. https://doi.org/10.1016/j.watres.2011.08.034
87. Shen R, Andrews SA (2013) Formation of NDMA from ranitidine and sumatriptan: the role of pH. Water Res 47:802–810. https://doi.org/10.1016/j.watres.2012.11.004
88. Shen R, Andrews SA (2013) NDMA formation from amine-based pharmaceuticals – impact from prechlorination and water matrix. Water Res 47:2446–2457. https://doi.org/10.1016/j.watres.2013.02.017
89. Hanigan D, Thurman EM, Ferrer I et al (2015) Methadone contributes to *N*-Nitrosodimethylamine formation in surface waters and wastewaters during Chloramination. Environ Sci Technol Lett 2:151–157. https://doi.org/10.1021/acs.estlett.5b00096
90. Ding S, Deng Y, Bond T et al (2019) Disinfection byproduct formation during drinking water treatment and distribution: a review of unintended effects of engineering agents and materials. Water Res 160:313–329. https://doi.org/10.1016/j.watres.2019.05.024
91. Park S-H, Wei S, Mizaikoff B et al (2009) Degradation of amine-based water treatment polymers during Chloramination as N-Nitrosodimethylamine (NDMA) precursors. Environ Sci Technol 43:1360–1366. https://doi.org/10.1021/es802732z
92. Krasner SW, Westerhoff P, Mitch WA et al (2018) Behavior of NDMA precursors at 21 full-scale water treatment facilities. Environ Sci Water Res Technol 4:1966–1978. https://doi.org/10.1039/C8EW00442K
93. Morran J, Whittle M, Fabris RB et al (2011) Nitrosamines from pipeline materials in drinking water distribution systems. J AWWA 103:76–83. https://doi.org/10.1002/j.1551-8833.2011.tb11549.x
94. Teefy S, Chan CC, Wong W, Work L (2014) NDMA formation from gaskets used in water storage tanks. J AWWA 106:E408–E417. https://doi.org/10.5942/jawwa.2014.106.0081
95. Wang C (2012) Occurrence, precursors and controlling of N-nitrosamine in drinking water. Ph.D. dissertation. Tsinghua University
96. Liao X, Bei E, Li S et al (2015) Applying the polarity rapid assessment method to characterize nitrosamine precursors and to understand their removal by drinking water treatment processes. Water Res 87:292–298. https://doi.org/10.1016/j.watres.2015.09.040
97. Wang C, Zhang X, Wang J et al (2013) Effects of organic fractions on the formation and control of N-nitrosamine precursors during conventional drinking water treatment processes. Sci Total Environ 449:295–301. https://doi.org/10.1016/j.scitotenv.2013.01.080
98. Rosario-Ortiz FL, Snyder S, Suffet IH(M) (2007) Characterization of the polarity of natural organic matter under ambient conditions by the polarity rapid assessment method (PRAM). Environ Sci Technol 41:4895–4900. https://doi.org/10.1021/es062151t
99. Pehlivanoglu-Mantas E, Sedlak DL (2008) Measurement of dissolved organic nitrogen forms in wastewater effluents: concentrations, size distribution and NDMA formation potential. Water Res 42:3890–3898. https://doi.org/10.1016/j.watres.2008.05.017
100. Wilczak A, Assadi-Rad A, Lai HH et al (2003) Formation of NDMA in Chloraminated water coagulated with DADMAC cationic polymer. J AWWA 95:94–106. https://doi.org/10.1002/j.1551-8833.2003.tb10457.x
101. Chu W, Yao D, Gao N et al (2015) The enhanced removal of carbonaceous and nitrogenous disinfection by-product precursors using integrated permanganate oxidation and powdered activated carbon adsorption pretreatment. Chemosphere 141:1–6. https://doi.org/10.1016/j.chemosphere.2015.05.087
102. Chen Z, Valentine RL (2008) The influence of the pre-oxidation of natural organic matter on the formation of N-nitrosodimethylamine (NDMA). Environ Sci Technol 42:5062–5067. https://doi.org/10.1021/es8006673

103. Lee C, Schmidt C, Yoon J, von Gunten U (2007) Oxidation of N-Nitrosodimethylamine (NDMA) precursors with ozone and chlorine dioxide: kinetics and effect on NDMA formation potential. Environ Sci Technol 41:2056–2063. https://doi.org/10.1021/es062484q
104. Liao XB, Zhao L, Shen LL et al (2019) Effect of pre-ozonation on N-nitrosodimethylamine (NDMA) formation from four drinking-water sources during subsequent chloramination. Water Supply 19:1816–1822. https://doi.org/10.2166/ws.2019.060
105. Hanigan D, Zhang J, Herckes P et al (2012) Adsorption of N-nitrosodimethylamine precursors by powdered and granular activated carbon. Environ Sci Technol 46:12630–12639. https://doi.org/10.1021/es302922w
106. Woods GC, Sadmani AHMA, Andrews SA et al (2016) Rejection of pharmaceutically-based *N*-nitrosodimethylamine precursors using nanofiltration. Water Res 93:179–186. https://doi.org/10.1016/j.watres.2016.02.002
107. Chen C, Zhang X-J, He W-J et al (2006) Simultaneous control of microorganism, disinfection by-products and bio-stability by sequential chlorination disinfection. Huan Jing Ke Xue 27: 74–79
108. Li S, Shu Y, Tang X et al (2018) Reaction patterns of NDMA precursors during the sequential chlorination process of short-term free chlorination and monochloramination. Sep Purif Technol 204:196–204. https://doi.org/10.1016/j.seppur.2018.04.055
109. Shah AD, Krasner SW, Lee CFT et al (2012) Trade-offs in disinfection byproduct formation associated with precursor preoxidation for control of N-Nitrosodimethylamine formation. Environ Sci Technol 46:4809–4818. https://doi.org/10.1021/es204717j
110. Fujioka T, Khan SJ, Poussade Y et al (2012) N-nitrosamine removal by reverse osmosis for indirect potable water reuse – a critical review based on observations from laboratory-, pilot- and full-scale studies. Sep Purif Technol 98:503–515. https://doi.org/10.1016/j.seppur.2012.07.025
111. Fujioka T, Khan SJ, McDonald JA et al (2013) N-nitrosamine rejection by reverse osmosis membranes: a full-scale study. Water Res 47:6141–6148. https://doi.org/10.1016/j.watres.2013.07.035
112. McCurry DL, Ishida KP, Oelker GL, Mitch WA (2017) Reverse osmosis shifts chloramine speciation causing re-formation of NDMA during potable reuse of wastewater. Environ Sci Technol 51:8589–8596. https://doi.org/10.1021/acs.est.7b01641
113. Song M, Roback SL, Ishida KP et al (2023) Contribution of dimethylamine to N-Nitrosodimethylamine formation at reverse osmosis water reclamation facilities. Environ Sci Technol Lett 10:66–71. https://doi.org/10.1021/acs.estlett.2c00852
114. Chuang Y-H, Szczuka A, Shabani F et al (2019) Pilot-scale comparison of microfiltration/reverse osmosis and ozone/biological activated carbon with UV/hydrogen peroxide or UV/free chlorine AOP treatment for controlling disinfection byproducts during wastewater reuse. Water Res 152:215–225. https://doi.org/10.1016/j.watres.2018.12.062
115. World Health Organization (2021) A global overview of national regulations and standards for drinking-water quality. WHO
116. DVGW-Technologiezentrum Wasser (TZW) (2007) Analysis, toxicity, occurrence, fate, and removal of nitrosamines in the water cycle. Global Water Research Coalition

Inorganic Disinfection By-Products (DBPs)

Qingqing Kong, Mengge Fan, Jingmeng Guan, Liuqing Zhang, and Xin Yang

Contents

1 Introduction 232
2 Occurrence and Generation Mechanisms 233
 2.1 Chlorite 233
 2.2 Chlorate 235
 2.3 Bromate 237
3 Control Strategy 240
 3.1 Precursor Removal 240
 3.2 Disinfection/Oxidation Process Optimization 241
 3.3 Direct Treatment of Chlorite, Chlorate, and Bromate 243
4 Summary 248
References 249

Abstract Chlorite, chlorate, and bromate are three typical inorganic disinfection by-products (DBPs), exhibiting risks to animals and humans. Once generated, these inorganic DBPs are very stable in treated waters. To date, a large amount of work has been carried out to examine their occurrence and formation mechanisms. Regulations exist in many jurisdictions to restrict their levels in drinking water. To decrease the exposure of these inorganic DBPs, extensive research has focused on controlling them. This chapter introduces the properties, toxicity, and regulatory standards for chlorite, chlorate, and bromate, summarizes their reported occurrence, formation mechanisms, and control strategies, and proposes measures that can be applied to control their formation.

Keywords Bromate, Chlorate, Chlorite, Control, Generation

Q. Kong, M. Fan, J. Guan, L. Zhang, and X. Yang (✉)
Guangdong Provincial Key Laboratory of Environmental Pollution Control and Remediation Technology, School of Environmental Science and Engineering, Sun Yat-sen University, Guangzhou, China
e-mail: kongqq6@mail.sysu.edu.cn; fanmg3@mail2.sysu.edu.cn; guanjm3@mail2.sysu.edu.cn; zhanglq63@mail2.sysu.edu.cn; yangx36@mail.sysu.edu.cn

Chao Chen, Susan Andrews, and Yuefeng Xie (eds.), *Disinfection By-Products in Water*, Hdb Env Chem (2024) 135: 231–254, DOI 10.1007/698_2024_1106,

Published online: 17 September 2024

1 Introduction

Chlorite, chlorate, and bromate are three inorganic DBPs that are stable once they are generated and exhibit risks to animals and humans. Chlorite has a chlorine oxidation state of +3. Exposure to elevated concentrations of chlorite, whether ingested or inhaled, may induce irritation in the gastrointestinal tract, respiratory system, and various organs [1]. The chlorate ion exists in a triangular cone structure, with a chlorine oxidation state of +5. Chlorate exhibits a higher level of toxicity compared to chlorite. Ingestion or inhalation of chlorate can result in severe health implications, impacting the kidneys, liver, and the respiratory system [1]. This heightened toxicity is primarily attributed to the robust oxidizing properties of chlorate, disrupting cellular function and inducing oxidative stress. Bromate is also an ion with a triangular conical structure and contains bromine in an oxidation state of +5. The oxidation activity of bromate is considered to be an important factor in the mechanism of its toxic action. Bromate can cause oxidative damage to cellular DNA and is potentially carcinogenic to animals and humans [2]. Bromate is genotoxic to liver cells and causes problems such as oxidative damage and chromosomal mutations in the kidney.

Chlorite, chlorate, and bromate have received global attention since the 1990s due to their toxicity and occurrence in treated waters. Regulatory standards exist worldwide to restrict the occurrence of these inorganic DBPs in drinking water. In 1998, the United States Environmental Protection Agency (EPA) firstly regulated chlorite and bromate with maximum contaminant levels of 1 mg L^{-1} and 10 μg L^{-1} [3]. Chlorate was added to the Third Contaminant Candidate List in 2010 [4], making chlorate a potential candidate for regulation under the Safe Drinking Water Act, and it was also evaluated as a candidate for regulation under the Regulatory Determinations 3 program in 2014. Chlorite and chlorate are regulated in Canada with limiting concentrations of 1.0 mg L^{-1} [5], while 0.7 mg L^{-1} was recommended by the World Health Organization and enforced by the Chinese Government [6, 7]. Bromate is controlled with the same concentration, i.e., 10 μg L^{-1}, in most parts of the world.

Chlorite, chlorate, and bromate can be formed in pre-oxidation, disinfection, or some advanced oxidation processes (AOPs) including UV/chlorine, UV/ClO_2, $SO_4^{\bullet-}$, and O_3-based processes. To reduce the risks of exposure to these inorganic DBPs, a number of strategies have been investigated, including precursor removal, disinfection/oxidation process optimization, and direct treatment based on photolysis and (advanced) reduction. This section summarizes the occurrence and formation mechanisms of chlorite, chlorate, and bromate in pre-oxidation, disinfection/oxidation, or some advanced oxidation processes, as well as the reported control strategies to decrease or remove the formation of these inorganic DBPs.

2 Occurrence and Generation Mechanisms

2.1 Chlorite

2.1.1 Formation in Chlorine Dioxide (ClO_2) Oxidation Process

ClO_2 is a chemical often used in water treatment for pre-oxidation or disinfection [8, 9]. It is highly soluble in water (about 3 g L^{-1} at 25°C) with insignificant hydrolysis at pH 2.0–10.0, but it can disproportionate to chlorite and chlorate in alkaline solution [10]. It's reported that chlorite generated in ClO_2 oxidation treatment can result in a drinking water concentration of 1,600 μg L^{-1} in Canada and the United States, and up to 2,000 μg L^{-1} in Italy [11–13]. Chlorite was also observed with concentrations varying from 13 to 436 μg L^{-1} in samples collected from desalination plants, reservoirs, and mosques in Qatar, where ClO_2 was used as the disinfectant [14].

Chlorite generated in ClO_2 treatment is mainly due to the reaction of ClO_2 with inorganic or organic compounds through electron transfer reactions as illustrated in Fig. 1. I^-, CN^-, NO_2^-, SO_3^{2-}, Fe(II), and Mn(II) are representative inorganic compounds that can be readily oxidized by ClO_2. ClO_2 reacts with these compounds with the apparent second-order rate constants (k_{app}) in the range of 10^2–10^6 M^{-1} s^{-1} at pH 7.0 [15], while these rate constants can be affected by pH. For example, the reaction rate constants of ClO_2 oxidizing Fe(II) and Mn(II) increase greatly with increasing pH from 5.5 to 9.0 [16]. Organic compounds with electron-rich moieties are prone to being oxidized by ClO_2. Generally, electron-rich compounds like phenols, anilines, and thiols can reduce ClO_2 with the apparent rate constants at 10^3–10^9 M^{-1} s^{-1}, 10^5–10^8 M^{-1} s^{-1}, and >10^8 M^{-1} s^{-1} [17–19]. Also, aliphatic

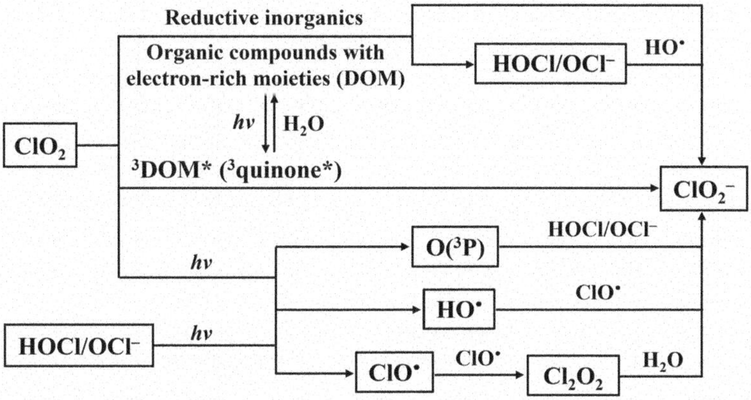

Fig. 1 Pathways of chlorite production in ClO_2 oxidation and UV/chlorine (ClO_2) AOPs (Note: ^{3}DOM* (^{3}quinone*): Triplet state DOM or quinone; HO•: Hydroxyl radical; O(^{3}P): Triplet states of oxygen atom; ClO•: Chloroxy radical; Cl_2O_2: Dichlorine dioxide; ClO_2^-: Chlorite)

tertiary amines and heterocyclic nitrogenous compounds (i.e., indoles and piperidines) show high reactivity toward ClO_2 ($k_{app} = 10^1$–10^6 M^{-1} s^{-1}) [20, 21].

Due to the high reactivity of ClO_2 with those inorganic and organic compounds, chlorite formed in ClO_2 oxidation has been detected frequently [15, 22, 23]. For example, the chlorite yield varying in 30–70% has been found in batch or pilot tests of different waters [22, 23]. The ClO_2 reduction product species can be influenced by the molar ratio of inorganics and ClO_2, with a maximum chlorite yield of more than 70% [24]. For instance, chlorite is the dominated reduction product at a molar ratio of SO_3^{2-} to ClO_2 of 1:2, while chlorate and chloride are generated through some competitive reactions if the molar ratio of SO_3^{2-} and ClO_2 varies between 0.5 and 1.5 [25]. When ClO_2 is in excess, the product of ClO_2 oxidation of Fe(II) is mainly chlorite, but chlorite can be further reduced by Fe(II) to chloride if ClO_2 is consumed [26]. Chlorite yields in ClO_2 reacting with organic compounds depend on the distribution of functional groups [27–29]. For instance, ClO_2 oxidation of amines, di- and trihydroxybenzenes at pH 7.0 shows chlorite yields higher than 50%, while oxidation of olefins, thiols, and benzoquinones exhibits chlorite yields lower than 50% [27]. Furthermore, hydroxyl- and amino-substituents in ortho- and para-positions of the aromatic ring of phenol were found to be the only functional groups strongly influencing chlorite formation, which could cause almost 100% chlorite formation [28]. The widespread dissolved organic matter (DOM) in real waters, consisting of electron-rich moieties, can also react with ClO_2 and lead to the chlorite yield of around 60% in water treatment [30–32].

2.1.2 Formation in AOPs

The photolysis of ClO_2 upon UV radiation (UV/ClO_2) can produce reactive species like $HO^\bullet$, $ClO^\bullet$, and triplet oxygen ($O(^3P)$) and has been proposed as an AOP to use in water treatment [33, 34]. However, chlorite can be formed in this AOP through multiple pathways as described in Fig. 1. For example, $ClO^\bullet$ can convert to chlorite after the hydrolysis of its addition product, dichlorine dioxide (Cl_2O_2) [35], and also, reactions between $HO^\bullet$ and $ClO^\bullet$ can induce chlorite formation (Fig. 1). Chlorite can also form from $O(^3P)$ oxidizing free chlorine ($HOCl/OCl^-$), which can be generated from ClO_2 reacting with phenolic compounds or DOM in real waters treated by the UV/ClO_2 AOP [22, 28, 32, 36]. Additionally, triplet states of DOM ($^3DOM^*$) can be produced from DOM absorbing photons upon UV irradiation. It can reduce ClO_2 directly through one-electron transfer reaction to form chlorite, and also, the triplet states of quinone-like components can transfer to phenols in water, which can reduce ClO_2 to chlorite through one-electron transfer reaction [37].

All of these pathways mentioned above contribute to chlorite formation in the UV/ClO_2 AOP. However, the chlorite yield from radicals including pathways containing $HO^\bullet$ and $ClO^\bullet$ is relatively lower than those from phenolic compounds or DOM involved pathways. It's reported that chlorite produced in the UV/ClO_2 AOP exhibited a yield of 6–15% in DOM-free water, but increased to 56–58% when DOM and other organic compounds were present [33, 38, 39].

The photolysis of free chlorine (UV/chlorine), as an emerging AOP, can also produce reactive species including $HO^•$, $ClO^•$, and $O(^3P)$ and promote micropollutant removal and microbial inactivation. Thus, reactions related to $HO^•$, $ClO^•$, and $O(^3P)$ can lead to chlorite formation in UV/chlorine AOP similarly as that in UV/ClO_2 AOP (Fig. 1). However, the chlorite yield due to these reactions is always very low in the UV/chlorine AOP, even lower than the chromatographic detection limit, because of the lower concentration of these reactive species in comparison with that dosed with ClO_2 in this process [38, 40, 41].

2.2 *Chlorate*

2.2.1 Formation in ClO_2 and Free Chlorine Oxidation Processes

Chlorate can form in a poorly-manufactured (or improperly stored) hypochlorite solution and ClO_2 oxidation processes. Sometimes, chlorate generated in the former can be higher than that in the latter. For example, the maximum concentrations of chlorate in drinking water disinfected by hypochlorite solutions were reported to be up to 4,340 μg L^{-1} in Spain, 4,082 μg L^{-1} in Egypt, and 1,600 μg L^{-1} in Canada [42–44], while the chlorate levels in water treated by ClO_2 seem to be lower at 11–283 μg L^{-1} in Qatar, 30–318 μg L^{-1} in Canada, and 20–1,500 μg L^{-1} in Italy [11, 12, 14]. Also, the chlorate concentrations in natural and hypochlorite solutions treated water were reported to range from 1.2 to 270 μg L^{-1} in the United States [45].

Generally, chlorate generated in the hypochlorite solution is attributed to the disproportionation of $HOCl/OCl^-$ [46], and that formed in ClO_2 oxidation process originates from reactions between ClO_2 or chlorite and $HOCl/OCl^-$, which is used to generate ClO_2 or formed from ClO_2 reacting with organic compounds as shown in Fig. 2 [47]. Furthermore, the oxidation of different model compounds by ClO_2

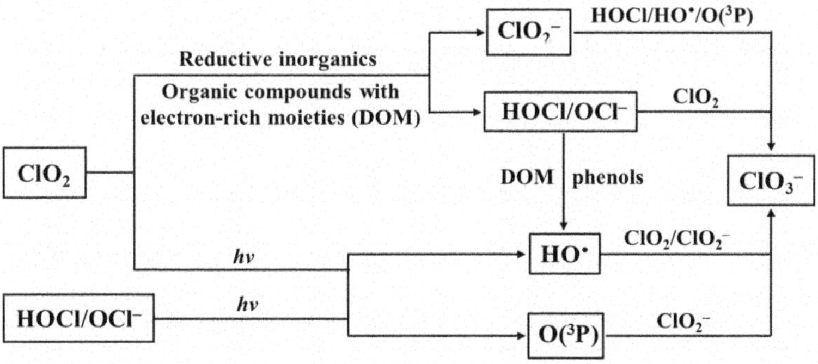

Fig. 2 Pathways of chlorate formation in ClO_2 oxidation and UV/chlorine (ClO_2) AOPs (Note: $HO^•$: Hydroxyl radical; $O(^3P)$: Triplet state oxygen atom; ClO_2^-: Chlorite; ClO_3^-: Chlorate)

produces chlorate with varying yields. For example, phenols, amines, and thiols were reported to transfer ClO_2 with the chlorate yields below 10%, while benzoquinones and olefins induced the chlorate yields around 50% and 15%, respectively [27]. Apart from the reaction between ClO_2 or chlorite and HOCl/ OCl^-, other oxidative species produced during ClO_2 oxidation, such as $HO^•$, may also contribute to chlorate formation through reactions with ClO_2 or chlorite (Fig. 2).

2.2.2 Formation in AOPs

AOPs that can induce chlorate generation mainly include UV/chlorine, UV/ClO_2, and those based on O_3 (e.g., O_3 alone, UV/O_3, O_3/H_2O_2, etc.) and $SO_4^{•-}$ (e.g., O_3/ peroxymonosulfate (PMS), Co(II)/PMS, UV/persulfate (PDS), UV/PMS, etc.). Specifically, chlorate formation in UV/chlorine and UV/ClO_2 AOPs results from the oxidation of chlorite or ClO_2 by reactive species (e.g., $HO^•$ and O (^{3}P)) generated from free chlorine and ClO_2 photolysis (Fig. 2). As reported by Chuang et al., the molar yield of chlorate was around 40–49% and 8%, respectively, in the UVA/ClO_2 and UVC/ClO_2 AOPs [48–50], while it was 1.7% in the UV/chlorine AOP according to Zhao's study [38]. In AOPs based on $SO_4^{•-}$ and O_3, chloride (Cl^-) is the source leading to chlorate formation [51]. As provided in Fig. 3, the interaction of Cl^- with PMS and O_3 induces free chlorine formation, which transforms to chlorate due to further oxidation by $SO_4^{•-}$, $HO^•$, or O_3. Meanwhile, the direct oxidation of Cl^- by $SO_4^{•-}$ and $HO^•$ can induce chlorate formation through the series of reactions listed in Fig. 3. Notably, Co(III) was reported to play a significant role in the free chlorine formation in the Co(II)/PMS system, which was an important intermediate of chlorate generation in this AOP [49].

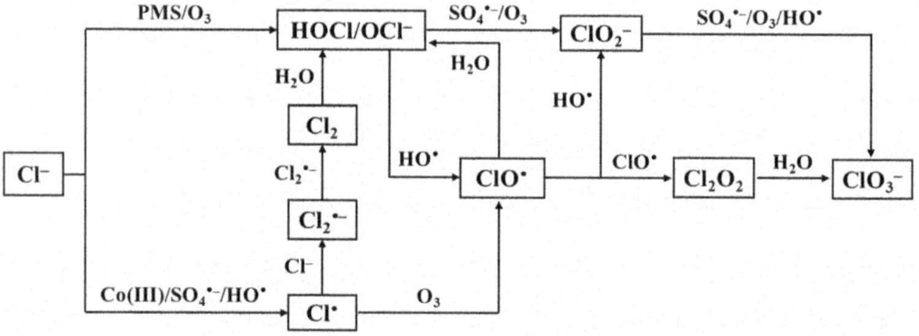

Fig. 3 Pathways of chlorate generation in AOPs based on $SO_4^{•-}$ and O_3 (Note: PMS: Peroxymonosulfate; Co(III): Co^{3+}; $HO^•$: Hydroxyl radical; O(^{3}P): Triplet states of oxygen atom; $Cl^•$: Chlorine radical; $Cl_2^{•-}$; Dichlorine radical anions; $ClO^•$: Chloroxy radical; Cl_2O_2: Dichlorine dioxide; ClO_2^-: Chlorite; ClO_3^-: Chlorate)

2.3 Bromate

2.3.1 Formation in O_3 and O_3-Based AOPs

O_3 Bromate is commonly formed in treating bromide (Br^-) containing waters by O_3. It's reported that ozonation of natural water containing Br^- could lead to bromate formation at concentrations up to 150 μg L^{-1} [52]. Lee et al. found significant bromate formation ranging from ~4 to 370 μg L^{-1} in ozonated secondary effluent samples from 10 wastewater treatment plants in the USA, Switzerland, and Australia [53].

Bromate formation in O_3 oxidation results from a multistep process based on the reaction of O_3 and $HO^•$ with Br^-. As shown in Fig. 4, Br^- can be oxidized by O_3 to form $HOBr/BrO^-$, a key intermediate, and then continue to be oxidized by O_3 to form BrO_2^-. BrO_2^- reacting with O_3 can form bromate (O transfer pathway) directly. BrO_2^- can also react with O_3 to form $BrO_2^•$, which combines to form Br_2O_4 which eventually splits into bromate and BrO_2^- (electron transfer pathway). Meanwhile, Br^- can also be oxidized by $HO^•$ to form $Br^•$ (Fig. 4). $Br^•$ then reacts with Br^- to produce $HOBr/BrO^-$ and further transfers into $BrO^•$ by $HO^•$ oxidation. $BrO^•$ can be hydrolyzed to BrO_2^-, which will be oxidized to $BrO_2^•$ by $HO^•$, which converts into bromate as an end-product after the intermediate Br_2O_4 formation. Additionally, $HO^•$ can be involved in bromate formation from O_3 oxidation pathways, and similarly O_3 can be involved in bromate formation from $HO^•$ oxidation. For example, $Br^•$, generated from $HO^•$ oxidizing Br^-, can be oxidized to $BrO^•$ by O_3, and BrO_2^-, formed from $HO^•$ pathways, can also be oxidized to $BrO_2^•$ or bromate directly by O_3 as listed in Fig. 4.

UV/O_3 The combination of UV and O_3 (UV/O_3) is a green photochemical oxidation method widely used in water treatment, and bromate formation can be promoted from the elevated production of $HO^•$ in this process. It's reported that 35.8 μg L^{-1} of bromate was detected in the effluent of a pilot scale treatment by UV/O_3 process,

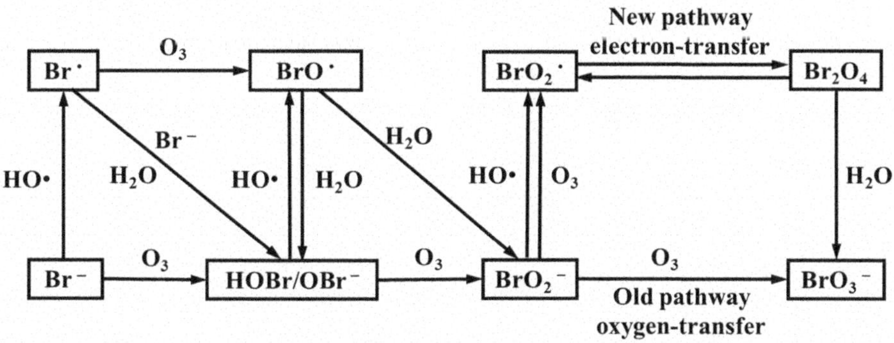

Fig. 4 Pathway for the BrO_3^- formation in O_3-based AOPs (Note: $Br^•$: Bromine radical; $BrO^•$: Bromoxy radical; $HOBr/OBr^-$: Hypobromous acid; $HO^•$: Hydroxyl radical; BrO_2^-: Bromite; $BrO_2^•$: Bromine dioxide; BrO_3^-: Bromate; Br_2O_4: Dibromo tetraoxide)

where the initial concentration of Br^- was 150–180 μg L^{-1} in the influent [54]. Generally, UV irradiation accelerates the decomposition rate of O_3, generating O_2 and $O^•$, and $O^•$ will convert into $HO^•$ (Eqs. 1 and 2). O_3 can also be converted into H_2O_2 under UV irradiation (Eq. 3), which will decompose into $HO^•$ upon UV irradiation (Eq. 4), or induces $HO^•$ formation through further reaction with O_3 (Eq. 5). As mentioned above, $HO^•$ is an important reactive species involved in bromate information and can induce bromate formation through multiple mechanisms as displayed in Fig. 4.

$$O_3 + h\nu \rightarrow O_2 + O^• \tag{1}$$

$$O^• + H_2O \rightarrow 2HO^• \tag{2}$$

$$O_3 + H_2O + h\nu \rightarrow H_2O_2 + O_2 \tag{3}$$

$$H_2O_2 + h\nu \rightarrow 2HO^• \tag{4}$$

$$2O_3 + H_2O_2 \rightarrow 2HO^• + 3O_2 \tag{5}$$

O_3/H_2O_2 H_2O_2 oxidation technology has the characteristics of simple operation and no secondary pollution. The combined use of O_3 and H_2O_2 can accelerate the decomposition of O_3 into $HO^•$ as in Eq. 5, which can achieve the purpose of degrading micropollutants. However, the elevated formation $HO^•$ in Br^- containing waters can also promote the conversion of Br^- to bromate as illustrated in Fig. 4.

2.3.2 Formation in $SO_4^{•-}$ Based AOPs

O_3/PMS In the O_3/PMS process, $SO_4^{•-}$ and $HO^•$ can be generated simultaneously according to Eqs. 6–9, which may play significant roles in bromate formation. Similar to other O_3-based processes, HOBr/BrO^- acts as the key intermediate in bromate formation in the O_3/PMS process. Br^- can be transferred to HOBr/BrO^- through the oxidation of $SO_4^{•-}$ and PMS in the O_3/PMS process as shown in Fig. 5,

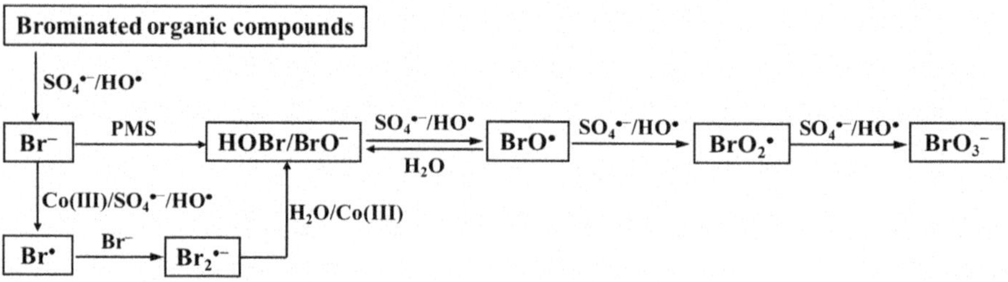

Fig. 5 Pathway for the BrO_3^- formation in $SO_4^{•-}$ based AOPs (Note: $HO^•$: Hydroxyl radical; PMS: Peroxymonosulfate; Co(III): Co^{3+}; $Br^•$: Bromine radical; $Br_2^{•-}$: Dibromine radical anions; HOBr/OBr^-: Hypobromous acid; $BrO^•$: Bromoxy radical; $BrO_2^•$: Bromine dioxide; BrO_3^-: Bromate)

as well as through the oxidation pathways of O_3 and $HO^\bullet$ as shown in Fig. 4. The subsequent series of reactions by $SO_4^{\bullet-}$ and $HO^\bullet$, along with intermediate $Br^\bullet$, $BrO^\bullet$, and $BrO_2^\bullet$ formation, lead the generated HOBr/BrO^- to form bromate as shown in Fig. 5.

$$SO_5{}^{2-} + O_3 \rightarrow O_3SO_5{}^{2-} \; k = (2.12 \pm 0.03) \times 10^4 \; M^{-1} \; s^{-1} \tag{6}$$

$$O_3SO_5{}^{2-} \rightarrow SO_5{}^{\bullet -} + O_3{}^{\bullet -} \tag{7}$$

$$O_3{}^{\bullet -} + H_2O \leftrightarrow HO^\bullet + HO^- + O_2 \tag{8}$$

$$SO_5{}^{\bullet -} + O_3 \rightarrow SO_4{}^{\bullet -} + 2O_2 \; k = 1.6 \times 10^5 M^{-1} s^{-1} \tag{9}$$

UV/PDS Upon UV irradiation, PDS can produce $SO_4^{\bullet-}$ as the primary radical (Eq. 10), and it hydrolyzes to form $HO^\bullet$ as in Eq. 11. Thus, bromate can be formed in the UV/PDS process due to the oxidation of Br^- or brominated organic compounds by $SO_4^{\bullet-}$ and $HO^\bullet$ as provided in Fig. 5 [55, 56]. As reported by Fang et al., 100% of the bromine atoms in dibromo-acetamide can be converted to bromate when UV/PDS was used to treat dibromo-acetamide [55, 56]. Again, HOBr/OBr^- were the key intermediate products for bromate formation in this system. Additionally, bromate formation when using UV/PDS to treat Br^- containing waters can be influenced by pH. It's reported that the bromate formation was similar at pH 4–7 but decreased over 90% with increasing pH from 7 to above 9 [55].

$$S_2O_8{}^{2-} + hv \rightarrow 2SO_4{}^{\bullet -} \tag{10}$$

$$SO_4{}^{\bullet -} + H_2O \rightarrow HO^\bullet + SO_4{}^{2-} \; k < 60 \; M^{-1} \; s^{-1} \tag{11}$$

UV/PMS Upon UV irradiation, PMS can be photolyzed similarly to PDS and generate $SO_4^{\bullet-}$ and $HO^\bullet$ (Eq. 12). Therefore, the same pathways occur in the UV/PMS process to lead Br^- to bromate. That is, Br^- undergoes a series of oxidation reactions by $SO_4^{\bullet-}$ and $HO^\bullet$ and transfers to bromate along with the formation of intermediate HOBr/BrO^-, $Br^\bullet$, $BrO^\bullet$, and $BrO_2^\bullet$ as exhibited in Fig. 5. Br^- can also be oxidized by PMS directly via two-electron transfer to form HOBr/BrO^- with a rate constant of 0.7 M^{-1} s^{-1} [57, 58], and then, a series of subsequent reactions by $SO_4^{\bullet-}$ and $HO^\bullet$ lead the generated HOBr/BrO^- to bromate as the final product (Fig. 5) [59].

$$HSO_5{}^- + hv \rightarrow HO^\bullet + SO_4{}^{\bullet -} \tag{12}$$

Transition Metal/PMS In addition to UV irradiation, metal materials (e.g., transition metal oxides, transition metal compounds, transition metal composites, etc.) are often used to activate PMS. After transferring one electron from the transition metal (i.e., the oxygen-containing transition metal itself) to PMS, $SO_4^{\bullet-}$ as well as bivalent metal ions can be produced according to Eq. 13, and $SO_4^{\bullet-}$ can hydrolyze to form $HO^{\bullet}$ as in Eq. 11 [60]. These highly oxidative species become the key cause of bromate production in treating Br^- containing waters. For example, Co(II) has been confirmed to offer the best activation performance for PMS in homogeneous transition metal activation, and the generated Co(III) can contribute to bromate formation together with $SO_4^{\bullet-}$ [30, 49]. Furthermore, Co(III) contributes to the formation of bromate by the same mechanism as $SO_4^{\bullet-}$ as shown in Fig. 5. That is, Br^- is first oxidized to $Br^{\bullet}$, which further reacts with Br^- to form $HOBr/BrO^-$, and then to bromate after a series of reactions [30].

$$M^{n+} + HSO_5^- \rightarrow M^{(n+1)+} + SO_4^{\bullet-} + OH^- \tag{13}$$

3 Control Strategy

3.1 Precursor Removal

One feasible approach to mitigate the formation of chlorite during ClO_2 oxidation is to remove its main precursors. As confirmed previously, phenolic moieties in DOM are the primary fast-reacting precursors of chlorite in the first 5 min of ClO_2 oxidation [27]. Consequently, eliminating phenolic moieties in DOM by activated carbon adsorption, ultrafiltration, or peroxidation process such as $KMnO_4$ and ferrate oxidation before ClO_2 oxidation should be able to reduce chlorite. Apart from DOM, reductive inorganics also play an important role in the formation of chlorite despite their lower contents in groundwater environment. The application of pre-oxidation processes should be considered to remove reductive inorganics and control chlorite formation. Additionally, controlling the dosage of ClO_2 is another strategy that can be applied to decrease chlorite yield. The United States has set the maximum ClO_2 at 0.8 mg L^{-1} to control chlorite formation.

Br^- acts as the precursor of bromate in all the processes mentioned above, and thus, the available control strategy for precursors of bromate mainly focus on removing Br^- from water, such as membranes, electrochemical and adsorptive techniques. Membrane techniques comprise reverse osmosis (RO), nanofiltration (NF), ion-exchange membranes, electrodialysis (ED), and electrodialysis reversal (EDR). RO, NF and ion-exchange membranes exhibit advantage of selectivity due to the difference in aperture size, shape, or charge, and ED/EDR technology has a low requirement for pretreatment and long membrane lifespan. However, all of them can be hindered by factors such as fouling, frequent cleaning requirements, and high costs. Electrochemical techniques comprise electrolysis and capacitive deionization (CDI). It's reported that electrochemical methods can remove about 35% of bromide from natural waters in laboratory batch and continuous flow systems. Sorption techniques comprise hydrous oxides, activated carbons, silver-doped activated carbons and carbon aerogels, ion-exchange resins, aluminum-based adsorbents and soils. These methods are low-cost and easy to operate, but are susceptible to be interfered by the co-existed DOM or other ions.

3.2 Disinfection/Oxidation Process Optimization

Chlorite and chlorate formed during ClO_2 and free chlorine oxidation can be partly controlled by optimizing the oxidation process. One option is to optimize ClO_2 production. Proper control of chemical concentrations and mixing conditions in the generator can improve ClO_2 yields and minimizes the unreacted chlorite and formation of chlorate in the reactor [61]. When using ClO_2 as a disinfectant, chlorite and chlorate can also be controlled by optimizing the pre-treatment process, because processes such as coagulation and flocculation before ClO_2 disinfection can influence chlorite and chlorate concentration. For example, pre-oxidation with $KMnO_4$ followed by coagulation/flocculation with $Al_2(SO_4)_3$ would lead to the lowest chlorite formation during ClO_2 disinfection among all the tested chemical pre-oxidation plus coagulation/flocculation plus filtration treatment. Correspondingly, the use of $Al_2(SO_4)_3$ instead of $FeCl_3$ can reduce chlorate production due to the decreased chlorite formation [23]. Therefore, replacing $FeCl_3$ with $Al_2(SO_4)_3$ might be a potential solution for water treatment plant to control chlorite and chlorate concentration.

In ClO_2 pre-oxidation-chlorination treatment, the formed chlorite during ClO_2 oxidation can be oxidized by chlorine to chlorate in post chlorination process. The reaction between chlorite and free chlorine is highly dependent on pH, DOM, and other water quality conditions. For example, a lower pH will facilitate this reaction, causing the concentration of chlorite to decrease and the concentration of chlorate to increase, while high pH shows an opposite trend [62]. Therefore, adjusting the pH of chlorination process can be an alternative strategy to change the ratio of the generated chlorate and chlorite and minimize the toxicity of the treated water. However, the operation may decrease the disinfection efficiency, which should be noted in real application. Additionally, kinetic models might be practical tools to predict the formation of chlorite and chlorate, and then help control their formation. Sorlini et al. have established a kinetic model to simulate the reaction between chlorine and chlorite [23], so as to optimize the ClO_2 pre-oxidation-chlorination process. However, the presently established model could just accurately predict chlorite, chlorate, and chlorine concentration in the prepared water, but may be ineffective in real waters due to the existence of complex matrix [23]. Therefore, comprehensive models are still needed to be built in future work.

Methods to prevent bromate formation in the O_3 process mainly include pH reduction, addition of ammonia or hydrogen peroxide, optimization of reactor configuration, and so on [63]. By reducing pH, the equilibrium between $HOBr/OBr^-$ shifts toward HOBr, which is less susceptible to be oxidized by O_3. Ammonia

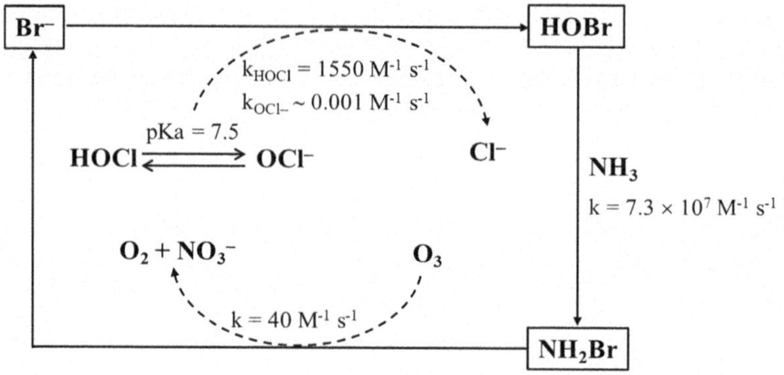

Fig. 6 Bromate minimization by Cl_2-NH_3 process ozonation (Note: NH_2Br: Monobromamine; NH_3: Ammonia; BrO_3^-: Bromate; NO_3^-: Nitrate) [64]

is commonly added to hinder bromate formation in ozonation, through rapidly reacting with $HOBr/OBr^-$ and forming NH_2Br [64, 65]. Besides, prechlorination is often conducted to improve the effects of the ammonia-only process (Fig. 6), becoming the chlorine-ammonia process (Cl_2-NH_3). It's confirmed that Cl_2-NH_3 process could cause a fourfold decrease in bromate formation compared to the ammonia-only process [64]. However, ammonia addition can't effectively impede bromate formation through the $HO^\bullet$ oxidation pathway [65]. In this instance, (bi)-carbonate (e.g., $CaCO_3$) can be chosen to render bromate formation through scavenging $HO^\bullet$ [65]. Additionally, H_2O_2 addition is another alternative method for controlling the formation of bromate during ozonation, especially when disinfection is not a concern [7]. Excess H_2O_2 reacts with $HOBr/OBr^-$ rapidly to form Br^-, preventing its further oxidation by O_3 or $HO^\bullet$, thereby inhibiting bromate formation [3].

In addition to chemical approaches, optimizing reactor configuration has been found effective in reducing bromate formation during ozonation [5]. Contact reactors often suffer from flow irregularities such as short-circuiting and recirculation that lead to lower hydraulic efficiencies [4]. Therefore, it is necessary to optimize reactor design to improve hydraulic efficiency. An optimized reactor configuration allows for lower oxidant dosage or shorter reaction time while achieving the same disinfection or oxidation efficiency. This optimization strategy has successfully reduced bromate formation in several cases involving ozonation. For instance, an optimized contact reactor demonstrated a 43–56% reduction in bromate formation compared to a reactor with a conventional design [5].

Apart from these methods described above, DOC removal treatment may also effectively minimize the level of bromate. DOC is a consumer of O_3 in ozonation, and DOC removal can help control the high O_3 exposures, so as to reduce the formation of bromate. This method might be especially important for fluids like wastewater that have greater DOC content. Generally, DOC can be removed by using powdered activated carbon (PAC) adsorption. It's demonstrated that this pre-treatment reduced the yield of bromate at relatively high PAC doses (50–100 mg L^{-1}, > 40% DOC elimination), most likely as a result of the lower O_3 doses required to achieve the micropollutant abatement goal. However, the bromate yield rose for PAC dosages between 10 and 20 mg L^{-1}, probably as a result of changes in electron-donating capacity of the DOM [6]. More experiments should be conducted before this strategy is used broadly. Also, DOC can be removed through enhanced coagulation, with optimal conditions based on coagulants and the dosage, pH, hydraulic conditions, etc. [66]. As reported previously, coagulation enhanced by 10–30 mg L^{-1} of $FeCl_3$ can decrease 10–47% of the DOC, which, in the study, subsequently reduced the dosed O_3 by a similar proportion to reach the treatment goals [67]. This reduction in O_3 dosage would likely reduce the production of bromate, but further evaluation is still required.

3.3 *Direct Treatment of Chlorite, Chlorate, and Bromate*

3.3.1 UV Photolysis

Chlorite absorbs photons with wavelength less than 320 nm, which means that it could be degraded by UV or solar irradiation. It is reported that the removal of chlorite is the highest under UV irradiation with wavelength of 255 nm (96%), followed by that of 285 nm (33%), but nominal (1%) for that of 365 nm [68]. The major reactions in chlorite photolysis are summarized in the following Eqs. 14–21 [69]. According to these reactions, several reactive radicals including $HO^\bullet$, $ClO^\bullet$, and $Cl^\bullet$ can be generated from chlorite photolysis. Hence, UV/chlorite has been proposed to destroy organic micropollutants in water while reducing chlorite concentration at the same time. It has been proved that the UV/chlorite process degraded carbamazepine effectively due to the contribution of reactive chlorine species [68, 70]. However, one problem exists is that chlorate would be generated as by-product via the UV photolysis of chlorite with maximum yields of ~35% [70].

$$ClO_2^- + h\nu \rightarrow (ClO_2^-)^* \tag{14}$$

$$(ClO_2^-)^* + ClO_2^- \rightarrow ClO_2\ (g) + ClO^- + O^{\bullet -} \tag{15}$$

$$ClO_2^- + h\nu \rightarrow ClO^\bullet + O^{\bullet -} \tag{16}$$

$$ClO_2 + h\nu \rightarrow ClO^\bullet + O^\bullet \tag{17}$$

$$ClO^- + h\nu \rightarrow Cl^\bullet + O^{\bullet -} \tag{18}$$

$$O^{\bullet -} + H_2O \leftrightarrow HO^\bullet + OH^- \tag{19}$$

$$O^\bullet + H_2O \rightarrow H_2O_2 \tag{20}$$

$$H_2O_2 + h\nu \rightarrow 2HO^\bullet \tag{21}$$

Photolysis utilizing UV light at different wavelengths has also been explored as a potential method for reducing bromate. Both low-pressure (UV-L) and medium-pressure (UV–M) lamps have been commonly used in these studies. UV-L primarily emits light at 254 nm while UV-M covers a wider range (~200–400 nm) [71]. Jung et al. investigated the degradation of bromate using UV irradiation at various wavelengths and observed that removal with UV–M was much higher than that with UV–L. The observed degradation rate constant of bromate varied between 0.051 min^{-1} and 0.017 min^{-1} due to the different absorbance characteristics of bromate [72]. However, the efficiency of bromate degradation under UV–L irradiation is relatively higher, with a quantum yield of approximately 0.9 ± 0.1 mol $Einstein^{-1}$, possibly due to the low molar absorption coefficient at 254 nm [73]. Therefore, when designing a UV treatment for bromate degradation, it is necessary to find a balance between process kinetics and efficiency. UV–M provides better kinetics while UV–L offers greater efficiency.

3.3.2 Direct Reduction

In order to eliminate the undesirable chlorite, chlorate, or bromate formed during ClO_2 or O_3 oxidation, or $SO_4^{\bullet-}$ based process, multiple reducing agents including ferrous ions (Fe^{2+}), zero valent iron (Fe^0), sulfite compounds (i.e., Na_2SO_3), and granular activated carbon (GAC) have been investigated.

Fe^{2+} Fe^{2+} can efficiently lower chlorite and bromate concentration by reducing them to chloride and bromide, respectively (Eqs. 22–23), and the theoretical ferrous ions required for the complete reduction of 1 mol chlorite and bromate are 3 and 6 mol according to Eqs. 22–23. The use of Fe^{2+} has been explored extensively and applied in some water treatment plants due to its simplicity and economy. However, several operational conditions would affect the efficiency of chlorite or bromate being reduced by Fe^{2+}. The presence of oxygen can hinder chlorite removal due to the oxidation of Fe^{2+} to Fe^{3+} [74]. pH changes can influence the removal efficiency of chlorite due to the conversion of Fe^{2+} (pKa = 10.1) to ferric hydroxide ($Fe(OH)_3$) at pH > 8.0. As reported by Sorlini et al., an additional 25% amount of Fe^{2+} with respect to the stoichiometric one is required to achieve a complete chlorite removal at pH 8.5 [75].

$$4Fe^{2+} + ClO_2{}^- + 10H_2O \rightarrow 4Fe(OH)_3 + Cl^- + 8H^+ \tag{22}$$

$$BrO_3{}^- + 6Fe^{2+} + 6H^+ \rightleftarrows Br^- + 6Fe^{3+} + 3H_2O \tag{23}$$

Additionally, the generated solid $Fe(OH)_3$ may affect the subsequent water treatment process. According to Henderson's study, the formed $Fe(OH)_3$ is an effective supplement to the coagulant, but significant negative impacts are found when Fe^{2+} is applied in post-flocculation or post-settling [26]. Another study found that dosing Fe^{2+} into coagulants not only eliminates chlorite, but also benefits the post UV/chlorine process in radical production and chlorate reduction in a ClO_2 oxidation followed by coagulation-flocculation-sedimentation and advanced oxidation processes system [76]. These two studies provide reference to the practical application of ferrous ions to control chlorite.

Fe^0 Fe^0 electrochemical reduction is a viable process for controlling chlorate and bromate, and it can directly reduce chlorate and bromate to innocuous chloride and bromide. Researches exploring the reduction of chlorate by Fe^0 are scarce compared to other reducing agents. Some researchers found that higher solid (Fe^0) to liquid ratios are conducive to chlorate electrochemical reduction [77]. Westerhoff investigated reduction rates of chlorate and bromate using seven different sources of Fe^0 through laboratory-scale batch and column tests [77]. The reduction of bromate occurred at a faster rate than that of chlorate, nearly 90% of bromate (10 mM) reduced to bromide within an empty bed contact time of 20 min during Fe^0 column trials. The source of Fe^0 was identified as a critical factor in overall reduction rate, but the mechanisms leading to these differences were not further investigated [77].

Sulfite Sulfite is another reported reducing agent to remove chlorite and bromate in water. The oxidation product of sulfite is sulfate, which is a common anion in water and won't cause adverse effects to water treatment systems. One early study found that sulfite can reduce chlorite to chloride efficiently, and the stoichiometric ratio of SO_3^{2-}/ClO_2^- is around 2, consistent with Eq. 24, with pH ranging from 4.0 to 7.5 and air being absent [61]. However, more sulfite is needed at pH > 8 under aerobic conditions, due to the increasing rate of dissolved oxygen competing for sulfite. Another study even found chlorate formation in the SO_3^{2-}/ClO_2^- system in the presence of oxygen [78], and this phenomenon is speculated to be related to sulfur (IV)-ClO_2^- intermediate formation, which can be either reduced by sulfur (IV) to form chloride or oxidized by oxygen to form chlorate. Additionally, reactive species including $SO_3^{\bullet -}$ and $SO_4^{\bullet -}$ were found in applying SO_3^2 to remove chlorite and bromate, which played significant roles in the degradation of contaminants, e.g., penicillin G. Moreover, the removal rates of chlorite and penicillin G were found to decrease with increasing pH in this system.

$$2SO_3^{2-} + ClO_2^- \rightarrow 2SO_4^{2-} + Cl^- \qquad (24)$$

For bromate removal, the addition of sulfite resulted in a rate ranging from 16% to 63% at pH 5–9.5 and an initial bromate concentration of 7.6 mg L^{-1}. The removal rate increased with increasing pH, reaching approximately 50% at neutral pH (pH = 7) [79]. Another trial using sulfite found that although up to 99% bromate removal could be achieved with initial contamination levels as low as 0.1 mg L^{-1}, it took up to 4 days for complete reaction under drinking-water conditions at pH 7 [80].

GAC GAC is another extensively studied reductant to reduce chlorite generation in a ClO_2-related oxidation system, and has been occupied in some water treatment plants. GAC can eliminate chlorite through adsorption and chemical reduction, and chemical reduction becomes the most important mechanism once the absorptive sites are occupied. On the GAC surface, chlorite reacts with GAC radicals to generate chloride as the final product. However, it should be noted that chlorite breakthrough may occur after long-term operation of GAC, because of the decreased carbon capacity for chlorite reduction. Regenerating GAC with acid or base-acid treatment is a viable choice to partially recover its chlorite removal efficiency [81]. Also, operational variances can influence the chlorite removal efficiency by GAC. For example, the presence of organic matter in water would reduce GAC's ability to eliminate chlorite through taking up those reactive sites [81, 82]; nitrate in treated water can inhibit chlorite removal by GAC via high affinity with GAC sites [81]; increasing pH, increasing hydraulic loading, increasing GAC particle size, and decreasing temperature may cause a lower removal efficiency as well [82]. In addition, the pre-loading of GAC can significantly affect the efficiency of the chlorite removal process [83].

3.3.3 Advanced Reduction

Advanced reduction processes (ARPs), including photocatalysis and combination of UV irradiation with reducing agents (i.e., UV/sulfite and UV/dithionite), can efficiently degrade chlorite, chlorate, and bromate while forming chloride and bromide as the final product. Specifically, photocatalysis can reduce chlorite, chlorate, and bromate into chloride and bromide applying electrons from the photocatalyst (e.g., $BiVO_4$, TiO_2, etc.) that are excited to a free state by light energy; UV-based ARPs mainly remove chlorite, chlorate, and bromate dependent on the generation of radicals with highly reducing capacity (e.g., $SO_2^{\bullet-}$, $SO_3^{\bullet-}$, etc.).

Photocatalysis Photocatalysis is a sustainable treatment technology that can reduce chlorite, chlorate, and bromate into chloride and bromide. Heterogeneous photocatalysts, such as $BiVO_4$, TiO_2, ZnO, Fe_3O_4, CdS, SnO_2, and WO_3, have been used to remove chlorite, chlorate, or bromate. Among these photocatalysts, $BiVO_4$ catalyst has been used to degrade chlorite upon irradiation with a wavelength over 360 nm. Chloride was formed as the main product, and oxalate can be added as a hole scavenger to avoid chlorate formation from hole oxidizing chlorite [84].

TiO_2 is a frequently utilized photocatalyst to reduce chlorite, chlorate, and bromate. Researchers have proved that TiO_2 photocatalysis can eliminate chlorate and bromate in water efficiently, following the multi-electron reduction half reaction and reduction potential as in Eq. 25. Chloride and bromide are produced as the final products [85, 86]. The size of the particles plays a crucial role in determining the migration time of electrons (e^-) and holes (h^+) from the crystal surface to reduce e^- and h^+ recombination. This results in better efficiency for reduced BrO_3^- to Br^- by small particles than larger ones [87]. Therefore, nanoparticles made from TiO_2 have great potential for removing contaminants due to their small size and large surface area. An alternative method proposed by Mills et al. uses a semiconductor photocatalyst, Pt/TiO_2, which was effective in reducing bromate due to noble metal doping that promoted e^- transport [88].

$$ClO_3^-/BrO_3^- + 6H^+ + 6e^- \rightarrow Cl^-/Br^- + 3H_2O \tag{25}$$

Chlorite, chlorate, and bromate removal efficiency during photocatalysis may be affected by various experimental conditions including solution pH, hole scavenger types, photocatalyst dose, and light intensity. For example, previous studies showed that TiO_2 photocatalysis could only reduce chlorate efficiently in acidic conditions, with little effect in neutral and basic solutions. This was attributed to the changed surface charge of TiO_2 as a function of pH, which interfered with the adsorption of chlorate on catalyst [85, 86]. Among three common hole scavengers (i.e., methanol, ethanol, and formate), methanol was confirmed to be the best for chlorate removal [85]. Additionally, the removal of chlorate can be enhanced by increasing TiO_2 dose, as a higher dosage of catalysis provides more surface area and produces more electrons [85].

In addition, the introduction of a semiconductor coating resulted in a significant enhancement in the stability of the photocurrent. The most optimal photo-stability for quantum-sized particles was attained by utilizing CdS on TiO_2 that had been prepared beforehand [87]. For colloidal CdS particles, an effective approach to enhance photo-stability could involve applying thin layers of wide bandgap materials to block hole states [89]. Consequently, a continuous reduction of bromate could be achieved due to the stability of the photocurrent.

UV/Sulfite and UV/Dithionite Combining UV irradiation and reducing agents to produce highly reducing radicals is also a workable choice to eliminate chlorite, chlorate, and bromate. Generally, irradiation of sulfite by UV light generates the sulfite anion radical ($SO_3^{\bullet-}$), hydrated electron and hydrogen atom, which can reduce chlorite and chlorate to chloride with the formation of free available chlorine as an intermediate. Also, previous research showed that acidic pH, high sulfite dose, and high UV light intensity promoted the reducing process of UV/sulfite, while the presence of radical scavengers such as dissolved oxygen, nitrite, and nitrate restrained it [90]. Dithionite possesses a long and weak S-S bond, and it can reversibly dissociate to produce highly reducing sulfur dioxide radicals ($SO_2^{\bullet-}$) upon UV irradiation, which can reduce chlorite and chlorate to chloride directly [91]. Therefore, UV/dithionite is another ARP that can be used to efficiently treat chlorite and chlorate.

In the context of bromate removal, the distribution of S(IV) species is influenced by pH. Lowering the pH reduces the fraction of SO_3^{2-}, leading to a significant inhibition of monochloroacetic acid degradation [92]. Similarly, increasing pH within a narrow range of 5.0–10.0 enhances bromate reduction in the UV/sulfite process, with optimal efficiency observed at pH 10.0 where a higher percentage of S (IV) exists as SO_3^{2-} [93]. Moreover, higher concentrations of sulfite promote bromate degradation through this process. For instance, Botlaguduru et al.

demonstrated that the observed first-order rate constants for SO_3^{2-} doses of 6.4, 12.8, and 25.6 mg L^{-1} were 0.026 ± 0.002, 0.033 ± 0.002, and 0.056 ± 0.001 min^{-1}, respectively [73]. Therefore, the increased SO_3^{2-} dosage can promote the degradation of BrO_3^- through the UV/sulfite process. Liu et al. also found that a high SO_3^{2-} dose can improve the decay rates of BrO_3^- without a significant inhibitory effect, probably due to the high formation of e_{aq}^- at high SO_3^{2-} doses [93]. Qiao et al. proposed a reductive strategy to transform BrO_3^- to Br^- with sulfite [94]. Unlike an e_{aq}^-based reductive strategy, sulfite could effectively reduce BrO_3^- over a wide pH range of 3.0–6.0 in pure water.

4 Summary

Chlorite, chlorate, and bromate are three typical inorganic DBPs regulated worldwide. These inorganic DBPs can be formed in water treatment including pre-oxidation, disinfection, or some advanced oxidation processes (AOPs), e.g., UV/chlorine, UV/ClO_2, $SO_4^{\bullet-}$, and O_3-based processes, and so on. In these scenarios, chlorite is mainly generated due to ClO_2 accepting one electron from reductive organics and inorganics, and also trace from the oxidation of $ClO^\bullet$ or hypochlorous by the generated $HO^\bullet$ or $O(^3P)$; the occurrence of chlorate is mainly attribued to the disproportionation of HOCl/OCl^- in the poorly-manufactured or improperly stored hypochlorite solution, and also, from the oxidation of Cl^-, chlorite, ClO_2 or hypochlorite by the dosed oxidants like O_3 and PMS, and the generated reactive species such as $HO^\bullet$, $SO_4^{\bullet-}$, $O(^3P)$ and Co(III); bromate is mainly produced from the oxidation of organic and inorganic bromine by the dosed oxidants and generated reactive species as those involved in chlorate formation.

To reduce risks of the exposure of these inorganic DBPs, a number of strategies have been investigated. Generally, adding a form of pre-treatment, such as activated carbon adsorption, ultrafiltration, or a peroxidation process by permanganate or ferrate, to remove a precursor (e.g., DOM and reductive inorganics) can lead to a reduction in chlorite and chlorate formation during ClO_2 treatment, as well as reducing bromate formation in O_3 and O_3-based processes. Optimization treatment conditions (e.g., chemical concentrations, pH, and other water quality conditions) or processes (e.g., replacing $FeCl_3$ with $Al_2(SO_4)_3$ as the coagulant) may reduce chlorite, chlorate, and bromate formation in disinfection/oxidation. Decreasing pH, adding NH_4^+ and H_2O_2, or optimizing O_3 dosing methods and reactor configuration have been applied in real engineering to control bromate formation. Additionally, UV photolysis, reduction by ferrous ions, zero valent iron, sulfite compounds and granular activated carbon, and advanced reduction processes based on photocatalysts (i.e., $BiVO_4$ and TiO_2) and reductants (i.e., sulfite and dithionite) were found to be efficient in directly reducing chlorite, chlorate, and bromate into chloride and bromide.

Overall, using $Al_2(SO_4)_3$ as the coagulant instead of $FeCl_3$ to remove precursors is more applicable to control chlorite and chlorate formation in ClO_2 oxidation treatment, and this measure does not need any costly additional technology or process. In ClO_2 pre-oxidation-chlorination treatment, the concentration of chlorite and chlorate can be controlled by adjusting the pH of the chlorination process. If conditions allow, permanganate and ferrate pre-oxidation before coagulation and the application of activated carbon adsorption or ultrafiltration are also viable approaches to reduce chlorite formation by removing the precursors (e.g., DOM). For bromate control, decreasing pH, adding NH_4^+ and H_2O_2, the chlorine-ammonia process, or optimizing O_3 dosing methods and reactor configuration are commonly applied in treatment practices. Also, activated carbon adsorption can be used to remove the precursor of bromate (i.e., Br^-) or bromate directly. As for the other strategies (i.e., photolysis, direct and advanced reduction, etc.), the costs and applicable conditions should be considered.

Acknowledgment The financial support from National Natural Science Foundation of China (No. 22306207) and Basic and Applied Basic Research Foundation of Guangdong Province (No. 2023A1515012113) is greatly appreciated.

References

1. World Health Organization (2022) Guidelines for drinking-water quality. 4th edn
2. Catherine CP, Richard MG, Michael DF et al (2010) Anomalous genotoxic responses induced in mouse *Lymphoma L5178Y* cells by potassium bromate. Toxicology 1–3(267):45–53
3. von Gunten U, Oliveras Y (1997) Kinetics of the reaction between hydrogen peroxide and hypobromous acid: implication on water treatment and natural systems. Water Res 4(31): 900–906
4. Wols BA, Uijttewaal WSJ, Rietveld LC et al (2008) Residence time distributions in ozone contactors. Ozone Sci Eng 1(30):49–57
5. Yang J, Li J, Zhu J et al (2017) A novel design for an ozone contact reactor and its performance on hydrodynamics, disinfection, bromate formation and oxidation. Chem Eng J 328:207–214
6. Soltermann F, Abegglen C, Tschui M et al (2017) Options and limitations for bromate control during ozonation of wastewater. Water Res 116:76–85
7. Merle T, Pronk W, von Gunten U (2017) MEMBRO3X, a novel combination of a membrane contactor with advanced oxidation (O_3/H_2O_2) for simultaneous micropollutant abatement and bromate minimization. Environ Sci Technol Lett 5(4):180–185
8. Huang J, Wang L, Ren N et al (1997) Disinfection effect of chlorine dioxide on bacteria in water. Water Res 3(31):607–613
9. Chauret CP, Radziminski CZ, Lepuil M et al (2001) Chlorine dioxide inactivation of *Cryptosporidium parvum* oocysts and bacterial spore indicators. Appl Environ Microbiol 7(67): 2993–3001
10. Aieta EM, Berg JD (1986) A review of chlorine dioxide in drinking water treatment. J Am Water Works Ass 6(78):62–72
11. Aggazzotti G, Righi E, Fantuzzi G et al (2004) Chlorination by-products (CBPs) in drinking water and adverse pregnancy outcomes in Italy. J Water Health 4(2):233–247
12. Rodriguez RA, Koudjonou B, Jay B et al (2008) Disinfection by-products (DBPs) in drinking water from eight systems using chlorine dioxide. Water Qual Res J 1(43):11–22
13. Wagner HP, Pepich BV, Hautman DP et al (2000) Performance evaluation of a method for the determination of bromate in drinking water by ion chromatography (EPA Method 317.0) and validation of EPA Method 324.0. J Chromatogr A 884:201–210

14. Al-Otoum F, Al-Ghouti MA, Ahmed TA et al (2016) Disinfection by-products of chlorine dioxide (chlorite, chlorate, and trihalomethanes): occurrence in drinking water in Qatar. Chemosphere 164:649–656
15. Gan W, Ge Y, Zhong Y et al (2020) The reactions of chlorine dioxide with inorganic and organic compounds in water treatment: kinetics and mechanisms. Environ Sci Water Res Technol 9(6):2287–2312
16. Knocke WR, Van Benschoten JE, Kearney MJ et al (1991) Kinetics of manganese and iron oxidation by potassium permanganate and chlorine dioxide. J Am Water Works Ass 6(83): 80–87
17. Hoigné J, Bader H (1994) Kinetics of reactions of chlorine dioxide (ClO_2) in water – I. Rate constants for inorganic and organic compounds. Water Res 1(28):45–55
18. Hull LA, Davis GT, Rosenblatt DH et al (1967) Oxidations of amines. III. Duality of mechanism in the reaction of amines with chlorine dioxide. J Am Chem Soc 5(89):1163–1170
19. Kutchin AV, Rubtsova SA, Loginova IV (2001) Reactions of chlorine dioxide with organic compounds. Selective oxidation of sulfides to sulfoxides by chlorine dioxide. Russ Chem Bull 50:432–435
20. Huie RE, Neta P (1986) Kinetics of one-electron transfer reactions involving chlorine dioxide and nitrogen dioxide. J Phys Chem 6(90):1193–1198
21. Merenyi G, Lind J, Shen X (1988) Electron transfer from indoles, phenol, and sulfite (SO_3^{2-}) to chlorine dioxide (ClO_2). J Phys Chem 1(92):134–137
22. Rougé V, Allard S, Croué J-P et al (2018) In situ formation of free chlorine during ClO_2 treatment: implications on the formation of disinfection byproducts. Environ Sci Technol 22(52):13421–13429
23. Sorlini S, Gialdini F, Biasibetti M et al (2014) Influence of drinking water treatments on chlorine dioxide consumption and chlorite/chlorate formation. Water Res 54:44–52
24. Yang B, Fang H, Chen B et al (2021) Effects of reductive inorganics and NOM on the formation of chlorite in the chlorine dioxide disinfection of drinking water. J Environ Sci 104:225–232
25. Suzuki K, Gordon G (1978) Stoichiometry and kinetics of the reaction between chlorine dioxide and sulfur (IV) in basic solutions. Inorg Chem 11(17):3115–3118
26. Henderson R, Carlson K, Gregory D (2001) The impact of ferrous ion reduction of chlorite ion on drinking water process performance. Water Res 18(35):4464–4473
27. Gan W, Huang S, Ge Y et al (2019) Chlorite formation during ClO_2 oxidation of model compounds having various functional groups and humic substances. Water Res 159:348–357
28. Jütte M, Wilbert JA, Reusing M et al (2023) Reaction mechanisms of chlorine dioxide with phenolic compounds—influence of different substituents on stoichiometric ratios and intrinsic formation of free available chlorine. Environ Sci Technol 47(57):18846–18855
29. Hupperich K, Mutke XAM, Abdighahroudi MS et al (2020) Reaction of chlorine dioxide with organic matter–formation of inorganic products. Environ Sci Water Res Technol 9(6): 2597–2606
30. Zhou Y, Jiang J, Gao Y et al (2015) Activation of peroxymonosulfate by benzoquinone: a novel nonradical oxidation process. Environ Sci Technol 21(49):12941–12950
31. Ranieri E, Swietlik J (2010) DBPs control in European drinking water treatment plants using chlorine dioxide: two case studies. J Environ Eng Landsc 2(18):85–91
32. Terhalle J, Kaiser P, Jütte M et al (2018) Chlorine dioxide—pollutant transformation and formation of hypochlorous acid as a secondary oxidant. Environ Sci Technol 17(52):9964–9971
33. Kong Q, Fan M, Yin R et al (2021) Micropollutant abatement and byproduct formation during the co-exposure of chlorine dioxide (ClO_2) and UVC radiation. J Hazard Mater 419:126424
34. Peng J, Yin R, Yang X et al (2022) A novel UVA/ClO_2 advanced oxidation process for the degradation of micropollutants in water. Environ Sci Technol 2(56):1257–1266
35. Buxton GV, Subhani MS (1972) Radiation chemistry and photochemistry of oxychlorine ions. Part 2.-photodecomposition of aqueous solutions of hypochlorite ions. J Chem Soc Faraday Trans 1(68):958–969

36. Wajon JE, Rosenblatt DH, Burrows EP (1982) Oxidation of phenol and hydroquinone by chlorine dioxide. Environ Sci Technol 7(16):396–402
37. Kong Q, Pan Y, Lei X et al (2022) Reducing properties of triplet state organic matter (^{3}DOM*) probed via the transformation from chlorine dioxide to chlorite. Water Res 225:119120
38. Zhao J, Peng J, Yin R et al (2022) Multi-angle comparison of UV/chlorine, UV/monochloramine, and UV/chlorine dioxide processes for water treatment and reuse. Water Res 217:118414
39. Wang Y, Sun W, Dong H et al (2023) Accelerated degradation of micro-pollutant by combined UV and chlorine dioxide: unexpected inhibition of chlorite formation. Environ Pollut 337:122600
40. Chuang Y, Wu K, Lin W et al (2022) Photolysis of chlorine dioxide under UVA irradiation: radical formation, application in treating micropollutants, formation of disinfection byproducts, and toxicity under scenarios relevant to potable reuse and drinking water. Environ Sci Technol 4(56):2593–2604
41. Zhao J, Shang C, Zhang X et al (2021) The multiple roles of chlorite on the concentrations of radicals and ozone and formation of chlorate during UV photolysis of free chlorine. Water Res 190:116680
42. Garcia-Villanova RJ, Oliveira Dantas Leite MV, Hernández Hierro JM et al (2010) Occurrence of bromate, chlorite and chlorate in drinking waters disinfected with hypochlorite reagents. Tracing their origins. Sci Total Environ 12(408):2616–2620
43. Radwan EK, Barakat MH, Ibrahim MBM (2021) Hazardous inorganic disinfection by-products in Egypt's tap drinking water: occurrence and human health risks assessment studies. Sci Total Environ 797:149069
44. Aranda-Rodriguez R, Lemieux F, Jin Z et al (2017) (Yet More) Challenges for water treatment plants: potential contribution of hypochlorite solutions to bromate, chlorate, chlorite and perchlorate in drinking water. J Water Supply Res T 8(66):621–631
45. Snyder SA, Vanderford BJ, Rexing DJ (2005) Trace analysis of bromate, chlorate, iodate, and perchlorate in natural and bottled waters. Environ Sci Technol 12(39):4586–4593
46. Adam LC, Fabian I, Suzuki K et al (1992) Hypochlorous acid decomposition in the pH 5–8 region. Inorg Chem 17(31):3534–3541
47. Suh M, Simpson AMA, Mitch WA (2023) Purified chlorine dioxide as an alternative to chlorine disinfection to minimize chlorate formation during postharvest produce washing. Environ Sci Technol 32(57):12063–12071
48. Zuo Z, Katsumura Y, Ueda K et al (1997) Reactions between some inorganic radicals and oxychlorides studied by pulse radiolysis and laser photolysis. J Chem Soc Faraday Trans 10(93):1885–1891
49. Hou S, Ling L, Dionysiou DD et al (2018) Chlorate formation mechanism in the presence of sulfate radical, chloride, bromide and natural organic matter. Environ Sci Technol 11(52):6317–6325
50. Bulman DM, Mezyk SP, Remucal CK (2019) The impact of pH and irradiation wavelength on the production of reactive oxidants during chlorine photolysis. Environ Sci Technol 8(53):4450–4459
51. Manna F, Cherry JA, McWhorter DB et al (2016) Groundwater recharge assessment in an upland sandstone aquifer of southern California. J Hydrol 541:787–799
52. Siddiqui MS, Amy GL, Rice RG (1995) Bromate ion formation: a critical review. J Am Water Works Ass 10(87):58–70
53. Lee Y, Gerrity D, Lee M et al (2016) Organic contaminant abatement in reclaimed water by UV/H_2O_2 and a combined process consisting of O_3/H_2O_2 followed by UV/H_2O_2: prediction of abatement efficiency, energy consumption, and byproduct formation. Environ Sci Technol 7(50):3809–3819
54. Yao WK, Rehman SWU, Wang HJ et al (2018) Pilot-scale evaluation of micropollutant abatements by conventional ozonation, UV/O_3, and an electro-peroxone process. Water Res 138:106–117

55. Fang JY, Shang C (2012) Bromate formation from bromide oxidation by the UV/persulfate process. Environ Sci Technol 16(46):8976–8983
56. Hou S, Ling L, Shang C et al (2017) Degradation kinetics and pathways of haloacetonitriles by the UV/persulfate process. Chem Eng J 320:478–484
57. Lente G, Kalmar J, Baranyai Z et al (2009) One-versus two-electron oxidation with peroxomonosulfate ion: reactions with iron(II), vanadium(IV), halide ions, and photoreaction with cerium(III). Inorg Chem 4(48):1763–1773
58. Zhou Y, Jiang J, Gao Y et al (2018) Oxidation of steroid estrogens by peroxymonosulfate (PMS) and effect of bromide and chloride ions: kinetics, products, and modeling. Water Res 138:56–66
59. Guan YH, Chen J, Chen LJ et al (2020) Comparison of UV/H_2O_2, UV/PMS, and UV/PDS in destruction of different reactivity compounds and formation of bromate and chlorate. Front Chem 8:581198
60. Pan Y, Zhang Y, Zhou M et al (2019) Enhanced removal of emerging contaminants using persulfate activated by UV and pre-magnetized Fe^0. Chem Eng J 361:908–918
61. Gordon G, Slootmaekers B, Tachiyashiki S et al (1990) Minimizing chlorite ion and chlorate ion in water treated with chlorine dioxide. J Am Water Works Ass 4(82):160–165
62. Rougé V, Lee Y, von Gunten U et al (2022) Kinetic and mechanistic understanding of chlorite oxidation during chlorination: optimization of ClO_2 pre-oxidation for disinfection byproduct control. Water Res 220:118515
63. Pinkernell U, von Gunten U (2001) Bromate minimization during ozonation: mechanistic considerations. Environ Sci Technol 12(35):2525–2531
64. Buffle MO, Galli S, von Gunten U (2004) Enhanced bromate control during ozonation: the chlorine-ammonia process. Environ Sci Technol 19(38):5187–5195
65. Hofmann R, Andrews RC (2006) Impact of H_2O_2 and (bi) carbonate alkalinity on ammonia's inhibition of bromate formation. Water Res 18(40):3343–3348
66. Sun Y, Zhou S, Chiang P-C et al (2019) Evaluation and optimization of enhanced coagulation process: water and energy nexus. Water-Energy Nexus 1(2):25–36
67. Wert EC, Gonzales S, Dong MM et al (2011) Evaluation of enhanced coagulation pretreatment to improve ozone oxidation efficiency in wastewater. Water Res 16(45):5191–5199
68. Wang J, Zhu S, Wu Y et al (2022) Insights into the wavelength-dependent photolysis of chlorite: elimination of carbamazepine and formation of chlorate. Chemosphere 288:132505
69. Hao R, Mao X, Wang Z et al (2019) A novel method of ultraviolet/$NaClO_2$-NH_4OH for NO removal: mechanism and kinetics. J Hazard Mater 368:234–242
70. Wang J, Wu Y, Bu L et al (2021) Simultaneous removal of chlorite and contaminants of emerging concern under UV photolysis: hydroxyl radicals vs. chlorate formation. Water Res 190:116708
71. Huang X, Wang L, Zhou J et al (2014) Photocatalytic decomposition of bromate ion by the UV/P25-graphene processes. Water Res 57:1–7
72. Jung B, Nicola R, Batchelor B et al (2014) Effect of low-and medium-pressure Hg UV irradiation on bromate removal in advanced reduction process. Chemosphere 117:663–672
73. Botlaguduru VSV, Batchelor B, Abdel-Wahab A (2015) Application of UV–sulfite advanced reduction process to bromate removal. J Water Process Eng 5:76–82
74. Katz A, Narkis N (2001) Removal of chlorine dioxide disinfection by-products by ferrous salts. Water Res 1(35):101–108
75. Sorlini S, Collivignarelli C (2005) Chlorite removal with ferrous ions. Desalination 1(176): 267–271
76. Cassol GS, Shang C, Li J et al (2022) Dosing low-level ferrous iron in coagulation enhances the removal of micropollutants, chlorite and chlorate during advanced water treatment. J Environ Sci 117:119–128
77. Westerhoff P (2003) Reduction of nitrate, bromate, and chlorate by zero valent iron (Fe^0). J Environ Eng 1(129):10–16

78. Griese MH, Hauser K, Berkemeier M et al (1991) Using reducing agents to eliminate chlorine dioxide and chlorite ion residuals in drinking water. J Am Water Works Ass 5(83):56–61
79. Jack R (2013) Comparison of regulated methods using ion chromatography for bromate analysis in drinking water, water supply. Thermo Fisher Scientific, Sunnyvale
80. Gordon G, Gauw RD, Emmert GL et al (2002) Chemical reduction methods for bromate ion removal. J Am Water Works Ass 2(94):91–98
81. Sorlini S, Collivignarelli C (2005) Chlorite removal with granular activated carbon. Desalination 1(176):255–265
82. Gonce N, Voudrias EA (1994) Removal of chlorite and chlorate ions from water using granular activated carbon. Water Res 5(28):1059–1069
83. Sorlini S, Collivignarelli MC, Canato M (2015) Effectiveness in chlorite removal by two activated carbons under different working conditions: a laboratory study. J Water Supply Res T 4(64):450–461
84. Marks R, Doudrick K (2019) Photocatalytic reduction of chlorite in water using bismuth vanadate ($BiVO_4$): effect of irradiance conditions and presence of oxalate on the reactivity and by-product selectivity. Environ Sci Water Res Technol 11(5):2015–2026
85. Jung B, Abu-Rub F, El-Ghenymy A et al (2021) Photocatalytic reduction of chlorate in aqueous TiO_2 suspension with hole scavenger under simulated solar light. Emerg Mater 2(4):435–446
86. Marks R, Yang T, Westerhoff P et al (2016) Comparative analysis of the photocatalytic reduction of drinking water oxoanions using titanium dioxide. Water Res 104:11–19
87. Vogel R, Hoyer P, Weller H (2002) Quantum-sized PbS, CdS, Ag_2S, Sb_2S_3, and Bi_2S_3 particles as sensitizers for various nanoporous wide-bandgap semiconductors. J Phys Chem 12(98): 3183–3188
88. Mills A, Belghazi A, Rodman D (1996) Bromate removal from drinking water by semiconductor photocatalysis. Water Res 9(30):1973–1978
89. Haase M, Weller H, Henglein A (1988) Photochemistry of colloidal semiconductors 28. Photoelectron emission from cadmium phosphide particles in aqueous solution. J Phys Chem 10(92): 1103–1107
90. Bensalah N, Abdel-Wahab A (2021) UV-L/sulfite advanced reduction process as an efficient method for chlorate reduction in water. Desalin Water Treat 234:58–67
91. Jung B, Sivasubramanian R, Batchelor B et al (2017) Chlorate reduction by dithionite/UV advanced reduction process. Int J Environ Sci Technol 1(14):123–134
92. Li X, Ma J, Liu G et al (2012) Efficient reductive dechlorination of monochloroacetic acid by sulfite/UV process. Environ Sci Technol 13(46):7342–7349
93. Liu X, Zhang T, Shao Y (2014) Aqueous bromate reduction by UV activation of sulfite. Clean Soil Air Water 10(42):1370–1375
94. Qiao J, Feng L, Dong H et al (2019) Overlooked role of sulfur-centered radicals during bromate reduction by sulfite. Environ Sci Technol 17(53):10320–10328

Part III
DBPs in our Lives

DBPs in Drinking Water Treatment Plants

Emma H. Goslan and Chao Chen

Contents

1 DBP Formation During Treatment 258
2 DBP Precursor Control/Removal by Various Processes 259
3 DBP in Distribution Systems 263
3.1 Impact of Water Blending 265
4 DBP in Small WTPs 265
5 Conclusion 266
References 267

Abstract Disinfection by-products (DBPs) can form in drinking water treatment plants when pre-chlorination is implemented to inhibit microbial growth and/or for oxidation. When this occurs, the formation of DBPs can increase throughout the treatment train although some removals by treatment processes are observed. Where pre-chlorination is not used, the focus is on removal of DBP precursors and how this is achieved by different advanced treatment processes intending to achieve better outcomes. The formation of DBPs can continue in distribution and is further complicated when waters are blended, particularly if they contain different residual disinfectants. The chapter concludes with a look at challenges faced by small water treatment plants when minimizing DBPs.

Keywords Disinfection by-products, Drinking water treatment, Haloacetic acids, Natural organic matter, Trihalomethanes

E. H. Goslan (✉)
Cranfield University, Cranfield, UK
e-mail: e.h.goslan@cranfield.ac.uk

C. Chen
Tsinghua University, Beijing, China
e-mail: chen_water@tsinghua.edu.cn

Chao Chen, Susan Andrews, and Yuefeng Xie (eds.), *Disinfection By-Products in Water*, Hdb Env Chem (2024) 135: 257–272, DOI 10.1007/698_2024_1161,

Published online: 8 October 2024

1 DBP Formation During Treatment

Chlorinated DBPs will form in drinking water treatment plants (DWTPs) when pre-chlorination is used. Pre-chlorination is widely used in DWTPs to inactivate algal cells, degrade toxins, and improve the resilience of the treatment process for changing source water quality [1, 2]. Additionally, pre-chlorination can reduce organics, suspended solids, viruses, bacteria, and algae in water [3, 4]. However, it has the potential to significantly increase the formation of DBPs due to the higher levels of precursors present before treatment and longer exposure time to chlorine. The impact of pre-chlorination on the formation of four trihalomethanes (THM_4) and five haloacetic acids (HAA_5) at a DWTP in China is shown ([5], Fig. 1).

This seasonal sampling campaign [5] was conducted in one conventional DWTP in northern China, along with testing of the concentrations of four THMs and five HAAs in the effluents of each unit process (Fig. 1). Due to the practice of pre-chlorination (1–2 mg/L), which aims to restrain algae growth and enhance coagulation, high concentrations of THMs and HAAs were detected in the effluent of the coagulation process. The average yields of THMs and HAAs accounted for 77% and 52% of the final yields in the effluent. Subsequently, the residual free chlorine continued to react with DBP precursor along the treatment train, i.e., sedimentation, filtration, and clear well, which led to the continuous increases of THM and HAA concentrations.

The concentration of DBPs also exhibited significant seasonal variations. The concentrations of THMs and HAAs in spring and winter were relatively low, partly due to the low concentration of organic matter and precursors in the source water. Moreover, due to the low water temperature at that time, the amount of disinfectant needed was low, and the reaction rate between the disinfectant and the precursor was also low. In summer and autumn, the concentrations of DBP precursors increased due to the growth and reproduction of aquatic organisms. In addition, the required

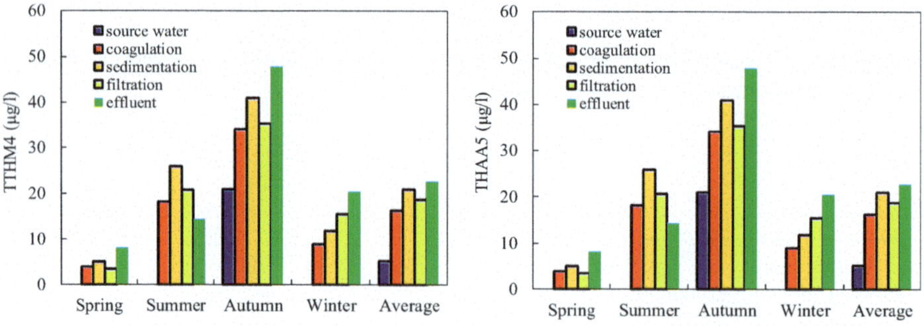

Fig. 1 The seasonal dynamic of total THMs and total HAAs along the treatment train. (Note: $TTHM_4$ means the sum of 4 THMs, including chloroform, bromodichloromethane, dibromochloromethane, bromoform; $THAA_5$ means the sum of 5 HAAs, including monochloroacetic acid, dichloroacetic acid, trichloroacetic acid, monobromoacetic acid, dibromoacetic acid)

dosage of disinfectant and reaction rate also increased with the increase of temperature, so that the concentrations of THMs and HAAs in these two seasons were the highest. In addition, in autumn, the DWTP generally applied pre-chlorination in the water intake point to restrain algae growth, so high concentrations of DBPs were also detected in the source water.

The formation of and transformation of pre-chlorination-formed DBPs in drinking water treatment processes have been investigated [6] to determine process favorable for reduction of DBPs. These processes include pre-chlorination, pre-ozonation, mechanical coagulation & clarification, post-ozonation, and biological activated carbon filtration (BAC), with samples taken after each process for DBP and element analyses. It was found that coagulation/clarification significantly reduced natural organic matter (NOM) but poorly controlled chlorine-containing DBPs. Pre-ozonation and post-ozonation/BAC processes were mostly favored for preformed chlorine-containing DBP removal. The researchers could elucidate the species of NOM and DBPs formed and treated by each process.

2 DBP Precursor Control/Removal by Various Processes

The removal of precursors reported in the literature is variable, even when the same treatment is applied. This is because the precursors continually vary among water sources and seasons [7, 8]. Many studies have examined DBP precursor removal, particularly THM precursors, with limited research examining HAA precursor removal [9].

Bond [10] has analyzed literature data regarding removals of NOM, THM precursors, and HAA precursors, and highlighted circumstances in which various treatment processes can be effectively used for precursor removal. First of all, they summarized the mechanisms and selectivity of NOM removal processes. They confirmed that the removal of NOM is dependent on physical properties, especially size, charge, and hydrophobicity (Table 1).

Bond [10] also made a table of the formation, physical properties, and treatability of NOM groups (Table 2). To assess the treatability of NOM groups by different treatment processes, NOM species were separated into groups of humic substances, carboxylic acids, amino acids, proteins, and carbohydrates according to their physical properties: charge, size, and hydrophobicity. Several assumptions were made while compiling Table 2 [34–38], because of the uncertainty about characteristics of NOM. The work did not consider N-DBPs although it was acknowledged that these are generated from nitrogen-enriched water sources, such as those impacted by wastewater and algae [39].

While the review by Bond [10] focused on THMs, HAAs, and types of NOM, a more recent review [43] focusing on removal of THMs, HAAs, and emerging DBPs has been carried out to report removals of DBPs at pilot and full scales. A selection of the data on removal of DBPs is summarized in Table 3.

Table 1 Selectivity of NOM removal processes (from [10])

Process	Mechanism/s	Selectivity	Least treatable	References
Coagulation	Adsorption onto flocs and charge neutralization/colloid destabilization. Sweep flocculation	Large, anionic molecules	Neutral molecules	[11–15]
Anion exchange	Ion exchange (electrostatic), also adsorption (hydrophobic) and hydrogen bonding	Small, anionic molecules	Neutral molecules	[11, 16–19]
Membranes	Size exclusion, differing diffusion rates across membrane. Electrostatics for charged membranes	Species >MWCO	Hydrophobic molecules <MWCO	[11, 14, 20–22]
Activated carbon	Reversible physical adsorption by non-specific forces	Small, neutral, hydrophobic molecules	Hydrophilic charged molecules	[11, 14, 15, 23–25]
Biotreatment	Enzyme-controlled microbial degradation and adsorption	Low MW polar molecules (e.g., amino acids, aldehydes)	Large and hydrophobic molecules	[19, 26–28]
AOPs	·OH reactions: electron transfer, H abstraction, and OH addition	Relatively unselective		[14, 29, 30]
Ozone	Electrophilic addition: oxidation and bond cleavage. Also OH radical reactions	Aromatic compounds and amines	Saturated compounds	[11, 14, 15, 31–33]

MW – Molecular weight, MWCO – molecular weight cut-off, •OH – hydroxide radical, OH – hydroxide, H – hydrogen

Papageorgiou et al. [49] reported that although dichloroacetic acid (DCAA) and trichloroacetic acid (TCAA) were reduced, other DBP species, including brominated HAAs (BrHAAs), increased by almost threefold within the DWTP. This was attributed to the alteration of NOM nature during treatment [49]. Gough et al. [50] attributed the increase of brominated THMs (BrTHMs) during water treatment to the removal of hydrophobic and aromatic fractions and the tendency of remaining hydrophilic fraction to the formation of brominated compounds. However, the change in bromide to dissolved organic carbon (DOC) ratio is also likely to have an impact as treatment processes removed NOM but not bromide.

The advanced treatment using ozonation and activated carbon adsorption is regarded as one of the best available technologies for minimization of THM and HAA and has been applied at large scale in the United States, Europe, Japan, and China. According to one pilot test with direct comparison of multiple processes, ozonation could reduce $29 \pm 67\%$ of total THM formation potential (TTHMFP) and $16 \pm 44\%$ of total HAA formation potential (THAAFP), while granular activated carbon could reduce $22 \pm 43\%$ of TTHMFP and $9 \pm 20\%$ of THAAFP [51].

Table 2 Proposed DBP formation, physical properties, and treatability of NOM groups (from [10])

Group	Humic species	Carboxylic acids	Amino acids	Proteins	Carbohydrates	Reference
Abundance	50–76% of DOC	Uncertain	2–5%	Variable. 1 mg L^{-1} during algal bloom	5–50% of DOC	[8, 36–38]
THMFP	Major source	β-Dicarbonyl species important for THMs and HAAs	Low	Variable	Significant at pH 8	[34, 35, 37, 40–42]
HAAFP	Major source		Significant	Uncertain	Low	
Physical property						
Charge	[a]	[a]	[b]	[b]		
Size	[a]	[c]	[c]	[a]	[b]	
Hydrophobicity	[a]	[b]	[b]	[b]	[b]	
Treatability						
Coagulation	[a]	[c]	[b]	[b]		
Ion exchange	[a]	[a]	[b]	[b]		
Ozone	[a]	[b]	[b]	[b]	[b]	
Biotreatment	[b]	[c]	[a]	[c]	[c]	
Activated carbon	[a]	[c]	[c]	[c]	[c]	
Membranes	[c]	[c]	[a]	[a]	[a]	
AOPs	[a]	[a]	[a]	[a]	[a]	

DOC – dissolved organic carbon, AOPs – advanced oxidation processes

[a] High

[b] Low

[c] Medium

Table 3 Removals of DBPs from different treatment processes at pilot and full-scale water treatment works (adapted from [43])

Treatment process	Scale	Removal efficiency (%)	References
Pre-oxidation (Cl_2), coagulation/flocculation; sand and anthracite filtration, granular activated carbon (GAC) filtration	Pilot scale	THMs 20%; Halonitroalkanes 59%; Haloaldehydes 34%; HAAs 28%; Haloalkanes 20%; Haloketones 13%; Nitrosamines 10%; Halonitriles 9%; Haloalkenes 10%; N-DBPs 5%	[44]
Membrane nanofiltration: ESNA 1-LF2; TS80; NF270	Pilot scale	NDMA 57–83%; HNM 48–87% and THM 72–97%	[45]
Pre-ozonation, coagulation, sand filtration, ozonation, and GAC filtration	Full scale	THMs: 79% for $CHCl_3$; 21% for BDCM; HAAs: 71% for DCAA and TCAA	[46]
Conventional (coagulation, sedimentation, sand filtration, chlorination), advanced (conventional plus ozonation, biologically active GAC filtration, chlorination)	Full scale	DCAcAm 9% in conventional plant and 76% in advanced plant	[47]
Coagulation, ozonation, ceramic membrane ultra-filtration and GAC filtration, chlorination	Full scale	73% for 4 THMs and 75% for 6 HAAs	[48]

DOC – dissolved organic carbon, THMFP – THM formation potential, HAAFP – HAA formation potential, NDMA – N-nitrosodimethylamine, HNM – halonitromethane, $CHCl_3$ – trichloromethane, BDCM – bromodichloromethane, DCAA – dichloroacetic acid, TCAA – trichloroacetic acid, DCAcAm – dichloroacetamide

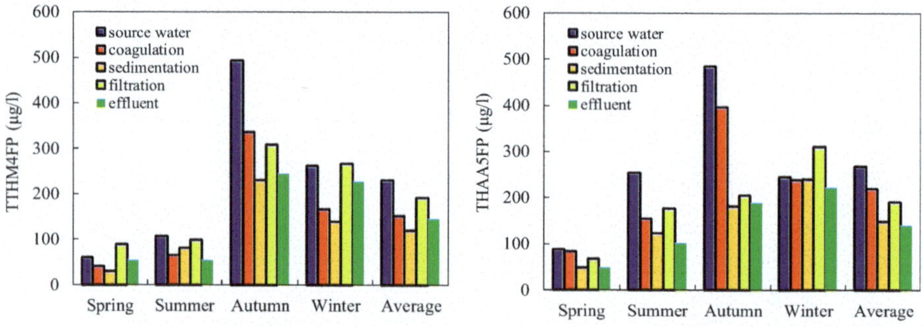

Fig. 2 The seasonal dynamic of total THMs and total HAAs along the treatment train. (Note: TTHM4FP means the sum of formation potential of 4 THMs, including chloroform, bromodichloromethane, dibromochloromethane, bromoform; THAA5FP means the sum of formation potential of 5 HAAs, including monochloroacetic acid, dichloroacetic acid, trichloroacetic acid, monobromoacetic acid, dibromoacetic acid)

During the sampling campaign reported above (Sect. 1) and carried out in 2005 [5], the formation potentials of four THMs and five HAAs were also measured, as shown in Fig. 2. Along the treatment train of conventional processes, the concentration of TTHM4FP and THAA5FP generally decreased. Coagulation and sedimentation removed about 40% of THM and HAA precursors on average. However, there was an obvious increase of TTHM4FP (59%) and THAA5FP (28%) after filtration, which means that the retained organic matters in the filter released THM and HAA precursors continually.

The concentrations of THM and HAA precursors varied greatly with seasons. The TTHM4FP and THAA5FP in source water in spring (March) and summer (May) were relatively low. The concentrations of DBP precursors reached a peak in Autumn (September) due to the growth and reproduction of aquatic organisms. Algae blooms occurred in Autumn in this case with the peak population being over 20 million cells per liter, which could contribute greatly to THM and HAA precursors. According to another investigation [52], the contribution of algae was quantified as 3.96 μgTHAA9/million algae and 1.67 μg TTHM/million algae, which could account for about 20% to 50% of the total formation potentials of THMs and HAAs.

3 DBP in Distribution Systems

There have been many full-scale surveys carried out that have demonstrated that THM levels generally increased with prolonged residence time in distribution system when free chlorine residual was maintained. On the contrary, concentrations of HAAs have been reported to decrease in distribution in some instances [53, 54] but increase in others [55]. Increases of HAAs were ascribed to higher free chlorine concentrations. Although most utilities that used chlorine in the distribution system found the highest concentrations of THMs at the location with the maximum detention time, it has been observed that DCAA as well as other HAAs were significantly reduced in concentration in many samples collected from the locations with the maximum residence time [56]. A further study [57] also found that bacteria at locations with maximum residence time (in a portion of the distribution system where the chlorine residual had greatly decreased) can degrade DCAA and other HAAs, which may help explain why HAAs were reduced to some extent at the maximum residence time.

When chloramination was used to provide a residual, there was no significant increase in HAA9 concentrations observed in distribution systems [58, 59]. Although the behavior of HAAs in distribution can be different from THMs, parameters including chlorine or chloramine residual, retention time, bacterial counts, NOM, pH, and temperature are believed to be important factors affecting the fate of DBPs. Chlorine residual is considered one of the most important parameters controlling DBP concentrations in distribution system with its dual impacts of increasing formation rates and decreasing biological activity. When DBPs were observed to increase, the chlorine residuals were reported to be elevated (3–7 mg/L) [55, 60, 61]; while lower chlorine residuals (0.5–1.5 mg/L) were concurrent with decreases in DBP concentrations [53, 62, 63]. In a distribution system without residual disinfectant, DCAA was stable only in cold water conditions while it was degraded under warm water conditions [64].

Additionally, pipe material is of importance when considering DBP degradation in distribution systems. The considerations are:

- Abiotic reduction of halogenated DBPs in unlined steel and iron pipes [65, 66]
- Corroded iron pipes harboring high densities of bacteria that are protected from the biocidal activity of chlorine by pipe wall and tubercles [67]
- The presence of biofilms occurring in any pipes (e.g., concrete, PVC), although they may be favored on corroded iron

The changes in HAA speciation can give information on the mechanism of degradation. If the reduction only occurs in iron pipes, and trihalogenated HAAs (TXAAs) are preferentially degraded to dihalogenated HAAs (DXAAs) and monohalogenated HAAs (MXAAs), abiotic reduction of HAAs dominates. Alternatively, if degradation of MXAAs and DXAAs causes an overall reduction of HAAs, aerobic biological degradation process dominates [65, 66].

The behavior of N-DBPs in distribution is less researched. A study on regulated and unregulated DBPs [68] measured haloacetonitriles (HANs) and haloacetamides (HAcAms) in drinking water distribution systems in the UK. In chlorinated systems, HANs and HAcAms occurred at median concentrations of 4.8 μg/L and 4.0 μg/L respectively while in chloraminated systems the median yields were 1.5 μg/L and 3.8 μg/L respectively. These values are similar to values reported previously in the USA and UK [69–71] where total HAN and HAcAms concentrations were similar or slightly higher at the end of the distribution system than the DWTP outlet.

Serrano et al. [72] investigated 46 DBPs and looked into their occurrence in DWTP and at two points in distribution system (24 and 48 h) over four seasons. They concluded that:

- Distribution system samples contained five aliphatic aldehydes at similar concentrations to the finished water. Two other aldehydes' concentrations increased slightly with time (total concentration of 10 and 12 μg/L for 24 and 48 h, respectively), while another aldehyde decreased in parallel (from 0.85 to 0.77 μg/L after 24 and 48 h respectively) [72].
- The concentrations of the four THMs in summer were approximately 1.5 and ~3 times higher at 24 h (13 μg/L) and 48 h (22 μg/L), respectively, than those in the finished water at the DWTP exit (8.1 μg/L) [72].
- The total average concentration of volatile species (including THMs, halonitromethanes (HNMs), and HANs) at 24 and 48 h from the DWTP exit increased more in summer than that in other seasons, indicating that warm water temperature favors the formation of volatile DBPs which are consistent with previous studies focusing on the effects of temperature and seasonal conditions on DBP formation.
- An increase in HAA levels in the distribution network was observed in all seasons, which suggests that they increased with increasing residence time, by reactions of residual disinfectant with remaining DBP precursors in water.

Similar to volatile DBPs, the total HAA concentrations found at 48 h were higher in the warm seasons than those in the cool seasons. However, the concentrations of DCAA and TCAA in the distribution system were similar in all seasons and TCAA prevailed in the summer. The seasonal variations of the two prevalent HAAs (DCAA and TCAA) is consistent with the results of other studies focusing on seasonal and spatial changes in a DWTP [72].

3.1 Impact of Water Blending

A recent review of the impact of blending of chlorinated water and chloraminated water was carried out [73]. There are certain circumstances that lead to blending of these types of water. Examples of this include when DWTPs or distribution systems switch from free chlorine to chloramines, or when distribution systems using different disinfectants are interconnected to increase water volume. In these cases, free chlorine and chloramines might meet and interact to form mixed chlorine/chloramine [74]. Additionally, unintended mixed chlorine/chloramine species might occur when re-chlor(am)ination is applied to counteract loss of disinfectant residual without adequate control of chlorine concentration and species [75, 76].

Total chlorine in finished drinking water after chlorination/chloramination includes free chlorine, mono-, di- and tri-chloramines, and organic chloramines [77]. The occurrence of mixed chlorine/chloramine species in water has been described [73] but the focus on the impact of water blending and on DBP formation remains to be studied. It has been reported that the presence of chlorine/chloramine species due to blending of waters is reactive with NOM and the subsequent DBP formation is less controlled and more unpredictable than those treated with chlorine or chloramines alone [59, 78–80].

4 DBP in Small WTPs

Most studies on the occurrence of regulated and non-regulated DBPs have been conducted in large-scale systems [81]. However, small water systems (i.e., serving 5,000 or fewer people) with surface water sources can be more susceptible to higher DBP formation due to: financial constraints, a relatively low capacity to implement adequate treatment technologies to remove contaminants, in particular DBP precursors, and inability to hire qualified operators [82–84]. There is little information on the seasonal occurrence of DBPs in the water from small DWTPs and their distribution networks. Although a number of studies on the occurrence of regulated DBPs have been published [61, 82, 85–88], the data available for non-regulated DBPs is sparse.

With respect to THM occurrence, a study on remote Scottish Water sites gave two examples between 2013 and 2019. The first one was a small island coagulation-based plant (1 MLD) which was replaced with a nanofiltration and reverse osmosis (RO) plant. Results confirmed that although the final water TOC was in average 1.2 mg/L, the THM formation potential of the NF permeate was over 100 μg/L, indicating that the risk of THM was high and that DBPs were not minimized. Installation of an RO yielded a significant drop of THMs with the final water concentrations <5 μg/L. The second one relates to another small island plant installing nanofiltration with a lot of organic carbon passing through it. At this site, chloramination was applied to minimize DBP formation within distribution, which eventually led to significant drops of THMs from values between 40 and 120 μg/L to below 40 μg/L [85]. Studies at Scottish Water [86–88] confirmed that a small but statistically significant increase in reactivity (i.e., increased DBPFP without a corresponding increase in the DOC level) was evident for the Scottish islands plants which are generally small (serving <5,000 people).

The study by Guilherme and Rodriguez [81] reported regulated and unregulated DBPs in 25 small systems in Canada with samples taken monthly for over 1 year. The average measured DBP concentrations were reported to be much higher than those reported in the literature for medium and large systems. The measured average values for THMs was 75 ± 69 μg/L; HAAs, 77 ± 75 μg/L; HANs, 2.5 ± 1.8 μg/L; CPK, 0.4 ± 0.3 μg/L, and HKs, 6.0 μg/L ± 4.5 μg/L. The findings support the hypothesis that small DWTPs serving small populations find it more difficult to minimize DBP precursors in water during disinfection. This finding carried over to the distribution system with higher concentrations of DBPs found than in larger systems.

A potential treatment solution for small systems (sometimes referred to as non-consecutive systems) is localized water treatment [89]. This includes storage tank aeration, GAC adsorption, and biologically active carbon (BAC) filtration. These localized treatments can be used to remove specific groups of DBPs that have similar characteristics [89]. Examples of these include:

- Aeration which is effective in removing volatile DBPs such as THMs [90]
- GAC and biologically active carbon which can be effective in removal of HAAs, chloral hydrate, aldehydes, and keto acids [90]

5 Conclusion

Levels of DBPs were higher when pre-chlorination was used compared to DWTPs that chlorinated only at the end of the treatment train. The removal of precursors varies according to the precursors present, the type of treatment implemented, and seasons. Investigations into DBP formation in distribution system reveal that most DBPs increase in the presence of free chlorine residual but others (HAAs in particular) can be degraded. Small DWTPs were identified to be more likely to produce higher concentrations of DBP than large-scale centralized DWTPs.

References

1. Huang X, Liang H, Yu Y, Shi B (2022) The enhanced treatment of algae-laden water by combination of powdered activated carbon and chlorine. Sep Purif Technol 280:119922. https://doi.org/10.1016/j.seppur.2021.119922
2. Lin JL, Ika AR (2022) Pre-oxidation of microcystis aeruginosa-laden water by intensified chlorination: impact of growth phase on cell degradation and in-situ formation of carbonaceous disinfection by-products. Sci Total Environ 805:150285. https://doi.org/10.1016/j.scitotenv.2021.150285
3. Solsona F (2002) Disinfection. In: Smet J, van Wijk C (eds) Small community water supplies. Delft
4. Niu ZG, Hu XP, Zhang Y, Sun YY (2017) Effect of chlorine dosage in prechlorination on trihalomethanes and haloacetic acids during water treatment process. Environ Sci Pollut Res 24: 5068–5077. https://doi.org/10.1007/s11356-016-8265-x
5. Chen C (2005) Studies on the development of sequential chlorination disinfection process and the optimization of water purification processes to control disinfection by-products. Ph.D. thesis. Tsinghua University, China
6. Cheng X, Dong H, Qiang Z (2023) Formation and transformation of pre-chlorination-formed disinfection byproducts in drinking water treatment process. Sci Total Environ 904:166241. https://doi.org/10.1016/j.scitotenv.2023.166241
7. Croué J-P, Debroux J-C, Amy GL, Aiken GR, Leenheer JA (1999) Natural organic matter: structural characteristics and reactive properties. In: Singer PC (ed) Formation and control of disinfection by-products in drinking water. American Water Works Association, Denver, pp 65–93
8. Sharp EL, Parsons SA, Jefferson B (2006) Seasonal variations in natural organic matter and its impact on coagulation in water treatment. Sci Total Environ 363:183–194. https://doi.org/10.1016/j.scitotenv.2005.05.032
9. Singer PC (ed) (1999) Formation and control of disinfection by-products in drinking water. American Water Works Association, Denver, pp 65–93
10. Bond T. Treatment of disinfection by-product precursors. PhD thesis, Cranfield University, UK. https://dspace.lib.cranfield.ac.uk/server/api/core/bitstreams/15571bcf-079f-474a-b382-c7291f7ab326/content
11. Parsons SA, Jefferson B (2006) Introduction to potable water treatment processes. Blackwell, Oxford
12. Sharp EL, Jarvis P, Parsons SA, Jefferson B (2006) Impact of fractional character on the coagulation of NOM. Colloids Surf A Physicochem Eng Asp 286:104–111. https://doi.org/10.1016/j.colsurfa.2006.03.009
13. Randtke SJ (1999) Disinfection by-product precursor removal by coagulation and precipitate softening. In: Singer PC (ed) Formation and control of disinfection by-products in drinking water. American Water Works Association, Denver, pp 237–258
14. Crittenden JC, Rhodes Trussell R, Hand DW, Howe KJ, Tchobanoglous G (2006) MWH's water treatment: principles and design.2nd edn. Wiley, New York
15. Pontius FW, American Water Works Association (eds) (2006) Water quality and treatment.4th edn. McGraw-Hill, New York
16. Mergen MRD, Jefferson B, Parsons SA, Jarvis P (2008) Magnetic ion-exchange resin treatment: impact of water type and resin use. Water Res 42:1977–1988. https://doi.org/10.1016/j.watres.2007.11.032
17. Johnson CJ, Singer PC (2003) Impact of a magnetic ion exchange resin on ozone demand and bromate formation during drinking water treatment. Water Res 38:3738–3750. https://doi.org/10.1016/j.watres.2004.06.021
18. Bolto B, Dixon D, Eldridge R, King S, Linge K (2002) Removal of natural organic matter by ion exchange. Water Res 36:5057–5065. https://doi.org/10.1016/S0043-1354(02)00231-2

19. Metcalf and Eddy Inc, Tchobanoglous G, Burton FL, Stensel HD (2006) Wastewater engineering treatment and reuse.4th edn. McGraw-Hill, New York
20. Judd S, Judd C (eds) (2006) The MBR book. Elsevier, Amsterdam
21. Chellam S (2000) Effects of nanofiltration on trihalomethane and haloacetic acid precursor removal and speciation in waters containing low concentrations of bromide ion. Environ Sci Technol 34:1813–1820, https://pubs.acs.org. https://doi.org/10.1021/es991153t
22. Van der Bruggen B, Vandecasteele C (2003) Removal of pollutants from surface water and groundwater by nanofiltration: overview of possible applications in the drinking water industry. Environ Pollut 122:435–445. https://doi.org/10.1016/S0269-7491(02)00308-1
23. Weber WJ, McGinley PM, Katz LE (1991) Sorption phenomena in subsurface systems: concepts, models and effects on contaminant fate and transport. Water Res 25:499–528. https://doi.org/10.1016/0043-1354(91)90125-A
24. Kilduff JE, Karanfil T, Chin Y-U, Weber WJ (1996) Adsorption of natural organic electrolytes by activated carbon: a size exclusion chromatography study. Environ Sci Technol 30:1336–1343. https://pubs.acs.org. https://doi.org/10.1021/es950547r
25. Karanfil T, Kitis M, Kilduff JE, Wigton A (1999) Role of granular activated carbon surface chemistry on the adsorption of organic compounds. 2. Natural organic matter. Environ Sci Technol 33:3325–3333. https://pubs.acs.org. https://doi.org/10.1021/es002008e
26. Shukairy HM, Summers RS (1992) The impact of preozonation and biodegradation on disinfection by-product formation. Water Res 26:1217–1227. https://doi.org/10.1016/0043-1354(92)90182-4
27. Hwang CJ, Krasner SW, Schlimenti MJ, Amy GL, Dickenson E, Bruchet A, Prompsy C, Filippi G, Croué J-P, Violleau D (2006) Polar NOM: characterization, DBPs, treatment. American Water Works Association Research Foundation, Denver. Report number 98077
28. Leisinger T (ed) (2006) Microbial degradation of xenobiotics and recalcitrant compounds. Academic Press, London
29. Parsons SA (ed) (2006) Advanced oxidation processes for water and wastewater treatment. IWA, London
30. Westerhoff P, Aiken G, Amy G, Debroux J (1999) Relationships between the structure of natural organic matter and its reactivity towards molecular ozone and hydroxyl radicals. Water Res 33:2265–2276. https://doi.org/10.1016/S0043-1354(98)00447-3
31. Speitel GE, Symons JM, Diehl AC, Sorenson HW, Cipparone LA (1993) Effect of ozone dosage and subsequent biodegradation on removal of DBP precursors. J Am Water Works Ass 85:89–95. https://doi.org/10.1002/j.1551-8833.1993.tb05989.x
32. Miltner RJ, Shukairy HM, Summers RS (1992) Disinfection by-product formation and control by ozonation and biotreatment. J Am Water Works Ass 84:53–62. https://doi.org/10.1002/j.1551-8833.1992.tb05882.x
33. Goel S, Hozalski RM, Bouwer EJ (1995) Biodegradation of NOM: effect of NOM source and ozone dose. J Am Water Works Ass 87:90–105. https://doi.org/10.1002/j.1551-8833.1995.tb06304.x
34. Hureïki L, Croué JP, Legube B (1994) Chlorination studies of free and combined amino acids. Water Res 28:2521–2531. https://doi.org/10.1016/0043-1354(94)90070-1
35. Navalon S, Alvaro M, Garcia H (2008) Carbohydrates as trihalomethanes precursors. Influence of pH and the presence of Cl− and Br− on trihalomethane formation potential. Water Res 42:3990–4000. https://doi.org/10.1016/j.watres.2008.07.011
36. Navalon S, Alvaro M, Alcaina I, Garcia H (2010) Multi-method characterization of DOM from the Turia river (Spain). Appl Geochem 25(11):1632–1643. https://doi.org/10.1016/j.apgeochem.2010.08.011
37. Scully FEJ, Howell GD, Kravitz R, Jewell JT, Hahn V, Speed M (1988) Proteins in natural waters and their relation to the formation of chlorinated organics during water disinfection. Environ Sci Technol 22:537–542. https://doi.org/10.1021/es00170a009
38. Thurman EM (2006) Organic geochemistry of natural waters. Nijhoff/Junk Publishers, Dordrecht

39. Mitch WA, Krasner SW, Westerhoff P, Dotson A (2009). Report number 91250) Occurrence and formation of nitrogenous disinfection by-products. Water Research Foundation, Denver
40. Croué J-P, Korshin GV, Benjamin MB (2006). Report number 90780) Characterization of natural organic matter in drinking water. American Water Works Association Research Foundation, Denver
41. Dickenson ER, Summers RS, Croué J-P, Gallard H (2008) Haloacetic acid and trihalomethane formation from the chlorination and bromination of aliphatic β-dicarbonyl acid model compounds. Environ Sci Technol 42:3226–3233. https://doi.org/10.1021/es0711866
42. Reckhow DA, Kim J (2008) Role of proteins and amino acids in DBP formation, IWA conference on NOM: from source to tap, Bath
43. Gilca AF, Teodosiu C, Fiore S, Musteret CP (2020) Emerging disinfection byproducts: a review on their occurrence and control in drinking water treatment processes. Chemosphere 259: 127476. https://doi.org/10.1016/j.chemosphere.2020.127476
44. Fu J, Lee WN, Coleman C, Nowack K, Carter J, Huang CH (2017) Removal of disinfection byproduct (DBP) precursors in water by two-stage biofiltration treatment. Water Res 123:224–235. https://doi.org/10.1016/j.watres.2017.06.073
45. Ersan MS, Ladner DA, Karanfil T (2016) The control of N-nitrosodimethylamine, halonitromethane, and trihalomethane precursors by nanofiltration. Water Res 105:274–281. https://doi.org/10.1016/j.watres.2016.08.065
46. Papageorgiou A, Papadakis N, Voutsa D (2016) Fate of natural organic matter at a full-scale drinking water treatment plant in Greece. Environ Sci Pollut Res 23:1841–1851. https://doi.org/10.1007/s11356-015-5433-3
47. Lin T, Zhou D, Yu S, Chen W (2016) The removal process of 2,2-dichloroacetamide (DCAcAm), a new disinfection by-product, in drinking water treatment process and its toxicity on zebrafish. Chemosphere 159:403–411. https://doi.org/10.1016/j.chemosphere.2016.06.029
48. Fan X, Tao Y, Wang L, Zhang X, Lei Y, Wang Z, Noguchi H (2014) Performance of an integrated process combining ozonation with ceramic membrane ultra-filtration for advanced treatment of drinking water. Desalination 335:47–54. https://doi.org/10.1016/j.desal.2013.12.014
49. Papageorgiou A, Voutsa D, Papadakis N (2014) Occurrence and fate of ozonation by-products at a full-scale drinking water treatment plant. Sci Total Environ 481:392–400. https://doi.org/10.1016/j.scitotenv.2014.02.069
50. Gough R, Holliman PJ, Willis N, Freeman C (2014) Dissolved organic carbon and trihalomethane precursor removal at a UK upland water treatment works. Sci Total Environ 468–469: 228–239. https://doi.org/10.1016/j.scitotenv.2013.08.048
51. Chen C, Zhang XJ, He WJ, Han HD (2007) Comparison of seven kinds of drinking water treatment processes to enhance organic material removal: a pilot test. Sci Total Environ 382:93–102. https://doi.org/10.1016/j.scitotenv.2007.04.012
52. Chen C, Zhang XJ, Zhu L, Liu J, He W, Han H (2008) Disinfection by-products and their precursors in a water treatment plant in North China: seasonal changes and fraction analysis. Sci Total Environ 397:140–147. https://doi.org/10.1016/j.scitotenv.2008.02.032
53. LeBel GL, Benoit FM, Williams DT (1997) A one-year survey of halogenated disinfection by-products in the distribution system of treatment plants using three different disinfection processes. Chemosphere 34(11):2301–2317. https://doi.org/10.1016/S0045-6535(97)00042-8
54. Williams DT, Lebel GL, Benoit FM (1997) Disinfection by-products in Canadian drinking water. Chemosphere 34(2):299–316. https://doi.org/10.1016/S0045-6535(96)00378-5
55. Singer PC, Obolensky A, Greiner A (1995) DBPs in chlorinated North Carolina drinking waters. J Am Water Works Ass 87(10):83–92. https://doi.org/10.1002/j.1551-8833.1995.tb06437.x
56. Williams SL, Rindfleisch DF, Williams RL (1994) Deadend on haloacetic acids (HAA). Proceedings of AWWA water quality technology conference, San Francisco, CA, USA

57. Williams SL, Williams RL, Gordon AS (1996) The impact of bacterial degradation of Haloacetic acids (HAA) in the distribution system. Proceedings of AWWA water quality technology Conf., Boston, MA, USA
58. Goslan EH, Krasner SW, Bower M, Rocks SA, Holmes P, Levy LS, Parsons SA (2009) A comparison of disinfection by-products found in chlorinated and chloraminated drinking waters in Scotland. Water Res 43:4698–4706. https://doi.org/10.1016/j.watres.2009.07.029
59. Bougeard C, Goslan EH, Jefferson B, S.A. (2010) Parsons comparison of the disinfection by-product formation potential of treated waters exposed to chlorine and monochloramine. Water Res 44(3):729–740. https://doi.org/10.1016/j.watres.2009.10.008
60. Krasner SW, McGuire MJ, Jacangelo JG, Patania NL, Reagan KM, Aieta EM (1989) The occurrence of disinfection by-products in US drinking water. J Am Water Works Ass 81(8): 41–53. https://doi.org/10.1002/j.1551-8833.1989.tb03258.x
61. Nieminski CE, Chaudhuri S, Lamoreaux T (1993) The occurrence of DBPs in Utah drinking waters. J Am Water Works Ass 85(9):98–105. https://doi.org/10.1002/j.1551-8833.1989.tb03258.x
62. Chen WJ, Weisel CP (1998) Halogenated DBP concentrations in a distribution system. J Am Water Works Ass 90(4):151–163. https://doi.org/10.1002/j.1551-8833.1998.tb08418.x
63. Tung H, Xie YF (2009) Association between haloacetic acid degradation and heterotrophic bacteria in water distribution systems. Water Res 43:971–978. https://doi.org/10.1016/j.watres.2008.11.041
64. Baribeau H, Krasner SW, Chinn R, Singer PC (2000) Impact of biomass on the stability of haloacetic acids and trihalomethanes in a simulated distribution system. Proceedings of AWWA water quality technology conference, Salt Lake City, UT, USA
65. Hozalski RM, Zhang L, Arnold WA (2001) Reduction of haloacetic acids by Fe(0): implications for treatment and fate. Environ Sci Technol 35(11):2258–2263. https://doi.org/10.1021/es001785b
66. Zhang L, Arnold WA, Hozalski RM (2004) Kinetics of haloacetic acid reactions with Fe(0). Environ Sci Technol 38:6881–6889. https://doi.org/10.1021/es049267e
67. LeChevallier MW, Welch NJ, Smith DB (1996) Full-scale studies of factors related to coliform regrowth in drinking water. Appl Environ Microbiol 62(7):2201–2211. https://doi.org/10.1128/aem.62.7.2201-2211.1996
68. Sfynia C, Bond T, Kanda R, Templeton M (2022) Simultaneous prediction of trihalomethanes, haloacetic acids, haloacetonitriles and haloacetamides using simulated distribution system tests. Environ Sci Water Res Technol 8:742–756. https://doi.org/10.1039/D1EW00824B
69. Krasner SW, Weinberg HS, Richardson SD, Pastor SJ, Chinn R, Sclimenti MJ, Onstad GD, Thruston AD (2006) Occurrence of a new generation of disinfection byproducts. Environ Sci Technol 40(23):7175–7185. https://doi.org/10.1021/es060353j
70. Templeton MR, Kanda R, Graham N, Kamal HM, Bond T (2012) Monitoring of nitrogenated DBPs in drinking water, report for the drinking water inspectorate (DWI), UK
71. Bond T, Templeton MR, Kamal NHM, Graham N, Kanda R (2015) Nitrogenous disinfection byproducts in English drinking water supply systems: occurrence, bromine substitution and correlation analysis. Water Res 85:85–94. https://doi.org/10.1016/j.watres.2015.08.015
72. Serrano M, Montesinos I, Cardador MJ, Silva M, Gallego M (2015) Seasonal evaluation of the presence of 46 disinfection by-products throughout a drinking water treatment plant. Sci Total Environ 517:246–258. https://doi.org/10.1016/j.scitotenv.2015.02.070
73. Pan R, Zhang T-Y, He H, Zheng Z-X, Dong Z-Y, Zhao H-X, Xu M-Y, Luo Z-N, Hu C-Y, Tang Y-L, Gamal El-Din M, Xu B (2023) Mixed chlorine/chloramines in disinfected water and drinking water distribution systems (DWDSs): a critical review. Water Res 247:120736. https://doi.org/10.1016/j.watres.2023.120736
74. Farley CG, Gentzler R, Flythe CW (2012) What happens when chloramines and chlorine meet? Opflow 38(3):22–26. https://www.jstor.org/stable/opflow.38.3.22

75. Karthik NB, Krishna K, Sathasivan A (2022) Factors controlling the effectiveness of rechlor (am)ination to recover chloramine from nitrification. Sci Total Environ 806(3):151322. https://doi.org/10.1016/j.scitotenv.2021.151322
76. Li RA, McDonald JA, Sathasivan A, Khan SJ (2021) A multivariate Bayesian network analysis of water quality factors influencing trihalomethanes formation in drinking water distribution systems. Water Res 190:116712. https://doi.org/10.1016/j.watres.2020.116712
77. Zhang TY, Lin YL, Xu B, Hu CY, Wang Z, Tian FX, Gao NY (2015) Identification and quantification of ineffective chlorine by $NaAsO_2$ selective quenching method during drinking water disinfection. Chem Eng J 277:295–302. https://doi.org/10.1016/j.cej.2015.04.126
78. Huang G, Blackstock L, Jiang P, Liu ZS, Lu XF, Li XF (2019) Formation, identification, and occurrence of new bromo- and mixed halo-tyrosyl dipeptides in chloraminated water. Environ Sci Technol 53(7):3672–3680. https://doi.org/10.1021/acs.est.8b06831
79. Li GW, Ma X, Chen RY, Yu Y, Tao H, Shi BY (2019) Field studies of manganese deposition and release in drinking water distribution systems: insight into deposit control. Water Res 163: 114897. https://doi.org/10.1016/j.watres.2019.114897
80. Mitch WA, Sedlak DL (2002) Formation of N-nitrosodimethylamine (NDMA) from dimethylamine during chlorination. Environ Sci Technol 36(4):588–595. https://doi.org/10.1021/es010684q
81. Guilherme S, Rodriguez MJ (2014) Occurrence of regulated and non-regulated disinfection by-products in small drinking water systems. Chemosphere 117:425–432. https://doi.org/10.1016/j.chemosphere.2014.08.002
82. Charrois JWA, Graham D, Hrudey SE, Froese KL (2004) Disinfection by-products in small Alberta community drinking water supplies. J Toxicol Environ Health A Curr Issues 67(20–22): 1797–1803. https://doi.org/10.1080/15287390490492494
83. Coulibaly HD, Rodriguez MJ (2004) Development of performance indicators for small Quebec drinking water utilities. J Environ Manage 73:243–255. https://doi.org/10.1016/j.jenvman.2004.07.003
84. Edwards JE, Henderson SB, Struck S, Kosatsky T (2012) Characteristics of small residential and commercial water systems that influence their likelihood of being on drinking water advisories in rural British Columbia, Canada: a cross-sectional study using administrative data. J Water Health 10(4):629–649. https://doi.org/10.2166/wh.2012.046
85. Murray CA, Booker N, Parsons SA (2020) Editorial perspectives: Scottish water case study. Environ Sci Water Res Technol 6:2257–2258. https://doi.org/10.1039/D0EW90038A
86. Golea DM, Upton A, Jarvis P, Moore G, Sutherland S, Parsons SA, Judd SJ (2017) THM and HAA formation from NOM in raw and treated surface waters. Water Res 112:226–235. https://doi.org/10.1016/j.watres.2017.01.051
87. Valdivia-Garcia M, Weir P, Frogbrook Z, Graham DW, Werner D (2016) Climatic, geographic and operational determinants of trihalomethanes (THMs) in drinking water systems. Sci Rep 6: 35027. https://doi.org/10.1038/srep35027
88. Sutherland S, Parsons SA, Daneshkhah A, Jarvis P, Judd SJ (2015) THM precursor rejection by UF membranes treating Scottish surface waters. Sep Purif Tecnol 149:381–388. https://doi.org/10.1016/j.seppur.2015.06.009
89. McGuire MJ, Karanfil T, Krasner SW, Reckhow DA, Roberson JA, Summers RS, Westerhoff P, Xie PY (2014) Not your granddad's disinfection by-product problems and solutions. J Am Water Works Ass 106(8):54–73. https://doi.org/10.5942/jawwa.2014.106.0128
90. Wang X, Mao Y, Tang S, Yang H, Xie Y (2014) Disinfection byproducts in drinking water and regulatory compliance: a critical review. Front Environ Sci Eng 9(1):3–15. https://doi.org/10.1007/s11783-014-0734-1

DBP Formation and Control in Water Reuse

Qianyuan Wu, Liu He, Xiao Xiao, DeXiu Wu, and Ron Hofmann

Contents

1 Potential Risk of DBPs in Reclaimed Water 274
1.1 Utilization and Exposure of Reclaimed Water 274
1.2 Chemical Disinfection and Oxidation to Treat Reclaimed Water 275
1.3 Characterization of DBP Precursors in Reclaimed Water 276
1.4 Formation and Potential Risks of DBP in Reclaimed Water 278
2 Chlorine Disinfection Byproducts 278
2.1 DBP Formation and Toxicity Change During Chlorine Disinfection 278
2.2 Control of Precursors of Chlorine Disinfection Byproducts 281
2.3 Control of Chlorine Disinfection Byproducts 283
3 Ultraviolet-Based Advanced Oxidation Byproducts 285
3.1 Biostability 286
3.2 Organohalide Byproducts 287
3.3 Impact on Downstream DBP Formation Potential 288
3.4 Transformation Products 290
3.5 Nitrosamines 290
3.6 Inorganics: Perchlorate, Chlorite, Chlorate, and Bromate 291
4 Ozonation Byproducts 292
4.1 Ozone DBP Formation 292
4.2 Toxicity Changes 293
References 294

Abstract DBP formation in water reuse treatment may follow trends that do not necessarily follow what is commonly observed in drinking water treatment. This is because organic precursors in treated wastewater can exhibit different characteristics

Q. Wu, L. He, X. Xiao, and D. Wu
Shenzhen International Graduate School, Tsinghua University, Shenzhen, People's Republic of China
e-mail: wu.qianyuan@sz.tsinghua.edu.cn

R. Hofmann (✉)
Department of Civil and Mineral Engineering, University of Toronto, Toronto, ON, Canada
e-mail: ron.hofmann@utoronto.ca

Chao Chen, Susan Andrews, and Yuefeng Xie (eds.), *Disinfection By-Products in Water*, Hdb Env Chem (2024) 135: 273–302, DOI 10.1007/698_2024_1149,

Published online: 24 September 2024

than organics found in nature, and also the treatment/oxidation methods employed in water reuse may be different – typically more powerful – such as ozonation or advanced oxidation. As more parts of the world experience water stress due to increased urbanization and climate change, using recycled wastewater as a source of drinking water or grey water will become increasingly common. This chapter highlights current knowledge about DBP formation in this context.

Keywords Advanced oxidation, EFOM, Recycling, Reuse, Wastewater

1 Potential Risk of DBPs in Reclaimed Water

1.1 Utilization and Exposure of Reclaimed Water

Water reclamation (also called reuse or recycling) is becoming increasingly important as a strategy to reduce water stress in many parts of the world. Reclaimed water can be used in industry, agriculture, landscaping, and even for potable reuse [1–3]. Due to the growing utilization of reclaimed water, it is important to explore the potential risk for contaminants in the water to adversely affect both the environment and human health [4]. For example, when using reclaimed water for agriculture, plants can be affected by contaminants such as residual chlorine and heavy metals, and reclaimed water for replenishing rivers can lead to accumulation of pollutants in large fish through biomagnification up the food chain, increasing the risk to human health when eating the fish [5]. Furthermore, as reclaimed water enters the watershed, various harmful chemicals may pose a risk to the bacterial community of the soil [6].

From the perspective of human health, there are three main routes of exposure to contaminants that may be present in reclaimed water, including oral intake, skin absorption, and inhalation [7, 8]. Oral intake means exposure through drinking or eating. Since most reclaimed water is used for non-potable purposes, exposure through oral intake does not occur frequently [5]. This may change in the future as direct potable reuse becomes more common. Skin absorption usually occurs when people are in direct contact with reclaimed water such as through swimming or occupational exposure (e.g., landscaping) [9]. Inhalation refers to routes through breathing, and is considered one of the most important exposure routes for reclaimed water due to vectors such as mists from cooling towers, or lawn irrigation spraying [10]. A list of examples of human exposure to reclaimed water is shown in Fig. 1, along with predicted magnitudes of exposure as estimated by He et al. [1].

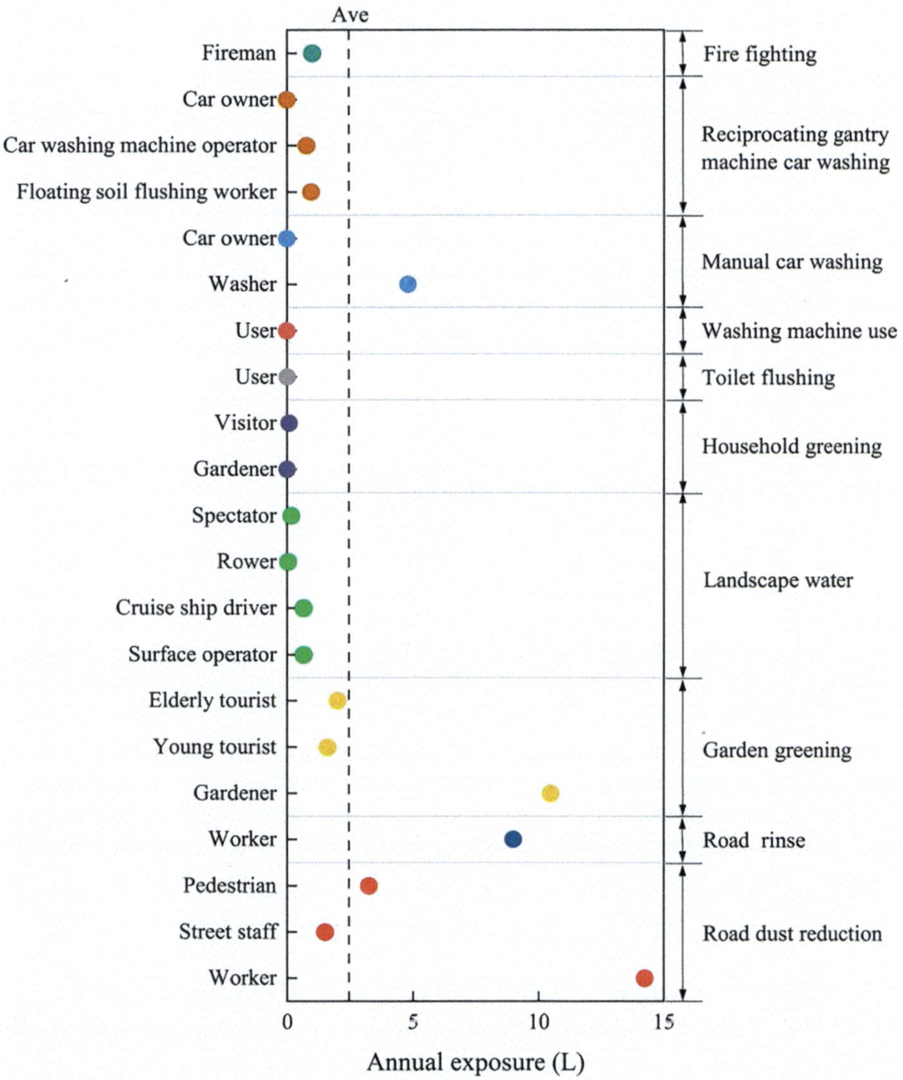

Fig. 1 Annual exposure of reclaimed water in different reused scenarios. Adapted from He et al. with permission [1]

1.2 Chemical Disinfection and Oxidation to Treat Reclaimed Water

There are four common stages in reclaimed water treatment, including primary treatment (such as screening, filtration, centrifugation, and sedimentation to remove the suspended materials from water), secondary treatment (generally biological methods such as activated sludge process and biological membrane treatment), tertiary treatment (including chemical disinfection and oxidation, precipitation, membrane filtration and reverse osmosis), and natural processes (such as the natural purification process of wetlands) [6]. Among these treatment methods, disinfection is critical for inactivating pathogenic organisms and preventing the spread of waterborne diseases to the environment. Common disinfectants for reclaimed water treatment include chlorine, chlorine dioxide, ozone, and ultraviolet light [11].

Disinfection with chlorine is the technology most extensively utilized to ensure the safety of reclaimed water [11]. Chlorine can effectively inactivate pathogens and can be useful in controlling the regrowth of microorganisms in distribution systems. Additionally, chlorine can effectively destroy biofilms during backflushing of biological activated carbon [12].

Chlorine dioxide (ClO_2) is known as an alternative disinfectant to chlorine, which has been widely used to control waterborne pathogens in wastewater treatment [13]. Compared with chlorine, ClO_2 mainly produces chlorite and chlorate rather than toxic halogenated disinfection products [14, 15].

Ozone is a well-known disinfectant that has been applied in water oxidation for decades [16]. Ozone reacts relatively fast with microorganisms and can be effective in inactivating chlorine-resistant microorganisms such as *Cryptosporidium parvum* [17].

UV light is also widely used in the treatment of reclaimed water. It is a very economical disinfectant, and at the doses applied for disinfection, it is unlikely to cause significant changes to the chemistry of the water (i.e., virtually no DBP formation). However, higher doses of UV can be applied together with an oxidant such as hydrogen peroxide to form hydroxyl radicals which are very strong oxidants that can destroy many chemical contaminants. In this application, the potential for DBP formation is more significant. More is discussed on this topic later in this chapter.

1.3 Characterization of DBP Precursors in Reclaimed Water

Dissolved organic matter (DOM) is a mixture of aromatic and aliphatic carbon-containing compounds that are widely found in reclaimed water sources. Common DOM in reclaimed water sources includes soluble microbial products (SMPs), transformation products from biotic or abiotic treatment processes, and emerging contaminants such as pharmaceuticals and personal care products (PPCPs) [18]. Numerous studies have been conducted to show that DOM is an important precursor of disinfection byproducts (DBPs) [19–21].

Dissolved organic carbon (DOC) is the most common indicator of DOM concentration [21]. The DOC level in reclaimed water has been reported by Hu et al. to range from 3.6 to 14.6 mg-C/L, with a median of 7.5 mg-C/L, based on a survey of hundreds of sites worldwide. In contrast, the same survey reported that concentrations of DOC in drinking water sources were typically lower, ranging from

1.5 to 11.2 mg/L, with a median of 3.9 mg/L [21]. The high DOC in reclaimed water would inevitably lead to an increased potential for DBP formation.

Dissolved organic nitrogen (DON) is an important indicator of the presence of organic nitrogen compounds in DOM [22]. The survey of reclaimed waters by Hu et al. found that DON ranged from 0.29 to 4.3 mg-N/L, and showed a clear trend of increasing with the concentration of DOC [21]. Nitrogenous proteins or amino acids produced by microbial utilization of DOC are the main pathway for the accumulation of DON in reclaimed water sources [23]. Studies have shown that DON is a precursor not only to trihalomethanes (THMs) and haloacetic acids (HAAs), but also to potentially more toxic nitrogen-containing DBPs (N-DBPs) such as haloacetonitriles, halonitromethanes, nitrosamines, and haloamides [24–27]. Increased DON concentrations in reclaimed water sources are expected to promote the formation of these N-DBPs.

The DON/DOC ratio can provide some indication of the proportion of N-DBPs that may be generated during the disinfection process. For example, during chlorination in one study, there was a positive correlation between the DON/DOC ratio of reclaimed water and the ratio of dichloroacetonitrile (DCAN) to chloroform in the treated water [27]. *N*-nitrosodimethylamine (NDMA) formation also increased with the increase in DON/DOC ratio [28]. The average of the DON/DOC ratio in reclaimed water was 0.18 mg-N/mg-C, while the average DON/DOC in potable water was only 0.04 mg-N/mg-C [21]. Therefore, a higher percentage of N-DBPs might be generated after disinfection of reclaimed water sources compared to drinking water sources.

Specific ultraviolet absorbance at 254 nm ($SUVA_{254}$) is defined as the UV absorbance of a water sample at 254 nm divided by the DOC concentration. $SUVA_{254}$ is an important parameter to characterize the density of aromatic carbon [29]. Hu et al. showed that the average of $SUVA_{254}$ in 117 samples of reclaimed water sources was 1.73 L/(mg-C-m) [21]. The density of aromatic carbon can affect the formation of DBPs during the subsequent disinfection of water samples. For example, it was found that $SUVA_{254}$ was positively correlated with the formation of HAAs [30]. However, as a generic parameter, $SUVA_{254}$ will not provide specific information on functional groups in DOM from different sources. In some cases, different DOM with similar $SUVA_{254}$ values showed different reactivity to subsequent disinfection [29].

The excitation-emission-matrix (EEM) spectra can provide information on the type, structure, and abundance of functional groups of organic molecules with chromophore and fluorophore fractions in DOM by matching excitation and emission wavelengths [31, 32]. The fluorescent signals of EEM spectra were used to demonstrate the presence of humic and fulvic acids in reclaimed water sources, and they were identified as the precursors of chlorine DBPs [30]. Through characterization of EEM spectra, humic and fulvic acids were also successfully demonstrated to be potential precursors of anti-estrogenic DBPs in reclaimed water sources [33]. The proteins were likewise identified by EEM spectra in reclaimed water sources, and they were considered as major precursor sources of unknown N-DBPs [34].

Disinfection is a critical process in reclaimed water treatment. Given the rich content of disinfection precursors in reclaimed water sources, the generation of DBPs cannot be ignored. The adoption of necessary pre-treatment or post-treatment processes to reduce the content of DBPs in reclaimed water is essential to guarantee the safety of reclaimed water.

1.4 Formation and Potential Risks of DBP in Reclaimed Water

The generally higher concentration of DOC and DON in reclaimed water compared to drinking water suggests that there may be a higher potential formation of DBPs and higher percentage of N-DBPs in reclaimed water [11, 21, 25, 35]. For example, some reports have shown that several classes of DBPs such as chlorohydrate (TCA) and *N*-nitrosodimethylamine (NDMA), which exhibit an elevated level of genotoxity and carcinogenity, are more likely to be generated during wastewater disinfection compared to drinking water disinfection [24, 26]. Despite the large amount of research on the formation of DBPs during reclaimed water disinfection, it is not possible to detect every byproduct because of limitations in sample preparation, analytical methods, and the cost of labour and materials. Instead, biotoxicity tests have been used to provide a more direct assessment of health risk of DBPs in reclaimed water. Biotoxicity assays widely used for water risk assessment include those of acute toxicity, such as the luminescent bacteria test and *Daphnia magna* test, evaluation of genotoxicity by the umu-test, and assays to investigate endocrine disruption and tests of cytotoxicity [36]. Notably, different bioassay methods have exhibited different trends so it may be prudent to perform multiple different bioassays to obtain different end-points to better access the overall risk of DBPs.

2 Chlorine Disinfection Byproducts

2.1 DBP Formation and Toxicity Change During Chlorine Disinfection

2.1.1 Formation of DBPs During Chlorine Disinfection

Figure 2 shows the levels of DBPs found in reclaimed water after chlorine disinfection using chlorination or chloramination reported by Du et al. (2017b), which included 156 wastewater samples [11]. Chlorinated DBPs are the main DBPs produced by chlorine disinfection, including regulated DBPs like THMs and HAAs, and different classes of emerging DBPs like haloketones, haloacetonitriles, and haloacetamides [30, 37]. Brominated DBPs were also generated to varying degrees when bromide ions were present in the reclaimed water. Chloramine could

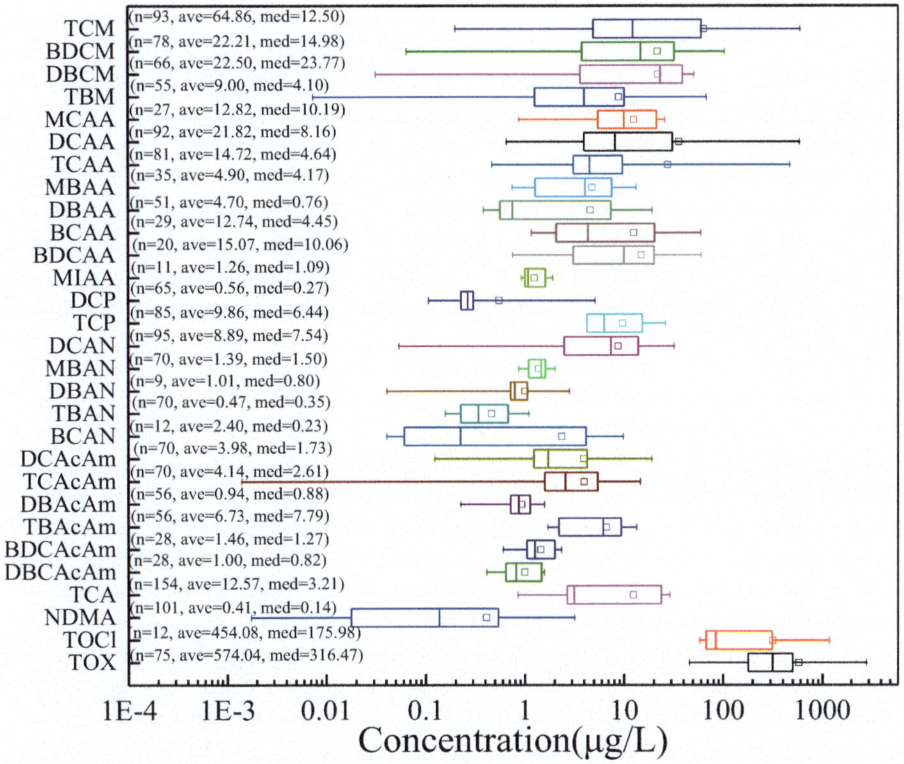

Fig. 2 Levels of DBPs found in reclaimed water during chlorination. The two whiskers in the box chart represent the 5th and 95th percentiles. The three lines in the box chart represent the 25th, 50th, and 75th percentiles. Adapted from Du et al. with permission [11]

also generate iodoacetic acid at the μg/L level while chlorine usually oxidizes iodide to iodate [38, 39].

2.1.2 Changes of Toxicity During Chlorine Disinfection

When applying a disinfectant or oxidant to a wastewater for the purposes of reclamation, toxicity can be formed due to the generation of DBPs, but it can also be reduced through the destruction of toxic species that were in the water to begin with. Some studies have explored the overall effect of disinfection/oxidation on toxicity of such waters. For example, it has been reported that chlorination led to a significant increase in the relative toxicity according to the luminescent bacteria test (defined as the change of ratio of toxicity change during chlorination), with generally up to 2 to 3 times increase in the level of the original toxicity [13, 40]. This indicated the generation of toxic DBPs during chlorination as the residual chlorine was quenched before the bioassays [41]. The same result has been seen in toxicity assays using *Daphnia magna*, while some studies have reported an opposite trend [42, 43]. In 2017, Du et al. collected the toxicity data of 24 wastewater samples from 17 studies, including secondary effluent and tertiary effluent [11]. According to the data, 78% of reports indicate an increase in toxicity to *Daphnia magna*, while a decreasing trend was reported by other researchers. In previous research, Blatchley et al. [42] claimed that there is little change in the toxicity of wastewater during chlorination. In contrast, some reports emphasized that a significant increase in *Daphnia magna* toxicity could usually be detected. Notably, several studies reported a mortality of 100% despite variations in chlorine dosage and contact time, which suggested the formation of some DBPs with high *Daphnia magna* toxicity during chlorination [43].

Compared with other bioassay methods, that of genotoxicity is more standardized, but the results of genotoxicity changes during chlorination are more complex than the changes according to other bioassays [11]. About 35% of reports showed that the genotoxicity was reduced by more than a half during chlorination. However, different reports showed very different results. The toxicity values could be reduced by more than 80% compared with the original water samples, or increase by 2–4 times compared to that before chlorination in other studies [44–46]. These results indicate that water quality of water samples might result in the variations in genotoxicity changes in different studies [46].

As for the estrogenic activity, a decreasing trend was always seen after chlorination [11]. Generally, a 40–90% decrease could be detected in estrogenic activity after chlorination, which indicated that chlorine had an effective effect on transforming the estrogenic chemicals to products with less estrogenic activity [13, 47–50]. In contrast, both increases and decreases were detected in estrogenic activity in the same wastewater effluent on different occasions in a study, with one sample showing an estrogenic activity increase of almost 4 times after chlorination [13]. It was possible that there was a combined action of industrial estrogen mimics, natural estrogens, and anti-estrogens in the samples of these studies which caused the resulting increasing estrogenic activity upon chlorination [11].

Previous research has shown that chlorination can also produce anti-estrogenic DBPs, with a reported increase in anti-estrogenic activity by 1 to 7.5 times [38, 48].

Cytotoxicity assays using mammalian cells are widely used in the assessment of health risk of reclaimed water because they evaluate toxic effects on cells that are presumably more representative of human health risk than other bioassay methods [51]. Generally, the cytotoxicity level of reclaimed water using mammalian cells (Chinese hamster ovary cells) shows an increase of 1 to 6 times following chlorination according to the review [11].

Another method of predicting the potential toxicity of DBPs in reclaimed water is to use computational modelling. Computational modelling, like quantitative structure-activity relationships (QSARs), are efficient in evaluating the biological effects of chemicals, especially those lacking standard compounds, and doing so quickly and at low cost [52–54]. QSAR models use the structure and composition of chemicals to predict their toxic endpoints by comparing the chemical structure of predicted compounds with the experimental data [55, 56]. QSAR modelling has predicted that chlorination of water for reclamation can increase the water's toxicity [57]. For example, Lu et al. claimed that some Cl-DBPs with high acute toxicity could be formed during tannic acid chlorination. The toxicity to fathead minnows of the newly discovered Cl-DBPs was much higher than that of controlled Cl-DBPs like trichloromethane and monochloroacetic acid, and some of them might be mutagenic according to the QSAR models [57].

2.2 *Control of Precursors of Chlorine Disinfection Byproducts*

2.2.1 Coagulation and Sedimentation

Coagulation and sedimentation are considered to be the most economical and efficient methods for the removal of DBP precursors of chlorination. Such treatment reduces DBP formation indicators such as DON, DOC, and UV_{254} of reclaimed water. This treatment also has been reported to reduce fluorescence intensity, especially for SMP-like compounds [58].

As a pre-treatment for chlorination, coagulation and sedimentation removed the formation potential of trihalomethanes (THMs), haloacetic acids (HAAs), and developmental toxicity to zebra fish embryos with removal rates of 45–86%, 3–61%, and 39–64%, respectively [43, 59–62]. In addition, Liu et al. (2019) reported that coagulation and sedimentation reduced DBP formation potential (DBPFP) of bromoacetonitrile (BCAN), dichloroacetonitrile (DCAN), dibromoacetonitrile (DBAN), trichloroacetonitrile (TCAN), and trichloroacetone (1,1,1-TCP) by about 17–22% [58].

However, *N*-nitrosodimethylamine formation potential (NDMAFP) of chloramination may be elevated after coagulation and sedimentation. Some dimethylamine-containing coagulants can serve as precursors to NDMA formation during chloramination [63]. Another study showed that the coagulant polyacrylamide could act as a DBP precursor of chlor(am)ination, increasing the formation of haloacetamides in the treated water [64]. When selecting coagulants, the effectiveness of the agent in controlling the original precursors in the effluent needs to be balanced against the risks when the agent itself acts as a precursor.

2.2.2 Activated Carbon Adsorption

Activated carbon (AC) adsorption can be an effective DBP precursor removal technique. Several studies showed that AC preferentially removed UV_{254} of reclaimed water sources compared to DOC, while it was ineffective in removing DON [65, 66].

The ratio of the concentration of trihalomethanes (THMs), haloacetic acids (HAAs), and haloacetonitriles (HANs) to the concentration of DOC decreased to varying degrees after chlorination of AC-treated water samples [67].

Bromide, compared to DOC, is reportedly not effectively removed by AC, resulting in a higher bromide/DOC ratio in the treated water than in the influent. This can result in an elevated concentration of brominated DBPs in the chlorinated effluent, which have been found to be more genotoxic than chlorinated analogues [66]. However, Peterson et al. (2022) found that AC pretreatment still reduced the cumulative toxicity of typical DBPs in the presence of bromide compared to controls with no AC pretreatment [67]. In other words, AC pretreatment may shift the DBP speciation towards more brominated compounds that are inherently more toxic, but Peterson et al. found that the AC reduced the total DBP formation to a point where there was nevertheless a net benefit in terms of toxicity reduction.

2.2.3 Ozonation

In recent years, ozone has attracted attention in reclaimed water treatment because of its efficient oxidizing ability to control micropollutants and destroy DBP precursors. Several studies have explored the effect of ozone on DBPFP when used as a pretreatment for chlorination. The removal efficiency of ozone on precursor-related indicators of reclaimed water sources followed UV_{254} > DON > DOC [58]. Another study similarly reported a sustained decrease in $SUVA_{254}$ of reclaimed water sources during ozonation [68]. Analysis of the EEM spectra revealed that ozone was able to reduce the fluorescence intensity in all regions. As with coagulation and precipitation, the fluorescence of SMP-like compounds was more easily removed by ozone [58].

Wert et al. [69], in a study of a tertiary municipal wastewater, reported that through degradation of precursors, ozone pretreatment reduced the total organic halide formation potential (TOXFP) of chlorination by at least 20%. However, ozone increased THMFP and HAAFP of chlorination [43]. Additionally, Liu et al. (2019) reported that the formation of trichloronitromethane (TCNM), 1,1-DCP, and 1,1,1-TCP increased when ozonation was used as a pretreatment for chlorination [58]. Research has also shown that more hydrophilic organics with smaller molecular weights containing -C=O may be formed during ozonation, which could be used to explain the significant increase in dichloroacetic acid formation potential (DCAAFP) upon chlorination [68].

2.2.4 O_3-BAC Treatment

In recent years, wastewater reuse treatment using O_3-BAC has been increasingly explored. O_3 has been reported to be able to destroy stubborn pollutants that are difficult to degrade by biological treatment [70]. BAC with biodegradation and physical adsorption can also significantly remove toxic pollutants and oxidation byproducts produced during the ozonation process [71].

It was found that O_3-BAC promoted the mineralization of DOC and the degradation of chromophores compared to BAC alone. In addition, it was difficult to degrade fluorophores using BAC alone, but O_3 could transfer hydrophobic or hydrophilic protein-like fluorophores to less fluorescently converted products that were more easily degraded by BAC treatment [72].

Ozonation increased bromate and TCNM in chlorinated reclaimed water, whereas these DBPs were greatly reduced when O_3-BAC was used as a pretreatment [73]. Therefore, the combination of BAC helps to further reduce the risk of increased DBPs. Generally, the total concentration of typical DBPs generated by chlorination after O_3-BAC pretreatment, including bromate, four THMs, six iodinated THMs, four HALs, four HAMs, two HKs, and one TCNM, was reduced compared to controls, with a similar decline in calculated toxicity [73]. However, as the BAC empty bed contact time increased, it was difficult to observe a further decrease in the concentration of typical DBPs. This can be explained by the low dissolved oxygen concentration limiting biological activity [73].

Currently, treatment with O_3-BAC is more focused on controlling DBPFPs in drinking water sources, with limited research on reclaimed water sources. In drinking water treatment, it has been shown that O_3-BAC can reliably reduce TOXFP and AOXFP [74], but organic matter in reclaimed water sources is typically more complex than that in drinking water sources. The effects of O_3-BAC on TOXFP or AOXFP in reclaimed water sources have not been reported and more research is needed.

2.3 *Control of Chlorine Disinfection Byproducts*

2.3.1 Sunlight Degradation

Sunlight is a "green" energy source that can degrade DBPs. Numerous studies have reported that sunlight irradiation could degrade most of the model chlorinated DBPs to varying degrees [75–77].

The composition of wastewater can affect the degradation of DBPs by sunlight irradiation. Indoor laboratory experiments using simulated sunlight have shown that when nitrosamines were added to secondary effluent from a wastewater plant, the degradation of nitrosamines was approximately 50% slower than in pure water and 11% slower in unfiltered water than in filtered water [75]. The effect of the substrate of the wastewater on the degradation of TOX was also reported in another study [78]. This study showed a reduced photolysis rate of TOX in the wastewater and the same phenomenon of being slower in unfiltered water than in filtered water. These phenomena occurred because organic or particulate matter in the wastewater could affect the photolysis rate through competitive reactions or by reducing the actual photolytic flux.

Although the composition of the wastewater may interfere to some extent with solar degradation of DBPs, some degradation is certain to occur. It was found that sunlight irradiation significantly reduced the TOX of chlorinated reclaimed water, accompanied by a concomitant reduction in the CHO cytotoxicity [79]. Ultraviolet light from sunlight was the main cause of this decrease. UV light from sunlight reduced cytotoxicity from 34.9 mg-phenol/L to 7.2 mg-phenol/L, whereas visible light from sunlight reduced cytotoxicity from 34.1 mg-phenol/L to 27.1 mg-phenol/L [79].

The photosensitizing transformation of free chlorine remaining in the reclaimed water under sunlight irradiation is of concern. It was shown that sunlight-driven photolysis of free chlorine would produce higher DBP levels [15]. Free chlorine in reclaimed water also tended to increase cytotoxicity, whereas the synergistic effect of sunlight and free chlorine did not enhance the reduction of cytotoxicity [79].

2.3.2 Ultraviolet Photolysis

In ultrapure water, photodegradation of DBPs by UV_{254} followed the order of TOI > TOBr > TOCl [80, 81]. Several common chlorinated DBPs, including TCM, MCAA, DCAA, TCAA, DCAL, CH, CAN, and TCAN, had removal efficiencies of less than 5% under irradiation with UV_{254} of 1,000 mJ/cm^2 in ultrapure water [80]. This is due to the fact that the carbon-chlorine bond has a low molar extinction coefficient and it requires high energy to break [82].

An interesting comparison is that while one study found that dichloroacetonitrle (DCAN) in ultrapure water is negligibly degraded by UV irradiation [83], another study showed that the k_{OBS} of DCAN in chlorinated reclaimed water was greater than $0.1 \times 10^{-3}\ s^{-1}$ [84]. It has been shown that UV irradiation of DOM in reclaimed water could produce photoreactive substances such as 3DOM*, ·OH and $O_2^{\cdot -}$ [85, 86]. These active species can be used to account for the degradation of UV irradiation of DCAN in reclaimed water. Furthermore, in chlorinated reclaimed water, the k_{OBS} of DBPs under UV irradiation increased with increasing Br substitution [87]. This can be explained by the fact that Br substitution increased the reaction rate of DBPs with UV and ·OH.

For real wastewaters, it was revealed that after performing UV irradiation at 2400 mJ/cm^2, the TOCl content of all chlorinated reclaimed water was significantly reduced (35.1–53.6%), and the TOBr degradation also exceeded 60%. Meanwhile, the cytotoxicity and genotoxicity of the water samples by CHO cells assay decreased simultaneously with the reduction of TOX [84]. A UV dose of 2,400 mJ/cm^2 is very high compared to current practices, suggesting that direct DBP photolysis may not be practical or economical at present.

2.3.3 Soil Aquifer Treatment

Soil aquifer treatment (SAT) is a natural treatment process that utilizes biotic and abiotic mechanisms in the soil to remove contaminants from reclaimed water, with the advantages of no waste generation and low operating costs.

Natural soil with high adsorption capacity was found to be ineffective in removing adsorbable organic halogen (AOX) in reclaimed water [88]. The removal of DBPs by SAT must rely on the action of microorganisms.

A pilot test found that aerobic soil columns showed the best AOX removal efficiency compared to coagulation and sedimentation or sand filtration [89]. In addition, other studies showed that anaerobic soil aquifer removed AOX in reclaimed water better than aerobic or anoxic soil aquifers [90, 91]. Observation of the degradation behaviour of different species of AOX formed by chlorination revealed that adsorbable organic chlorine (AOCl) and adsorbable organic bromine (AOBr) could be degraded under both aerobic and anaerobic conditions, but adsorbable organic iodine (AOI) was resistant to degradation under aerobic conditions [88]. Therefore, anaerobic degradation of AOX is more effective.

Co-metabolism was proposed as a possible mechanism by which organic matter that was difficult for microorganisms to utilize as an energy source in anaerobic environments could be efficiently degraded through a series of reactions when other compounds were present to provide energy for the microorganisms. It was found DOC may play a role as a co-substrate in the co-metabolism of AOX, as evidenced by the fact that the degradation rate of AOX in reclaimed water under anaerobic conditions increased with the increase of DOC content [88]. The addition of glucose and glutamate was proposed as a protocol to simulate the increase of co-substrate DOC in reclaimed water [90, 92]. When glucose and glutamate were added to the reclaimed water, the removal of AOX by the anaerobic soil aquifer was significantly improved up to 55% [90].

Overall, the study showed that the uppermost 8 cm of the soil accomplished 57% of the AOX removal [91]. Another test in the field found that 80% of total organic halogen (TOX) was removed by filtration through a 37 m unsaturated zone [93]. Therefore, SAT can be considered as a potential scheme for the degradation of DBPs.

3 Ultraviolet-Based Advanced Oxidation Byproducts

There is growing interest in using ultraviolet light (UV) together with an oxidant, normally hydrogen peroxide (H_2O_2) or chlorine, to generate powerful oxidants that can destroy contaminants that might otherwise be resistant to more conventional treatment. The popularity of these UV-based advanced oxidation processes (UV-AOPs) is tied to the growing popularity of UV alone as a very powerful and economical disinfectant. There can be economic savings by combining both UV disinfection and UV-AOP chemical oxidation in a single treatment step, compared to treatment trains where these two goals are accomplished using two separate processes. UV-AOPs are gaining particular interest in the United States for potable water reuse treatment, but they have also historically been used for treating contaminated groundwater, and are gaining popularity for surface water treatment where taste and odour is seasonal. The ability to operate the UV system in disinfection mode when taste and odour is absent, and then to increase the UV dose and add an oxidant when taste and odour is present to operate in an AOP mode offers a very effective practical way to optimize treatment.

This section will focus on the two most common UV-AOPs: UV-H_2O_2 and UV-chlorine. In these processes, H_2O_2 or chlorine is applied upstream of the UV system, and the oxidants undergo photolysis by the UV light to form reactive oxygen species (ROS) and (in the case of chlorine) reactive chlorine species (RCS). The main reactive species for the UV-H_2O_2 process is the hydroxyl radical ($^{\bullet}OH$), whereas UV-chlorine generates both $^{\bullet}OH$ and several RCS species, such as $Cl^{\bullet}$, $Cl_2^{-\bullet}$, and $ClO^{\bullet}$.

Using reactive chemicals such as $^{\bullet}OH$ and RCS to treat water can lead to undesirable byproducts (the term "DBP" is used here, despite UV-AOPs not strictly being used solely to disinfect). There are several general mechanisms of formation, and types of DBPs, that can be formed by UV-AOP. The AOP can lead to changes in the characteristics of organic matter in the water to enhance the biological activity downstream. AOP reaction with organic matter can make that organic matter more or less reactive to chlorine that may be applied downstream in a way that alters subsequent chlorination DBPs; advanced oxidation can react with specific contaminants or general organic matter to form transformation products that, rather than being rendered non-toxic, could potentially maintain or enhance toxic properties; and finally, UV-AOPs can react to form traditional DBPs, such as trihalomethanes.

3.1 Biostability

UV-AOPs are designed to generate powerful oxidants. This is analogous to ozonation, which is known to break large organic molecules down into smaller, more oxygenated compounds such as carboxylic acids and aldehydes, that are often more bioassimilable than the parent organic matter. For example, when ozonation was widely adopted for drinking water disinfection in North America in the 1990s, it was sometimes accompanied by challenges in maintaining a distribution system chlorine residual [94, 95]. The cause of this problem was identified as the reaction between ozone and organic matter to form bioassimilable organic matter (BOM), which served as a food source for microorganisms in the distribution system, leading to more biofilm formation on pipes, and a corresponding increase in chlorine demand. There is some likelihood that UV-AOPs would behave similarly to ozone in terms of BOM formation, although the extent is not well known. The oxidation of the organic matter can be directly by the oxidant (chlorine or H_2O_2), or by the radical species. It has been reported that UV-AOP treatment may lead to changes to the organic matter that reduces the molecular weight and the aromaticity, and makes the organics more biodegradable [96–100]. UV-AOP doses generally are not high enough to lead to any significant mineralization of organic carbon to CO_2, which would eliminate concerns about biodegradability. Studies have reported virtually no change in total organic carbon (TOC) across UV-AOP at doses used in practice [101, 102].

The potential for UV-AOP treatment to form BOM has only been briefly studied. Metz et al. and Bazri et al. conducted UV-H_2O_2 testing at the laboratory and pilot scales and reported an increase in BOM ranging from 30 to 500% [103, 104]. These studies therefore demonstrate a potential for a deterioration in the biostability of water following UV-AOP treatment, but there is no evidence in practice that has been reported to suggest that UV-AOP causes such operational problems. More research is needed, particularly given that UV-AOP treatment would typically be near the end of a treatment process, implying that it may have a particularly significant impact on downstream biostability in a distribution system.

3.2 Organohalide Byproducts

UV-H_2O_2 cannot directly form organohalide byproducts since there are no halogen atoms used in the process. However, there are some theoretical reaction pathways whereby the hydroxyl radical from the AOP can react with naturally occurring chloride, bromide, or iodide, to form reactive versions of these species that can subsequently react with organic matter to form organohalides. In practice, however, this has not been observed, with UV-H_2O_2 treatment being associated with a lack of conventional halogenated DBPs.

UV-chlorine AOP, however, can form organohalogen DBPs by virtue of either direct free chlorine reaction with organic precursors, or by the reactive chlorine species (RCS) that are formed as part of the AOP. In a UV-chlorine AOP, the free chlorine doses can be at least as high, if not higher, than in conventional drinking water treatment, such as in the 5 mg/L range compared to the 1–2 mg/L normally found in tap water. This might imply a much greater likelihood of DBP formation. The high chlorine dose, however, is mitigated by a very small reaction time. In the UV-chlorine AOP, the chlorine is likely added immediately upstream of the UV reactor, and the majority of the chlorine typically undergoes photolysis across the UV reactor, such that the overall chlorine reaction time can be less than 1 min. Therefore, while relatively high chlorine doses are used in the UV-chlorine AOP, they exist for only a very brief period. The other main consideration, however, is the potential role of the reactive chlorine species that are formed during the AOP. Unfortunately, the chemistry of RCS is relatively poorly understood at present, both in terms of its ability to destroy contaminants, as well as its likelihood to react with organic matter to form halogenated DBPs. There has been some work to demonstrate that the chlorine atom from RCS can transfer onto organic molecules under some circumstances to form organohalide DBPs, but the evidence is still largely theoretical and very case-specific at present, and the more widespread implications remain unknown [105, 106].

There are very few laboratory- or field-scale investigations into the ability of UV-chlorine AOPs to form organohalogen DBPs. A study by Wang et al. can offer perspective on DBPs that might form upon water reuse treatment, even though it was from research on a surface water treatment plant [107]. The plant had a TOC of 1.5 mg-C/L and used UV-chlorine at chlorine doses ranging from 2 to 10 mg/L. There was less than 5 μg/L formation of THMs, HAAs, and other common non-regulated DBPs across the UV system [107]. The water at this plant had already been prechlorinated, potentially eliminating many of the DBP precursors, but the same study reported a bench-scale test of UV-chlorine treatment of unchlorinated raw lake water (TOC 3.5 mg-C/L) that revealed only up to 13 μg/L of HAA formation, and lower amounts of THMs and other non-regulated DBPs. Another full-scale study of UV-chlorine AOP at a surface water drinking water treatment plant (TOC 2.2 mg-C/L) at chlorine doses of 5–10 mg/L reported less than 6 μg/L of formation of HAAs, THMs, and other common non-regulatory organohalide DBPs across the UV-chlorine AOP [108]. An interesting observation from this study was that adsorbable organohalides (AOX – similar to "total" organohalides) were enhanced by UV-chlorine, but primarily at pH 6.5 and not pH 8, where the dominant mode of oxidation was due to $^{\bullet}OH$ and not reactive chlorine species. This is preliminary evidence that while UV-chlorine AOP might enhance some DBP precursors which can subsequently react with free chlorine to form organohalide DBPs, it may be mostly due to the action of the $^{\bullet}OH$ and not the RCS. This implies that a UV-H_2O_2 system would be similarly likely to form these DBP precursors as the UV-chlorine system if downstream chlorination is applied.

One study has examined the formation of DBPs when using UV-chlorine to treat secondary wastewater effluent [109]. The chlorine dose in this study was 14 mg/L, and the resulting increase in the formation of DBPs, including THMs, HAAs, chloral hydrate, haloacetonitriles, halonitromethanes, and haloacetamides, was in the range of 90–500% when compared to a chlorinated control with no UV. As part of this study, the toxicity of the UV-chlorine treated water was estimated by summing the nominal CHO comet assay toxicity (a measure of DNA damage and cytotoxicity) of each of the individual DBPs, as reported by Wagner and Plewa [110]. The UV-chlorine treatment demonstrated an approximate 100% increase in the calculated toxicity when compared to a parallel sample receiving chlorine-only treatment. However, when a different toxicity metric was used that is specific to genotoxicity, the UV-chlorine was estimated to reduce toxicity by about 75% relative to a chlorine-only control.

3.3 *Impact on Downstream DBP Formation Potential*

While a UV-AOP might directly form DBPs during treatment, it may also alter DBP precursors that could either enhance or reduce downstream DBP formation when a secondary disinfectant, such as chlorine, is applied.

UV at disinfection doses, often in the order of 40 mJ/cm^2, is unlikely to cause much chemical change to DBP precursors that would affect downstream DBP formation. One possible exception is when applying medium pressure (polychromatic) UV light as a disinfectant in waters that contain elevated nitrate. Medium-pressure lamps emit significant light at wavelengths below 250 nm, which can be absorbed by nitrate to lead to nitrite radicals [111]. There is evidence that these nitrite radicals can react with organic precursors to form several organohalide nitrogen-containing byproducts, including chloropicrin, chloral hydrate, and 1,1,1-trichloropropanone [112].

When UV is applied as part of an advanced oxidation process, the doses are often in the range of 500–1,000 mJ/cm^2, where there is much more potential for direct chemical photolysis compared to disinfection that can lead to changes to downstream DBP formation potential. One study applied UV to a synthetic water containing Suwannee River organic matter at doses up to 1,000 mJ/cm^2 [109]. At this dose, there was no observable change in THM or HAA formation potential when the water was subsequently chlorinated, although there was some formation of aldehydes and carboxylic acids at doses as low as 500 mJ/cm^2, which are bioassimilable and could contribute to biostability issues. A much higher dose of 6,360 mJ/cm^2 was then applied, and THM and HAA formation potential were observed to decrease by 15–45%, suggesting that the organic precursors to these DBPs were being destroyed. In contrast, Dotson et al. conducted UV-H_2O_2 experiments on Ohio River water, applying a UV dose of 1,000 mJ/cm^2 with 10 mg/L of H_2O_2, and reported a 150% increase in THM formation potential, but no impact on HAA or AOX formation potentials [113]. Sarathy and Mohseni conducted similar tests with UV-H_2O_2 on a surface water and reported that at UV doses in the order of 1,000 mJ/cm^2, there was little change to THM formation potential but a 30% reduction in HAA formation potential [99]. Pantin evaluated the impact of UV-H_2O_2 at a full-scale treatment plant treating a surface water for taste and odour, and observed a 20–30% variation, both higher and lower, in THM formation potential over a period of 1 year, implying that perhaps the impact of AOP on DBP formation potential may vary with changes to the organic DBP precursor composition in the water [114]. While this work was on a drinking water plant, it is possible that similar conclusions could be drawn about the impact of UV-H_2O_2 on DBP formation when used for water reuse.

Wang et al. conducted parallel tests on two surface waters using both UV-H_2O_2 and UV-chlorine AOPs at similar doses (UV dose of 1,800 mJ/cm^2), and reported increases to THM and HAA formation potential of 20–110%, as well as similar increases to haloacetonitrile and AOX potentials [107]. The similarity between UV-H_2O_2 and UV-chlorine impacts on DBP formation potential suggests that the change to the DBP precursors may be controlled by the hydroxyl radical, which occurs in both AOPs, and not by the reactive chlorine species, which are formed only by UV-chlorine.

3.4 Transformation Products

When treating wastewater, there is some risk that the concentration of specific contaminants with toxic properties is high enough that, if the UV-AOP were to somehow enhance the toxic properties of the parent compound through reaction, there could be an overall net increase in overall toxicity. This concern is likely unique to wastewater treatment or possibly potable reuse: in drinking water treatment plants with well-protected sources, the majority of DBPs are likely to originate from natural organic matter and not from specific anthropogenic contaminants with pre-existing toxicity.

There is some evidence that DBPs can form via UV-AOP treatment due to the transformation of specific pre-existing contaminants. Huang et al. used UV-chlorine treatment to oxidize dodecylbenzyldimethylammonium chloride (DDBAC) in ultrapure water, and found some formation of dichloropropanol, trichloropropanol, and chloral hydrate [115]. When UV-chlorine and UV-H_2O_2 were applied to solutions containing carbamazepine and atrazine, it was observed that carbamazepine degradation led to similar DBP formation and changes to genotoxicity for both forms of treatment, but that cytotoxicity was lower when treating with UV-chlorine [116]. UV-chlorine treatment of atrazine led to slightly higher DBP formation than UV-H_2O_2, but reduced the cytotoxicity and genotoxicity by a greater amount than UV-H_2O_2.

Wang et al. used modelling to predict the potential for UV-H_2O_2 to generate toxic products upon the oxidation of a large number of different micropollutants, and reported that more than 80% of the predicted reactions could form such products, with an estimated 51% chance of products of the oxidation reactions being more toxic than the parent compounds [117].

3.5 Nitrosamines

N-nitrosodimethylamine (NDMA) is a particularly important DBP in the context of potable water reuse in California, where the concentration in the treated water must be below 10 ng/L. NDMA can be efficiently destroyed by high-dose UV irradiation (e.g., a 1 log destruction at about 1,000 mJ/cm^2). This is one of the principal reasons behind the use of UV-AOP in potable water reuse plants in California: the UV destroys NDMA, and the radicals, which are unreactive with NDMA, destroy other potential contaminants. But while the UV light in a UV-AOP process can destroy NDMA, NDMA can be formed during other parts of the treatment process. A good understanding of the chemistry of NDMA formation and destruction is required to ensure that NDMA concentrations are minimized.

In a potable reuse treatment train consisting of prefiltration, reverse osmosis, and then UV-AOP, chloramines can be applied for fouling control of the prefilters and RO membranes. Chloramine rejection by the RO is incomplete, leading to chloramines entering the UV-AOP process. The unique conditions of this treatment train (especially pH depression to below 6 to reduce RO fouling) typically lead to a large fraction of dichloramine among the chloramines. Dichloramine is the main precursor to NDMA, so these conditions may lead to significant NDMA formation prior to the UV reactor. Furthermore, while UV-AOP treatment may destroy some of the NDMA precursors [118], enough may survive to continue to react with dichloramine downstream of the AOP to form NDMA. While the speed of this reaction is slow (e.g., hours), dichloramine can persist for long enough post-treatment to allow for such formation to occur when using UV-H_2O_2 (H_2O_2 does not react significantly dichloramines: if UV-chlorine AOP is used, this phenomenon won't occur because the chlorine will destroy the dichloramine) [119, 120]. The increased concentration of NDMA from UV-H_2O_2 treatment is called NDMA "rebound". The rate of this rebound has been reported to be in the order of 0.7 ng/L/h under some conditions [121]. One way to minimize this rebound is to raise the pH to 8.5 as soon as possible after the UV-H_2O_2, to shift the chloramine equilibrium away from dichloramine [119]. Szczuka et al. have reported that NDMA formation can also occur in a UV reactor by photolysis of monochloramine to form NDMA, although at UV doses above 800 mJ/cm^2 they report that the rate of NDMA destruction by UV photolysis should exceed the rate of formation [122]. Furst et al. also report a strategy to minimize NDMA formation if using UV-chlorine AOP by applying the free chlorine in multiple smaller doses [123]. This keeps the chlorine:nitrogen ratio smaller at each point of chlorine application than would occur if chlorine were applied all at once, which minimizes the local formation of dichloramine.

3.6 Inorganics: Perchlorate, Chlorite, Chlorate, and Bromate

Chloride and bromide in water sources can be oxidized to form inorganic byproducts, including chlorite (ClO_2^-), chlorate (ClO_3^-), perchlorate (ClO_4^-), and bromate (BrO_3^-). These species have been reported to have adverse health effects at high enough concentrations [124–126]. In the United States, the limits for chlorite and bromate in drinking water are given as 1.0 mg/L and 10 mg/L, respectively [127]. The World Health Organization has set maximum drinking water chlorate and perchlorate concentrations of 0.7 mg/L and 0.07 mg/L [128]. Sodium hypochlorite solutions that are used for UV-chlorine AOP can contain chlorate, bromate, and perchlorate, particularly if the hypochlorite solution is stored for extended periods at warm temperatures, or contains heavy metal impurities [107, 129, 130]. Chlorate is a particular risk in improperly stored hypochlorite solutions, so when using hypochlorite for UV-chlorine AOP, or for quenching H_2O_2 with liquid chlorine, chlorate concentrations should be assessed.

In a UV-chlorine AOP, chlorate appears to be an inevitable and stable end product of chlorine photolysis. The mechanism and kinetics of this process are not fully understood, but from experience, the mass ratio of photolyzed chlorine that becomes chlorate varies from approximately 2 to 20% [107, 108, 124, 131]. This might lead to a de facto limit on the amount of chlorine that can be used in a UV-chlorine AOP to avoid exceeding any regulatory limits on chlorate in the treated water, if they exist. Studies indicate that chlorate yields tend to be higher at increasing pH, with a higher chlorine concentration, and with greater applied UV power [108, 132, 133]. In contrast to chlorate, evidence suggests that while perchlorate is a theoretical product of UV photolysis of chlorine [129], its actual formation in practice might be negligible, with a yield of less than 0.01% by mass of chlorine photolyzed [133].

Bromide can be oxidized to bromate during the UV-chlorine AOP. The mechanism requires bromide to first be oxidized by free chlorine to free bromine (HOBr and OBr^-) [134]. The free bromine is then oxidized by the radicals of the AOP to lead to bromate. The yield of this process varies widely in the literature. In one water containing 30 mg/L of bromide, UV-chlorine treatment only resulted in 2 mg/L of bromate formation [107]. In a different water containing 50–80 mg/L bromide, and with similar UV-chlorine doses, 10–20 mg/L bromate formation was reported [135]. The Terminal Island Water Reclamation Plant in Los Angeles treats RO permeate with 90 mg/L bromide using UV-chlorine. To suppress bromate formation, 0.3 mg-N/L ammonia is injected before chlorine. In this way, when chlorine oxidizes the bromide to free bromine, the free bromine quickly reacts with ammonia to form bromamines instead of bromate, which blocks the bromate formation pathway [136]. Wu et al., however, report that the bromamines may react with organics to form other toxic byproducts, such as brominated nitromethanes and brominated acetamides [137].

4 Ozonation Byproducts

Ozonation applied in water reuse is very effective at inactivating pathogens, but requires a high ozone dose [138]. This high dose can subsequently lead to problematic inorganic and organic DBPs. Importantly, if bromide ion exists in the source water, ozone can oxidize the bromide to free bromine, which can lead to the formation of hazardous bromine-containing byproducts.

4.1 Ozone DBP Formation

Organic ozonation byproducts include aldehydes, ketones, carboxylic acids, *N*-nitrosodimethylamine (NDMA), and other regulated DBPs. Recent advances in analytical techniques, such as GC × GC-qMS, have also revealed new ozonation byproducts.

In a study by Wu et al. of ozonated municipal wastewater, regulated chlorine-containing DBPs were below detection limits (<0.1 μg/L), including trichloromethane, bromodichloromethane, dibromochloromethane, tribromomethane, trichloronitromethane, trichloroacetonitrile, dichloroacetonitrile, bromochloroacetonitrile, dibromoacetonitrile, trichloroacetamide, chloral hydrate, bromodichloroacetaldehyde, tribromoacetaldehyde, 1,1-dichlororopanone, and 1,1,1-trichloropropanone [139]. Formaldehyde, acetaldehyde, and glyoxal at 10–60 μg/L were produced when ozone was dosed at a O_3:DOC ratio of 1. Formaldehyde was the major species at 62 μg/L.

NDMA, a probable human carcinogen, has been reported in raw wastewater and/or formed during ozonation [140, 141]. NDMA can react with ozone and hydroxyl radical (•OH) with rate constants of reactions of $0.052 \pm 0.0016\ M^{-1}\ s^{-1}$ and $(4.5 \pm 0.21) \times 10^8\ M^{-1}\ s^{-1}$, respectively [49]. Molecular ozone dominates NDMA formation as compared to hydroxyl radicals [142]. To avoid the formation of NDMA, Marti et al. reported that multiple injections of ozone of the same overall dose somewhat improved the oxidant utilization efficiency by minimizing •OH scavenging contribution of oxidants.

In the presence of 2 mg/L bromide ion (a high bromide concentration), bromate, tribromomethane, and dibromoacetonitrile were the primary DBPs detected in three ozonated secondary effluent samples in a study by [139]. The concentration ranges were 20.2–77.7 μg/L for tribromomethane, 1.9–6.6 μg/L for dibromoacetonitrile, and bromate reached a concentration of 80 μg/L. Heeb et al. reported that in the presence of bromide, formaldehyde formation during ozonation decreased due to the competition between bromide ions and formaldehyde precursors [143]. Zhang et al. (2021) have used GCXGC-qMS methods to identify new bromine-containing byproducts containing benzene rings in ozonated second effluent [144].

4.2 *Toxicity Changes*

For wastewater ozonation, the driver for toxicity among aldehyde formation appears to be formaldehyde when considering both the LC_{50} and the concentration of the aldehydes [139]. When there is bromide ion present, Wu et al. also identified DBAN as a significant contributor to toxicity. Other species that may contribute to toxicity in ozonated wastewater, despite low concentrations (around tens of ng/L), include 2,4,6-tribromophenol and bromophenylacetonitriles. These two species were predicted to contribute to the cytotoxicity and genotoxicity index, and cause oxidative damage to mammalian cells [145].

Ozonation of wastewater, however, also has a detoxifying effect. Wu et al. reported that adding 5 mg/L O_3 significantly reduced the cytotoxicity equivalents in a secondary effluent from 13.8 mg-phenol/L to 7.2 mg-phenol/L ($p < 0.05$) exposed to mammalian Chinese hamster cells [146]. Increasing the ozone dose to 20 mg/L further decreased the cytotoxicity to 6.2 mg-phenol/L. In the presence of bromide ion, however, the toxicity to mammalian cells increased in the ozonated wastewater, suggesting a contribution of AOBr products and bromate to toxicity.

References

1. Sun Y, Chen Z, Wu GX, Wu QY, Zhang F, Niu ZB et al (2016) Characteristics of water quality of municipal wastewater treatment plants in China: implications for resources utilization and management. J Clean Prod 131:1–9. https://doi.org/10.1016/j.jclepro.2016.05.068
2. Bagatin R, Klemes JJ, Reverberi AP, Huisingh D (2014) Conservation and improvements in water resource management: a global challenge. J Clean Prod 77:1–9. https://doi.org/10.1016/j.jclepro.2014.04.027
3. Griffin G, Hussain S, Natarajan A, Pannirselvam M, Philip L, Shu L (2019) Water scarcity and ways to reduce the impact: management strategies and Technologies for zero liquid discharge and future smart cities. Applied environmental science and engineering for a sustainable future1st edn. Springer, Cham. https://doi.org/10.1007/978-3-319-75199-3
4. Rizzo L, Gernjak W, Krzeminski P, Malato S, McArdell CS, Perez JAS et al (2020) Best available technologies and treatment trains to address current challenges in urban wastewater reuse for irrigation of crops in EU countries. Sci Total Environ 710:136312. https://doi.org/10.1016/j.scitotenv.2019.136312
5. He L, Wu DX, Zhang XY, Peng L (2023) The exposure dose of reclaimed water in daily life: a mini review. Water Cycle 4:120–126. https://doi.org/10.1016/j.watcyc.2023.05.003
6. Deng S, Yan X, Zhu Q, Liao C (2019) The utilization of reclaimed water: possible risks arising from waterborne contaminants. Environ Pollut 254(Pt A):113020. https://doi.org/10.1016/j.envpol.2019.113020
7. Gao P, Deng F, Chen WS, Zhong YJ, Cai XL, Ma WM et al (2022) Health risk assessment of inhalation exposure to airborne particle-bound nitrated polycyclic aromatic hydrocarbons in urban and suburban areas of South China. Int J Environ Res Public Health 19(23):15536. https://doi.org/10.3390/ijerph192315536
8. Richardson SD, Plewa MJ, Wagner ED, Schoeny R, DeMarini DM (2007) Occurrence, genotoxicity, and carcinogenicity of regulated and emerging disinfection by-products in drinking water: a review and roadmap for research. Mutat Res-Rev Mutat 636(1–3):178–242. https://doi.org/10.1016/j.mrrev.2007.09.001
9. Hesam G, Shekarloo MV, Atamaleki A, Jalali M, Hajipour-Verdom B, Moradpour Z (2022) Health risk assessment of inhalation exposure to dry fogging of hydrogen peroxide in a dental clinic during the COVID-19 pandemic. Environ Sci Pollut Res Int 29(50):75338–75343. https://doi.org/10.1007/s11356-022-21174-1
10. Liu YQ, Sakaguchi N, Iijima M, Islam MRR, Zhang JY, Islam R et al (2023) Effects of short-term exposure of chloramine-T solution on the characteristics of light-cured and chemical-cured adhesives. Polymers-Basel 15(19):3995. https://doi.org/10.3390/polym15193995
11. Du Y, Lv XT, Wu QY, Zhang DY, Zhou YT, Peng L et al (2017) Formation and control of disinfection byproducts and toxicity during reclaimed water chlorination: a review. J Environ Sci (China) 58:51–63. https://doi.org/10.1016/j.jes.2017.01.013
12. Wang C, Liggio J, Wentzell JJB, Jorga S, Folkerson A, Abbatt JPD (2023) Chloramines as an important photochemical source of chlorine atoms in the urban atmosphere. Proc Natl Acad Sci U S A 120(30):e2220889120. https://doi.org/10.1073/pnas.2220889120
13. Watson K, Shaw G, Leusch FD, Knight NL (2012) Chlorine disinfection by-products in wastewater effluent: bioassay-based assessment of toxicological impact. Water Res 46(18):6069–6083. https://doi.org/10.1016/j.watres.2012.08.026
14. Hao R, Zhang Y, Du T, Yang L, Adeleye AS, Li Y (2017) Effect of water chemistry on disinfection by-product formation in the complex surface water system. Chemosphere 172:384–391. https://doi.org/10.1016/j.chemosphere.2016.12.034
15. Young TR, Li W, Guo A, Korshin GV, Dodd MC (2018) Characterization of disinfection byproduct formation and associated changes to dissolved organic matter during solar photolysis of free available chlorine. Water Res 146:318–327. https://doi.org/10.1016/j.watres.2018.09.022

16. Jia A, Escher BI, Leusch FDL, Tang JYM, Prochazka E, Dong BF et al (2015) In vitro bioassays to evaluate complex chemical mixtures in recycled water. Water Res 80:1–11. https://doi.org/10.1016/j.watres.2015.05.020
17. Zhu Y, Ling Y, Peng Z, Zhang N (2020) Formation of emerging iodinated disinfection by-products during ballast water treatment based on ozonation processes. Sci Total Environ 743:140805. https://doi.org/10.1016/j.scitotenv.2020.140805
18. Chen W, Xu J, Lu S, Jiao W, Wu L, Chang AC (2013) Fates and transport of PPCPs in soil receiving reclaimed water irrigation. Chemosphere 93(10):2621–2630. https://doi.org/10.1016/j.chemosphere.2013.09.088
19. Xu X, Kang J, Shen J, Zhao S, Wang B, Zhang X et al (2021) EEM–PARAFAC characterization of dissolved organic matter and its relationship with disinfection by-products formation potential in drinking water sources of northeastern China. Sci Total Environ 774:145297. https://doi.org/10.1016/j.scitotenv.2021.145297
20. Bougeard CMM, Goslan EH, Jefferson B, Parsons SA (2010) Comparison of the disinfection by-product formation potential of treated waters exposed to chlorine and monochloramine. Water Res 44(3):729–740. https://doi.org/10.1016/j.watres.2009.10.008
21. Hu HY, Du Y, Wu QY, Zhao X, Tang X, Chen Z (2016) Differences in dissolved organic matter between reclaimed water source and drinking water source. Sci Total Environ 551:133–142. https://doi.org/10.1016/j.scitotenv.2015.12.111
22. Pehlivanoglu-Mantas E, Sedlak DL (2008) Measurement of dissolved organic nitrogen forms in wastewater effluents: concentrations, size distribution and NDMA formation potential. Water Res 42(14):3890–3898. https://doi.org/10.1016/j.watres.2008.05.017
23. Liu JL, Li XY, Xie YF, Tang H (2014) Characterization of soluble microbial products as precursors of disinfection byproducts in drinking water supply. Sci Total Environ 472:818–824. https://doi.org/10.1016/j.scitotenv.2013.11.139
24. Reckhow D (2003) Reactivity of organic nitrogen and formation of nitrogenous DBPs. AWWA ACE, Santa Ana
25. Mitch WA, Sedlak DL (2004) Characterization and fate of N-nitrosodimethylamine precursors in municipal wastewater treatment plants. Environ Sci Technol 38(5):1445–1454. https://doi.org/10.1021/es035025n
26. Zheng A, Dzombak DA, Luthy RG (2004) Effects of nitrosation on the formation of cyanide in publicly owned treatment works secondary effluent. Water Environ Res 76(3):197–204. https://doi.org/10.2175/106143004X141735
27. Lee W, Westerhoff P, Croué JP (2007) Dissolved organic nitrogen as a precursor for chloroform, dichloroacetonitrile, N-nitrosodimethylamine, and trichloronitromethane. Environ Sci Technol 41(15):5485–5490. https://doi.org/10.1021/es070411g
28. Wang C, Zhang X, Wang J, Chen C (2013) Characterization of dissolved organic matter as N-nitrosamine precursors based on hydrophobicity, molecular weight and fluorescence. J Environ Sci (China) 25(1):85–95. https://doi.org/10.1016/S1001-0742(12)60029-1
29. Weishaar JL, Aiken GR, Bergamaschi BA, Fram MS, Fujii R, Mopper K (2003) Evaluation of specific ultraviolet absorbance as an indicator of the chemical composition and reactivity of dissolved organic carbon. Environ Sci Technol 37(20):4702–4708. https://doi.org/10.1021/es030360x
30. Zhang H, Qu J, Liu H, Zhao X (2009) Characterization of isolated fractions of dissolved organic matter from sewage treatment plant and the related disinfection by-products formation potential. J Hazard Mater 164(2–3):1433–1438. https://doi.org/10.1016/j.jhazmat.2008.09.057
31. Coble PG (2014) Aquatic organic matter fluorescence. Cambridge University Press, Cambridge
32. Wang Z, Wu Z, Tang S (2009) Characterization of dissolved organic matter in a submerged membrane bioreactor by using three-dimensional excitation and emission matrix fluorescence spectroscopy. Water Res 43(6):1533–1540. https://doi.org/10.1016/j.watres.2008.12.033

33. Tang X, Wu QY, Du Y, Yang Y, Hu HY (2014) Anti-estrogenic activity formation potential assessment and precursor analysis in reclaimed water during chlorination. Water Res 48:490–497. https://doi.org/10.1016/j.watres.2013.10.009
34. Liao Y, Ji W, Wang Z, Tian Y, Peng J, Li W et al (2024) Effects of alternative disinfection methods on the characteristics of effluent organic matter and the formation of disinfection byproducts. Environ Pollut 340:122796. https://doi.org/10.1016/j.envpol.2023.122796
35. Chang EE, Chiang PC, Ko YW, Lan WH (2001) Characteristics of organic precursors and their relationship with disinfection by-products. Chemosphere 44(5):1231–1236. https://doi.org/10.1016/S0045-6535(00)00499-9
36. Barceló D, Žonja B, Ginebreda A (2020) Toxicity tests in wastewater and drinking water treatment processes: a complementary assessment tool to be on your radar. J Environ Chem Eng 8(5):104262. https://doi.org/10.1016/j.jece.2020.104262
37. Krasner SW, Westerhoff P, Chen B, Rittmann BE, Amy G (2009) Occurrence of disinfection byproducts in United States wastewater treatment plant effluents. Environ Sci Technol 43(21): 8320–8325. https://doi.org/10.1021/es901611m
38. Lee Y, Escher BI, Von Gunten U (2008) Efficient removal of estrogenic activity during oxidative treatment of waters containing steroid estrogens. Environ Sci Technol 42(17): 6333–6339. https://doi.org/10.1021/es7023302
39. Plewa MJ, Wagner ED, Richardson SD, Thruston AD, Woo YT, McKague AB (2004) Chemical and biological characterization of newly discovered iodoacid drinking water disinfection byproducts. Environ Sci Technol 38(18):4713–4722. https://doi.org/10.1021/es049971v
40. Wei D, Wang L, Wei J, Hu HY (2006) Toxicity screening and evaluating in chlorination disinfection of wastewater reclamation processes. Water Sci Technol 53(9):239–246. https://doi.org/10.2166/wst.2006.281
41. Wang LS, Hu HY, Wang C (2007) Effect of ammonia nitrogen and dissolved organic matter fractions on the genotoxicity of wastewater effluent during chlorine disinfection. Environ Sci Technol 41(1):160–165. https://doi.org/10.1021/es0616635
42. Blatchley ER, Hunt BA, Duggirala R, Thompson JE, Zhao JG, Halaby T et al (1997) Effects of disinfectants on wastewater effluent toxicity. Water Res 31(7):1581–1588. https://doi.org/10.1016/S0043-1354(96)00396-X
43. Cao N, Miao T, Li K, Zhang Y, Yang M (2009) Formation potentials of typical disinfection byproducts and changes of genotoxicity for chlorinated tertiary effluent pretreated by ozone. J Environ Sci (China) 21(4):409–413. https://doi.org/10.1016/s1001-0742(08)62284-6
44. Wu QY, Li Y, Hu HY, Sun YX, Zhao FY (2010) Reduced effect of bromide on the genotoxicity in secondary effluent of a municipal wastewater treatment plant during chlorination. Environ Sci Technol 44(13):4924–4929. https://doi.org/10.1021/es100152j
45. Wei D, Tan Z, Du Y (2012) Toxicity-based assessment of the treatment performance of wastewater treatment and reclamation processes. J Environ Sci (China) 24(6):969–978. https://doi.org/10.1016/s1001-0742(11)60860-7
46. Wang LS, Wei DB, Wei J, Hu HY (2007) Screening and estimating of toxicity formation with photobacterium bioassay during chlorine disinfection of wastewater. J Hazard Mater 141(1): 289–294. https://doi.org/10.1016/j.jhazmat.2006.07.001
47. Schiliro T, Pignata C, Rovere R, Fea E, Gilli G (2009) The endocrine disrupting activity of surface waters and of wastewater treatment plant effluents in relation to chlorination. Chemosphere 75(3):335–340. https://doi.org/10.1016/j.chemosphere.2008.12.028
48. Wu QY, Hu HY, Zhao X, Sun YX (2009) Effect of chlorination on the estrogenic/antiestrogenic activities of biologically treated wastewater. Environ Sci Technol 43(13): 4940–4945. https://doi.org/10.1021/es8034329
49. Lee C, Yoon J, Von Gunten U (2007) Oxidative degradation of N-nitrosodimethylamine by conventional ozonation and the advanced oxidation process ozone/hydrogen peroxide. Water Res 41(3):581–590. https://doi.org/10.1016/j.watres.2006.10.033

50. Zeng S, Huang Y, Sun F, Li D, He M (2016) Probabilistic ecological risk assessment of effluent toxicity of a wastewater reclamation plant based on process modeling. Water Res 100: 367–376. https://doi.org/10.1016/j.watres.2016.05.039
51. Fukushima T, Hara-Yamamura H, Urai M, Kasuga I, Kurisu F, Miyoshi T et al (2014) Toxicity assessment of chlorinated wastewater effluents by using transcriptome-based bioassays and Fourier transform mass spectrometry (FT-MS) analysis. Water Res 52:73–82. https://doi.org/10.1016/j.watres.2014.01.006
52. Klopman G, Zhu H, Fuller MA, Saiakhov RD (2004) Searching for an enhanced predictive tool for mutagenicity. SAR QSAR Environ Res 15(4):251–263
53. Richard AM, Benigni R (2002) AI and SAR approaches for predicting chemical carcinogenicity: survey and status report. SAR QSAR Environ Res 13(1):1–19
54. Yang C, Benz RD, Cheeseman MA (2006) Landscape of current toxicity databases and database standards. Curr Opin Drug Discov Devel 9(1):124–133
55. Zhu H, Martin TM, Young DM, Tropsha A (2009) Combinatorial QSAR modeling of rat acute toxicity by oral exposure. Chem Res Toxicol 22(12):1913–1921
56. Martin TM, Young DM (2001) Prediction of the acute toxicity (96-h LC_{50}) of organic compounds in the Fathead Minnow (*Pimephales Promelas*) using a group contribution method. Chem Res Toxicol 14(10):1378–1385
57. Lu Y, Song ZM, Wang C, Liang JK, Xu N, Hu Q, Wu QY (2021) Combination of high resolution mass spectrometry and a halogen extraction code to identify chlorinated disinfection byproducts formed from aromatic amino acids. Water Res 190:116710
58. Liu B, Gu L, Li Q, Yu G, Zhao C, Zhai H (2019) Effect of pre-ozonation-enhanced coagulation on dissolved organic nitrogen in municipal wastewater treatment plant effluent. Environ Technol 40(20):2684–2694. https://doi.org/10.1080/09593330.2018.1449897
59. Musikavong C, Wattanachira S, Marhaba TF, Pavasant P (2005) Reduction of organic matter and trihalomethane formation potential in reclaimed water from treated industrial estate wastewater by coagulation. J Hazard Mater 127(1–3):58–67. https://doi.org/10.1016/j.jhazmat.2005.06.037
60. Han Q, Yan H, Zhang F, Xue N, Wang Y, Chu Y et al (2015) Trihalomethanes (THMs) precursor fractions removal by coagulation and adsorption for bio-treated municipal wastewater: molecular weight, hydrophobicity/hydrophily and fluorescence. J Hazard Mater 297: 119–126. https://doi.org/10.1016/j.jhazmat.2015.04.070
61. Wei LL, Zhao QL, Xue S, Chang CC, Tang F, Liang GL et al (2009) Reduction of trihalomethane precursors of dissolved organic matter in the secondary effluent by advanced treatment processes. J Hazard Mater 169(1–3):1012–1021. https://doi.org/10.1016/j.jhazmat.2009.04.045
62. Liu Y, Duan J, Li W, Beecham S, Mulcahy D (2016) Effects of organic matter removal from a wastewater secondary effluent by aluminum sulfate coagulation on haloacetic acids formation. Environ Eng Sci 33(7):484–493. https://doi.org/10.1089/ees.2015.0446
63. Park SH, Piyachaturawat P, Taylor AE, Huang CH (2009) Potential N-nitrosodimethylamine (NDMA) formation from amine-based water treatment polymers in the reactions with chlorine-based oxidants and nitrosifying agents. Water Supply 9(3):279–288. https://doi.org/10.2166/ws.2009.452
64. Ding S, Chu W, Bond T, Cao Z, Xu B, Gao N (2018) Contribution of amide-based coagulant polyacrylamide as precursors of haloacetamides and other disinfection by-products. Chem Eng J 350:356–363. https://doi.org/10.1016/j.cej.2018.06.002
65. Chili CA, Westerhoff P, Ghosh A (2012) GAC removal of organic nitrogen and other DBP precursors. J AWWA 104(7):E406–EE15. https://doi.org/10.5942/jawwa.2012.104.0090
66. Krasner SW, Lee TCF, Westerhoff P, Fischer N, Hanigan D, Karanfil T et al (2016) Granular activated carbon treatment may result in higher predicted genotoxicity in the presence of bromide. Environ Sci Technol 50(17):9583–9591. https://doi.org/10.1021/acs.est.6b02508

67. Peterson ES, Johnson S, Shiokari S, Yu Y, Cook SM, Summers RS (2022) Impacts of carbon-based advanced treatment processes on disinfection byproduct formation and speciation for potable reuse. Water Res 220:118643. https://doi.org/10.1016/j.watres.2022.118643
68. Qi W, Zhang H, Hu C, Liu H, Qu J (2018) Effect of ozonation on the characteristics of effluent organic matter fractions and subsequent associations with disinfection by-products formation. Sci Total Environ 610-611:1057–1064. https://doi.org/10.1016/j.scitotenv.2017.08.194
69. Wert EC, Rosario-Ortiz FL, Drury DD, Snyder SA (2007) Formation of oxidation byproducts from ozonation of wastewater. Water Res 41(7):1481–1490. https://doi.org/10.1016/j.watres.2007.01.020
70. Gomes J, Costa R, Quinta-Ferreira RM, Martins RC (2017) Application of ozonation for pharmaceuticals and personal care products removal from water. Sci Total Environ 586:265–283. https://doi.org/10.1016/j.scitotenv.2017.01.216
71. Scott JP, Ollis DF (1995) Integration of chemical and biological oxidation processes for water treatment: review and recommendations. Environ Prog 14(2):88–103. https://doi.org/10.1002/ep.670140212
72. Wang WL, Cai YZ, Hu HY, Chen J, Wang J, Xue G et al (2019) Advanced treatment of bio-treated dyeing and finishing wastewater using ozone-biological activated carbon: a study on the synergistic effects. Chem Eng J 359:168–175. https://doi.org/10.1016/j.cej.2018.11.059
73. Chuang YH, Mitch WA (2017) Effect of ozonation and biological activated carbon treatment of wastewater effluents on formation of N-nitrosamines and halogenated disinfection byproducts. Environ Sci Technol 51(4):2329–2338. https://doi.org/10.1021/acs.est.6b04693
74. de Vera GA, Keller J, Gernjak W, Weinberg H, Farré MJ (2016) Biodegradability of DBP precursors after drinking water ozonation. Water Res 106:550–561. https://doi.org/10.1016/j.watres.2016.10.022
75. Chen B, Lee W, Westerhoff PK, Krasner SW, Herckes P (2010) Solar photolysis kinetics of disinfection byproducts. Water Res 44(11):3401–3409. https://doi.org/10.1016/j.watres.2010.03.014
76. Lekkas TD, Nikolaou AD (2004) Degradation of disinfection byproducts in drinking water. Environ Eng Sci 21(4):493–506. https://doi.org/10.1089/1092875041358566
77. Jones CE, Carpenter LJ (2005) Solar photolysis of CH_2I_2, CH_2ICl, and CH_2IBr in water, saltwater, and seawater. Environ Sci Technol 39(16):6130–6137. https://doi.org/10.1021/es050563g
78. Abusallout I, Hua GH (2016) Natural solar photolysis of total organic chlorine, bromine and iodine in water. Water Res 92:69–77. https://doi.org/10.1016/j.watres.2016.01.047
79. Lv XT, Zhang X, Du Y, Wu QY, Lu Y, Hu HY (2017) Solar light irradiation significantly reduced cytotoxicity and disinfection byproducts in chlorinated reclaimed water. Water Res 125:162–169. https://doi.org/10.1016/j.watres.2017.08.043
80. Erdem CU, Liu C, Karanfil T (2023) Photodegradation of halogenated organic disinfection by-products: decomposition and reformation. Water Res 245:120565. https://doi.org/10.1016/j.watres.2023.120565
81. Lei X, Lei Y, Zhang X, Yang X (2021) Treating disinfection byproducts with UV or solar irradiation and in UV advanced oxidation processes: a review. J Hazard Mater 408:124435. https://doi.org/10.1016/j.jhazmat.2020.124435
82. Xiao Y, Fan R, Zhang L, Yue J, Webster RD, Lim TT (2014) Photodegradation of iodinated trihalomethanes in aqueous solution by UV_{254} irradiation. Water Res 49:275–285. https://doi.org/10.1016/j.watres.2013.11.039
83. Hou S, Ling L, Shang C, Guan Y, Fang J (2017) Degradation kinetics and pathways of haloacetonitriles by the UV/persulfate process. Chem Eng J 320:478–484. https://doi.org/10.1016/j.cej.2017.03.042
84. Huang WC, Du Y, Liu M, Hu HY, Wu QY, Chen Y (2019) Influence of UV irradiation on the toxicity of chlorinated water to mammalian cells: toxicity drivers, toxicity changes and toxicity surrogates. Water Res 165:115024. https://doi.org/10.1016/j.watres.2019.115024

85. Wu DX, Huang WC, Liang ZF, Wang WL, Xiang T, Wang G et al (2022) Essential role of sunlight irradiation in aqueous micropollutant transformations: influence of the water matrix and changes in toxicities. Environ Sci Water Res Technol 8(8):1619–1638. https://doi.org/10.1039/D2EW00082B
86. Jacobs LE, Fimmen RL, Chin YP, Mash HE, Weavers LK (2011) Fulvic acid mediated photolysis of ibuprofen in water. Water Res 45(15):4449–4458. https://doi.org/10.1016/j.watres.2011.05.041
87. Chuang YH, Parker KM, Mitch WA (2016) Development of predictive models for the degradation of halogenated disinfection byproducts during the UV/H_2O_2 advanced oxidation process. Environ Sci Technol 50(20):11209–11217. https://doi.org/10.1021/acs.est.6b03560
88. Drewes JE, Jekel M (1998) Behavior of DOC and AOX using advanced treated wastewater for groundwater recharge. Water Res 32(10):3125–3133. https://doi.org/10.1016/S0043-1354(98)00064-5
89. Cheng XZ, Yang L, Wu TB, Gan YP, Hu J (2002) Pilot test of advanced treatments combination of wastewaterfor groundwater recharge. Tsinghua Sci Technol 05:539–543
90. Jian XC (1999) Studies on the processes and mechanisms of removing organic compounds in municipal wastewater by advanced treatment Technologies for Artificial Groundwater Recharge. Tsinghua University, Beijing
91. Quanrud DM, Arnold RG, Wilson LG, Gordon HJ, Graham DW, Amy GL (1996) Fate of organics during column studies of soil aquifer treatment. J Environ Eng 122(4):314–321. https://doi.org/10.1061/(ASCE)0733-9372(1996)122:4(314)
92. Cheng XZ (1999) Studies on design and optimal operation of groundwater recharge plant and advanced sewage treatment technologies. Tsinghua University, Beijing
93. Wilson LG, Quanrud DM, Arnold RG, Amy G, Gordon H, Conroy AD (1995) Field and laboratory observations on the fate of organics in sewage effluent during soil aquifer treatment. Proceedings of the 2nd international symposium on artificial recharge of ground water. Florida, USA, pp 529–538
94. Cipparone LA, Diehl AC, Speitel GE (1997) Ozonation and BDOC removal: effect on water quality. J AWWA 89(2):84–97. https://doi.org/10.1002/j.1551-8833.1997.tb08181.x
95. Jadas-Hécart A, Ventresque C, Legube B, Doré M (1991) Effect of ozonation on the chlorine demand of a treated surface water and some macromolecular compounds. Ozone Sci Eng 13(2):147–160. https://doi.org/10.1080/01919519108552423
96. Kleiser G, Frimmel FH (2000) Removal of precursors for disinfection by-products (DBPs) — differences between ozone- and OH-radical-induced oxidation. Sci Total Environ 256(1):1–9. https://doi.org/10.1016/s0048-9697(00)00377-6
97. Sarathya SR, Stefan MI, Royce A, Mohseni M (2011) Pilot-scale UV/H_2O_2 advanced oxidation process for surface water treatment and downstream biological treatment: effects on natural organic matter characteristics and DBP formation potential. Environ Technol 33(15–16):1709–1718. https://doi.org/10.1080/09593330.2011.553843
98. Wang X, Zhang H, Zhang Y, Shi Q, Wang J, Yu J et al (2017) New insights into trihalomethane and haloacetic acid formation potentials: correlation with the molecular composition of natural organic matter in source water. Environ Sci Technol 51(4):2015–2021. https://doi.org/10.1021/acs.est.6b04817
99. Sarathy S, Mohseni M (2010) Effects of UV/H_2O_2 advanced oxidation on chemical characteristics and chlorine reactivity of surface water natural organic matter. Water Res 44(14):4087–4096. https://doi.org/10.1016/j.watres.2010.05.025
100. Sarathy SR, Mohseni M (2007) The impact of UV/H_2O_2 advanced oxidation on molecular size distribution of chromophoric natural organic matter. Environ Sci Technol 41(24):8315–8320. https://doi.org/10.1021/es071602m
101. Pisarenko AN, Stanford BD, Snyder SA, Rivera SB, Boal AK (2013) Investigation of the use of chlorine based advanced oxidation in surface water: oxidation of natural organic matter and formation of disinfection byproducts. J Adv Oxid Technol 16(1):137–150. https://doi.org/10.1515/jaots-2013-0115

102. Ike IA, Lee Y, Hur J (2019) Impacts of advanced oxidation processes on disinfection byproducts from dissolved organic matter upon post-chlor(am)ination: a critical review. Chem Eng J 375:121929. https://doi.org/10.1016/j.cej.2019.121929
103. Metz DH, Reynolds K, Meyer M, Dionysiou DD (2011) The effect of UV/H_2O_2 treatment on biofilm formation potential. Water Res 45(2):497–508. https://doi.org/10.1016/j.watres.2010.09.007
104. Bazri MM, Barbeau B, Mohseni M (2012) Impact of UV/H_2O_2 advanced oxidation treatment on molecular weight distribution of NOM and biostability of water. Water Res 46(16): 5297–5304. https://doi.org/10.1016/j.watres.2012.07.017
105. Guo K, Wu Z, Shang C, Yao B, Hou S, Yang X et al (2017) Radical chemistry and structural relationships of PPCP degradation by UV/chlorine treatment in simulated drinking water. Environ Sci Technol 51(18):10431–10439. https://doi.org/10.1021/acs.est.7b02059
106. Xiang Y, Fang J, Shang C (2016) Kinetics and pathways of ibuprofen degradation by the UV/chlorine advanced oxidation process. Water Res 90:301–308. https://doi.org/10.1016/j.watres.2015.11.069
107. Wang D, Bolton JR, Andrews SA, Hofmann R (2015) Formation of disinfection by-products in the ultraviolet/chlorine advanced oxidation process. Sci Total Environ 518-519:49–57. https://doi.org/10.1016/j.scitotenv.2015.02.094
108. Wang C, Moore N, Bircher K, Andrews S, Hofmann R (2019) Full-scale comparison of UV/H_2O_2 and UV/Cl_2 advanced oxidation: the degradation of micropollutant surrogates and the formation of disinfection byproducts. Water Res 161:448–458. https://doi.org/10.1016/j.watres.2019.06.033
109. Hua Z, Li D, Wu Z, Wang D, Cui Y, Huang X et al (2021) DBP formation and toxicity alteration during UV/chlorine treatment of wastewater and the effects of ammonia and bromide. Water Res 188:116549. https://doi.org/10.1016/j.watres.2020.116549
110. Wagner ED, Plewa MJ (2017) CHO cell cytotoxicity and genotoxicity analyses of disinfection by-products: an updated review. J Environ Sci (China) 58:64–76. https://doi.org/10.1016/j.jes.2017.04.021
111. Sharpless CM, Seibold DA, Linden KG (2003) Nitrate photosensitized degradation of atrazine during UV water treatment. Aquat Sci 65(4):359–366. https://doi.org/10.1007/s00027-003-0674-5
112. Reckhow DA, Linden KG, Kim J, Shemer H, Makdissy G (2010) Effect of UV treatment on DBP formation. J AWWA 102(6):100–113. https://doi.org/10.1002/j.1551-8833.2010.tb10134.x
113. Dotson AD, Keen VS, Metz D, Linden KG (2010) UV/H_2O_2 treatment of drinking water increases post-chlorination DBP formation. Water Res 44(12):3703–3713. https://doi.org/10.1016/j.watres.2010.04.006
114. Pantin SRH (2008) Impact of UV/H_2O_2 on subsequent chlorine stability and DBP formation. Proceedings of the 13th Canadian national conference on drinking water. Quebec, Canada
115. Huang N, Wang T, Wang WL, Wu QY, Li A, Hu HY (2017) UV/chlorine as an advanced oxidation process for the degradation of benzalkonium chloride: synergistic effect, transformation products and toxicity evaluation. Water Res 114:246–253. https://doi.org/10.1016/j.watres.2017.02.015
116. Sun JL, Kong DZ, Aghdam E, Fang JY, Wu QY, Liu JJ et al (2019) The influence of the UV/chlorine advanced oxidation of natural organic matter for micropollutant degradation on the formation of DBPs and toxicity during post-chlorination. Chem Eng J 373:870–879. https://doi.org/10.1016/j.cej.2019.05.096
117. Wang WL, Wu QY, Huang N, Xu ZB, Lee MY, Hu HY (2018) Potential risks from UV/H_2O_2 oxidation and UV photocatalysis: a review of toxic, assimilable, and sensory-unpleasant transformation products. Water Res 141:109–125. https://doi.org/10.1016/j.watres.2018.05.005
118. Roback SL, Kodamatani H, Fujioka T, Plumlee MH (2020) Validation of a novel direct-injection chemiluminescence-based method for N-nitrosamine analysis in advanced-treated

recycled water, drinking water, and wastewater. Environ Sci Water Res Technol 6(4): 1106–1115. https://doi.org/10.1039/c9ew00943d
119. McCurry DL, Ishida KP, Oelker GL, Mitch WA (2017) Reverse osmosis shifts chloramine speciation causing re-formation of NDMA during potable reuse of wastewater. Environ Sci Technol 51(15):8589–8596. https://doi.org/10.1021/acs.est.7b01641
120. Sgroi M, Roccaro P, Oelker GL, Snyder SA (2015) N-nitrosodimethylamine (NDMA) formation at an indirect potable reuse facility. Water Res 70:174–183. https://doi.org/10.1016/j.watres.2014.11.051
121. Roback SL, Ishida KP, Plumlee MH (2019) Influence of reverse osmosis membrane age on rejection of NDMA precursors and formation of NDMA in finished water after full advanced treatment for potable reuse. Chemosphere 233:120–131. https://doi.org/10.1016/j.chemosphere.2019.05.259
122. Szczuka A, Huang N, MacDonald JA, Nayak A, Zhang Z, Mitch WA (2020) N-nitrosodimethylamine formation during UV/hydrogen peroxide and UV/chlorine advanced oxidation process treatment following reverse osmosis for potable reuse. Environ Sci Technol 54(23):15465–15475. https://doi.org/10.1021/acs.est.0c05704
123. Furst KE, Pecson BM, Webber BD, Mitch WA (2018) Tradeoffs between pathogen inactivation and disinfection byproduct formation during sequential chlorine and chloramine disinfection for wastewater reuse. Water Res 143:579–588. https://doi.org/10.1016/j.watres.2018.05.050
124. Feng Y, Smith DW, Bolton JR (2010) A potential new method for determination of the fluence (UV dose) delivered in UV reactors involving the photodegradation of free chlorine. Water Environ Res 82(4):328–334. https://doi.org/10.2175/106143009x447920
125. Yang J, Dong Z, Jiang C, Wang C, Liu H (2019) An overview of bromate formation in chemical oxidation processes: occurrence, mechanism, influencing factors, risk assessment, and control strategies. Chemosphere 237:124521. https://doi.org/10.1016/j.chemosphere.2019.124521
126. WHO (2005) Chlorite and chlorate in drinking-water background document for development of WHO guidelines for drinking-water quality. World Health Organization
127. EPA (2021) National Primary Drinking Water Regulations U.S. Accessed 6 Feb 2024
128. WHO (2017) Guidelines for Drinking-water Quality, 4th edn. Accessed 6 Feb 2024
129. Asami M, Kosaka K, Kunikane S (2009) Bromate, chlorate, chlorite and perchlorate in sodium hypochlorite solution used in water supply. J Water Supply Res Technol 58(2):107–115. https://doi.org/10.2166/aqua.2009.014
130. Stanford BD, Pisarenko AN, Snyder SA, Gordon G (2011) Perchlorate, bromate, and chlorate in hypochlorite solutions: guidelines for utilities. J AWWA 103(6):71–83. https://doi.org/10.1002/j.1551-8833.2011.tb11474.x
131. Buxton GV, Subhani MS (1972) Radiation chemistry and photochemistry of oxychlorine ions. Part 2. Photodecomposition of aqueous solutions of hypochlorite ions. J Chem Soc Faraday Trans 68:958–969. https://doi.org/10.1039/F19726800958
132. Wu Z, Guo K, Fang J, Yang X, Xiao H, Hou S et al (2017) Factors affecting the roles of reactive species in the degradation of micropollutants by the UV/chlorine process. Water Res 126:351–360. https://doi.org/10.1016/j.watres.2017.09.028
133. Rao B, Estrada N, McGee S, Mangold J, Gu B, Jackson WA (2012) Perchlorate production by photodecomposition of aqueous chlorine solutions. Environ Sci Technol 46(21): 11635–11643. https://doi.org/10.1021/es3015277
134. Lee TY, Notari RE (1987) Kinetics and mechanism of captopril oxidation in aqueous solution under controlled oxygen partial pressure. Pharm Res 4(2):98–103. https://doi.org/10.1023/a:1016406716989
135. Huang X, Gao N, Deng Y (2008) Bromate ion formation in dark chlorination and ultraviolet/chlorination processes for bromide-containing water. J Environ Sci (China) 20(2):246–251. https://doi.org/10.1016/s1001-0742(08)60038-8

136. Hofmann R, Andrews RC (2001) Ammoniacal bromamines: a review of their influence on bromate formation during ozonation. Water Res 35(3):599–604. https://doi.org/10.1016/s0043-1354(00)00319-5
137. Wu QY, Yang LL, Zhang XY, Wang WL, Lu Y, Du Y et al (2020) Ammonia-mediated bromate inhibition during ozonation promotes the toxicity due to organic byproduct transformation. Environ Sci Technol 54(14):8926–8937. https://doi.org/10.1021/acs.est.0c02984
138. von Gunten U (2003) Ozonation of drinking water: part II. Disinfection and by-product formation in presence of bromide, iodide or chlorine. Water Res 37(7):1469–1487. https://doi.org/10.1016/S0043-1354(02)00458-X
139. Wu QY, Yang LL, Du Y, Liang ZF, Wang WL, Song ZM et al (2021) Toxicity of ozonated wastewater to HepG2 cells: taking full account of nonvolatile, volatile, and inorganic byproducts. Environ Sci Technol 55(15):10597–10607. https://doi.org/10.1021/acs.est.1c02171
140. Zhang Y, Zhang B, Xu T, Yang X, Wei S, Yin D (2021) Developmental and neurobehavioral assessment of low-dose N-nitrosodimethylamine (NDMA) using zebrafish embryo bioassay. Sci Total Environ 770:144748. https://doi.org/10.1016/j.scitotenv.2020.144748
141. Taniguchi M, Yasutake A, Takedomi K, Inoue K (1999) Effects of N-nitrosodimethylamine (NDMA) on the oxidative status of rat liver. Arch Toxicol 73(3):141–146. https://doi.org/10.1007/s002040050598
142. Marti EJ, Pisarenko AN, Peller JR, Dickenson ERV (2015) N-nitrosodimethylamine (NDMA) formation from the ozonation of model compounds. Water Res 72:262–270. https://doi.org/10.1016/j.watres.2014.08.047
143. Heeb MB, Criquet J, Zimmermann-Steffens SG, von Gunten U (2014) Oxidative treatment of bromide-containing waters: formation of bromine and its reactions with inorganic and organic compounds – a critical review. Water Res 48:15–42. https://doi.org/10.1016/j.watres.2013.08.030
144. Zhang XY, Lu Y, Du Y, Wang WL, Yang LL, Wu QY (2021) Comprehensive GC×GC-qMS with a mass-to-charge ratio difference extraction method to identify new brominated byproducts during ozonation and their toxicity assessment. J Hazard Mater 403:124103. https://doi.org/10.1016/j.jhazmat.2020.124103
145. Zhang XY, Du Y, Lu Y, Wang WL, Wu QY (2022) Characteristics of the formation and toxicity index of nine newly identified brominated disinfection byproducts during wastewater ozonation. Sci Total Environ 824:153924. https://doi.org/10.1016/j.scitotenv.2022.153924
146. Wu QY, Zhou YT, Li WX, Zhang XR, Du Y, Hu HY (2019) Underestimated risk from ozonation of wastewater containing bromide: both organic byproducts and bromate contributed to the toxicity increase. Water Res 162:43–52. https://doi.org/10.1016/j.watres.2019.06.054

Disinfection Byproducts in Daily Life

Lei Wang, H. Larry Tang, and Baiyang Chen

Contents

1 Introduction 304
2 Boiling 307
3 Adsorption 310
4 Membrane Filtration 312
4.1 Membrane Characteristics 313
4.2 DBP Properties 313
4.3 Impacts of Operating Conditions on DBPs Removal 315
5 Photolysis 316
5.1 Kinetics of DBPs Photodegradation 316
5.2 Factors Impacting DBP Photodegradation Efficiency 317
5.3 Mechanisms for DBPs Photodegradation 319
6 Combined Technologies 320
7 Tips for Consumers 321
7.1 How to Boil Tap Water 321
7.2 How to Choose a Suitable POU Facility 322
8 DBPs in Swimming Pools 322
8.1 Formation Characteristics 323
8.2 Modeling Efforts 323
8.3 Mitigation Choices 324

L. Wang
Key Laboratory of Aquatic Botany and Watershed Ecology, Wuhan Botanical Garden, Chinese Academy of Sciences, Wuhan, China

State Key Laboratory of Urban Water Resource and Environment, Harbin Institute of Technology, Shenzhen, China
e-mail: wanglei@wbgcas.cn

H. L. Tang
Environmental Engineering Program, Indiana University of Pennsylvania, Indiana, PA, USA
e-mail: htang@iup.edu

B. Chen (✉)
State Key Laboratory of Urban Water Resource and Environment, Harbin Institute of Technology, Shenzhen, China
e-mail: chen.baiyang@hit.edu.cn

Chao Chen, Susan Andrews, and Yuefeng Xie (eds.), *Disinfection By-Products in Water*, Hdb Env Chem (2024) 135: 303–332, DOI 10.1007/698_2024_1157,

Published online: 10 November 2024

9 Summary 325
References 327

Abstract Disinfection byproducts (DBPs), formed as secondary contaminants during water disinfection processes, are frequently detected in domestic settings such as drinking water and swimming pools, posing considerable health risks to humans. Consequently, controlling DBPs is crucial for water consumers. This chapter provides a comprehensive overview of various methods for removing DBPs through household facility processes, also known as point-of-use facilities, including boiling, activated carbon adsorption, membrane filtration, and ultraviolet photolysis. In addition to discussing the performance and mechanisms of these processes, the chapter explores intrinsic factors (such as the physical-chemical properties of DBPs) and extrinsic factors (including pH and temperature). Accordingly, a detailed procedure for boiling water to achieve DBPs removal is presented. Furthermore, given that swimming is a widely recognized health-boosting activity suitable for all age groups, the formation and control of DBPs in swimming pools has also been discussed.

Keywords DBPs, Household water treatment, Point-of-use, Swimming pool water

1 Introduction

Since the advent of chlorination in the early twentieth century, disinfection technology has played a vital role in inactivating pathogens and preventing waterborne diseases [1]. However, chemical disinfection can also generate disinfection byproducts (DBPs) as a side effect. To mitigate their risks, many countries and organizations have enacted regulations or guidelines to control DBPs in drinking water. For example, the national standards for Drinking Water Quality of China (GB/T5749-2022) have established maximum contaminant levels (MCLs) for over 10 DBPs in 2023 (Fig. 1). The US EPA has also set MCLs for four trihalomethanes (THMs) and five haloacetic acids (HAAs) since 2006 [2] (Fig. 1).

To comply with these DBPs rules, many studies have explored various DBP control strategies, including removal of DBP precursors, optimization of disinfection processes at water treatment plants, and elimination of already-formed DBPs [1]. Although the first two strategies are cost-effective, they cannot eliminate all DBPs formed in water treatment plants and distribution systems, which often contain considerable residual disinfectants and organic matter, leaving DBPs inevitably present in consumers' tap [3]. Thus, controlling DBPs from tap water via point-of-use (POU) treatment (also namely household water treatment processes) becomes the last resort to improve drinking water quality [4].

Table 1 summarizes the concentrations of common DBPs in tap water around the world. Generally, the concentrations of DBPs in tap water fall within the range of ng L^{-1} to μg L^{-1}. Among these DBPs, the mass concentrations of two dominant types, THMs and HAAs, were observed to be the highest. The mass concentrations

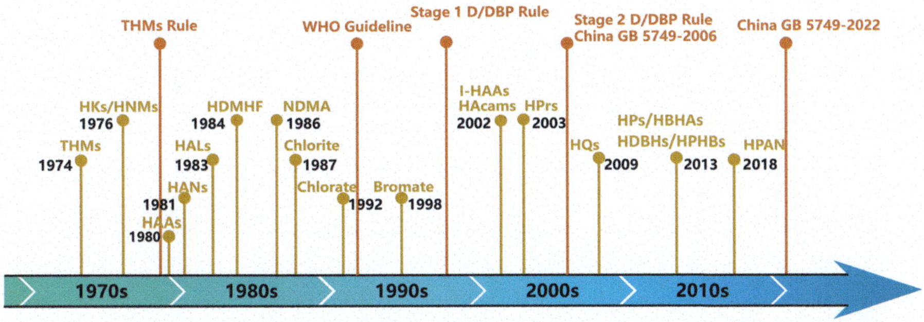

Fig. 1 Discovery and legislation processes of DBPs in drinking water from the 1970s to 2020s. Abbreviations: THMs, trihalomethanes; HAAs, haloacetic acids; HALs, haloacetaldehydes; HAcAms, haloacetamides; HNMs, halonitromethanes; NDMA, N-nitrosodimethylamine; HKs, haloketones; HDMHF, halofuranones; I-HAAs, iodinated haloacetic acids; HPrs, halopyrroles; HQs, haloquinones; HPs, halophenols; HBHAs, halosalicylic acids; HDBHs, halohydroxybenzoic acids; HPHBs, halohydroxybenzaldehydes; HPANs, halophenylacetonitriles; WHO, World Health Organization

of haloacetaldehydes (HALs) occupied the third position, with trichloroacetaldehyde (TCAL) exhibiting the highest concentration of all HALs. Among nitrogenous DBPs, such as haloacetonitriles (HANs), haloacetamides (HAcAms), halonitromethanes (HNMs), and nitrosamines (NAs), those with high concentrations and high detection frequencies included dichloroacetonitrile (DCAN), dibromoacetonitrile, dichloroacetamide, dibromoacetamide, trichloronitromethane (TCNM), and N-nitrosodimethylamine (NDMA).

POU provides a supplementary approach to mitigate DBPs beyond municipal treatment strategies [1] (Fig. 2). For example, recent researches indicate that boiling water can effectively control various types of DBPs [22, 23], although it is often perceived as a pathogen inactivation method only [24]. Besides boiling, POU technologies like membrane filtration, advanced oxidation, and adsorption also merit in-depth investigations [4]. For example, large particles, rust, and debris are removed by primary filtration through string-wound sediment or polypropylene (PP) cotton filters first [7, 25]. Subsequently, redox media, activated carbon, and ion exchange may be employed to remove micropollutants, toxic ions, and heavy metals [4]. Reverse osmosis (RO) and nanofiltration (NF) membrane technologies have been commercialized in POU facilities and widely used nowadays due to their ability to remove almost all contaminants from water [4]. Ultraviolet (UV) not only inactivates pathogenic microorganisms but might also degrade micropollutants; so, some POU products are also equipped with UV devices. Overall, there exist various technologies with the potential to remove DBPs robustly at domestic scales.

Hereafter, we present some common POU methods known to be efficient in controlling DBPs, including boiling/heating, activated carbon adsorption, membrane filtration, and ultraviolet photolysis (Fig. 2). This chapter starts with an introduction of the technology or device along with its DBPs removal efficiencies, proceeds with discussions of relevant mechanisms, and then ends with the evaluation of influencing factors. In addition to drinking water, swimming pool water also contains abundant DBPs, which are to be discussed too.

Table 1 Occurrence of DBPs in tap water in different regions or countries

Location		THMs (μg/L)	HAAs (μg/L)	HALs (μg/L)	HANs (μg/L)	HAcAms (μg/L)	HNMs (μg/L)	NAs (ng/L)	HBQs[a] (ng/L)	HPANs[b] (ng/L)	References
Asia	China	[n.d., 197.5][c]	[n.d., 60.4][c]	[4.9, 46.2]	[n.d., 16.2][c]	[0.1, 3.1]	[0.2, 2.9]	[n.d., 224.7][c]	–	[170, 530]	[5–8]
	Japan	[21.7, 61.6]	[17.1, 83.9]	[n.d., 5.4][d]	–	–	–	–	[8, 51]	–	[9, 10]
	Korea	[3.9, 53.5]	[n.d., 49.5]	[n.d., 33.2][d]	[n.d., 22.5]	–	[n.d., 4.2][e]	–	–	–	[11, 12]
	Thailand	[104.2, 198.0]	–	–	[5, 20]	–	–	–	–	–	[13]
	Saudi Arabia	[0.1, 52.4]	[6.4, 9.9]	–	–	–	–	–	–	–	[14]
Europe	UK	–	–	–	[n.d., 12.1]	[n.d., 7.0]	[n.d., 7.0]	–	–	–	[1]
	Spain	[17, 83]	[n.d., 86.9]	[n.d., 12.1][d]	[1.3, 5.0]	–	n.d.	–	–	–	[1, 15]
	Greece	[5.0, 106.0]	–	–	–	–	–	–	–	–	[1]
North America	USA	[37, 55]	[34, 44]	[0.5, 4.0]	[0.5, 5.0]	–	–	–	[4.5, 274.5]	–	[16, 17]
	Canada	[12.4, 99.6]	–	[7.2, 26.8]	–	–	–	[0.9, 3.3][f]	[4.5, 274.5]	–	[16, 18, 19]
Oceania	Austria	[6.0, 191.0]	–	[0.2, 19.0][d]	–	–	–	–	–	–	[1]
Africa	Tunisia	[46.0, 186.7]	[47.3, 105.7]	–	–	–	–	–	–	–	[14]

Notes: The data in the bracket indicate the range of concentrations of DBPs. For example, [20, 21] for the USA in the column for THMs means that the minimum concentration of THMs detected in the USA in the cited literature is 37 μg/L and the maximum concentration is 55 μg/L

n.d., not detected; –, unmeasured

Abbreviations: THMs, trihalomethanes; HAAs, haloacetic acids; HALs, haloacetaldehydes; HAcAms, haloacetamides; HNMs, halonitromethanes; NAs, nitrosamines; HBQs, halobenzonicquinones; HPANs, halophenylacetonitriles

[a] 2,6-chloro-benzoquinone only

[b] 2-chloro-benzeneacetonitrile

[c] National-scale survey

[d] TCAL only

[e] TCNM only

[f] NDMA only

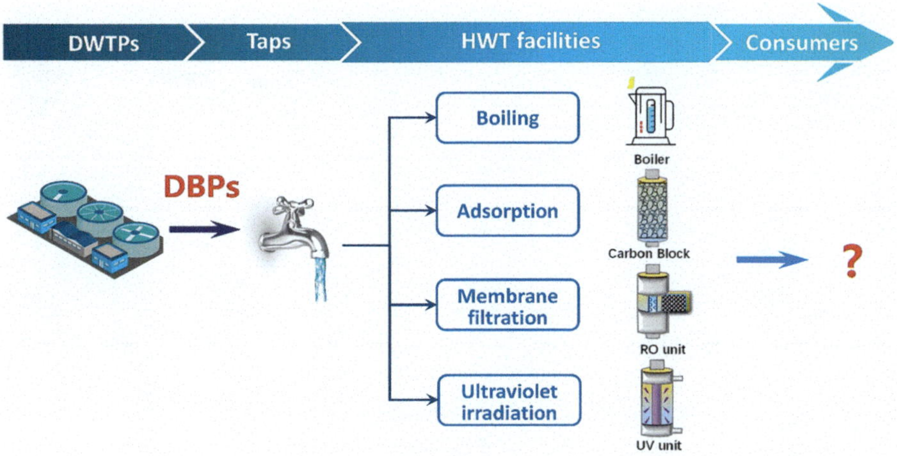

Fig. 2 A summary of POU methods for DBP mitigation in this chapter. Abbreviations: DWTPs, drinking water treatment plants; HWTs, household water treatment

2 Boiling

Boiling, a widely adopted POU technique, particularly in China, not only inactivates bacteria but also effectively reduces total organic halides (TOX) [23, 26, 27]. Within identical DBPs groups, the removal efficiency through boiling follows the order: chlorinated > brominated > iodinated DBPs, with lowly halogenated DBPs showing greater removal than highly halogenated DBPs [28].

During the boiling process, two major mechanisms are involved: volatilization and hydrolysis (Fig. 3). The volatilization potential of DBPs is determined by Henry's law constant or hydrophobicity. It was reported that DBPs with higher Henry's law constants or greater hydrophobicity were more likely to volatilize compared to DBPs with hydrophilic properties or lower Henry's law constants [28]. So far, the tested DBPs include THMs, which were removed by 69–87% [29, 30], HANs, which were removed by 80% on average [31], and HALs, which were removed by 32% on average [22].

Hydrolysis at elevated temperatures is another DBPs removal mechanism during boiling process [32], which may cause either addition or decomposition of functional groups to target compounds [33]. For example, the electron-rich group of water, HO–, may form a bond with the electron-deficient substituent of halogenated DBPs, leading to breakages of carbon–carbon and/or carbon–halogen bonds and modifying halogenated DBP molecules [34–36]. Despite so, decomposition of DBPs was more frequently observed, decreasing the concentrations of parent DBPs but increasing the formation of other DBPs [30, 32, 37]. For example, the primary hydrolysis products of HAAs and HALs were THMs (Reactions 2 and 3), while the major hydrolysis products of haloketones were THMs and acetic acids (Reaction 6).

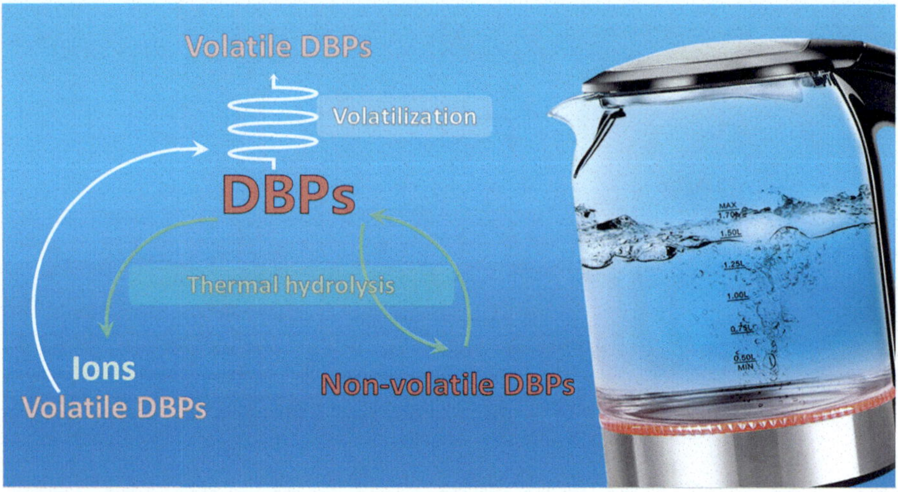

Fig. 3 Mechanisms for DBPs mitigation by boiling tap water

Although HAcAms were formed through HANs hydrolysis (Reaction 4), they were subsequently hydrolyzed into HAAs (Reaction 5). The only exception was THMs, which dehalogenated stepwise and yielded halide anions mostly (Reaction 1) during hydrolysis process [35, 37, 38]. For aromatic DBPs like halohydroxybenzaldehyde, halosalicylic acid, halohydroxybenzoic acid, and halophenol, their brominated analogues (i.e., 3,5-dibromo-4-hydroxybenzaldehyde, 3,5-dibromosalicylic acid, 3,5-dibromo-4-hydroxybenzoic acid, and 2,4,6-tribromophenol) increased by 25–90% after boiling for 10 min [20, 23]. This is due to the decarboxylation pathway undergone by DBPs bearing one or more carboxylic acid groups attached to the benzene ring [20, 23]. This process results in the formation of smaller aromatic DBPs, such as 2,4,6-trihalophenol, as well as THMs and HAAs [4].

$$CX_nH + OH^- \rightarrow CX_{n-1}OH + X^- \quad (1)$$

$$CX_nCOO^- + H_2O \rightarrow CHX_n + HCO_3{}^- \quad (2)$$

$$X_nCC(=O)H + OH^- \rightarrow CHX_n + HCO_3{}^- \quad (3)$$

$$X_nCC \equiv N + H_2O \rightarrow X_nC(OH) = N^- \rightarrow X_nCC(=O)NH_2 \quad (4)$$

$$X_nCC(=O)NH_2 + OH^- \rightarrow CX_nCOO- + NH_3 \quad (5)$$

$$X_nCC(=O)C + OH^- \rightarrow CCOO^- + CHX_n \quad (6)$$

Water temperature is a key parameter during the heating process as it significantly influences the formation and removal of DBPs. The Arrhenius equation has demonstrated that the hydrolysis rate constants for THMs at room temperature ranged from 10^{-8} to 10^{-6} h^{-1}. However, these rates increased significantly to 10^{-2} to 10^{-1} h^{-1} at the boiling point of water [35], indicating that the half-life of trichloromethane (TCM) is approximately 3,407 years at 25°C but diminishes to approximately 3 h at 100°C [33]. Moreover, elevated water temperature can notably increase the concentration of hydroxide ions and significantly influence the chemical stability and reaction rates of DBPs. This impact is particularly pronounced in the hydrolysis of THMs [37], where higher temperature (65 and 95°C) within this range markedly accelerated their degradations, thereby reducing their concentrations in drinking water. Boiling real tap water samples exhibited similar effects [33]. Additionally, elevated temperatures, particularly up to 70°C, significantly enhanced the hydrolysis of N-nitrosamine precursors, facilitating their removal during water treatment processes [35]. For chlorinated HALs, increased temperatures significantly improved their degradation rates, especially when chlorine was present [22]. In summary, increasing water temperature can enhance hydrolysis of various DBPs, including THMs and HALs, while the presence of chlorine can further amplify such effect. In addition, these findings suggest that prolonging heating time can significantly enhance DBP removal.

To optimize DBP removal by boiling, several temperature-associated parameters can be adjusted. Increasing heating power, lowering water volume, prolonging heating time, lowering cooling time, and keeping the lid open during boiling can all enhance DBP removal [28, 31, 39]. For instance, boiling tap water and storing it covered or uncovered for 4 h resulted in 84% and 90% reductions in total THMs, respectively [30]. If stored uncovered after boiling, THM concentrations may continue to decline over time, with up to a 97% reduction observed after 48 h [30]. So, distinctive THMs removals (97% and 82%) were observed between uncovered or covered conditions, respectively [40]. However, it is not advisable to store boiled water for extended periods as bacteria may rapidly grow after 12 h [41].

Different heating tools, such as electric kettles or microwave oven, demonstrated variable effectiveness in removing DBPs [28, 31, 39]. Various studies indicate that the removals of THMs by electric kettles are more effective than microwave-oven boiling [28]. This observation was largely attributed to the uneven heat distribution and subsequent increased turbulence in electric boilers relative to microwave ovens [28]. Nevertheless, microwave irradiation has been found to remove HANs more efficiently, achieving a 7% average higher removal than electric kettles within a 4-min heating period [31]. This contrasts with the trends observed in the removal of THMs, which can be attributed to the differences in how the containers are capped: while the containers in microwave oven is fully open, which is favorite for volatilization of HANs, electric kettle is almost closed with only a small outlet, which restricts volatilization [31].

Similar differences have also been observed in the impact of initial DBPs concentrations on boiling performance. Increasing the initial THM concentration from 50 to 200 μg L^{-1} led to an average increase in THMs removal from 80% to 92% [28]. However, an increase in initial HAN concentration from 20 to 100 μg L^{-1} resulted in only minor increases (2%) in HANs removals [31]. Similarly, increasing the initial concentrations of HALs, such as monochloroacetaldehyde, dichloroacetaldehyde, and TCAL, from 5 to 100 μg L^{-1} decreased their removals from 14.6%, 13.8%, and 18.7% to 9.4%, 4.3%, and 6.3% respectively [39]. So far, the underlying reasons for these observations remain elusive.

Coexisting chlorine is another significant factor that impacts DBPs transformation during boiling process. For example, increasing the chlorine dosage from 0 to 1.0 mg L^{-1} apparently enhanced the removals of HALs from ~15% to >80% [22]. Furthermore, boiling water for 4.2 min removed 33–100% of HANs from real tap water containing 0.4 mg L^{-1} of residual chlorine, much more sufficient than those when chlorine is absent [31]. Notably, residual chlorine might promote the formation of thermally stable DBPs, such as TCM and dichloroacetic acid (DCAA) [39]; so its side effect also needs attention.

Interestingly, Liu et al. found that adding ascorbic acid (vitamin C) and carbonate (baking soda) before boiling could reduce the levels of TOX by up to 36% and 28%, respectively [42]. The reason is that ascorbate could quench residual chlorine, thereby avoiding formation of DBPs during subsequent boiling process [42], and carbonate could elevate pH, thereby enhancing hydrolysis (dehalogenation) of DBPs [42]. Ultimately, additions of these additives before boiling significantly reduced the developmental toxicity of tap water [43].

3 Adsorption

Adsorption is a widely employed technology in POU with activated carbon being the predominant adsorbent. Owing to its low cost, high efficiency, and ease of operation, it was already recognized by the US EPA as the best available technology for drinking water treatment [4]. Presently, activated carbon was found effective in treating various DBPs, including THMs, HAAs, HALs, HANs, HAcAms, HNMs, and halogenated aromatic DBPs. Some studies have also explored the uses of biochar (from 20 to 200 mg L^{-1}) for removing iodinated DBPs (100% in 4 min with initial concentrations from 0.2 to 1.2 mM) and NDMA (61.7% in 2 days with 5 mg L^{-1} as the initial concentration and 1.0 g L^{-1} as the biochar dosage) [44, 45].

The mechanisms responsible for DBP removal by activated carbon adsorption include physical adsorption via van der Waals forces and ion exchange:

1. Physical adsorption via van der Waals forces. Physical adsorption mainly occurs in activated carbon micropores with low reversibility or desorption potential [46, 47]. Scanning electron microscope images of activated carbon before and after adsorption of THMs indicate that non-polar DBPs were mostly adsorbed on AC's surface and micropore channels [48]. In comparison, Fourier Transform Infrared Spectra (FTIS) of activated carbon indicate that AC's functional groups minimally influence THM adsorption [48]. Similar outcomes were observed in removing NAs via activated carbon [45].

2. Ion exchange between DBPs and oxygen-containing functional groups of activated carbon [49]. For activated carbon with oxygen-containing functional groups, its surface electronegativity is high due to hydroxyl dissociation effect

[50], which induces electrostatic repulsion of HAAs and results in low adsorption of HAAs. In contrast, activated carbon with phenolic hydroxyl and carboxyl groups or highly basic groups is prone to undergo ion exchange with HAAs [48]. A comparison of FTIS of activated carbon before and after HAAs adsorption showed that the FTIS peaks shifted from 1,604 to 1,613 cm^{-1} [48]. As vibration peak at this frequency indicates the presence of C=O bond, such shift means that a chemical reaction occurred between HAAs and activated carbon functional groups [49].

The performances of activated carbon adsorption of DBPs in water are more complicated than those in ultrapure water because many factors, including physicochemical properties of activated carbon [49], properties of DBPs [51], and environmental conditions [49, 52], might impose additional impacts. The physicochemical properties of activated carbon are governed by the raw material, type, quantity of activating agent, combustion temperature and duration, and heating methods used during the activated carbon manufacturing process. Given that activated carbons manufactured by different methods feature distinctive physical structures and chemical properties [53], the adsorption performances of activated carbons are reasonably different. For instance, He et al. assessed the adsorption of THMs by four activated carbons with similar pore sizes [54] and found that the coconut shell-based activated carbon displayed superior efficacy for THMs adsorption (1.6–3.8 $\mu g\ g^{-1}\ min^{-1}$) over those wood (1.2–2.6 $\mu g\ g^{-1}\ min^{-1}$) and nutshell (1.0–1.6 $\mu g\ g^{-1}\ min^{-1}$) -based ACs [54]. In addition, both new and used carbon blocks reduced HAAs by approximately 72% and 60%, respectively [30], meaning that the performance of activated carbon could deteriorate with time.

Different DBPs exhibit varying adsorption susceptibilities. In general, activated carbon preferentially adsorbs DBPs with weak polarity, as seen in the selective adsorption of THMs over HAAs [48], and in the preferential removals of HANs over HALs and HAcAms by activated carbon [47]. In addition, halogenation degree and molecular weight influence adsorption, with higher halogenation and molecular weight leading to increased adsorption rate but decreased adsorption capacity. For example, the adsorption rates of DBPs were ranked as TCM (0.03 h^{-1}) < bromodichloromethane (BDCM, 0.09 h^{-1}) < dibromochloromethane (DBCM, 0.14 h^{-1}) in synthetic water [47, 55], whereas the order of adsorption capacity in full-scale GAC bed test within 200 days reversed as TCM (458.0 mg kg^{-1} AC) > BDCM (149.4 mg kg^{-1} AC) > DBCM (130.4 mg kg^{-1} AC) in real water [21]. As for ionic DBPs like HAAs, the sequence of adsorption potentials was found as monobromoacetic acid (MBAA, 42.7 mg kg^{-1} AC) > bromochloroacetic acid (BCAA, 35.0 mg kg^{-1} AC) > dibromoacetic acid (DBAA, 28.5 mg kg^{-1} AC) among the bromine substituted HAAs and trichloroacetic acid (TCAA, 585.7 mg kg^{-1} AC) > DCAA (516.0 mg kg^{-1} AC) > monochloroacetic acid (174.6 mg kg^{-1} AC) among the chlorine substituted HAAs [21].

Environmental and operation conditions, including the presence of coexisting organics, water temperature, and pH, can also significantly influence DBPs' removal through activated carbon adsorption. One influential factor is the competitive adsorption between organics and DBPs. For example, DOM was preferentially adsorbed to activated carbon first, followed by THMs and HAAs [56]. Hence, the presence of humic substances in water led to decreased removals of TCM, BDCM, and DCAA [48].

The influence of water temperature on DBPs adsorption by activated carbon is noteworthy. Using THMs as an example, the pseudo-first order adsorption rate constants for TCM, BDCM, and DBCM were 0.06, 0.18, and 0.22 h^{-1}, respectively, at 10°C but they dropped to 0.02, 0.03, and 0.10 h^{-1}, respectively, at 35°C (AC dosage: 0.4 to 3 g L^{-1}; DBPs initial concentration: 2 mg L^{-1}) [54]. Conversely, increasing pH tends to increase the adsorption capacity of activated carbon for DBPs removal [49]. Li et al. reported that the adsorptions of four regulated THMs increased by 50–100% with a rise of pH from 3 to 11 [57].

In addition to activated carbon, recent studies have reported the effectiveness of hydrophobic zeolites and magnetic composites in the adsorption-based removal of DBPs [58, 59]. Silicalite-1 has demonstrated superior performance compared to dealuminated zeolite Y in adsorbing DBPs, particularly TCM, even in the presence of humic acid. Furthermore, it can be regenerated through advanced oxidation processes using Fe^0/H_2O_2 [58]. Similarly, Sun and colleagues developed a magnetic composite, $Fe_3O_4@SiO_2@PDA@NH_2$-MIL-53(Fe), which exhibited significant adsorptive capacity for 2,6-dichloro-1,4-benzoquinone, an emerging aromatic DBP [59]. The adsorptive mechanism of the magnetic composites involves π-π interactions, hydrogen bonding, and pore-filling, differing from the strategy of hydrophobic zeolites, which rely on hydrophobic interactions and molecular sieving properties [58, 59]. Both methods are highly pH-sensitive, with optimal adsorption occurring under neutral to mildly basic conditions. These findings underscore the adaptability of these materials in advancing sustainable water purification technologies [58, 59].

4 Membrane Filtration

High pressure membrane filtration technologies, such as Nanofiltration (NF) and Reverse Osmosis (RO), offer effective removal of organic and inorganic pollutants, including DBPs [60]. Because the molecular sizes of DBPs are much smaller than the pore sizes of microfiltration and ultrafiltration membranes, these two membranes are not suitable for removing DBPs. The global POU market is growing rapidly and is anticipated to reach a value of 34.1 billion US dollars by 2025, with RO dominating the share (43%) [4]. In 2019, China alone witnessed the sale of over 18 million POU units, with 80% of them incorporating RO cartridges [4]. The efficiency of DBPs removal by NF/RO depends on membrane properties, DBP characteristics [61], and various operational factors [62].

4.1 Membrane Characteristics

Membrane characteristics mainly include molecular weight cut-off (MWCO), salt rejection ability, surface charge, and membrane materials [63]. MWCO signifies the pore size of a membrane that retains >90% of the smallest non-charged solute [64]. As typical halogenated DBPs have molecular weights exceeding 100 Da, membranes with MWCO of 100 Da ~ 300 Da are effective for their removal [65, 66]. Comparatively, Chalatip et al. found that tight NF membranes (branded as ES10 and NTR729HF) with lower MWCO (100 Da and 200 Da) outperformed a loose NF membrane (branded as NTR7410 with MWCO 20,000 Da) in removing five HAAs [67].

Given that MWCO measurement is challenging, salt rejection efficiency is often used as an alternative parameter by membrane manufacturers for evaluating membrane performance. For example, an NF membrane branded as NF200 rejected >97% of $MgSO_4$ and 79–96% THMs, significantly higher than a membrane branded as DS5, which removed <80% $MgSO_4$ and 65–82% THMs [68]. Similarly, NF90 rejected more NaCl (82.7% of 2000 mg/L) than NF270 (30.6%), which reasonably explain why NF90 performed better in removing nine HAAs (80–99%) than NF270 (20–50%) [69].

Surface charge, another crucial property, plays a role in electrostatic repulsion effects. To the best of our knowledge, most commercial NF/RO membranes are negatively charged, thereby rejecting negative solutes well [70, 71]. Sometimes, the effect of electrostatic repulsion might supersede size exclusion. For instance, although a tight NF membrane (branded as NTR729HF with 200 Da MWCO) has a much lower MWCO than a loose NF membrane (branded as NTR7410 with 20,000 Da MWCO), the former gave lower rejection (66% ~ 72%) of five HAAs than the latter (74% ~ 82%) [67] because the former was uncharged whereas the latter was charged.

Membrane materials influence DBPs removal significantly. For example, a thin (10–500 nm), non-polar aromatic polyamide RO membrane removed more HAAs (80–100%) than a thick (1–100 μm), polar cellulose acetate RO membrane (50–80%) [69]. Polyether sulfone membrane performed similarly to polyamide membranes in salt rejection but was weaker for THMs removal [68].

4.2 DBP Properties

DBPs exhibit diverse characteristics, including molecular size, hydrophobicity, and acid dissociation constant (pK_a) values. All might determine their removals mechanisms, including size exclusion, hydrophobic interactions, and electrostatic repulsion (Fig. 4) [69, 73, 74].

Molecular size can be described as molecular weight, volume, length, width, and Stokes radius. Sometimes, the rejections of neutral DBPs (HAcAms, HANs, and

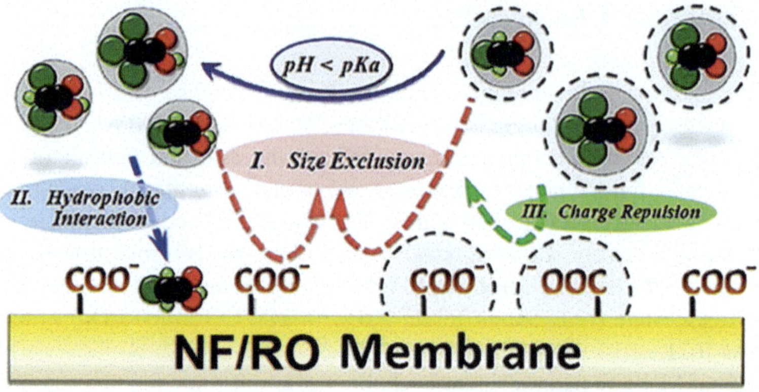

Fig. 4 Mechanisms for HAAs mitigation by NF/RO filtration [72]

THMs) by NF and RO membranes were positively correlated with molecular weights of DBPs [73], and molecular volume may better reflect size than molecular weight [73]. Some other times, Yang et al. correlated Stokes radius well with HAA transports [69]. Although certain studies have applied molecular length and width to estimate RO rejection of other compounds [75], their uses in DBPs studies remain limited.

Octanol-water partition coefficient (log K_{ow}) serves as an indicator of the hydrophobicity of DBPs, potentially reflecting their adsorption to membranes or hydrophobic interactions with membranes. The log K_{ow} of halogenated DBPs varies widely from 0–3, meaning that some hydrophobic DBPs might be removed efficiently by membrane adsorption [4, 74, 76] but some hydrophilic ones may not. As a result, while certain literature reported negative correlation between log K_{ow} values and the rejection levels of 20 DBPs [4], little correlations were found between 29 DBPs and their log K_{ow} values in other studies [73, 77]. Therefore, it is necessary to elucidate the role of hydrophobic interaction in DBP rejections in the future.

For charged DBPs such as HAAs, bromate, and perchlorate, pK_a is a key parameter determining not only the charge states of DBPs but also their treatability toward membranes [65]. Ionic DBPs at pH above their pK_a experience strong electrostatic repulsion from negatively charged NF/RO membranes [65, 72], resulting in often satisfactory removal. For example, the HAA removals were negatively correlated with pK_a at a given pH [69]. Although the properties of neutral DBPs remain unchanged with varying pH, increasing pH has been shown to increase the rejections of certain DBPs like HAcAms or NAs in NF/RO processes [62, 73] without compromising desalting ability. The reason is likely that increasing pH can interfere with the H-bonding between membrane polymer and DBPs, hence leading to increased rejections of DBPs [73].

4.3 Impacts of Operating Conditions on DBPs Removal

pH, DBPs' concentration, ionic strength, and other water quality factors greatly influence DBPs removal [72, 78–81]. pH primarily affects DBPs removal by altering DBP speciation and membrane surface charge and structure [69]. One study shows that raising pH from 3 to 7 lowered the zeta potential of membrane from 0 to −35 mV and increased HAAs rejections from 20–40% to 60–80%, but did not change water flux significantly [69]. In another case, increasing pH from 4.5 to 8.5 did not alter the rejection levels of non-charged DBPs (i.e., THMs) by NF but the pore size was contracted and permeate flux was decreased [73].

Concerning the impact of DBPs concentration, one study showed that increasing THMs concentration from 20 to 80 μg L^{-1} raised the rejection from 75 to 95% by NF200 [68]. Another study, nonetheless, found that increasing mono-HAAs concentrations from 10 to 90 μg L^{-1} reduced their removals from 20% to 10% by a NF membrane branded as VNF2 [65]. This indicates that the influence of DBPs concentration varies with varying types of compounds.

The influence of ionic strength on DBPs rejection varies among ionic and neutral DBPs. For ionic DBPs, one study found that the HAAs rejections increased by 5–15% when ionic strength increased from 7 to 70 mM NaCl [67]. Under high ionic strength, membrane electrostatic repulsion weakens, thus decreasing effective pore size and permeability [82]. Additionally, high ionic strength reduced the Debye length of ionic DBPs which enhanced their transport [71, 72]. Thus, water flux was reduced and DBPs rejection was increased when ionic strength increased. For neutral DBPs, increasing ionic strength moderately improved rejections of small DBPs like THMs but did not benefit large DBPs like HANs [73]. The ion strength effect on neutral DBPs rejection was attributed to solute dehydration [83], increase of dipole moment [4], salt-induced membrane swelling [62], and reduction of hydration layer on pore walls [62].

DBP rejection also depends on certain operating parameters like permeate flux, feed flow velocity, and pressure. As predicted by solution-diffusion theory, DBPs rejection occurred more likely at high flux and low salt recovery conditions, whereas low rejection occurred at low flux and high salt recovery conditions [62, 73]. This aligns with a general trend that improved DBP rejection corresponds to increasing permeate flux [84]. However, a pressure threshold exists in this case. For example, increasing RO pressure from 0.6 to 0.8 MPa severely damaged the performance of membrane [72], and increasing operating pressure from 0.2 to 1.0 MPa restrained HAAs rejection by 4–15% for NF [65]. These suggest that raising operating pressure and flux does not always enhance rejection for all DBPs, because membrane deformation may occur, enlarging the pore sizes of membranes. So, a proper membrane operating condition needs to balance both water flux and membrane preservation.

Although simple and convenient, single-stage NF/RO lacks energy and water use efficiency, which might waste over 50% feed as brine and cause energy loss [4, 85]. To address such issues, earlier studies once applied multi-stage RO process

[72] and flexible RO process [4] to recover water and energy. Results showed that the five-stage RO rejected >74.6% of HAAs and recovered 87.0% of water [72]; the flexible RO removed >80% of selected DBPs while recovered 80% of water recovery constantly [4]. Despite so, decreasing DBPs removal efficiencies were observed along with RO stages or time, meaning that the water/energy recovery benefits were obtained at the cost of compromised pollutant removal efficiency, although it is still within acceptable scope. To balance the needs of water quality, water quantity, energy, and cost, flexible RO was probably a promising technology to be explored in the future.

In summary, membrane technologies such as NF and RO are highly effective in removing DBPs and are becoming standard components in POU devices. Despite their prevalence, significant knowledge gaps persist, particularly concerning long-term efficacy, reusability, and the optimal timing for membrane replacement in practical applications. RO is often preferred for POU applications due to its desalination capabilities and its ability to reject DBPs. However, the efficiency of RO in rejecting DBPs can fluctuate based on various factors such as pH, ionic strength, membrane characteristics, and water components. The potential for smaller and neutral DBPs to permeate RO membranes is a significant concern, as it could result in the presence of these contaminants in treated water. It is crucial to differentiate adsorption rejection of DBPs during RO process. Future research is needed to ascertain the sustained effectiveness of RO in DBPs management and to develop innovative strategies to reduce the permeation of micropollutants over the operational lifetime of the membranes.

5 Photolysis

Previously, ultraviolet (UV) irradiation was one of the modules in POU facility to ensure water safety from microbial contaminants [25]. Currently, UV has been tested for degradation of many DBPs, including THMs [86], HAAs [87], HALs [88], HANs [88], HNMs [89], HAcAms [90], and NAs [91]. The following summarizes the efficiency, mechanisms, and influencing factors of DBPs photodegradation.

5.1 *Kinetics of DBPs Photodegradation*

The photodegradation of DBPs generally follows pseudo-first-order reaction kinetics with varying rate constants ($k_p < 0.05 \times 10^4$ to 29.84×10^4 cm^2 mJ^{-1}) and quantum yields (Φ, 0.11 mol $Einstein^{-1}$ ~ 0.87 mol $Einstein^{-1}$) [92]. Among DBPs, one database once reported that the k_p were ranked as: THMs > HAAs > HAcAms > HANs ≈ HNMs [93] with identical halogenation types and degrees. However, another study reported the order of k_p was: TCNM (2.6×10^{-3} min^{-1}) > TCAA (1.7×10^{-3} min^{-1}) > trichloroacetonitrile (TCAN, 1.4×10^{-3} min^{-}

^{1}) > trichloroacetamide (TCAcAm, 6.5×10^{-4} min^{-1}) > TCAL (4.1×10^{-4} min^{-1}) > TCM (1.7×10^{-4} min^{-1}) [88], which are significantly different from the earlier database.

Among DBPs with identical halogen types but different numbers of halogens, the $\boldsymbol{k_p}$ of monohalogenated DBPs ranges within 6×10^3 to 2.4×10^4 cm^2 mJ^{-1}, significantly lower than those of di- and tri-halogenated species [93]. For instance, the $\boldsymbol{k_p}$ for MBAA to DBAA and tribromoacetic acid (TBAA) were 6×10^3, 4.3×10^4, and 2.2×10^5 cm^2 mJ^{-1}, respectively [87]. Similarly, the $\boldsymbol{k_p}$ of monobromonitromethane, dibromonitromethane, and tribromonitromethane increased sequentially from 7×10^3 cm^2 mJ^{-1} to 1.1×10^5 cm^2 mJ^{-1} and 1.5×10^5 cm^2 mJ^{-1}, respectively [89]. Another HANs photolysis study also showed increasing photolytic stability with decreasing halogenation degrees: TCAN ($\boldsymbol{k_p} = 8.8 \times 10^{-5}$ s^{-1}) > DCAN (2.9×10^{-5} s^{-1}) > chloroacetonitrile (MCAN, not detected) [94]. These results all support that compounds with higher halogenation tend to be photolyzed faster than those with lower halogenation degrees [93]. This pattern may be ascribed to the steric and electronic effects of halogens, which can lead to different electron-withdrawing influences on photolysis [95].

For compounds with the same number of halogens, the degradation kinetics increased with rising molecular weights of halogens (i.e., iodinated > brominated > chlorinated DBPs). For instance, the direct photolysis rates exceeded 2.0×10^5 cm^2 mJ^{-1} for tribrominated and triiodinated DBPs, markedly higher than trichlorinated species such as TCNM (1.0×10^4 cm^2 mJ^{-1}) and TCAN (7×10^3 cm^2 mJ^{-1}) [93]. Similarly, the $\boldsymbol{k_p}$ of dihalogenated HAAs followed the order of diiodoacetic acid (3.0×10^5 cm^2 mJ^{-1}) > bromoiodoacetic acid (1.3×10^5 cm^2 mJ^{-1}) > chloroiodoacetic acid (4.7×10^4 cm^2 mJ^{-1}) > DBAA (4.3×10^4 cm^2 mJ^{-1}) > BCAA (2.1×10^4 cm^2 mJ^{-1}) [87]. These trends consist with the carbon-halogen bond dissociation energies of DBPs, which are 209 kJ mol^{-1}, 280 kJ mol^{-1}, and 397 kJ mol^{-1} respectively, for carbon-chlorine, carbon-bromine, and carbon-iodine bonds, respectively, all below the photon energy of 254 nm UV (472 kJ mol^{-1}) [86].

5.2 *Factors Impacting DBP Photodegradation Efficiency*

The photodegradation efficiency of DBPs strongly depends on both UV light and environmental conditions. Concerning UV light, the key light parameters include UV intensity of lamp, optical path length of UV reactor, and the wavelength of UV light. In an earlier study, the time-based photodegradation rate ($\boldsymbol{k_t}$) of TCAL increased from 0.4 to 1.0 h^{-1} with increases of UV intensity from 6 W to 12 W [96]. The $\boldsymbol{k_t}$ of TBAA also increased with increasing light intensity [97]. In contrast, shortening optical path length from 2.6 to 0.6 cm increased the $\boldsymbol{k_p}$ of TIM from 1.9×10^5 to 1.0×10^6 cm^2 mJ^{-1} [86, 98]. Although 185 nm vacuum UV light (VUV) usually accounts for ~10% of total power input energy, it is much more robust than 254 nm UV light in degrading, dehalogenating, and mineralizing DBPs [94, 99]. In a comparative study, researchers found almost complete degradation of MCAN by a VUV/UV lamp (emitting UV lights with wavelengths at both 254 and 185 nm) but <20% removal of MCAN for a UV lamp (emitting UV light with wavelength of 254 nm only) within 120 min [94]. The reason lies in the fact that VUV triggers formation of oxidative species like hydroxyl radicals (•OH) and ozone as well as reductive species like hydrate electron and hydrogen atom [99]. For typical drinking water, a common UV dosage required for disinfection (e.g., 40 mJ cm^{-2}) can reduce the contents of brominated and iodinated HAAs considerably but hardly decompose chlorinated HAAs [87]. In this regard, the performance of UV/VUV should consider the types and degrees of halogenation.

Regarding environmental conditions, several aqueous factors have been evaluated, including pH, DBPs' initial concentration (C_0), free chlorine (Cl_2), DOM, dissolved oxygen (DO), halides (X^-), carbonate (CO_3^{2-}), bicarbonate (HCO_3^-), and nitrate (NO_3^-). In one study, increasing pH from 5 to 10 exerted little effect on THMs photolysis kinetics [86]. Similar results were observed for HAAs photolysis [100]. However, in other studies, increasing pH enhanced reaction rates of TCAL, iodinated THMs, and iodinated HAAs clearly [96, 98, 101]. Increasing C_0 from 50 μg L^{-1} to 10 mg L^{-1} did not affect the photolysis kinetics of HAAs [100], which was consistent with the THMs photolysis [86]. However, in another study, increasing C_0 from 20 to 120 μg L^{-1} decelerated the reaction of TCAL [96], suggesting that the impact of C_0 was compound-specific.

In contrast, photolysis rates of THMs and HAAs were always greatly enhanced by addition of Cl_2, a typical disinfectant/oxidant, or significantly decelerated by addition of DOM or humic acids, a typical UV-absorbing and reducing agent [86, 100]. Although UV irradiation of DOM might trigger formation of •OH, a highly reactive oxidant that can facilitate contaminant degradation [102], selected DOM in earlier studies usually played inhibiting roles [86, 100]. Photolysis of chlorinated HAAs in oxygenated solutions led to higher quantum yields of oxidation products than deoxygenated water [103], suggesting that DO is a facilitating factor. The impacts of halide on DBPs photolysis were compound-specific. For example, although iodide has a higher UV-absorbing ability than bromide and chloride, the rates of DBPs (e.g., THMs, HAAs, and HALs) photolysis were unexpectedly enhanced by adding iodide [96, 100, 101], suggesting that the presence of halide has both inhibitive and facilitative effects. The reasons are attributed to: (1) the inner filter effect of halide, resulting in decrease of direct photolysis of DBPs (i.e., inhibitive effect) [100]; (2) the light-induced formation of solvated electron and consequent reaction with DBPs (i.e., facilitative effect) [101]. Spiking CO_3^{2-}, HCO_3^-, nitrite, and NO_3^- exhibited case-specific influences on DBPs photolysis [86, 87, 100]. In terms of real water matrix, the photodegradation rates of DBPs dosed into real tap water were mostly lower than those in ultrapure water, meaning that the combined effect of water matrix reduced the treatability of UV [86, 87, 100].

5.3 *Mechanisms for DBPs Photodegradation*

In terms of removal mechanisms, both direct photolysis and indirect photolysis exist (Fig. 5). Direct photolysis refers to a transformation process of DBPs mainly driven by direct irradiation of a UV lamp. The direct k_p depends on DBP optical properties (i.e., the 254 nm molar extinction coefficient, ε_{254}) and the phototransformation probability per absorbed photon (i.e., Φ). One study found that ε_{254} escalated with increasing bromine/iodine substitution, while chlorinated DBPs had lower ε_{254} (e.g., TCM, DCAN, and dichloroacetamide have ε_{254} < 35 M^{-1} cm^{-1}) than tribrominated/triiodinated DBPs (ε_{254}: 1,090 to 1,370 M^{-1} cm^{-1}) [88]. Photolysis rates positively correlate with the ε_{254} ($R^2 = 0.86$) of 37 DBPs [88]. In comparison, Φ was calculated by dividing photons by total absorbed photons [86, 87]. Although the Φ values of DBPs vary within one order of magnitude (0.12–0.59 mol Einstein^{-1}), their photolysis rates differed by up to three orders of magnitude (5×10^2 cm^{-2} mJ^{-1} for DCAN to 2.6×10^5 for dibromoiodomethane) [88]. Generally, DBPs with high Φ value underwent fast direct photodegradation under identical UV exposure conditions.

Besides direct photolysis, DBPs may undergo indirect photolysis too, which refers to degradation by reactive species derived from photoexcitation of coexisting compounds. Taking the most typical reactive species as an example, hydroxyl radicals (•OH) can rapidly oxidize many DBPs [88]. Among tri-halogenated DBPs, TCAA and TCAcAm exhibited the lowest degradation rates ($<10^6$ and 2.1×10^6 M^{-1} s^{-1}) while TCAL was the fastest (3.8×10^8 M^{-1} s^{-1}) [88]. •OH are usually produced by natural photosensitizers (e.g., NO_3^- and DOM) and engineering additives (e.g., Cl_2) in tap water [104]. However, an earlier study found that •OH was unexpectedly present during 254 nm UV irradiation of chlorinated HAAs

Fig. 5 Typical POU facility containing UV lamp and the removal mechanisms for HAAs by UV irradiation [95, 103]

in water without any photosensitizers [103]. This phenomenon was called as direct photolysis-induced indirect photolysis (Fig. 5), which might merit in-depth exploration in the future.

6 Combined Technologies

In general, a combination of activated carbon adsorption with other processes, such as boiling, RO filtration, and UV irradiation, is able to enhance the removal of DBPs. For example, RO filtration as a post-treatment of activated carbon adsorption has reduced TCAL from 20 to ~1 μg L^{-1} in tap water [105], which is far below the regulatory MCL (i.e., 10 μg L^{-1}). Combining UV irradiation technique with activated carbon adsorption enhanced the removal of THMs, which showed much higher $\boldsymbol{k_t}$ (573–942 min^{-1}) for the hybrid technology than photodegradation alone (327–656 min^{-1}) and adsorption alone (1.7–3.0 min^{-1}) [106]. In a one-year long survey of DBPs in tap water treated by a commercial POU system containing a PP cotton filter, two activated carbon cartridges, and a boiler, the THMs removal efficiency (40–60%) gradually declined over time, while HANs and HALs were constantly eliminated regardless of system age [7]. Meanwhile, the POU system reduced HAAs from 42.6 ± 11.8 to 7.0 ± 3.7 μg L^{-1} in summer and from 23.2 ± 5.9 to 9.9 ± 9.2 μg L^{-1} in autumn, both surpassing the THM removals in two seasons [7]. So, existing literature affirms that combining technologies like activated carbon adsorption, boiling, RO filtration, and UV photodegradation can all significantly reduce DBPs at residential POU.

Table 2 shows that water filtration technology varies significantly across different brands, each offering unique processes and costs [25]. APEC Water provides a diverse product line, including quintuple filtration systems with sediment filters, activated carbon blocks, RO technology, and granulated activated carbon (GAC) post-filters, along with options for remineralization and UV disinfection. Aquasana features a unique "Claryum" phase that combines activated carbon, sediment, and ion exchange (IX) technologies, while Culligan offers models ranging from tankless systems with RO membranes to four-stage filtration processes. Whirlpool is known for its three-stage filtration systems, including both tank-mounted and two-stage AC models with sediment phases. Pelican stands out with a six-stage model incorporating sediment filters, GAC pre-filters, RO membranes, dual GAC post-filters, and calcium remineralization. iSpring captures the market with systems integrating ultrafiltration membranes, hybrid copper-zinc filter/GAC, and carbon post-filters.

Costs vary among these brands. APEC Water's products range from $190 to $400, Aquasana's four-stage model is priced at $200, and Pelican's six-stage model costs $220. iSpring's systems range from $120 to $170, depending on the filtration stages. Additional features like UV disinfection and remineralization, offered by brands such as APEC Water and Pelican, may enhance water quality protection at increased costs. Other brands, like Aquasana, offer special filtration media, performance, and price. Brand-specific differences also highlight product strategies, with

Table 2 Number and cost of POU products with various treatment unit(s) [25]

	mfrs.	SED	KDF	CB	RO	UV	ALK	Other	Total	Cost ($)
Systems w. RO	APEC	11	0	11	11	3	3	0	11	190–400
	Aquasana	1	0	1	1	0	0	1	1	200
	Culligan	2	0	2	3	0	0	0	3	na
	iSpring	10	0	10	10	2	4	1	10	na
	Kinetico	1	0	1	1	0	1	0	1	180
	Pelican	1	0	1	1	0	1	0	1	220
	Whirlpool	3	0	3	3	0	0	0	3	na
Systems w.o. RO	Aquasana	3	0	3	na	0	0	3	3	99–175
	GE	0	0	3	na	0	0	0	3	70–130
	iSpring	3	2	4	na	0	0	2	4	120–175
	Kinetico				na				2	na
	Pelican	1	0	2	na	0	0	0	2	74–154
	Whirlpool	0	0	3	na	0	0	0	3	90

Notes: na: not available; SED: sediment prefilter; KDF: copper-zinc filter; CB: carbon blockers; ALK: remineralization media filters

GE and Kinetico offering limited choices, while iSpring provides customized solutions with various configurations. Overall, consumers must consider their water quality needs, health concerns, and budget when choosing a suitable water filtration system.

7 Tips for Consumers

7.1 How to Boil Tap Water

Based on the information mentioned above, here below we outline a protocol for boiling tap water to reduce volatile and hydrolyzable DBPs exposure:

Step 1: Turn on the exhaust fan or range hood. During the boiling process, volatile DBPs may escape into the air, posing a risk of respiratory exposure. Therefore, it is better to turn on the exhaust fan.

Step 2: Fill the kettle halfway with tap water and boil with lid open. Leaving headspace above the liquid and opening the lid can enhance the phase transition of volatile DBPs into air.

Step 3: Maintain boiling for ≥ 60 s before stop heating. This extended boiling period ensures optimal DBP removal.

Step 4: Allow water to cool to room temperature in a disclosed container before drinking. This step reduces the risk of burns of heat and ensures continuous removal of DBPs like THMs, HANs, HALs, and so on. Notably, storing boiled water for over 12 h is not recommended because bacteria may regrow. Also, it is not convenient to wait for so long to get potable water.

7.2 How to Choose a Suitable POU Facility

Selecting an effective POU facility begins with a thorough understanding of the source and quality of tap water. The type of water supply – whether from rural wells or municipal taps – significantly impacts the contaminants present and the required treatment method.

Water from rural wells typically comes from either deep or shallow aquifers, with the latter recharged by surface runoff. Compared with shallow well water, deep well water tends to have higher levels of dissolved metals but lower levels of pesticides, fertilizers, and organic contaminants derived from agricultural sources. To effectively remove heavy metals and organic toxins from both types of well water, people can employ NF/RO membrane and activated carbon adsorption cartridges. If total dissolved solids exceed 1,000 mg L^{-1}, pretreatment by desalination may be necessary to prolong the lifespan of NF/RO membrane.

In centralized water system, municipal tap water often undergoes rigorous treatments such as sedimentation, filtration, and disinfection before distribution. For tap water with pristine water source and treated without disinfection process, activated carbon adsorption and ultrafiltration can provide an extra barrier against common contaminants like suspended particles and high molecular weight organic contaminants. However, for waters facing aging infrastructure that can leach metals and harbor opportunistic pathogens, or if secondary water supply tank is not properly maintained, activated carbon adsorption and NF/RO membrane filtration are recommended for use.

8 DBPs in Swimming Pools

As water flows from a public water supply system into a swimming pool, its quality is no longer governed by drinking water regulations. This raises the question: is it necessary to mitigate the formation of DBPs in swimming pools? Given that swimming is a widely recognized health-boosting activity suitable for all age groups, the formation and control of DBPs in swimming pools have emerged as a compelling research topic over the past two decades. Bathers may inhale, ingest, or absorb these DBPs through their skin, which can be detrimental to health. In this section, we will briefly introduce the formation characteristics, modeling efforts, and mitigation strategies for DBPs in swimming pools.

8.1 Formation Characteristics

Swimming pools often contain a complex mix of chemicals, both natural and anthropogenic, complicating water chemistry. Swimmers introduce substances such as skin, hair, sweat, urine, saliva, pathogens, and personal care products like lotions and sunscreens. These substances increase toxicity and expose swimmers to potential biological and chemical health risks. Disinfecting pool water is a common practice to prevent the spread of infectious diseases, with chlorination being the most prevalent method for deactivating microbial pathogens. Chlorine, introduced in various forms, forms hypochlorous acid in the water, providing a residual effect to prevent microbial recontamination. Effective disinfection requires maintaining a specific disinfectant level, which varies by regulation. For instance, regulations in Canada differ by province [107], and a US study showed a median chlorine residual of 3 mg/L in 23 indoor pools [108].

The formation of DBPs in swimming pool water depends on anthropogenic inputs from bathers and several other factors, including the type and amount of disinfectant used and pool characteristics. Studies have focused on THMs and HAAs, two types of DBPs from chlorine-based disinfection. The levels of these DBPs vary widely. For example, THM levels ranged from 18 to 71 μg/L in Italian indoor pools [109], while HAAs typically have much higher concentrations [110]. Nitrogenous DBPs, such as HANs, are also found in swimming pool water and can be highly toxic, although they are present at lower concentrations compared to THMs and HAAs [111].

Quantifying the unintentional release of human wastes as the main source of organic precursors is challenging in interpreting DBP formation in swimming pools. It's estimated that a bather introduces 25–80 mL of urine and 200–1,000 mL of sweat per swimming session [112, 113]. As these wastes are continuously introduced and pools are consistently exposed to disinfectants, the disinfection process in swimming pools resembles a DBP Formation Potential (DBPFP) test.

The DBP profile in swimming pools is influenced by several factors: (1) Extended water retention time in pools, leading to high DBP levels due to the transformation of most DBP precursors; (2) Continuous introduction and removal of human wastes, correlating with varying DBP levels; and (3) Different chemical loss rates, resulting in varying profiles of HAAs and THMs.

8.2 Modeling Efforts

There is a need for models capable of predicting DBP levels in swimming pool water. These models are designed to identify the significance of various operational and water quality parameters that influence the formation of DBPs or to investigate the kinetics of their formation. Such models might incorporate empirical and mechanistic relationships between water quality and operational parameters and the prevailing levels of DBPs at different stages within the same pool. This section begins with our research on developing predictive models for DBPs, based on data from field and laboratory-scale studies [110, 112, 114].

Specifically, Tang and colleagues analyzed five types of anthropogenic inputs—namely, sweat, saliva, skin wash, hair wash, and urine—for their DBPFPs [110, 112, 114]. All these human wastes were found to contribute to the formation of DBPs. The yields of HAAs in chlorinated samples of sweat, saliva, skin wash, hair wash, and urine were 300, 390, 380, 160, and 160 μg/mg C, respectively. In contrast, the yields of THMs in the corresponding samples were 67, 130, 96, 50, and 74 μg/mg C, respectively. We used these DBP yield data in modeling an indoor pool monitored over a 1.5-year period. It turns out that these anthropogenic inputs were the primary cause of high DBP levels in the pool water. The model suggests that each person introduces 1,000 mg of carbon from sweat and 480 mg of carbon from urine into the pool daily. Prior research indicates that body fluids have a carbon content of approximately 6,000 mg/L [115]. This implies that each pool user releases about 170 mL of sweat and 80 mL of urine into the pool. The model also calculates DBP loss coefficients. It is believed that water agitation impacts the DBP levels in swimming pool water, though quantifying the degree of water agitation is challenging. According to the model, the loss coefficient for HAAs was 0.035 per day, while that for THMs was 0.23 per day in the studied pool. A lower coefficient suggests a longer half-life and, consequently, a higher concentration.

In general, modeling efforts for DBPs in swimming pools focus on understanding their formation, health risks, and effective management strategies. Key insights include: (1) Long-term monitoring and empirical modeling studies revealing significant variations in halogenated DBPs in chlorinated pools [110, 116]; (2) The critical role of body fluids in DBP formation in chlorinated water, highlighting the need for accurate predictive models [117, 118]; (3) The utility of physiologically based kinetic models for assessing intermittent chronic exposure to chemicals in pool water [119]; (4) Predictive modeling aiding in identifying health risks associated with DBPs and in developing safer swimming environments [120]; and (5) The importance of human health risk estimation models, especially for halogenated DBPs like chloroform, in assessing the impact of swimming pool water on bathers [121].

8.3 *Mitigation Choices*

Although the formation of DBPs is initiated by organic precursors during chlorination, applying a source reduction approach to mitigate DBP formation poses challenges, as bathers themselves are the primary source of these precursors. The typical treatment process for swimming pool water includes flocculation, sand filtration, and subsequent disinfection with chlorine. However, continuous chlorination, combined with the introduction of human waste and the recirculation of pool water, can lead to an accumulation of DBPs. Therefore, more advanced technologies are necessary to address this issue.

In swimming pool water, DBP formation may be influenced by three factors: residual level of disinfectant, amount of organic material in the water, water temperature, and recirculation process. The quantity of chlorine added to the pool water, whether in the form of granular powder or tablets, can be precisely controlled either manually or through a mechanical dosing system. Proper dosing ensures an optimal chlorine residual level that effectively eliminates pathogenic microorganisms while maintaining low DBP concentrations [117]. The amount of organic material can be reduced by minimizing the input of DBP precursors, such as NOM derived from tap water and anthropogenic inputs. The anthropogenic input can be significantly decreased through education to alter the behavior of bathers before and during swimming. Actions such as showering, using toilet facilities, washing off sunscreen lotions, and using water-tight diapers can reduce human contributions to DBP formation. It has been found that the duration of a shower greatly impacts the removal of anthropogenic inputs, with most being eliminated through a 60-s shower. While establishing a showering protocol is an important first step in educating pool users about their impact on water quality, it might not be sufficient on its own. Aquatic staff also need to emphasize the importance of pre-swim showering. Supportive policies, such as maintaining clean and well-stocked bathroom facilities to encourage both showering and toileting, should be established, implemented, and enforced. The recirculation of pool water can lead to a buildup of DBPs, necessitating improved treatment efficiency for reducing both DBP precursors and already-formed DBPs. This can be achieved through more effective filtration and oxidation methods, such as tight membrane (e.g., NF or RO) filtration, carbon filtration, biological filtration, and advanced oxidation processes [110, 122]. Removing the low-molecular-weight fraction is recommended due to its significant contribution to toxicity. Toxicity data also suggest maintaining low concentrations of bromide and iodide in pool water, as brominated and iodinated DBPs are more toxic than their chlorinated counterparts. Additionally, increasing air circulation in indoor pool settings can help reduce the levels of volatile DBPs.

9 Summary

DBPs are a special type of micropollutant formed primarily in water supply systems rather than in source water. Therefore, the control strategy for DBPs should somehow differ from that of other pollutants found in raw water sources. The implementation of POU devices has been rapidly increasing because they can significantly improve water quality before consumption. Therefore, the roles of POU methods in mitigating DBPs deserve evaluation.

In this chapter, we have summarized some information regarding the susceptibility of DBPs to four common POU methods, including boiling, activated carbon adsorption, membrane filtration, and UV photolysis (Fig. 6). Some advantages and

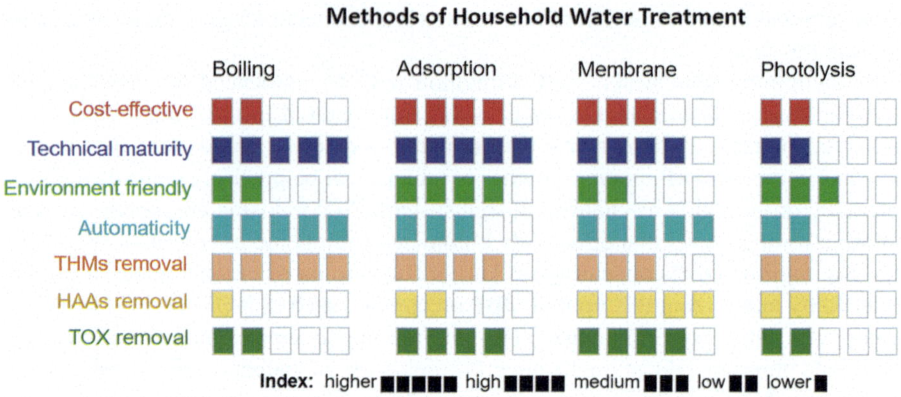

Fig. 6 A general comparison among four typical POU methods used for DBPs removal from tap water. The information are summarized based on earlier studies reported in literatures [1, 4, 105].

disadvantages were identified accordingly. Overall, a selection of treatment method should be tailored to the dominant DBPs of concern in a specific water system. For example, although boiling can effectively remove volatile DBPs like THMs, it may generate other toxic compounds during the boiling process [123, 124]. Therefore, although boiling process can overall reduce TOX and DBPs toxicity from a holistic perspective [123, 124], special care should be taken into account when treating water containing specific precursors.

Activated carbon filters generally exhibit preferential removal of hydrophobic DBPs like iodinated THMs, but are limited in their removal of hydrophilic DBPs like HAAs. Brominated DBPs exhibit greater removals by adsorption than their chlorinated analogues due to their higher hydrophobicity capacities. Combining activated carbon adsorption with other processes, such as boiling, RO filtration, and UV irradiation is very popular nowadays to exert synergistic effect on DBPs removal at residential POU.

The effectiveness of RO and NF membranes on DBPs removal varies significantly depending on the characteristics of DBPs. To enhance their performance, reducing membrane pore size, augmenting surface charge and roughness, and improving desalination capacity without compromising water flux are desired. Small, hydrophilic, neutral species (such as NDMA) are more difficult to be rejected than the large, hydrophobic, and negatively charged DBPs (such as HAAs), therefore the former might deserve more attention in POU treated water.

UV devices offer a promising approach for DBP control too in household settings although formation of intermediate products might comprise its practicability. The key barrier of this method is that the UV dosage required for removal of chlorinated DBPs is still too high.

Moreover, a comprehensive understanding of POU influencing factors is necessary to establish reliable dose-response relationships for epidemiologic studies, because DBP exposures after in-home treatment are more representative than

DBPs in water treatment plants. While a lot of attention had been put on regulated THMs and HAAs, emerging and unregulated/emerging DBPs like HANs and NAs should be investigated too [5]. Although POU systems provide benefits and promises in removing pollutants, their long-term performances, potential side effects, instrumental and maintenance costs, and convenience of replacement/refreshments should all be considered.

Acknowledgments The study is supported financially by the National Natural Science Foundation of China (42307102, 51978194), the Hubei Provincial Natural Science Foundation of China (2022CFB583), the Shenzhen Science and Technology Innovation Commission (JCYJ20210324121403010), and the Knowledge Innovation Program of Wuhan-Shuguang Project (2023020201020295). We would like to take this opportunity to acknowledge the enormous time and efforts of experts who helped improve the quality of this study.

References

1. Xiao R et al (2023) Disinfection by-products as environmental contaminants of emerging concern: a review on their occurrence, fate and removal in the urban water cycle. Crit Rev Environ Sci Technol 53(1):19–46
2. USEPA (2006) National primary drinking water regulations: stage 2 disinfectants and disinfection byproducts rule. Fed Regist 71(2):387–493
3. Hua G, Reckhow DA (2008) DBP formation during chlorination and chloramination: effect of reaction time, pH, dosage, and temperature. J AWWA 100(8):82–95
4. Chen B et al (2021) Roles and knowledge gaps of point-of-use technologies for mitigating health risks from disinfection byproducts in tap water: a critical review. Water Res 200:117265
5. Bei E et al (2016) Occurrence of nitrosamines and their precursors in drinking water systems around mainland China. Water Res 98:168–175
6. Huang H et al (2017) Occurrence of nitrogenous and carbonaceous disinfection byproducts in drinking water distributed in Shenzhen, China. Chemosphere 188:257–264
7. Wang L et al (2019) A one-year long survey of temporal disinfection byproducts variations in a consumer's tap and their removals by a point-of-use facility. Water Res 159:203–213
8. Du P et al (2022) Hydrogen atom abstraction mechanism for organic compound oxidation by acetylperoxyl radical in Co(II)/peracetic acid activation system. Water Res 212:118113
9. Kawamoto T, Makihata N (2004) Distribution of bromine/chlorine-containing disinfection by-products in tap water from different water sources in the Hyogo prefecture. J Health Sci 50(3):235–247
10. Nakai T et al (2015) Analysis and occurrence of 2,6-dichloro-1,4-benzoquinone in drinking water by liquid chromatography-tandem mass spectrometry. J Jpn Soc Water Environ 38(3): 67–73
11. Lee KJ et al (2001) A study on the distribution of chlorination by-products (CBPs) in treated water in Korea. Water Res 35(12):2861–2872
12. Lee J et al (2013) Occurrence of disinfection by-products in tap water distribution systems and their associated health risk. Environ Monit Assess 185(9):7675–7691
13. Ratpukdi T et al (2019) Occurrence of trihalomethanes and haloacetonitriles in water distribution networks of Khon Kaen municipality, Thailand. Water Supply 19(6):1748–1757
14. Vizioli BC, Hantao LW, Montagner CC (2022) Chapter 14 – disinfection byproducts in emerging countries. In: Dalu T, Tavengwa NT (eds) Emerging freshwater pollutants. Elsevier, pp 241–266

15. Redondo-Hasselerharm PE et al (2022) Insights to estimate exposure to regulated and non-regulated disinfection by-products in drinking water. J Expo Sci Environ Epidemiol
16. Zhao Y et al (2012) Occurrence and formation of chloro- and bromo-benzoquinones during drinking water disinfection. Water Res 46(14):4351–4360
17. Krasner SW et al (1989) The occurrence of disinfection by-products in US drinking water. J Am Water Works Ass 81(8):41–53
18. Brisson IJ et al (2013) Spatial and temporal occurrence of N-nitrosamines in seven drinking water supply systems. Environ Monit Assess 185(9):7693–7708
19. Koudjonou B, LeBel GL, Dabeka L (2008) Formation of halogenated acetaldehydes, and occurrence in Canadian drinking water. Chemosphere 72(6):875–881
20. Liu X et al (2020) The occurrence, characteristics, transformation and control of aromatic disinfection by-products: a review. Water Res 184:116076
21. Lekkas TD et al (2008) Removal of specific DBPs by GAC in Galatsi WTP, Athens. *Global NEST J* 11(3):349–356
22. Li B et al (2021) Removing chlorinated haloacetaldehydes from drinking water by household heating devices with and without chlorine: efficiency, influencing factors, and mechanisms. Chemosphere 284:131202
23. Pan Y et al (2014) Boiling of simulated tap water: effect on polar brominated disinfection byproducts, halogen speciation, and cytotoxicity. Environ Sci Technol 48(1):149–156
24. Bivins A et al (2019) Selecting household water treatment options on the basis of World Health Organization performance testing protocols. Environ Sci Technol 53(9):5043–5051
25. Wu J et al (2021) A critical review of point-of-use drinking water treatment in the United States. npj Clean Water 4(1):40
26. Sun X et al (2019) Research progress of disinfection and disinfection by-products in China. J Environ Sci 81:52–67
27. Carey CM, Lee H, Trevors JT (2004) Biology, persistence and detection of Cryptosporidium parvum and Cryptosporidium hominis oocyst. Water Res 38(4):818–862
28. Ma S et al (2017) Understanding and exploring the potentials of household water treatment methods for volatile disinfection by-products control: kinetics, mechanisms, and influencing factors. J Hazard Mater 321:509–516
29. Batterman S et al (2000) Reduction of ingestion exposure to trihalomethanes due to volatilization. Environ Sci Technol 34(20):4418–4424
30. Levesque S et al (2006) Effects of indoor drinking water handling on trihalomethanes and haloacetic acids. Water Res 40(15):2921–2930
31. Shi W, Wang L, Chen B (2017) Kinetics, mechanisms, and influencing factors on the treatment of haloacetonitriles (HANs) in water by two household heating devices. Chemosphere 172:278–285
32. Krasner SW, Wright JM (2005) The effect of boiling water on disinfection by-product exposure. Water Res 39(5):855–864
33. Liu B, Reckhow DA (2015) Disparity in disinfection byproducts concentration between hot and cold tap water. Water Res 70:196–204
34. Zhao J et al (2022) Revisiting the effect of boiling on halogenated disinfection byproducts, total organic halogen, and cytotoxicity in simulated tap water. Chemosphere 309:136577
35. Wang L, Chen B, Zhang T (2018) Predicting hydrolysis kinetics for multiple types of halogenated disinfection byproducts via QSAR models. Chem Eng J 342:372–385
36. Wang Y et al (2023) Removal and transformation of disinfection by-products in water during boiling treatment. Chemosphere 326:138426
37. Zhang X et al (2015) Trihalomethane hydrolysis in drinking water at elevated temperatures. Water Res 78:18–27
38. Chen B (2011) Hydrolytic stabilities of halogenated disinfection byproducts: review and rate constant quantitative structure–property relationship analysis. J Environ Eng 28(6):385–394

39. Ma X et al (2024) Impact of boiling on chemical and physical processes for reduction of halomethanes, haloacetonitriles, and haloacetic acids in drinking water. Sci Total Environ 906: 167657
40. Chowdhury S, Rodriguez MJ, Serodes J (2010) Model development for predicting changes in DBP exposure concentrations during indoor handling of tap water. Sci Total Environ 408(20): 4733–4743
41. Møretrø T et al (2021) Dishwashing sponges and brushes: consumer practices and bacterial growth and survival. Int J Food Microbiol 337:108928
42. Liu J et al (2021) Addition of lemon before boiling chlorinated tap water: a strategy to control halogenated disinfection byproducts. Chemosphere 263:127954
43. Liu J et al (2020) Effects of ascorbate and carbonate on the conversion and developmental toxicity of halogenated disinfection byproducts during boiling of tap water. Chemosphere 254: 126890
44. Fan Y et al (2023) Overcoming slow removal efficiency-induced highly toxic I-DBPs in water by oxygen vacancies enriched invasive plant biochar catalyst: experimental and theoretical studies. J Hazard Mater 459:132086
45. Chen C, Zhou W, Lin D (2015) Sorption characteristics of N-nitrosodimethylamine onto biochar from aqueous solution. Bioresour Technol 179:359–366
46. Oke EA et al (2023) Latest insights on technologies for halides and halogenated compounds extraction/abatement from water and wastewater: challenges and future perspectives. J Water Process Eng 53:103724
47. Li Y et al (2020) Removal and toxicity control of disinfection by-products in industrial wastewater by activated carbon. Ind Water Treat 40(11):36–40
48. Zhang P et al (2020) Study on adsorption competition on activated carbon between chloroform and other chlorine disinfectant in drinking water. J Zhejiang Univ Sci Ed 47(1):107–114
49. Castiglioni M et al (2022) Biochars intended for water filtration: a comparative study with activated carbons of their physicochemical properties and removal efficiency towards neutral and anionic organic pollutants. Chemosphere 288:132538
50. Cai M et al (2020) Characterizing property and treatability of dissolved effluent organic matter using size exclusion chromatography with an array of absorbance, fluorescence, organic nitrogen and organic carbon detectors. Chemosphere 243:125321
51. Jiang J et al (2017) Removal of intermediate aromatic halogenated DBPs by activated carbon adsorption: a new approach to controlling halogenated DBPs in chlorinated drinking water. Environ Sci Technol 51(6):3435–3444
52. He H et al (2022) Molecular transformation of dissolved organic matter and the formation of disinfection byproducts in full-scale surface water treatment processes. Sci Total Environ 838: 156547
53. Lu Y et al (2021) Combination of high resolution mass spectrometry and a halogen extraction code to identify chlorinated disinfection byproducts formed from aromatic amino acids. Water Res 190:116710
54. Ritt CL et al (2022) The open membrane database: synthesis–structure–performance relationships of reverse osmosis membranes. J Membr Sci 641:119927
55. Tung H-H, Unz RF, Xie YF (2006) HAA removal by GAC adsorption. J Am Water Works Ass 98(6):107–112
56. Babi KG et al (2007) Pilot study of the removal of THMs, HAAs and DOC from drinking water by GAC adsorption. Desalination 210(1):215–224
57. Li JW et al (1997) Trihalomethanes adsorption on activated carbon fiber and granular activated carbon. Water Air Soil Pollut 97(3):367–378
58. Koryabkina N et al (2007) Adsorption of disinfection byproducts on hydrophobic zeolites with regeneration by advanced oxidation. Microporous Mesoporous Mater 104(1):77–82
59. Sun X-H et al (2023) Synthesis of magnetic metal-organic framework for efficient adsorption of disinfection by-products in water. Colloids Surf A Physicochem Eng Asp 660:130797

60. Shin MG et al (2022) Critical review and comprehensive analysis of trace organic compound (TOrC) removal with polyamide RO/NF membranes: mechanisms and materials. Chem Eng J 427:130957
61. Castaño Osorio S et al (2022) Modeling micropollutant removal by nanofiltration and reverse osmosis membranes: considerations and challenges. Water Res 225:119130
62. Fujioka T et al (2012) Effects of feed solution characteristics on the rejection of N-nitrosamines by reverse osmosis membranes. J Membr Sci 409–410:66–74
63. Hu S et al (2022) Formation and cytotoxicity of halophenylacetamides: a new group of nitrogenous aromatic halogenated disinfection byproducts in drinking water. Environ Sci Technol 56(5):3181–3192
64. Fujioka T et al (2013) N-nitrosamine rejection by nanofiltration and reverse osmosis membranes: the importance of membrane characteristics. Desalination 316:67–75
65. Wang F et al (2023) Experiments and machine learning-based modeling for haloacetic acids rejection by nanofiltration: influence of solute properties and operating conditions. Sci Total Environ 883:163610
66. Xia S et al (2023) Pressure–driven membrane filtration technology for terminal control of organic DBPs: a review. Sci Total Environ 904:166751
67. Chalatip R, Chawalit R, Nopawan R (2009) Removal of haloacetic acids by nanofiltration. J Environ Sci 21(1):96–100
68. Uyak V et al (2008) Removal of trihalomethanes from drinking water by nanofiltration membranes. J Hazard Mater 152(2):789–794
69. Yang L et al (2017) Removal of haloacetic acids from swimming pool water by reverse osmosis and nanofiltration. Water Res 116:116–125
70. Yang L et al (2017) Role of calcium ions on the removal of haloacetic acids from swimming pool water by nanofiltration: mechanisms and implications. Water Res 110:332–341
71. Zhao H et al (2021) Magnesium-induced variation of polyamide membrane behavior for the treatment of haloacetic acids in swimming pool waters. ACS ES&T Water 1(2):346–355
72. Wang L, Sun Y, Chen B (2018) Rejection of haloacetic acids in water by multi-stage reverse osmosis: efficiency, mechanisms, and influencing factors. Water Res 144:383–392
73. Doederer K et al (2014) Rejection of disinfection by-products by RO and NF membranes: influence of solute properties and operational parameters. J Membr Sci 467:195–205
74. Fang C et al (2021) Rejection of chlorinated, brominated, and iodinated trihalomethanes by multi-stage reverse osmosis: efficiency and mechanisms. Chemosphere 268:129307
75. Xu R et al (2020) Influences of temperature on the retention of PPCPs by nanofiltration membranes: experiments and modeling assessment. J Membr Sci 599:117817
76. Kong F et al (2017) The role of solubility on the rejection of trace organics by nanofiltration membrane: exemplified with disinfection by-products. Environ Sci Pollut Res 24(22): 18400–18409
77. Fujioka T et al (2014) Rejection of small solutes by reverse osmosis membranes for water reuse applications: a pilot-scale study. Desalination 350:28–34
78. Agenson KO, Oh J-I, Urase T (2003) Retention of a wide variety of organic pollutants by different nanofiltration/reverse osmosis membranes: controlling parameters of process. J Membr Sci 225(1):91–103
79. Agus E, Sedlak DL (2010) Formation and fate of chlorination by-products in reverse osmosis desalination systems. Water Res 44(5):1616–1626
80. Bellona C et al (2008) Comparing nanofiltration and reverse osmosis for drinking water augmentation. J AWWA 100(9):102–116
81. Fujioka T et al (2013) N-nitrosamine rejection by reverse osmosis membranes: a full-scale study. Water Res 47(16):6141–6148
82. Fujioka T et al (2012) N-nitrosamine removal by reverse osmosis for indirect potable water reuse – a critical review based on observations from laboratory-, pilot- and full-scale studies. Sep Purif Technol 98:503–515

83. Fang C et al (2020) Effects of feed solution characteristics and membrane fouling on the removal of THMs by UF/NF/RO membranes. Chemosphere 260:127625
84. Sentana I et al (2010) Effect of pressure and pH over the removal of disinfection by-products using nanofiltration membranes in discontinuous systems. Desalin Water Treat 23(1–3):3–12
85. Elimelech M, Phillip WA (2011) The future of seawater desalination: energy, technology, and the environment. Science 333(6043):712–717
86. Xiao Y et al (2014) Photodegradation of iodinated trihalomethanes in aqueous solution by UV 254 irradiation. Water Res 49:275–285
87. Xiao Y et al (2016) Comparative evaluation of iodoacids removal by UV/persulfate and UV/H_2O_2 processes. Water Res 102:629–639
88. Chuang Y-H, Parker KM, Mitch WA (2016) Development of predictive models for the degradation of halogenated disinfection byproducts during the UV/H_2O_2 advanced oxidation process. Environ Sci Technol 50(20):11209–11217
89. Fang J-Y, Ling L, Shang C (2013) Kinetics and mechanisms of pH-dependent degradation of halonitromethanes by UV photolysis. Water Res 47(3):1257–1266
90. Chu W et al (2014) Impact of UV/H_2O_2 pre-oxidation on the formation of haloacetamides and other nitrogenous disinfection byproducts during chlorination. Environ Sci Technol 48(20): 12190–12198
91. Sharpless CM, Linden KG (2003) Experimental and model comparisons of low- and medium-pressure Hg lamps for the direct and H_2O_2 assisted UV photodegradation of N-nitrosodimethylamine in simulated drinking water. Environ Sci Technol 37(9):1933–1940
92. Wang L et al (2020) Some issues limiting photo(cata)lysis application in water pollutant control: a critical review from chemistry perspectives. Water Res 174:115605
93. Zhang Y et al (2019) UV direct photolysis of halogenated disinfection byproducts: experimental study and QSAR modeling. Chemosphere 235:719–725
94. Kiattisaksiri P et al (2016) Photodegradation of haloacetonitriles in water by vacuum ultraviolet irradiation: mechanisms and intermediate formation. Water Res 98:160–167
95. Wang L et al (2017) A comparison of photodegradation kinetics, mechanisms, and products between chlorinated and brominated/iodinated haloacetic acids in water. Chem Eng J 330: 1326–1333
96. Gan Y et al (2019) Photolysis of chloral hydrate in water with 254 nm ultraviolet: kinetics, influencing factors, mechanisms, and products. Chemosphere 218:104–109
97. Li XC, Ma J, Yue S (2012) Total and efficient removal of tribromoacetic acid by ultraviolet irradiation. Adv Mat Res 518–523:2939–2943
98. Hua Z et al (2019) DBP alteration from NOM and model compounds after UV/persulfate treatment with post chlorination. Water Res 158:237–245
99. Pan H et al (2022) Coexisting oxidation and reduction of chloroacetaldehydes in water by UV/VUV irradiation. Water Res 214:118192
100. Du Y et al (2018) Effects of typical water components on the UV254 photodegradation kinetics of haloacetic acids in water. Sep Purif Technol 200:255–265
101. Xu Y et al (2023) Photolysis of CH_3I under UV irradiation: effects of solvents, pH, and concentration. J Radioanal Nucl Chem 332(4):973–979
102. McKay G, Rosario-Ortiz FL (2015) Temperature dependence of the photochemical formation of hydroxyl radical from dissolved organic matter. Environ Sci Technol 49(7):4147–4154
103. Wang L et al (2021) Generation of hydroxyl radicals during photodegradation of chloroacetic acids by 254 nm ultraviolet: a special degradation process revealed by a holistic radical determination methodology. J Hazard Mater 404:124040
104. Wang L et al (2020) Photolysis and photocatalysis of haloacetic acids in water: a review of kinetics, influencing factors, products, pathways, and mechanisms. J Hazard Mater 391: 122143
105. Chen B et al (2015) Chloral hydrate control by point-of-use and household appliances. In: Recent advances in disinfection by-products. American Chemical Society, pp 363–379

106. Yakout SM (2010) Removal of trihalomethanes from aqueous solution through adsorption and photodegradation. Adsorpt Sci Technol 28(7):601–610
107. Dyck R et al (2011) Trihalomethane exposures in indoor swimming pools: a level III fugacity model. Water Res 45(16):5084–5098
108. Kanan A, Karanfil T (2011) Formation of disinfection by-products in indoor swimming pool water: the contribution from filling water natural organic matter and swimmer body fluids. Water Res 45(2):926–932
109. Fantuzzi G et al (2001) Occupational exposure to trihalomethanes in indoor swimming pools. Sci Total Environ 264(3):257–265
110. Tang HL, Ristau II RJ, Xie YF (2015) Disinfection by-products in swimming pool water: formation, modeling, and control. In: Recent advances in disinfection by-products. American Chemical Society, pp 381–403
111. Walse SS, Mitch WA (2008) Nitrosamine carcinogens also swim in chlorinated pools. Environ Sci Technol 42(4):1032–1037
112. Erdinger L, Kirsch F, Sonntag HG (1997) Potassium as an indicator of anthropogenic contamination of swimming pool water. Zentralbl Hyg Umweltmed 200(4):297–308
113. Gunkel K, Jessen H-J (1988) The problem of urea in bathing water. Zeitschrift für die Gesamte Hygiene 34:248–250
114. Berg AP, Fang T-A, Tang HL (2019) Unlocked disinfection by-product formation potential upon exposure of swimming pool water to additional stimulants. Front Environ Sci Eng 13(1): 10
115. Judd SJ, Black SH (2000) Disinfection by-product formation in swimming pool waters: a simple mass balance. Water Res 34(5):1611–1619
116. Genisoglu M et al (2023) Halogenated by-products in chlorinated indoor swimming pools: a long-term monitoring and empirical modeling study. ACS Omega 8(12):11364–11372
117. Berg AP, Fang T-A, Tang HL (2018) Variability of residual chlorine in swimming pool water and determination of chlorine consumption for maintaining hygienic safety of bathers with a simple mass balance model. J Water Health 17(2):227–236
118. Hua P, Chen Y, Zhang J (2022) Modeling the formation of disinfection byproducts in chlorinated swimming pool water: role of body fluid analog. Water Supply 22(9):7337–7351
119. Li Z, Xiong J, Guo Y (2022) Physiologically based kinetic model for assessing intermittent chronic internal exposure to chemicals: application for disinfection by-products in swimming pool water. Comput Toxicol 22:100227
120. Peng F et al (2020) Health risks and predictive modeling of disinfection byproducts in swimming pools. Environ Int 139:105726
121. Anchal P, Kumari M, Gupta SK (2020) Human health risk estimation and predictive modeling of halogenated disinfection by-products (chloroform) in swimming pool waters: a case study of Dhanbad, Jharkhand, India. J Environ Health Sci Eng 18(2):1595–1605
122. Tang HL, Xie YF (2016) Biologically active carbon filtration for haloacetic acid removal from swimming pool water. Sci Total Environ 541:58–64
123. Pan Y, Zhang X, Li Y (2016) Identification, toxicity and control of iodinated disinfection byproducts in cooking with simulated chlor(am)inated tap water and iodized table salt. Water Res 88:60–68
124. Zhang D et al (2020) Identification, formation and control of polar brominated disinfection byproducts during cooking with edible salt, organic matter and simulated tap water. Water Res 172:115526

DBPs in Food and Cooking

Yan Liu, Jia Ni, Kai Ye, Guiying He, Zan Sheng, Yang Pan, and Mengting Yang

Contents

1 DBPs in Agriculture Produce Processing and Industrial Processes 334
  1.1 Differences Between Food Industry and Drinking Water Disinfection 334
  1.2 Sample Pretreatment and Analytical Methods 335
  1.3 Occurrence and Formation of DBPs 338
  1.4 Control Strategies 345
2 DBP Formation During Household Cooking 348
  2.1 Cooking Method 348
  2.2 Cooking Food 348
  2.3 Detection and Identification 349
  2.4 Formation and Occurrence 350
  2.5 Toxicity 351
  2.6 Control Strategies 351
References 351

Abstract Although disinfection byproducts (DBPs) are well-known to be formed during the disinfection of drinking water and wastewater, a growing number of studies have also demonstrated that these unwanted contaminants are of great concern during agricultural produce processing, industrial processes, and household cooking. This chapter begins by examining the formation of DBPs during

Yan Liu and Jia Ni contributed equally to this work.

Y. Liu, G. He, Z. Sheng, and M. Yang (✉)
College of Chemistry and Environmental Engineering, Shenzhen University, Shenzhen, Guangdong, People's Republic of China
e-mail: liuyan2017@email.szu.edu.cn; heguiying2023@email.szu.edu.cn; shengzan2023@email.szu.edu.cn; yangmt@szu.edu.cn

J. Ni, K. Ye, and Y. Pan (✉)
State Key Laboratory of Pollution Control and Resource Reuse, School of the Environment, Nanjing University, Nanjing, Jiangsu, People's Republic of China
e-mail: 522023250075@smail.nju.edu.cn; 201830076@smail.nju.edu.cn; panyang@nju.edu.cn

Chao Chen, Susan Andrews, and Yuefeng Xie (eds.), *Disinfection By-Products in Water*, Hdb Env Chem (2024) 135: 333–356, DOI 10.1007/698_2024_1162,

Published online: 22 October 2024

agricultural produce processing and industrial processes, outlining the disparities between food industry and drinking water disinfection. It discusses various sample pretreatment techniques and analytical methods used for detecting DBPs, as well as their occurrence and formation, classifying them into inorganic DBPs, organic low molecular weight DBPs, and organic high molecular weight DBPs. Minimizing the formation of DBPs during food processing in the industry can be achieved via optimization of disinfection conditions and production processes. The chapter then shifts focus to DBP formation during household cooking, exploring the impacts of cooking methods and food types on DBP formation. Overviews of detection and identification, formation and occurrence, and toxicity profiles are provided. Finally, the chapter ends with a look at strategies for controlling DBP formation to mitigate their presence in cooking processes, which is beneficial for ensuring food safety and public health.

Keywords Agricultural produce processing, DBPs, Food industry disinfection, Food safety, Household cooking contaminants, Mitigation strategies

1 DBPs in Agriculture Produce Processing and Industrial Processes

1.1 Differences Between Food Industry and Drinking Water Disinfection

In food industry, the post-harvest processing of agricultural products plays a pivotal role in safeguarding food from microbial contamination. The use of disinfectants, primarily chlorine-based, has been a conventional approach to mitigate the risk of foodborne pathogens [1–3]. However, the interaction between these disinfectants and the organic matter present in food products can lead to the formation of DBPs, while initially identified in drinking water, is now recognized as a significant concern in food processing [4–6].

Generally, there are several differences between food industry and drinking water disinfection. Firstly, in drinking water treatment, disinfectants typically have longer contact times, usually measured in hours, whereas in food industry disinfection, the contact times are generally only measured in minutes [7–9]. Secondly, in comparison to drinking water, both the solid phase of food and the washing water used for food disinfection contain higher levels of dissolved organic matter [8, 10–12]. Furthermore, owing to established treatment technologies, drinking water facilities tend to adhere to uniform standards, while facilities in the food industry exhibit greater variability.

In the processing of agricultural products such as fruits and vegetables, chemical sanitizers are initially applied to deactivate pathogens and prevent cross-contamination. Subsequently, tap water is usually sprayed for rinsing to remove any residual disinfectant and to regulate the temperature of agricultural products [13–15]. Wash waters are often recycled within postharvest facilities to reduce water consumption, with effluents from later processing stages being reused in earlier ones [12, 16]. Typically, wash waters in agricultural processing contain high levels of dissolved organic matter (DOM), which significantly influences the demand for disinfectants. The chemical oxygen demand (COD), commonly used to measure DOM, varies greatly depending on the type of product and the processing method. For instance, chopping or shredding produce such as carrots, cabbage, and tomatoes releases juices that elevate COD levels in the wash water [12]. The baby vegetable leaves contribute 112 mg/L COD, while shredded vegetables contribute 7,092 mg/L COD to the washing water [12]. The rapid accumulation of COD, as observed in various facilities, poses a challenge to maintaining consistent disinfectant exposure due to its high and fluctuating demand, especially considering the short contact times typical of chlorine disinfectants. For instance, within 5 h, the COD of the washing water used for processing shredded carrots, onions, and tomatoes increased from 448 to 7,092 mg/L [12].

The processing of poultry, distinct from fruits and vegetables, involves initial stages like slaughtering, bleeding, scalding, defeathering, evisceration, and washing, crucial for reducing microbial load [17]. For instance, pigs are typically soaked at a temperature of approximately 60°C for several minutes post-slaughter for easier hair removal [18], followed by evisceration and washing, where scalding water and washing water may contain chemical disinfectants [19]. Seafood processing is more prone to spoilage and contamination than poultry, and usually begins processing immediately after capture [20]. Fish processing includes slaughtering, bleeding, scaling, evisceration, and washing [20], with potential chemical sanitizer use in washing water. Additionally, seafood processing and preservation typically require refrigeration and freezing, which are capable of keeping the compounds stable for longer periods at lower temperatures, potentially leading to higher DBP levels in the final products reaching consumers [21, 22].

1.2 Sample Pretreatment and Analytical Methods

In general, homogenization is the first pretreatment step for analysis of DBPs in solid food substances, aiming to crush solid samples and increase the efficiency of subsequent extraction procedures. It can be carried out using a food processor [5], a kitchen blender [22], a ball mill, or an agate mortar [4]. For example, a ball mill was used to homogenize freeze-dried vegetable samples before analyzing chlorate [23]. Subsequently, samples are often acidified before extraction, using either liquid-liquid extraction (LLE) or solid-phase extraction (SPE) methods.

Methyl tert-butyl ether (MTBE) is the prevailing solvent used for LLE to extract carbonaceous DBPs (C-DBPs) and nitrogenous DBPs (N-DBPs) from fresh-cut vegetables and meat products [24]. Other commonly used solvents include pentane and dichloromethane (such as for analyzing nitrosamines in food) [8]. Notably, for the detection of polar DBPs using gas chromatography (GC) techniques, derivatization is required after extraction. For haloacetic acids (HAA) analysis, typical derivatization reagents are sulfuric acid in methanol [8, 10], dimethylsulfate (DMS), and tetrabutylammonium hydrogen sulfate (TBA-HSO_4) [22, 25, 26]. Aldehydes can be derivatized with o-(2,3,4,5,6-pentafluorobenzyl)-hydroxylamine [8]. The extraction of samples can be accelerated by vortexing or vigorous shaking. In particular, ultrasonication or microwaving is effective for pretreating solid food substances [22], expediting extraction process after the addition of extraction solvent, and centrifugation is used to separate the solid and liquid phases easily [5]. SPE can be used to retain nitrosamines [8], halogenated benzoquinones [27], halogenated amino acids [28], etc., from the sample solution by adsorbing them onto solid-phase cartridges for extraction. In contrast to selecting commercially available SPE cartridges, a special case involved the synthesis of a new material—molecularly imprinted polymer—as SPE sorbents. This homemade SPE cartridge led to the development and successful application of a novel method for measuring nitrosamines in beverage samples [29]. Prior to instrumental determination, concentrating extracted samples is achieved via nitrogen blowing or rotary evaporation [11].

GC has been widely used in qualitative and quantitative analysis of low molecular weight DBPs in food, especially for those (semi-)volatile species. Commonly used separation columns include DB-5MS, HP-5MS, and SLB-5MS capillary columns. The isolated fractions were then detected by a coupled electron capture detector (ECD) or mass spectrometer (MS) [5, 30, 31]. For instance, Lee et al. utilized GC-ECD to quantify C-DBPs (including trihalomethanes [THMs] and HAAs), N-DBPs, and nitrosamines in lettuce and chlorinated wash water [5]. By using the static headspace GC-MS (SHS-GC-MS) method, 14 halogenated DBPs were comparably determined in frozen foods (fish, meats, and vegetables) [22]. In this technique, the sample is heated in a sealed headspace vial, allowing volatile compounds to transfer to the headspace. Its advantages include minimal sample preparation, high sensitivity for volatile compounds, and the ability to analyze complex matrices. Additionally, the SHS-GC-MS method has been reported for analyzing DBPs in fruit juices and soft drinks [26], cheeses [25], and canned vegetables [4].

Recently, high-resolution mass spectrometry (HRMS) coupled with GC has been increasingly utilized in the analysis of DBPs in disinfected waters. For instance, GC-HRMS was employed for suspect and non-target screening, resulting in the identification of two new haloketones and tribromo furoic acid in thermal water swimming pools [32]. This advanced technique offers higher resolution and precision. However, its application in food analysis remains limited. Future studies can benefit from it to analyze and screen a wider range of DBP categories. This will enable the comprehensive and accurate identification and quantitative analysis of various DBPs in food products.

Unlike GC, liquid chromatography (LC) operates at ambient or moderately elevated temperatures and utilizes a liquid mobile phase, thereby preserving the integrity of heat-sensitive analytes and facilitating the separation of a broad spectrum of polar and semi-polar compounds. Although less frequently adopted than GC,

LC-related techniques are suitable for the detection of non-volatile, thermally labile, and larger molecular weight DBPs, which cannot be effectively achieved via GC analyses. Therefore, they have been increasingly employed for analyzing DBPs in food and their associated liquid portions. Both coupled with MS as detector, ultra-performance liquid chromatography (UPLC) and high-performance liquid chromatography (HPLC) were used to determine the concentration of chlorate in fresh-cut produce wash water [16] and to explore the formation of chlorotyrosines in vegetables during disinfection washes [28], respectively. Based on the multiple reaction monitoring (MRM) mode, HPLC coupled with triple quadrupole MS (tqMS) was applied to develop a quantification method for determining nitrosamines in beverage samples [29].

In addition to target analysis, non-target screening is an important solution for identifying new emerging DBPs. By setting precursor ion scan (PIS) of *m/z* 35/37, 79/81, and 126.9, Zhang's group has developed a series of methods, which can selectively detect all electrospray ionized chlorinated, brominated, and iodinated DBPs in drinking water. Moreover, in conjunction with fragment information obtained from product ion scans and/or validating with standard compounds, more than 200 new emerging DBPs have been revealed and identified for the first time, including some highly toxic compounds such as 2,5-dibromohydroquinone and tetrabromopyrrole [33]. However, these powerful methods are rarely utilized in investigating DBPs in agriculture produce processing and industrial processes.

HRMS coupled to LC is another promising approach to characterize unknown polar halogenated DBPs, typical types include quadrupole-time-of-flight (Q-ToF) and Orbitrap mass spectrometers. By using HPLC-QToF-MS, eight halogenated DBPs derived from aspartame (a widely adopted sweetener in beverages) were reliably identified through observed accurate masses and tandem-mass spectrometry (MS/MS) spectra information [27]. Fourier transform ion cyclotron resonance (FT-ICR) mass spectrometry, although extremely expensive, can provide even higher accuracy in mass detection compared to the former two types of HRMS. However, to the best of our knowledge, there have been no studies using it to explore DBPs in food-related samples.

Other useful analytical methods include ion chromatography (IC) and total organic halogen (TOX) analysis. Inorganic DBPs can be effectively detected by IC. For example, IC was employed to explore the formation of chlorates during postharvest produce washing [23]. TOX is a collective parameter that doesn't require knowledge of each specific DBP concentration, allowing it to reflect the overall formation level of DBPs in a given water sample. The total concentration of identified DBPs found in drinking water only accounts for $<50\%$ of TOX, indicating that a large number of DBPs still remain unidentified [34]. As the composition of halogenated DBPs derived from foodstuffs may be more complex than those formed in drinking water, the use of TOX is encouraged to characterize food-related DBP formation. Based on TOX analysis, a recent study reported the effect of sugar on DBPs during the preparation of hot beverages, indicating that the TOX value in samples with sucrose (100 mg/L) increased by 35% compared to boiling water samples without sugar [35].

1.3 Occurrence and Formation of DBPs

1.3.1 Inorganic DBPs

The occurrence of chlorate is a significant concern due to its potential impact on thyroid function and overall health. Its formation is closely associated with the utilization of chlorine-based disinfectants [7]. Suh et al. reported that after exposing lettuce to 10 mg/L of free chlorine for 5 min, the concentration of chlorate significantly increased in the simulated washing water or lettuce itself, as contact with free chlorine enhanced its disproportionation to form ClO_3^- [23]. After treatment with free chlorine, the concentration of ClO_3^- in lettuce exceeded the limit currently suggested by the European Commission (EC) for food, which is 0.01 mg/kg [7, 36–38]. It was found that chlorate residues in fresh-cut lettuce correspond to approximately 10% of the concentration present in the wash water [7, 39]. Additionally, previous studies reported that chlorate was frequently detected in ready-to-eat carrots, and the highest content in tested processed carrots in the United States was found to be 0.54 mg/kg [37, 40].

Chlorate uptake from washing water into vegetables is commonly observed due to its high solubility. It was reported that shredded carrots exhibited the greatest chlorate uptake compared to baby lettuce, baby spinach, and chopped lettuce, all of which had intact epidermis [12]. Another study indicated that the uptake of chlorate in lettuce and kale was higher than that in broccoli [23]. Moreover, as cut grade increased, chlorate uptake increased. For example, shredded lettuce presented a higher level of chlorate uptake after washing compared to chopped lettuce [36]. Similarly, the chlorate uptake of shredded carrots was higher than that in disk or large sticks [36]. In addition to product type and cut piece size, the uptake of chlorate during the washing process can be influenced by washing time and washing water content; longer washing times and higher chlorate content in the washing water significantly increased the chlorate uptake [36]. Variations in the formation of chlorates are observed when employing different disinfectants. To assess the formation and accumulation of chlorate in postharvest produce treated with different disinfectants under conditions relevant to post-harvest facility washes, Suh et al. investigated the disinfection of kale, broccoli, butter lettuce, or romaine lettuce exposed to chlorine and chlorine dioxide (ClO_2) [23]. The results revealed that the use of ClO_2 significantly reduced the chlorate concentration compared to chlorination. Additionally, the introduction of ammonia during ClO_2 treatment further resulted in an approximately 40% reduction in chlorate formation. This reduction occurred by binding the chlorine generated during ClO_2 treatment with the produce. Thus, ClO_2 is a potential attractive alternative to chlorine for disinfecting wash water in post-harvest facilities [23]. Furthermore, Petri et al. compared the effects of peracetic acid (PAA) and chlorine in industrial-scale fresh-cut vegetable processing [41]. In general, the two disinfectants showed comparable antimicrobial effects during washings, but chlorination resulted in more substantial formation of chlorates and other DBPs than PAA treatment. Thus, PAA was found to be a promising alternative to chlorine-based disinfection approaches in the fresh-cut industry.

Bromate is a concerning compound due to its potential carcinogenic effects, as it can induce oxidative stress and damage deoxyribonucleic acid (DNA) [42]. The regulatory limit established by the US Environmental Protection Agency (EPA) for bromate in drinking water is 10 mg/L [43]. This inorganic DBP can be generated via oxidation of bromide during chlorination. Similar to the ozonation of drinking water rich in bromide, which favors the formation of bromate, higher concentrations of bromate are expected to be detected in produce wash waters with elevated bromide levels when ozone serves as the disinfectant. Lombardo et al. and Guzel-Seydim et al. both emphasized the importance of controlling ozone levels during food processing to minimize bromate formation while ensuring microbial safety [44, 45]. Interestingly, potassium bromate is also commonly used in commercial breadmaking as it is a slow-acting oxidant and can enhance gluten strength during baking. Sandhu et al. compared the performance of $KBrO_3$ and ozone in breadmaking and found that using ozone treatment on flour may potentially replace the necessity of using $KBrO_3$ in the breadmaking process [46].

1.3.2 Organic Low Molecular Weight DBPs

In current food studies, researchers have detected various types of organic low molecular weight DBPs in food after processing, including THMs, HAAs, and nitrosamines [12, 22, 25]. However, due to variations in research objectives and experimental conditions, the types and concentrations of DBPs in food reported in different studies can vary significantly.

In general, similar to drinking water samples, THMs and HAAs are the most prevalent groups of low molecular organic DBPs observed in processed food compared to other species. For example, based on the measurements of 32 commonly known DBPs in lettuce, cabbage, and strawberries, Lee et al. reported their concentration ranges following this trend: THMs (<limit of detection (LOD) – 490 ng/g) > HAAs (<LOD – 354 ng/g) > N-DBPs (halonitromethanes, haloacetonitriles, and haloacetamides) (<LOD – 204 ng/g) and other C-DBPs (halopropanones, haloethane, etc.) (<LOD – 37.3 ng/g) [5]. This trend is further supported by studies conducted by Cardador et al., Cardador and Gallego, and López-Gálvez et al. [12, 22, 25]. Table 1 shows the typical DBPs and their concentration levels detected in different types of food.

The formation levels of DBPs show significant differences among food types. According to the study conducted by Lee et al., among three types of vegetables and fruits, washed lettuce exhibited the highest concentration of THMs (490 ng/g), while strawberries and cabbage showed lower THM concentrations of 26.1 ng/g and 9.3 ng/g, respectively [5]. Huang and Batterman explored the generation of THMs while preparing foods and beverages with disinfected drinking water [47]. They found that tea had the highest THM levels (3–67 μg/L), followed by coffee (3–13 μg/L), rice (9 μg/L), soup (0.4–3.0 μg/L), vegetables (<1 μg/L), and infant food

Table 1 Summary of DBPs detected in different types of food

Food tape	DBPs	Concentration	Pretreatment	Detection method	Reference
Meat products (bacon, corned beef, etc.)	THMs (trichloromethane, bromodichloromethane) HAAs (dichloroacetic acid, trichloroacetic acid) Trichloronitromethane	n.d.–5.0 ng/g n.d.–2.5 ng/g n.d.–2.1 ng/g	Homogenization; derivatization with DMS and TBA-HSO_4 (for HAAs); microwave heating assisted extraction with MTBE and n-pentane	SHS-GC-MS	[70]
Fresh-cut lettuce (washed by peracetic acid)	Chloroform HAAs (chloroacetic acid, dichloroacetic acid, etc.) Nitrosamines (*N*-nitrosodimethylamine and N-nitrosodiethylamine) Aldehydes (formaldehyde, acetaldehyde, etc.) N-DBPs (2,2-dichloroacetamide) Others (chloral hydrate and 1,2-dibromo-3-chloropropane)	126–242 ng/g n.d.–12.2 ng/g n.d.–0.43 ng/g 0.89–137 ng/g 14.7–15.3 ng/g n.d.–3.77 ng/g	Homogenization; LLE with MTBE, dichloromethane (for nitrosamines) and hexane (for aldehydes); derivatization with 10% sulfuric acid in methanol (for HAAs) and o-(2,3,4,5,6-pentafluorobenzyl)-hydroxylamine (for aldehydes)	GC/ECD GC-MS	[70]
Fresh-cut lettuce (washed by sodium hypochlorite)	Chloroform HAAs (chloroacetic acid, bromoacetic acid, etc.) Nitrosamines (*N*-nitrosodimethylamine and *N*-nitrosomorpholine) Aldehydes (formaldehyde, acetaldehyde, etc.) N-DBPs (2,2-dichloroacetamide) Others C-DBPs (chloral hydrate, 1,1-dichloro-2-propanone, etc.)	490 ± 217 ng/g 4.00–354 ng/g 0.22–0.74 ng/g 1.36–106 ng/g 61.6 ± 6.93 ng/g 4.00–37.3 ng/g	Homogenization; LLE with MTBE, dichloromethane (for nitrosamines) and hexane (for aldehydes); derivatization with 10% sulfuric acid in methanol (for HAAs) and o-(2,3,4,5,6-pentafluorobenzyl)-hydroxylamine (for aldehydes)	GC/ECD GC-MS	[70]

Tap water (aspartame)	2,6-dichloro-1,4-benzoquinone	n.d.–15 ng/L	SPE with Waters oasis HLB cartridge; filtered with PVDF syringe filters (0.45 μm)	HPLC-MS/MS	[70]
Fruit juices (pear, pineapple, etc.)	HAAs (dichloroacetic acids, trichloroacetic acids, etc.)	n.d.–14 μg/L	Derivatization with TBA-HSO_4 and DMS; organic modified with n-pentane	SHS-GC-MS	[70]
Boiled tap water with the addition of sugar (sucrose, glucose, etc.)	TOBr TOCl TII	2.45–3.95 μg/L 7.05–10.13 μg/L 1.0%–86.8%	LLE with MTBE	AOX-IC UPLC/ESI-tqMS	[70]
Process wash water treated with electrochemical disinfection (lettuce, carrot, etc.)	Total THMs (trichloromethane, tribromomethane, etc.) Chlorates	210–1,300 μg/L n.d.–80 mg/L	n.a.	GC-MS UPLC-MS	[70]
Process wash water treated with sodium hypochlorite (lettuce, onion, etc.)	Total THMs (trichloromethane, bromodichloromethane, etc.) Total HAAs (dichloroacetic acid, trichloroacetic acid, etc.) Chlorates	41–504 μg/L 1,651–1802 μg/L (cabbage) 51–733 μg/L 2073–3,960 μg/L (cabbage) 2.2–62 mg/L	n.a.	GC-MS LC-MS UPLC-MS	[70]
Baby lettuce irrigated with electrolyzed water	Total THMs Chlorates	0.3–2.9 μg/kg 0.05–0.10 mg/kg	n.a.	GC-MS LC-MS	[70]
Fresh-cut produce (lettuce, strawberry, etc.)	Chloroform HAAs (chloroacetic acid, trichloroacetic acid, etc.) Other C-DBPs (1,1-dichloro-2-propanone, tetrachloroethylene, etc.) 2,2-dichloroacetamide	n.d.–26.1 ng/g n.d.–186 ng/g n.d.–28.1 ng/g 6.3–204 ng/g	Homogenization; LLE with MTBE; derivatization with 10% (v/v) H_2SO_4 in methanol (for HAAs)	GC/MS GC/ECD	[70]
Canned vegetable edible phase (beans, beets, etc.)	THMs (trichloromethane, bromodichloromethane, etc.) HAAs (dichloroacetic acid, trichloroacetic acid, etc.)	n.d.–25 μg/kg n.d.–5.7 μg/kg	Homogenization; extraction with 0.5 g/mL Na_2SO_4 aqueous solution; derivatization with DMS and TBA-HSO_4 (for HAAs)	SHS-GC-MS	[70]

(continued)

Table 1 (continued)

Food tape	DBPs	Concentration	Pretreatment	Detection method	Reference
Frozen foods (spinach, pork, etc.)	THMs (trichloromethane, bromodichloromethane, etc.) HAAs (dichloroacetic acid, trichloroacetic acid, etc.) Trichloronitromethane	n.d.–28 μg/kg n.d.–24 μg/kg n.d.–2.2 μg/kg	Homogenization; extraction with 0.5 g/ml Na_2SO_4 aqueous solution; derivatization with DMS and TBA-HSO_4 (for HAAs); microwave heating assisted extraction (for meats and fish)	SHS-GC-MS	[70]
Postharvest vegetables (spinach)	9,10-chlorohydrin	n.d.–53 nmol/g	LLE with MTBE; derivatization with silylating agent	GC/MS	[70]
Fresh-cut produce wash water (shredded lettuce, diced cabbages)	THMs (chloroform, bromoform, etc.) HAAs (chloroacetic acid, bromoacetic acid, etc.) Other C-DBPs (chloral hydrate, carbon tetrachloride, etc.) N-DBPs (dichloroacetonitrile, trichloroacetonitrile, etc.)	n.d.–467.9 μg/L n.d.–312.0 μg/L n.d.–249.2 μg/L n.d.–5794.4 μg/L	LLE with MTBE; derivatization with 10% (v/v) H_2SO_4 in methanol (for HAAs)	GC/ECD	[70]
Shredded vegetables and wash water (lettuce, onions, etc.)	Trichloromethane	2.8–22 μg/kg (solid phase) n.d.–284 μg/L (wash water)	Homogenization; filtration	GC-MSD	[70]
Cheese (cheddar, mozzarella, etc.)	THMs (trichloromethane and bromodichloromethane) HAAs (dichloroacetic acid, bromochloroacetic acid, etc.)	n.d.–11 μg/kg n.d.–13 μg/kg	Homogenization; vortex and centrifuge; derivatization with DMS and TBA-HSO_4	SHS-GC-MS	[70]
Chocolate (hot cocoa)	THMs ($CHCl_3$ and $CHCl_2Br$)	n.d.–3.1 μg/L	n.a.	SHS-GC-ECD	[70]

Soup (chicken-noodle soup, miso-tofu soup, etc.)	THMs ($CHCl_3$ and $CHCl_2Br$)	n.d.–2.1 μg/L	n.a.	SHS-GC-ECD	[70]
Starchy food (rice, pasta, etc.)	THMs ($CHCl_3$ and $CHCl_2Br$)	n.d.–9.4 μg/L	n.a.	SHS-GC-ECD	[70]
Baby food (baby cereal, infant formula)	THMs ($CHCl_3$ and $CHCl_2Br$)	n.d.–0.7 μg/L	n.a.	SHS-GC-ECD	[70]
Fresh produce (shredded vegetables, baby leaves, etc.)	Chlorates THMs (trichloromethane, bromodichloromethane, etc.)	n.d.–299.5 μg/kg n.d.–44.4 μg/kg	Extraction with formic acid (1%) in methanol; vortex and centrifuge; filtration	UPLC-MS GC-MS	[70]
Wash water (shredded vegetables, baby leaves, etc.)	Chlorates THMs (trichloromethane, bromodichloromethane, etc.)	0.44–47.4 mg/L 1.9–4,227 μg/L	n.a.	UPLC-MS GC-MS	[70]

n.d., not detected, n.a., not analyzed

(<0.7 μg/L). Interestingly, Cardador and Gallego comparably measured the DBP concentrations in frozen food and their fresh counterparts [22]. The average concentrations of THMs, HAAs, and N-DBPs in frozen food tend to be higher compared to fresh food, and the DBP formation level in vegetables was significantly higher than in pork and fish. It is most likely that the stability of residual chlorine and formed DBPs can be prolonged under low temperatures during freezing. The lower DBP concentrations observed in meat and fish samples may be due to the prohibition of disinfection for animal-derived foods according to European Union regulations. Consequently, these samples undergo only minimal treatment (such as rinsing or washing with chlorinated water) before freezing. Meanwhile, the adsorption of formed contaminants on sample surfaces is minimal.

The production of total THMs is observed to rise as the organic load and chlorine dose increase [48, 49]. Increased levels of organic matter result in a more rapid depletion of chlorine, therefore, elevated COD values require the addition of increased chlorine to sustain the desired residual concentration. For wash water with COD levels of 500 and 1,000 mg O_2/L, respectively, the total THM concentrations were detected to be 27.8 and 124.5 μg/L after 1 h of washing fresh-cut lettuce while maintaining a consistently comparable residual chlorine level [48].

Additionally, the type of washing system also impacts the DBP formation levels in food. Zhang et al. compared DBP formation in wash water and treated produce during the operation with two washing systems, i.e., the traditional flume washing system (where water is recirculated) and the novel single-pass washing system (where water is not recirculated) [31]. The results demonstrated that the levels of organic low molecular weight DBPs in washed products from the former system were significantly higher than those from the latter, and after disinfection, a post-rinse procedure with potable water proved effective in eliminating the levels of DBPs produced in flume-washed agricultural products.

1.3.3 Organic High Molecular Weight DBPs

In contrast to low molecular weight DBPs, investigations focusing on higher molecular weight DBPs in the context of food processing remain notably limited in current studies. Komaki et al. reported that using chlorine disinfectants in washing spinach and lettuce can produce a group of high molecular weight DBPs, i.e., chlorotyrosines, which form and retain in plant tissues, and their formation increases with the dosage of chlorine. In shredded vegetables, chlorotyrosine formation is elevated due to the enhanced contact between disinfectants and internal leaf proteins [28]. Moreover, a recent study revealed that aspartame, when reacted with residual chlorine in drinking water, forms an aromatic DBP, namely 2,6-dichloro-1,4-benzoquinone (DCBQ) [27]. This reaction primarily involves the free amino group on the aspartic acid residue of aspartame, which exhibits high reactivity with chlorine, leading to the formation of various N-Cl and C-Cl products. In water samples supplemented with aspartame, a significant increase in the concentration of DCBQ was observed compared to samples without aspartame. Interestingly, boiling tap water, which effectively reduces residual chlorine, subsequently decreases the production of DCBQ.

Given the increasing reports of high molecular weight DBPs in drinking water, and the complexity of food-related samples compared to drinking water, it is expected that numerous new high molecular weight DBPs will also be uncovered in agricultural produce processing and industrial processes in the future. The identification and detection of these potential DBPs can utilize methods commonly employed for DBPs in drinking water, particularly liquid chromatography-related techniques. For example, the PIS mode (*m/z* of 79/81, 35/37, and 126.9) in ultra-performance liquid chromatography/electrospray ionization-triple quadrupole mass spectrometry (UPLC/ESI-tqMS) enables selective detection of brominated, chlorinated, and iodinated DBPs [33]. Facilitated by this methodology, Zhang's group identified over 150 emerging DBPs, many of which are aromatic halogenated DBPs [33]. These higher molecular weight DBPs, especially aromatic ones, typically pose higher toxicity order. For instance, 2,6-dibromo-4-nitrophenol and 2,4,6-tribromophenol were demonstrated to exhibit significantly higher toxic potency compared to bromoacetic acid, by two to three orders of magnitude [50]. Therefore, it is crucial to further investigate the high molecular weight DBPs, especially aromatic and heterocyclic ones with elevated toxicity levels, in food-related samples. Future exploration is essential for enhancing our comprehension of the health risks associated with DBPs in food industry and for shaping pertinent policies.

1.4 Control Strategies

In general, minimizing the formation of DBPs during food processing in the industry involves optimizing disinfection conditions and production processes. Effective disinfection conditions are crucial for enhancing disinfection efficacy while minimizing the potential formation of DBPs. Strategies in this regard typically involve using appropriate chlorine doses, maintaining neutral pH conditions, and selecting suitable washing systems and disinfectant types. Production processes may entail alterations to the process sequence, as well as techniques such as boiling, adding ascorbic acid, etc.

At present, the widespread use of chlorine-containing disinfectants in food hygiene is common. Organic compounds can "consume" available chlorine, thus reducing its efficacy for thorough disinfection. Therefore, it is a widely employed practice in the food processing field to add excess chlorine to the treatment solution to achieve the desired level of disinfection efficacy. To maintain adequate free chlorine levels, demand-based automated injection systems or regular manual chlorine administration are commonly employed [51]. However, a previous study has shown that repeated addition of excess chlorine to washing solutions with high organic loads can lead to the formation of high levels of DBPs, such as THMs, HAAs, haloketones, etc. [52]. Thus, a logical approach is needed to supply the required chlorine dose. Interestingly, Chen and Hung developed an empirical model for predicting the chlorine requirement for specific treatment processes [53]. They established equations that factor in parameters related to the characteristics of washing water to predict the chlorine requirements for different agricultural products. Potential indicators of ultraviolet absorbance at 254 nm (UV_{254}) and the phenols-to-protein ratio can be used to predict the required chlorine levels for various agricultural products. By utilizing this predictive model, the process can avoid overuse of chlorine, thus minimizing the potential formation of DBPs.

Previous studies have shown that many chlorine reactions are pH-dependent; for instance, the concentration of THMs decreases while the concentration of HAAs increases as the pH decreases [54, 55]. Hung et al. reported the impact of pH on DBP formation during food processing, highlighting that extreme acidic and alkaline conditions were ineffective in achieving strong antimicrobial effects while minimizing DBP formation [56]. They emphasized that DBP generation can be minimized under near-neutral pH conditions (around 6–7). This is because chlorine's strongest antibacterial effect occurs within this pH range, maximizing hypochlorous acid production for optimal disinfection while reducing DBP formation. Therefore, maintaining a near-neutral pH during food disinfection can be considered a feasible DBP control strategy.

Choosing an appropriate washing system during food processing is another useful method to control the formation of DBPs. Currently, typical types commonly used for disinfecting fresh-cut food include sink washing systems and one-way washing systems. Zhang et al. found that the one-way washing system yielded considerably lower levels of total DBPs compared to the sink washing system, although the difference was less pronounced when examining washed food versus washing water [31]. Furthermore, the free chlorine concentration in the washing water exhibited significant fluctuations in the sink system compared to the one-way system, and the total DBP levels were notably higher in the sink system compared to the incoming water. The lower DBP formation levels observed in the one-way washing system may be attributed to its use of water only once, without any reuse or recycling. This finding is consistent with the study by Bornhorst et al. on the use of a single washing system in food processing [57]. Overall, the levels of DBPs in the samples from the sink washing system are relatively high and increase with longer treatment times. Accordingly, the one-way washing system may help maintain lower DBP levels in both wash water and produce.

Using alternative disinfectants may be effective in decreasing the formation of halogenated DBPs during food processing. Ozone is regarded as a promising disinfectant due to its superior antibacterial activity compared to chlorine, rapid action, and lack of by-product formation [58]. Rosenblum et al. treated process water contaminated with *Bacillus subtilis* with <2 mg/L ozone for 10 min achieving a 1.56 log CFU/mL reduction, while chlorination resulted in a 1.3 log CFU/mL reduction. In addition, weak organic acids such as citric acid, lactic acid, and acetic acid have been found to show good disinfection effects [59]. It was observed that dipping of iceberg lettuce in a 0.5% lactic acid or 0.5% citric acid solution for 2 min demonstrated efficacy comparable to chlorine in reducing microbial populations on fresh-cut iceberg lettuce [60]. Previous studies have also highlighted the superior disinfection potential of neutral electrolyzed oxidizing (NNEO) water within a pH range of 6–7 [61, 62]. Gómez-López et al. noted that wash water treated with sodium hypochlorite exhibited higher THM levels compared to electrolyzed oxidizing water treatment [63]. Thus, NNEO water emerges as a promising alternative to chlorine in order to minimize DBP formation while achieving the goal of disinfection [56]. Other typical alternative methods include hydrogen peroxide and the previously mentioned ClO_2 and PAA. Interestingly, the combination of UV and hydrogen peroxide to clean fresh-cut produce is a useful approach, and it can destroy refractory organics without leaving residues [64, 65].

Notably, these alternative disinfectants also come with certain drawbacks, which can hinder their widespread adoption in the food industry. Switching disinfectants to reduce the formation of certain DBP groups may inadvertently promote the formation of other groups. For instance, ozone reacts with bromide present in water, favoring the formation of brominated DBPs. Similarly, when iodide is present in water, ozone or ClO_2 forms more iodinated DBPs than chlorine. These brominated/iodinated DBPs are generally more toxic than their chlorinated analogs [66]. Currently, the use of ClO_2 is restricted to a maximum concentration of 3 mg/L when treating whole produce. Additionally, the US Code of Federal Regulations (CFR) mandates rinsing produce with potable water after ClO_2 treatment [9, 67]. Furthermore, ozone, as a strong oxidant, has the potential to cause physiological damage to products when used above a certain level. Kosek and Isobe observed rapid browning in lettuce treated with 10 mg/L ozonated water [68]. Similarly, research indicates that hydrogen peroxide treatment may also lead to significant browning in certain products, such as lettuce and mushrooms, unless supplemented with an anti-browning agent [9]. Therefore, the use of hydrogen peroxide in the food industry is limited to some products (e.g., milk, dried eggs, starch, tea, and wine) as an antimicrobial agent or bleaching agent within the dose range of 0.04–1.25% [9]. Moreover, the CFR requires the removal of residual hydrogen peroxide from food processing through appropriate physical and chemical methods. Additionally, organic acids may negatively affect the sensory quality of products. Wu et al. reported that high levels of acetic acid can lead to noticeable discoloration and a strong vinegar odor [69].

During food processing, it is common for food to be chopped and the impact of the chopping and disinfection order on DBPs in the washing water cannot be overlooked. For instance, a study by López-Gálvez et al. highlighted the disparity in DBP levels in the washing water of baby leaves and shredded vegetables; the average concentrations of THMs in the washing water of baby leaves and shredded vegetables were found to be 34.1 and 2635.2 μg/L, respectively [12]. Komaki's study also suggests that disinfection after chopping spinach increases the formation of chlorotyrosines, due to the greater accessibility of chlorine to proteins in the leaf interiors [28]. Therefore, performing the disinfection process before chopping the food is more conducive to controlling the formation of DBPs.

When processing canned food, if the food itself has a high moisture content, it is advisable to refrain from adding chlorinated water externally. A previous study demonstrated that canned tomato samples exhibit lower levels of DBPs as they were commonly preserved in their natural juices, bypassing the need for additional water [70]. Also, heat treatment can influence the formation of DBPs in canned food. Heating processes, typically around 80°C, during the canning of vegetables may result in the volatilization of THMs in the liquid phase, while HAAs remain stable [70]. Hence, endorsing high-temperature sterilization in the canned food industry could be beneficial, provided it does not alter the taste profile.

Furthermore, Wawryk et al. found that boiling or incorporating ascorbic acid into tap water sweetened with aspartame reduces residual chlorine levels, thereby preventing the formation of DCBQ [27]. Previous studies have also demonstrated that boiling tap water for 5 min effectively reduces the levels of DBPs and toxicity [71, 72]. Hence, it is recommended to maintain a certain boiling time and add ascorbic acid when preparing hot beverages, as these measures can help mitigate the formation of DBPs.

2 DBP Formation During Household Cooking

2.1 Cooking Method

A cooking process (e.g., soup making) involves combining tap water, organic food ingredients, and seasonings/sauces/oil in a cooking device and heating the mixture to boiling gently for a few minutes to a couple of hours [73], in which bromine/iodine/organic matter in the food react with disinfectants to form harmful DBPs. Cooking methods include boiling, shallow frying, roasting, pressure cooking, deep frying, microwave cooking, and so on. Up to now, the main method used in studies of DBPs in the cooking process is boiling.

2.2 Cooking Food

Salt, seaweed salt, kelp, sugar, and tea have all been proven to lead to DBPs formation during a household cooking process.

Bromide was found in 20 types of edible salts, ranging from 67 to 375 mg/kg of salt, with an average concentration of 173 mg/kg of salt [74]. Bromide is generally regarded as harmless by itself in edible salt, likely due to its relatively low toxicity to mammals [75] and slow rate of individual edible salt consumption (no more than 3.75–5.75 g/day depending on age) [76]. However, during household cooking, bromide in edible salts can react with disinfectants in tap water and organic matter, leading to the formation of polar Br-DBPs [74, 77].

Iodine is an essential micronutrient for synthesis of thyroid hormones to maintain normal functions of the human thyroid [78]. In many countries and regions, obtaining iodine from the normal diet and water is challenging, thus the iodine fortification of salt is considered as an effective method to guarantee a sufficient iodine intake [79]. However, similar to bromide, chlorine or chloramines in tap water during cooking react with the iodide in iodized table salt and organic matter to form I-DBPs [73, 77, 80, 81].

Therefore, other iodine-containing foods have also attracted the attention of the researchers. Seaweed salt and kelp are rich in iodine [82, 83], and they were all reported to promote forming high concentrations of I-DBPs during household cooking [84–87].

Sugar could also influence formation and transformation of DBPs in tap water because the carbohydrates present in table sugar, such as glucose and maltose, can enhance DBP formation [35]. And tea is the second most consumed nonalcoholic beverage next to drinking water [88]. Some of the compounds in tea (polyphenols, amino acids, caffeine, etc. [89]) have functional groups that can react with chlorine to form DBPs.

In addition, some additions have been proven to prevent DBP formation during household cooking. For example, adding sodium ascorbate (or lemon rich in ascorbic acid) could eliminate residual chlorine in tap water and prevent halogenated DBP formation during boiling [90]; adding carbonate could enhance the hydrolysis or dehalogenation of halogenated DBPs and lead to the increased formation of volatile THMs during the boiling processing [91].

2.3 Detection and Identification

The preparation procedure of cooking water sample is basically as follows: prepare tap water samples with different types and concentrations of disinfectant residual; add food and organic matter, etc.; heat, and keep cooking; filter the water sample.

Pretreatment procedure for the filtered samples often uses LLE, then pressure-reducing rotary evaporation or derivatization, etc. Finally, the sample is usually subjected to the analysis of GC/ECD [84, 85, 87], GC/MS [81, 92], UPLC/ESI-tqMS [35, 73, 74, 86, 92], or ultra-fast liquid chromatography-ion trap/time-of-flight mass spectrometry (UFLC/IT-TOF-MS) [73].

During household cooking, many new polar DBPs were generated with chloraminated/chlorinated tap water, and some reported DBPs (e.g., 3,5-diiodo-4-hydroxybenzaldehyde, 3-iodo-4-hydroxy-5-methylbenzoic acid, 2,4-diiodo-6-nitrophenol, etc.) were first found in cooking water [35, 73, 86, 93].

The presence of the food matrix may impact the DBP formation. Owing to the interference of matrix and the lack of corresponding DBP standards, some detected unknown DBPs couldn't be identified and others identified were not quantified with existing sample pretreatment methods, which require further optimization to accurately measure these trace unknown DBPs in cooking water.

Up to now, only Dong et al. studied the formation of I-DBPs during cooking real food with iodized salt and developed a new sensitive quantitative method (optimizing cleanup sorbents, extraction solvent, and GC-MS conditions) for the target I-DBPs in the real food (pasta) cooking water [81].

2.4 Formation and Occurrence

The main influencing factors of DBP formation and occurrence in cooking water are disinfectant species and levels of tap water, type and pretreatment of food, and cooking temperature and time.

2.4.1 Tap Water

The formation of DBPs varied with disinfectant species and levels. It has been reported that the total THM concentrations formed from chlorination, chloramination (monochloramine), and chloramination (organic chloramines) with seaweed salt rose by 228%, 295%, and 279%, respectively [85]. Increased residual chlorine levels in tap water resulted in a higher of DBPs in household cooking. Total I-THM concentrations increased by approximately 30% as the residual chlorine concentration increased from 0.05 to 1.0 mg L^{-1} [77].

2.4.2 Food

Type and dose of food affect the formation of DBPs during cooking, by offering different levels of bromide/iodine or precursors. Compared with KIO_3-fortified salt, cooking with KI-fortified salt significantly elevated both the number and concentrations of polar I-DBPs and I-THMs [73, 77]. Dong and Cao both proposed that a positive linear relationship was found between salt concentrations and total THMs ($R^2 > 0.90$) [81, 84]. In boiled tap water samples with added sugar at concentrations of 100 or 500 mg/L, the levels of TOX followed a descending rank order from highest to lowest: sucrose sample, lactose sample, maltose sample, glucose sample, xylose sample [35].

2.4.3 Temperature and Time

It was commonly known that heating the tap water to boiling effectively reduces the concentration of DBPs. As adding the food, the formation and reduction of DBPs became more complex. For example, higher temperature leads to increased formation of I-THMs in water samples with seaweed salt [84] but made lower level of I-THMs in water samples with KI salt and KIO_3 salt [77].

Prior researchers found that the formation of I-THMs increased with longer reaction time. However, the higher temperature in household cooking might modify the impact of reaction time. THM levels in the kelp soup and pasta with table salt presented a consistent decline as the reaction period increased [81, 87].

2.5 Toxicity

The toxicity detection methods of cooking water include toxicity assay (with Q67 or embryos of a polychaete *P. dumerilii*) and calculated method. Prior research confirmed the toxicity of cooking water increases with the formation of DBPs. The total calculated cytotoxicity of the cooked pasta sample was 126 times higher than that of tap water, and the genotoxicity of the pasta sample was 18 times higher than tap water [81]. After simulated household cooking, the soup containing 1.0 g/L of kelp was 7.3 times more toxic than tap water [87]. The toxicity of cooking water with seaweed salt increased by 4.6–8.7 times than tap water [84].

2.6 Control Strategies

The measures to control the formation of DBPs in the cooking process were proposed in the prior research.

(1) Use water with relatively low chlorine residual levels for cooking [73, 92]. (2) Choose food containing less iodine and bromine [73, 77, 81] or take pretreatment to reduce the iodine and bromine content (e.g., soak kelp and similar seafood) [87]. (3) Cook in open containers (without a lid) [81, 94] and with comparatively high temperature and long cooking time length [74, 94]. (4) Adding ascorbate or lemon slices (rich in vitamin C) to water before boiling, which could reduce the toxicity of boiling water [91, 95].

References

1. Beuchat LR, Nail BV, Adler BB, Clavero MRS (1998) Efficacy of spray application of chlorinated water in killing pathogenic bacteria on raw apples, tomatoes, and lettuce. J Food Prot 61(10):1305–1311
2. Cayanan DF, Zhang P, Liu W, Dixon M, Zheng Y (2009) Efficacy of chlorine in controlling five common plant pathogens. HortScience Horts 44(1):157–163
3. Nou X, Luo Y (2010) Whole-leaf wash improves chlorine efficacy for microbial reduction and prevents pathogen cross-contamination during fresh-cut lettuce processing. J Food Sci 75(5): M283–M290
4. Cardador MJ, Gallego M (2017) Simultaneous determination of 14 disinfection by-products in meat products using microwave-assisted extraction and static headspace coupled to gas chromatography–mass spectrometry. J Chromatogr A 1509:9–15
5. Lee W-N, Huang C-H, Zhu G (2019) Analytical methods for conventional and emerging disinfection by-products in fresh-cut produce. Food Chem 291:30–37
6. Miyahara MTYMUHKNSNSIY (1995) Volatile halogenated hydrocarbons in foods. J Agric Food Chem 43(2):320–326
7. Gil MI, Marín A, Andujar S, Allende A (2016) Should chlorate residues be of concern in fresh-cut salads? Food Control 60:416–421

8. Lee W-N, Huang C-H (2019) Formation of disinfection byproducts in wash water and lettuce by washing with sodium hypochlorite and peracetic acid sanitizers. Food Chem X 1:100003
9. Ölmez H, Kretzschmar U (2009) Potential alternative disinfection methods for organic fresh-cut industry for minimizing water consumption and environmental impact. LWT- Food Sci Technol 42(3):686–693
10. Gil MI, López-Gálvez F, Andújar S, Moreno M, Allende A (2019) Disinfection by-products generated by sodium hypochlorite and electrochemical disinfection in different process wash water and fresh-cut products and their reduction by activated carbon. Food Control 100:46–52
11. Lee W-N, Huang C-H, Zhu G (2018) Analysis of 40 conventional and emerging disinfection by-products in fresh-cut produce wash water by modified EPA methods. Food Chem 256:319–326
12. López-Gálvez F, Tudela JA, Allende A, Gil MI (2019) Microbial and chemical characterization of commercial washing lines of fresh produce highlights the need for process water control. Innov Food Sci Emerg Technol 51:211–219
13. Gombas D, Luo Y, Brennan J, Shergill G, Petran R, Walsh R, Hau H, Khurana K, Zomorodi B, Rosen J, Varley R, Deng K (2017) Guidelines to validate control of cross-contamination during washing of fresh-cut leafy vegetables. J Food Prot 80(2):312–330
14. Luo Y, Nou X, Yang Y, Alegre I, Turner E, Feng H, Abadias M, Conway W (2011) Determination of free chlorine concentrations needed to prevent Escherichia coli O157:H7 cross-contamination during fresh-cut produce wash†. J Food Prot 74(3):352–358
15. Gil MI, Selma MV, Suslow T, Jacxsens L, Uyttendaele M, Allende A (2015) Pre- and postharvest preventive measures and intervention strategies to control microbial food safety hazards of fresh leafy vegetables. Crit Rev Food Sci Nutr 55(4):453–468
16. Tudela JA, López-Gálvez F, Allende A, Hernández N, Andújar S, Marín A, Garrido Y, Gil MI (2019) Operational limits of sodium hypochlorite for different fresh produce wash water based on microbial inactivation and disinfection by-products (DBPs). Food Control 104:300–307
17. Guo M, Buchanan RL, Dubey JP, Hill DE, Lambertini E, Ying Y, Gamble HR, Jones JL, Pradhan AK (2015) Qualitative assessment for toxoplasma gondii exposure risk associated with meat products in the United States. J Food Prot 78(12):2207–2219
18. Liu Y, Gao J, Zhu Q, Zhou X, Chu W, Huang J, Liu C, Yang B, Yang M (2023) Zerovalent iron/Cu combined degradation of halogenated disinfection byproducts and quantitative structure–activity relationship modeling. Environ Sci Technol 57(30):11241–11250
19. Chen Y, Liang Q, Liang W, Li W, Liu Y, Guo K, Yang B, Zhao X, Yang M (2023) Identification of toxicity forcing agents from individual aliphatic and aromatic disinfection byproducts formed in drinking water: implications and limitations. Environ Sci Technol 57(3): 1366–1377
20. Liang W, Zhang Z, Zhu Q, Han Z, Huang C, Liang X, Yang M (2023) Molecular interactions between bovine serum albumin (BSA) and trihalophenol: insights from spectroscopic, calorimetric and molecular modeling studies. Spectrochim Acta A Mol Biomol Spectrosc 287: 122054
21. Hassoun A, Siddiqui SA, Smaoui S, Ucak İ, Arshad RN, Garcia-Oliveira P, Prieto MA, Aït-Kaddour A, Perestrelo R, Câmara JS, Bono G (2022) Seafood processing, preservation, and analytical techniques in the age of industry 4.0. Appl Sci
22. Cardador MJ, Gallego M (2018) Determination of several common disinfection by-products in frozen foods. Food Addit Contam Part A Chem Anal Control Expo Risk Assess 35(1):56–65
23. Suh M-J, Simpson AMA, Mitch WA (2023) Purified chlorine dioxide as an alternative to chlorine disinfection to minimize chlorate formation during postharvest produce washing. Environ Sci Technol 57(32):12063–12071
24. Yang Y, Liu Z, Zheng H, Zhu S, Zhang K, Li X, Ma X, Dietrich AM (2021) Sucralose, a persistent artificial sweetener in the urban water cycle: insights into occurrence, chlorinated byproducts formation, and human exposure. J Environ Chem Eng 9(4):105293
25. Cardador MJ, Gallego M, Cabezas L, Fernández-Salguero J (2016) Detection of regulated disinfection by-products in cheeses. Food Chem 204:306–313

26. Cardador MJ, Gallego M (2015) Haloacetic acids content of fruit juices and soft drinks. Food Chem 173:685–693
27. Wawryk NJP, Huang G, Craven C, Qiu J, Jmaiff Blackstock LK, Li X-F (2023) Aspartame-sweetened tap water: transformation products and 2,6-dichloro-1,4-benzoquinone formation. Environ Sci Technol 57(3):1332–1341
28. Komaki Y, Simpson AMA, Choe JK, Plewa MJ, Mitch WA (2018) Chlorotyrosines versus volatile byproducts from chlorine disinfection during washing of spinach and lettuce. Environ Sci Technol 52(16):9361–9369
29. Li Z, Wang J, Chen X, Hu S, Gong T, Xian Q (2019) A novel molecularly imprinted polymer-solid phase extraction method coupled with high performance liquid chromatography tandem mass spectrometry for the determination of nitrosamines in water and beverage samples. Food Chem 292:267–274
30. Coroneo V, Carraro V, Marras B, Marrucci A, Succa S, Meloni B, Pinna A, Angioni A, Sanna A, Schintu M (2017) Presence of trihalomethanes in ready-to-eat vegetables disinfected with chlorine. Food Addit Contam Part A Chem Anal Control Expo Risk Assess 34(12): 2111–2117
31. Zhang T, Lee W-N, Luo Y, Huang C-H (2022) Flume and single-pass washing systems for fresh-cut produce processing: disinfection by-products evaluation. Food Control 133:108578
32. Usman M, Kuckelkorn J, Kämpfe A, Zwiener C, Wintgens TA, Linnemann V (2023) Identification of disinfection by-products (DBP) in thermal water swimming pools applying non-target screening by LC-/GC-HRMS. J Hazard Mater 449:130981
33. Yang M, Zhang X, Liang Q, Yang B (2019) Application of (LC/)MS/MS precursor ion scan for evaluating the occurrence, formation and control of polar halogenated DBPs in disinfected waters: a review. Water Res 158:322–337
34. Zhang Y, Bu Y, Han J, Liu Y, Chen B, Zhang X, Yang M, Sui Y (2018) A facile and green pretreatment method for nonionic total organic halogen (NTOX) analysis in water – step I. Using electrodialysis to separate NTOX and halides. Water Res 145:631–639
35. Qiu C, He W, Li Y, Jiang F, Pan Y, Zhang M, Lin D, Zhang K, Yang Y, Wang W, Hua P (2022) Formation of halogenated disinfection byproducts in chlorinated real water during making hot beverage: effect of sugar addition. Chemosphere 305:135417
36. Garrido Y, Marín A, Tudela JA, Allende A, Gil MI (2019) Chlorate uptake during washing is influenced by product type and cut piece size, as well as washing time and wash water content. Postharvest Biol Technol 151:45–52
37. Kaufmann-Horlacher I, Kolberg DIS, Wildgrube C, Benkenstein A, Scherbaum E, Anastassiades M (2014) Chlorate residues in plant-based food. Eurl-Srm
38. Simpson AMA, Mitch WA (2022) Chlorine and ozone disinfection and disinfection byproducts in postharvest food processing facilities: a review. Crit Rev Environ Sci Technol 52(11): 1825–1867
39. Nitsopoulos A, Glauner T (2014) Chlorate-a contaminant in foodstuff and drinking water. Agilent Technologies, Labor Friedle GmbH
40. Kaufmann-Horlacher I (2015) Chlorate residues in plant-based food: an update. Report from a day in the lab
41. Petri E, Virto R, Mottura M, Parra J (2021) Comparison of peracetic acid and chlorine effectiveness during fresh-cut vegetable processing at industrial scale. J Food Prot 84(9): 1592–1602
42. Kawanishi S, Murata M (2006) Mechanism of DNA damage induced by bromate differs from general types of oxidative stress. Toxicology 221(2):172–178
43. Amarasiri M, Sano D, Suzuki S (2020) Understanding human health risks caused by antibiotic resistant bacteria (ARB) and antibiotic resistance genes (ARG) in water environments: current knowledge and questions to be answered. Crit Rev Environ Sci Technol 50(19):2016–2059
44. Lombardo S, Restuccia C, Pandino G, Licciardello F, Muratore G, Mauromicale G (2015) Influence of an O3-atmosphere storage on microbial growth and antioxidant contents of globe artichoke as affected by genotype and harvest time. Innov Food Sci Emerg Technol 27:121–128

45. Guzel-Seydim ZB, Greene AK, Seydim AC (2004) Use of ozone in the food industry. LWT-Food Sci Technol 37(4):453–460
46. Sandhu HPS, Manthey FA, Simsek S, Ohm J-B (2011) Comparison between potassium bromate and ozone as flour oxidants in breadmaking. Cereal Chem 88(1):103–108
47. Huang AT, Batterman S (2009) Formation of trihalomethanes in foods and beverages. Food Addit Contam Part A Chem Anal Control Expo Risk Assess 26(7):947–957
48. Van Haute S, Sampers I, Holvoet K, Uyttendaele M (2013) Physicochemical quality and chemical safety of chlorine as a reconditioning agent and wash water disinfectant for fresh-cut lettuce washing. Appl Environ Microbiol 79(9):2850–2861
49. López-Gálvez F, Allende A, Truchado P, Martínez-Sánchez A, Tudela JA, Selma MV, Gil MI (2010) Suitability of aqueous chlorine dioxide versus sodium hypochlorite as an effective sanitizer for preserving quality of fresh-cut lettuce while avoiding by-product formation. Postharvest Biol Technol 55(1):53–60
50. Yang M, Zhang X (2013) Comparative developmental toxicity of new aromatic halogenated DBPs in a chlorinated saline sewage effluent to the marine Polychaete Platynereis dumerilii. Environ Sci Technol 47(19):10868–10876
51. Shen C, Norris P, Williams O, Hagan S, Li K (2016) Generation of chlorine by-products in simulated wash water. Food Chem 190:97–102
52. Gil MI, Selma MV, López-Gálvez F, Allende A (2009) Fresh-cut product sanitation and wash water disinfection: problems and solutions. Int J Food Microbiol 134(1):37–45
53. Chen X, Hung Y-C (2016) Predicting chlorine demand of fresh and fresh-cut produce based on produce wash water properties. Postharvest Biol Technol 120:10–15
54. Obolensky A, Singer PC (2008) Development and interpretation of disinfection byproduct formation models using the information collection rule database. Environ Sci Technol 42(15): 5654–5660
55. Waters BW, Hung Y-C (2014) The effect of organic loads on stability of various chlorine-based sanitisers. Int J Food Sci Technol 49(3):867–875
56. Hung Y-C, Waters BW, Yemmireddy VK, Huang C-H (2017) pH effect on the formation of THM and HAA disinfection byproducts and potential control strategies for food processing. J Integr Agric 16(12):2914–2923
57. Bornhorst ER, Luo Y, Park E, Vinyard BT, Nou X, Zhou B, Turner E, Millner PD (2018) Immersion-free, single-pass, commercial fresh-cut produce washing system: an alternative to flume processing. Postharvest Biol Technol 146:124–133
58. Khadre MA, Yousef AE, Kim JG (2001) Microbiological aspects of ozone applications in food: a review. J Food Sci 66(9):1242–1252
59. Rosenblum J, Ge C, Bohrerova Z, Yousef A, Lee J (2012) Ozonation as a clean technology for fresh produce industry and environment: sanitizer efficiency and wastewater quality. J Appl Microbiol 113(4):837–845
60. Akbas MY, Ölmez H (2007) Inactivation of Escherichia coli and listeria monocytogenes on iceberg lettuce by dip wash treatments with organic acids. Lett Appl Microbiol 44(6):619–624
61. Abadias M, Usall J, Oliveira M, Alegre I, Viñas I (2008) Efficacy of neutral electrolyzed water (NEW) for reducing microbial contamination on minimally-processed vegetables. Int J Food Microbiol 123(1):151–158
62. Fang J, Cannon JL, Hung Y-C (2016) The efficacy of EO waters on inactivating norovirus and hepatitis a virus in the presence of organic matter. Food Control 61:13–19
63. Gómez-López VM, Marín A, Medina-Martínez MS, Gil MI, Allende A (2013) Generation of trihalomethanes with chlorine-based sanitizers and impact on microbial, nutritional and sensory quality of baby spinach. Postharvest Biol Technol 85:210–217
64. Hadjok C, Mittal GS, Warriner K (2008) Inactivation of human pathogens and spoilage bacteria on the surface and internalized within fresh produce by using a combination of ultraviolet light and hydrogen peroxide. J Appl Microbiol 104(4):1014–1024
65. Meireles A, Giaouris E, Simões M (2016) Alternative disinfection methods to chlorine for use in the fresh-cut industry. Food Res Int 82:71–85

66. Hua G, Reckhow DA (2007) Comparison of disinfection byproduct formation from chlorine and alternative disinfectants. Water Res 41(8):1667–1678
67. Banach JL, Sampers I, Van Haute S, Van der Fels-Klerx HJ (2015) Effect of disinfectants on preventing the cross-contamination of pathogens in fresh produce washing water. Int J Environ Res Public Health:8658–8677
68. Koseki S, Isobe S (2006) Effect of ozonated water treatment on microbial control and on Browning of iceberg lettuce (Lactuca sativa L.). J Food Prot 69(1):154–160
69. Wu FM, Doyle MP, Beuchat LR, Wells JG, Mintz ED, Swaminathan B (2000) Fate of Shigella sonnei on parsley and methods of disinfection. J Food Prot 63(5):568–572
70. Cardador MJ, Gallego M (2017) Control of disinfection by-products in canned vegetables caused by water used in their processing. Food Addit Contam Part A Chem Anal Control Expo Risk Assess 34(1):10–23
71. Liu J, Zhang X, Li Y (2015) Effect of boiling on halogenated DBPs and their developmental toxicity in real tap waters, recent advances in disinfection by-products. American Chemical Society, pp 45–60
72. Pan Y, Zhang X, Wagner ED, Osiol J, Plewa MJ (2014) Boiling of simulated tap water: effect on polar brominated disinfection byproducts, halogen speciation, and cytotoxicity. Environ Sci Technol 48(1):149–156
73. Pan Y, Zhang X, Li Y (2016) Identification, toxicity and control of iodinated disinfection byproducts in cooking with simulated chlor(am)inated tap water and iodized table salt. Water Res 88:60–68
74. Zhang D, Wu Y, Zhang X, Li W, Li Y, Li A, Pan Y (2020) Identification, formation and control of polar brominated disinfection byproducts during cooking with edible salt, organic matter and simulated tap water. Water Res 172:115526
75. Sangster B, Blom JL, Sekhuis VM, Loeber JG, Rauws AG, Koedam JC, Krajnc EI, Van Logten MJ (1983) The influence of sodium bromide in man: a study in human volunteers with special emphasis on the endocrine and the central nervous system. Food Chem Toxicol 21(4):409–419
76. Shelley MG (2011) U.S. Department of Agriculture and U.S. Department of Health and Human Services, Dietary Guidelines for Americans, 2010, vol 3, 7th edn, Advances in nutrition, U.S. Government Printing Office, Washington, DC, January 2011, pp 293–294
77. Yan M, Li M, Han X (2016) Behaviour of I/Br/Cl-THMs and their projected toxicities under simulated cooking conditions: effects of heating, table salt and residual chlorine. J Hazard Mater 314:105–112
78. Anke M, Seifert M (2004) Elements and their compounds in the environment
79. Leung A, Braverman L, Pearce E, History of U.S. Iodine fortification and supplementation (2012) Nutrients 4(11)
80. Krasner SW, Wright JM (2005) The effect of boiling water on disinfection by-product exposure. Water Res 39(5):855–864
81. Dong H, Nordhorn ID, Lamann K, Westerman DC, Liberatore HK, Forster ALB, Aziz MT, Richardson SD (2023) Overlooked iodo-disinfection byproduct formation when cooking pasta with iodized table salt. Environ Sci Technol 57(9):3538–3548
82. Ye N, Zhang X, Miao M, Fan X, Zheng Y, Xu D, Wang J, Zhou L, Wang D, Gao Y (2015) Saccharina genomes provide novel insight into kelp biology. Nat Commun 6:6986
83. Xl H, Qf Q, Xj Y, Fan X, Cf C (1997) Determination of chemical species of iodine in some Seaweeds (I). Sci Total Environ 3:204
84. Cao Z, Ding S, Deng Y, Luan X, Chu W (2022) Household cooking with seaweed salts and chloraminated tap water produce harmful iodinated disinfection by-products. Chem Eng J 431
85. Qu R, Ding S, Li H, Wang P, Pan L, Chu W (2022) Comparison of iodinated disinfection by-product formation from the reaction of chlorine, monochloramine, and organic chloramine with seaweed salt during a simulated household cooking process. Environ Sci Water Res Technol 8(8):1665–1674

86. Ding S, Deng Y, Wu M, Qu R, Du Z, Chu W (2023) Leaching of organic matter and iodine, formation of iodinated disinfection by-products and toxic risk from Laminaria japonica during simulated household cooking. J Hazard Mater 459:132241
87. Wu M, Ding S, Cao Z, Du Z, Tang Y, Chen X, Chu W (2022) Insights into the formation and mitigation of iodinated disinfection by-products during household cooking with Laminaria japonica (Haidai). Water Res 225:119177
88. Storelli MM, Giacominelli-Stuffler R, Barone G (2016) Evaluation of trace metal and polychlorinated biphenyl levels in tea brands of different origin commercialized in Italy, food and chemical toxicology: an international journal published for the British industrial. Biol Res 87:113–119
89. Jaziri I, Slama MB, Mhadhbi H, Urdaci MC, Hamdi M (2008) Effect of green and black teas (Camellia sinensis L.) on the characteristic microflora of yogurt during fermentation and refrigerated storage. Food Chem 112(3):614–620
90. Zhang LA. Removal of chlorine residual in tap water by boiling or adding ascorbic acid
91. Liu J, Li Y, Jiang J, Zhang X, Sayes CM (2020) Effects of ascorbate and carbonate on the conversion and developmental toxicity of halogenated disinfection byproducts during boiling of tap water. Chemosphere 254(1):126890
92. Li J, Aziz MT, Granger CO, Richardson SD (2021) Are disinfection byproducts (DBPs) formed in my cup of tea? Regulated, priority, and unknown DBPs. Environ Sci Technol 55(19): 12994–13004
93. Li B, Chen B, Chen S, Yang M, Wu Q (2021) Removing chlorinated haloacetaldehydes from drinking water by household heating devices with and without chlorine: efficiency, influencing factors, and mechanisms. Chemosphere 284:131202
94. Ma X, Cheng J, Zhang P, Wu Y, Deng J, Dong F, Li X, Dietrich AM (2023) Impact of boiling on chemical and physical processes for reduction of halomethanes, haloacetonitriles, and haloacetic acids in drinking water. Sci Total Environ 906:167657
95. Liu J, Sayes CM, Sharma VK, Li Y, Zhang X (2021) Addition of lemon before boiling chlorinated tap water: a strategy to control halogenated disinfection byproducts. Chemosphere 263:127954

Climate Change Implications for DBP Occurrence in Municipal Drinking Water

Monica B. Emelko, Alex Chow, Fariba Amiri, Huan Chen, and Micheal Stone

Contents

1 Global Warming, Hydroclimatic Change, and Climate Shocks 358
1.1 Weather, Climate, and Global Warming 358
1.2 Hydroclimatic Change and Climate Shocks 359
2 Landscape Characteristics 360
2.1 Land Cover, Land Use, and Land Management 360
2.2 Soil Characteristics, Geology, and Landform 361
3 Landscape Response to Climate Change 363
3.1 Increased Erosion and Slope Failure 363
3.2 Increased Atmospheric Deposition 364
3.3 Increased Saltwater Intrusion 364
4 DBP Precursor Implications of Landscape Responses to Climate Change 364

The original version of the chapter has been revised. A correction to this chapter can be found at https://doi.org/10.1007/978-3-031-74260-6_1169

M. B. Emelko (✉)
Department of Civil & Environmental Engineering, University of Waterloo, Waterloo, ON, Canada
e-mail: mbemelko@uwaterloo.ca

A. Chow and F. Amiri
The Chinese University of Hong Kong, EESC, Hong Kong, People's Republic of China
e-mail: achow@cuhk.edu.hk; f3amiri@uwaterloo.ca

H. Chen
Department of Environmental Engineering and Earth Sciences, Clemson University, Clemson, SC, USA
e-mail: huanc@clemson.edu

M. Stone
Department of Geography and Environmental Management, University of Waterloo, Waterloo, ON, Canada
e-mail: mstone@uwaterloo.ca

Chao Chen, Susan Andrews, and Yuefeng Xie (eds.), *Disinfection By-Products in Water*, Hdb Env Chem (2024) 135: 357–376, DOI 10.1007/698_2024_1168,

Published online: 19 October 2024

5 Evidence of Climate Change Impacts on DBP Precursors and Formation 365
5.1 Glacial Ice and Snowpack 365
5.2 Heavy Rainfall 366
5.3 Drought 367
5.4 Wildfire 368
5.5 Sea Level Rise 368
6 Key Messages 369
References 369

Abstract Disinfection by-product (DBP) precursors are inextricably linked to the landscapes and waters on which they originate and transform. Warming global temperatures, hydroclimatic change, and climate shocks alter key landscape characteristics (e.g., land cover and soil characteristics) and responses (e.g., erosion and runoff), altering the mobilization and delivery of disinfection by-product (DBP) precursors to water supplies. In this chapter, the inter-relationships between (1) global warming, (2) hydroclimatic change and climate shocks, (3) landscape characteristics, (4) landscape responses to changes in landscape characteristics, and (5) DBP precursor delivery and change in aquatic systems are discussed.

Keywords AnOM, AOM, Climate shocks, DBP precursors, Global warming, Hydroclimatic variability, Land disturbance, NOM, Treatability, Water quality, Wildfire

1 Global Warming, Hydroclimatic Change, and Climate Shocks

1.1 Weather, Climate, and Global Warming

Weather and climate are key factors that affect DBP precursor availability and transformation in aquatic systems. Weather refers to atmospheric conditions in a specific location. In contrast, climate refers to average weather patterns over an extended period, typically 30 years (i.e., a climate normal) or more [1]. Key factors influencing climate include solar radiation, latitude, longitude, landforms, elevation, proximity to oceans, ocean currents, and greenhouse gas concentrations. Changes in land cover – including those caused by anthropogenic disturbances (e.g., changes in land use and land management) can also significantly affect climate.

Weather and climate affect temperature, precipitation, evaporation/evapotranspiration, and the hydrologic cycle more broadly. Over the shorter term, temperature variations affect the state of water (i.e., higher temperatures can lead to increased evaporation/evapotranspiration, while lower temperatures may cause water to freeze) and influence the rates of chemical and biological reactions that occur within it. Over the longer term, average temperatures impact the frequency and duration of freezing and thawing (where relevant), thereby affecting water availability and aquatic ecosystem development and activity. Similarly, variations in precipitation (i.e., rainfall and snowfall) and evaporation affect the immediate availability of water over the shorter term while regional precipitation levels and evaporation rates are governed by longer-term climatic conditions. Accordingly, water resources are typically more abundant in regions with higher average precipitation while water scarcity is more common in arid regions. The hydrologic cycle describes the continuous movement of water in the earth-atmosphere system through evaporation, condensation, precipitation, interception, infiltration, percolation, transpiration, run-off, and storage. These processes are affected by both short-term weather events and long-term climate patterns that impact the intensity and frequency of weather events. Thus, water is influenced by climate, and also influences climate because it is integral to the large-scale exchanges of heat and mass between the atmosphere, the ocean, and the land surface [2, 3].

1.2 Hydroclimatic Change and Climate Shocks

Hydroclimate reflects feedback between climate patterns and hydrological processes [3]. Long-term changes in climate patterns affect hydrological processes and aquatic ecosystems – these shifts are often referred to as hydroclimatic change [4]. It is because of these feedbacks that global warming is leading to shifts in precipitation patterns, which are linked to atmospheric conditions. Elevated atmospheric greenhouse gas concentrations result in positive radiative forcing (i.e., greater incoming energy than outgoing energy because they absorb and re-emit heat from outgoing infrared radiation).

Deep snowpacks and melting glaciers maintain river, lake, and groundwater supplies in many parts of the world. Warmer air temperatures result in earlier and greater ice and snow melt during spring and summer months and shifts from snowfall to rainfall in some regions. Warming temperatures and changes in precipitation patterns can also lead to climate shocks, which are climate change-exacerbated and weather-associated landscape disturbances such as extreme precipitation, droughts, and wildfires [5, 6]. For example, warmer temperatures lead to increased humidity and affect atmospheric circulation patterns. Increased atmospheric water vapor content contributes to more intense and heavy rainfall during precipitation events. Altered weather systems and shifts in regional climates contribute to more intense and variable rainfall patterns that affect the distribution, intensity, frequency, and duration of precipitation [7]. Increasing temperatures can also increase evapotranspiration, however, thereby reducing soil moisture and shifting rainfall distributions to create conditions conducive to prolonged and severe drought events in some areas [8].

Wet springs can also promote vegetation growth while warm and dry summers dry out vegetation and landscapes, enhancing fuel availability and promoting the ignition and spread of wildfires. Additionally, climate-induced factors, like insect infestations and disease, weaken forests, providing more combustible material.

These compounding effects create a conducive environment for larger, more severe wildfires with greater burned areas [9]. Thus, over shorter- and longer-term time scales, climate shocks and shifts in hydroclimate can alter the mobilization and delivery of DBP precursors to water supplies.

2 Landscape Characteristics

2.1 Land Cover, Land Use, and Land Management

Land cover, land use, and land management collectively affect DBP precursor source availability, mobilization, and delivery to receiving waters. As described by [3], land cover includes the amount and type of vegetation present in watersheds. Land use reflects activities that occur within watersheds. Land management is closely related to land use; it involves policies and practices for mitigating the detrimental impacts of land use change on ecosystem services, including water. In general, decreases in vegetation cover result in increased delivery of DBP precursors from terrestrial to aquatic systems.

The loss of vegetation can reduce surface roughness and increase the delivery of these materials from hillslopes to aquatic systems during periods of runoff [10]. Conversely, vegetated areas tend to retain these materials. Nutrient assimilation rates can vary widely depending on soil type, slope, moisture content, and the vegetation type [11]. For example, forests remove more nitrogen from soils than grasslands and shrubs [12]. In contrast, engineered features (impervious surfaces and drainage structures) on residential, industrial, and agricultural areas are designed to route runoff from the landscape to receiving waters by increasing runoff velocities and volumes [13]. The resulting high energy conditions in these transport pathways increase the delivery of DBP precursors to surface waters thereby increasing their downstream transport at large river basin scales [14].

Land cover, land use, and land management are inter-connected in their effects on DBP precursors: natural organic matter (NOM) (which typically also reflects soil organic matter [SOM]), nitrogenous compounds (e.g., ammonia, amines, proteins, and amino acids), algae and microorganisms, and halides. NOM is a complex mixture of organic compounds (e.g., humic acids, fulvic acids, and phenolic compounds) [15]. Both dissolved and particulate forms of NOM are generally associated with increased potential for DBP formation [16]. Nitrogenous compounds, such as ammonia, amines, and amino acids, can be present in water due to the decomposition of organic matter and land use-associated delivery – algae and wastewater are the two most cited sources of dissolved organic nitrogen (DON). It has been estimated that DON typically comprises 0.5–10% of total NOM by weight, depending on watershed characteristics [17]. Thus, waters containing higher levels of algal organic matter (AOM) and effluent organic matter (EfOM) will typically have enhanced DON concentrations and consequently form higher levels of nitrogenous DBPs (N-DBPs) than non-impacted waters [18]. The nitrogen in N-DBPs can also be derived from chloramine disinfectants themselves [19]. Algae and microorganisms are ubiquitous in the aquatic environment and release organic matter (e.g., AOM) to water supplies as they grow and die [20, 21]. AOM can be a major source of DBP precursors during chlorination [22]. Increased production of bacterial extracellular polymeric substances is also generally associated with increased potential to form DBPs [23]. Halides (e.g., bromide and iodide) also affect DBP formation.

Different land cover types, land uses, and land management practices can deliver varying amounts and types of NOM and nitrogenous compounds to receiving waters. For example, forests and wetlands can release higher levels of NOM and DON to water from the microbially mediated decomposition of organic material, such as decaying leaves, plants, and algae and root exudation [24, 25]. Lakes and reservoirs, especially those that are nutrient-rich, can release higher levels of AOM because availability of limiting nutrients such as phosphorus and micronutrients such as iron stimulate algal growth [26, 27]. As algae proliferate, they produce organic matter through photosynthesis and subsequent decay [27]. Urban and peri-urban areas may introduce anthropogenic sources of organic matter such as EfOM from sewage and industrial discharges [28–30]. Stormwater can also deliver organic matter to water due to its interaction with impervious surfaces and urban environments. When rainwater runs off surfaces such as roads, parking lots, and rooftops, it can pick up various pollutants from these surfaces, including (1) dissolved and suspended natural organic matter, (2) pollutants like oil, grease, salts, pesticides, manure, and fertilizers, and (3) microbial contaminants, including bacteria, viruses, and protozoa from pet waste, sewage overflows, and other sources [31, 32]. The additional delivery of halides (e.g., bromide and iodide) and sediments can further contribute to the formation of certain DBPs [33–35] and the increased availability of key macro- and micronutrients (e.g., phosphorus and iron, respectively) that promote the proliferation of algae and the associated production of AOM [26, 27].

2.2 Soil Characteristics, Geology, and Landform

The relationship between soil characteristics, geology, and landforms is integral to understanding how climate change can impact DBP formation potential. Geology reflects landscape structure, composition, and processes [3]. It provides the parent material needed for soil formation. Different rock types contribute distinct mineral compositions to soils through weathering. Landforms (e.g., mountains, plains, and basins) are shaped by geological forces such as erosion, weathering, and tectonics. They dictate the spatial distribution and properties of soils. Landform attributes such as slope, aspect, and elevation impact drainage, sunlight exposure, and temperature, thereby influencing soil development.

Organic and inorganic mineral particles, air, water, and living organisms comprise soil by forming a relatively thin layer on the earth's crust where plants can grow. The chemical characteristics of soils and rocks affect nutrients [36, 37] and salt sources/availability [38]. SOM has been described as a continuum of progressively decomposing organic compounds in various forms, derived from the decomposition of plant and animal residues, microbial activity, and organic matter accumulation [39]. Soil humus, substances formed from the biological and chemical breakdown of plant and animal residues, account for 60–70% of the total organic carbon in soils, which includes fulvic acids, humic acids, and humins. Phosphorus, nitrogen, and salts are contained within sedimentary deposits [37]. Particulate nutrients and salts composing rocks and soils can be mobilized by weathering and erosion, while dissolved forms can be mobilized by dissolution, leaching, and desorption [3]. Soil characteristics, including texture, structure, and fertility, are inextricably linked to underlying geology and the landforms they create. For example, well-drained soils may dominate on slopes, while poorly drained soils might characterize low-lying areas. Land use, land management, and climate shocks can alter soil characteristics.

The amount and timing of particulate dissolved organic matter (POM and DOM) transferred from landscapes to receiving streams is governed by soil type and vegetation, soil biogeochemical processes that regulate soil solute concentrations, and moisture conditions that influence the transport of hillslope solute pools [40]. On most landscapes, POM and DOM concentrations in soils decrease with depth in the soil profile because of leaching, microbially mediated decomposition and transformation, and humification processes [41]. POM and DOM export from landscapes occurs during spring snow melt (where applicable) and precipitation events [42]. During high-flow conditions, rising water tables can promote the lateral transport of DOC pools into receiving surface waters [43]. Thus, stream DOC variability is linked to the concentrations, spatial distribution, and hydrologic connectivity of hillslope solute pool DOC [44].

The pathway of nitrogen export from landscapes is primarily through liquid leaching or in gaseous forms (denitrification and ammonia volatilization) lost to the atmosphere. The magnitude of wildfire effects on N concentrations varies greatly [45]. Once transported from hillslopes to receiving streams during precipitation events particulate forms of nitrogen can be exported from watersheds. Nitrogen dynamics have been studied extensively in agricultural landscapes (e.g., [46, 47]), urban settings [48], and forest disturbances such as harvesting (e.g., [49, 50]).

The major pathway of nutrient transfer is related to its form (dissolved, particulate). Delivery of particulate materials from terrestrial to aquatic systems generally occurs during storm events in overland flow, tile drains, or stormwater sewer systems [3]. Groundwater is an important pathway for dissolved nutrients, particularly nitrogen and to a lesser degree phosphorus. The textural composition of subsurface geological materials and overlying soils influences their hydraulic conductivity, residence time, and delivery to surface waters. Where aquifers and catchment soils have generally low hydraulic conductivities, the residence time of dissolved nutrients can increase. Nutrients in shallow groundwater can be assimilated by plants or be transformed in biogeochemical cycling.

3 Landscape Response to Climate Change

3.1 Increased Erosion and Slope Failure

Erosion is the process of wearing away soil, rock, and sediment. It can be driven by water (e.g., rivers and rainfall), wind, glacial, and gravitational forces. Surface runoff is the unconfined flow of water over the ground surface. Erosion mobilizes DBP precursors, suspended sediments, nutrients, and pollutants and enables their delivery to water bodies. This occurs predominantly through surface runoff and atmospheric deposition [51]. Mass wasting is gravitational erosion that involves the downslope movement of rock, soil, and debris under the influence of gravity. It occurs when the forces acting on a slope exceed the strength of the materials holding it in place. Landslides and debris flows are both mass-wasting processes. While landslides involve the downslope movement of rock, soil, and debris, debris flows are fast-moving mixtures of water, rock, soil, and debris that have been described as resembling flowing concrete [52].

One of the major pathways by which global warming results in increased erosion and slope failures is through changes in land cover [3]. Altered precipitation patterns and increased temperatures influence vegetation growth. Vegetation is essential to minimizing erosion and maximizing slope stability because it helps bind soil particles together, reducing erosion and enhancing slope cohesion [53]. Changes in vegetation cover, such as shifts in plant types, resulting from shifts in precipitation patterns, increased temperature, or climate shocks can lead to decreased root strength and increased water runoff [54, 55]. The loss of vegetation cover contributes to soil erosion, reduces slope stability, and increases the risk of landslides and debris flows, particularly during heavy rainfall events.

Increased temperatures also result in decreases in glacial ice and snowpacks. Glacial retreat can expose unstable, freshly deglaciated slopes composed of unconsolidated, loose sediments and rocks, making them susceptible to landslides and debris flows [56, 57]. Rapid melting of snow and ice can contribute to increased water runoff, leading to soil saturation and potential slope failures [58].

Altered precipitation patterns can lead to increases in both extreme rainfall and drought. Intense rainfall can create preferential flow paths, increasing water infiltration and increasing pore water pressure in soils, which can reduce slope stability, making slopes more prone to failure [59]. During drought, soil moisture content decreases, weakening the cohesion between soil particles, thereby making slopes more prone to instability [60]. Vegetation cover can also decline during drought, further compromising slope integrity [61]. After prolonged dry periods, friable soils may not effectively absorb water due to factors such as reduced moisture content and capillarity, compaction, hydrophobicity, and reduced soil structure leading to rapid surface runoff and an elevated risk of landslides and debris flows as water cascades over desiccated slopes [62]. Thus, rapid surface runoff and an elevated risk of landslides and debris flows are more likely because of rainfall on drought-impacted landscapes. Climate shocks such as wildfires, hurricanes, typhoons, and intense storms can further trigger slope failure flows through weakening or removal of vegetation, saturation of soils, erosion of soils, creation of hydrophobic soil layers that prevent infiltration and cause rapid surface runoff, and tree uprooting and windthrow [63, 64].

3.2 Increased Atmospheric Deposition

Atmospheric deposition is an important and understudied pathway by which DBP precursors are delivered from terrestrial to aquatic environments. It can occur through dry and wet deposition. Wet deposition involves the capture of these materials in precipitation (i.e., rain, snow, sleet, or hail) as it fall through the atmosphere while dry deposition involves direct settling. The melting of glaciers and snow exposes trapped materials, which can then become part of wet atmospheric deposition during precipitation events [65, 66].

Atmospheric deposition is a significant source of organic carbon [67–69]. For example, atmospheric organic carbon includes aerosolized particles and organic gases from natural and anthropogenic sources and volatile organic carbon molecules (VOCs) emitted by vegetation [70]. Sea spray, smoke from natural wildfires, and dust from soils are the predominant natural sources of organic aerosols, while fossil fuels and biomass burning are major anthropogenic sources [68]. The majority of atmospheric organic carbon deposition in watersheds is from natural, biogenic VOCs [70]. Sea spray, wildfire smoke, and dust can also contain inorganic halides (e.g., bromide and iodide) [71, 72]. Changes in atmospheric circulation, influenced by global temperature shifts, can redistribute pollutants over larger distances, impacting regions that have been historically less affected.

3.3 Increased Saltwater Intrusion

Rising global temperatures lead to the melting of glaciers and ice sheets and the thermal expansion of seawater. The combination of melting ice and thermal expansion increases seawater volume, causing higher sea levels [3]. Sea level rise is expected to alter natural groundwater flow patterns, causing an inland migration of the saltwater-freshwater interface [73]. The increased intrusion of seawater into freshwater supplies has resulted in increased halide (e.g., bromide and iodide) concentrations in coastal regions [74].

4 DBP Precursor Implications of Landscape Responses to Climate Change

DBP precursors are inextricably linked to the landscapes and waters on which they originate and transform. As discussed above, warming global temperatures, hydroclimatic change, and climate shocks alter key landscape characteristics,

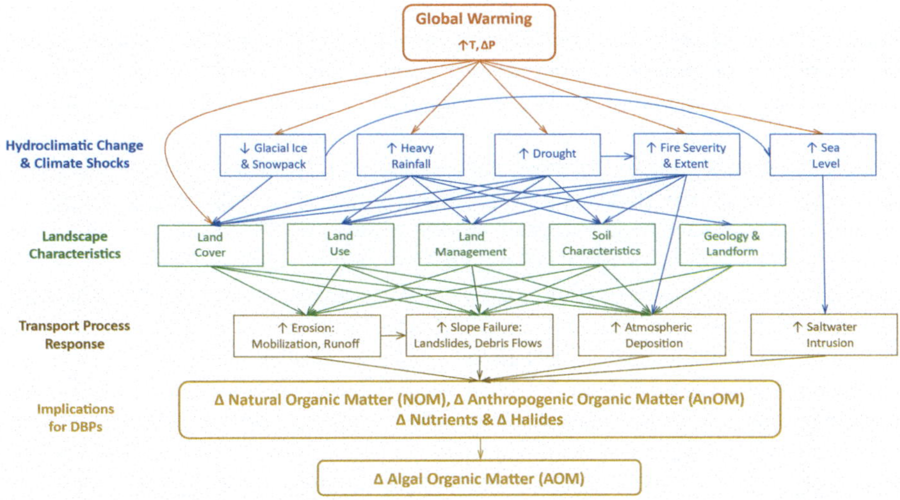

Fig. 1 Inter-relationships between global warming, hydroclimatic change and climate shocks, landscape characteristics, landscape responses to changes in landscape characteristics, and DBP precursors (modified from [3])

including land cover, land use, land management, soil characteristics, and geology and landforms; associated landscape responses include increases in erosion and associated runoff, slope failure, atmospheric deposition, and saltwater intrusion [3]. These processes alter mobilization and delivery of NOM (allochthonous organic carbon), anthropogenic organic carbon (AnOC; comprised EfOM and other anthropogenic carbon [e.g., from stormwater runoff]), and other aspects of water quality such as halides (e.g., bromide and iodide) and nutrients (e.g., bioavailable phosphorus) that serve as DBP precursors or affect the production of autochthonous organic carbon (i.e., AOM). The inter-relationships between global warming, hydroclimatic change and climate shocks, landscape characteristics, landscape responses to changes in landscape characteristics, and DBP precursors are shown in Fig. 1.

5 Evidence of Climate Change Impacts on DBP Precursors and Formation

5.1 Glacial Ice and Snowpack

It has been suggested that glacial NOM is more labile than that found in marine and freshwater environments [75]. NOM in the high-latitude cryosphere includes DOM and cryoconite organic matter (COM) from glaciers and SOM in permafrost. Glacier ice DOM and COM appear mainly composed of small, labile biomolecules associated with microbes, though COM may contain some plant-derived material [76]. In one investigation, higher dissolved organic carbon (DOC) concentrations have been found in snow samples as compared to glacial runoff samples, though higher concentrations of inorganic ions were found in the glacial runoff [77]. This suggested that the source of the DOM in the glacier runoff was the glacier. During transport, more than half of the molecules were transformed from aliphatic and peptide-like compounds in the snow samples into highly unsaturated and phenolic-like compounds in the glacial runoff; the transformation of DOM was attributed to biological and/or photolytic evolution during transport [77]. In another study, the content and characteristics of DOM and DBP formation and associated cytotoxicity index during chlorination of snow meltwater were investigated. Overall, meltwater exhibited high nitrogen content and DOM was predominantly composed of low molecular weight and low aromaticity organics. Trihalomethane (THM) and haloacetic acid (HAA) yields (i.e., DBP mass per unit mass of DOC) from these meltwater matrices were relatively low, while yields of highly cytotoxic DBPs, especially halonitromethanes were significantly higher than those found in surface water [78]. DBPs in these waters correlated positively with air quality index and were significantly impacted by snowfall stages, underscoring the importance of atmospheric deposition in DBP precursor delivery on these landscapes.

5.2 Heavy Rainfall

Increases in heavy rainfall can deliver large amounts of atmospheric organic matter to aquatic systems [79–81]. Rainwater DOM concentrations and character depend on atmospheric sources, anthropogenic activities, and rainfall patterns [80–90]. However, multiple investigations have shown that rainwater DOM can include appreciable amounts of DBP (e.g., THM, HAA, and trichloroacetonitrile) precursors [80, 81, 86, 87, 90–93]. Rainfall-associated increases in DOM concentration are not necessarily proportional to shifts in DOM character (e.g., aromaticity) [94]. In one investigation of European rainwater, more chromophoric dissolved organic matter (CDOM) was found in summer and fall than in other seasons [83]. Relative to pre-rainfall levels, small, but significant increases in treated water THMs (threefold) and HAAs (twofold) were observed during and after spring rainfall events in one study of four large, full-scale plants [95].

Heavy rainfall events also intensify connectivity between terrestrial and aquatic environments. Rainwater erosion and runoff can mobilize and transport NOM, especially during the early phases of rainfall events with differing impacts of rainfall intensity [96–98]. Higher-intensity rainfall tends to mobilize and transport organic carbon in particulate and colloidal forms [96, 99]. Exhaustion of humic compounds scoured from landscape surfaces during long duration, higher intensity rainfall events, and following successive rainfall events has been observed [100]. Similar relationships between rainfall intensity and DOM runoff and exhaustion have also been observed on urban landscapes [101] and in coastal regions experiencing subsurface flow flushing of organic matter from wetlands and organic-rich riparian zones [102]. In addition to NOM, on urban landscapes, anthropogenic organic matter (AnOM) composed of EfOM and stormwater containing anthropogenic organic micropollutants will typically have enhanced DON concentrations and consequently form higher levels of N-DBPs than unimpacted waters [18, 102].

In addition to direct delivery of NOM and AnOM to source water supplies, heavy rainfall events can also lead to the increased delivery of sediment and associated phosphorus to drinking water reservoirs [103]. As phosphorus is a key limiting nutrient for primary productivity in freshwaters, bioavailable particulate phosphorus forms that desorb from sediments to the water can promote the proliferation of algae and other microorganisms [104, 105] and associated AOM [20, 21, 23]. AOM can be a major source of DBP precursors, especially THMs and HAAs; it also contributes to the potential formation of haloacetonitriles (HANs) and haloacetamides (HAcAms) during chlorination [22]. Increased production of bacterial extracellular polymeric substances is also generally associated with increased potential to form DBPs [23].

5.3 *Drought*

Drought affects organic matter concentration, character, and the formation of disinfection by-product (DBP) precursors through reduced organic matter delivery to aquatic systems and increased retention time within them, which can at last initially enable more complete DOM degradation and mineralization, and alters DOM character [106, 107]. Increases in more recalcitrant and complex organic compounds are often observed in aquatic systems impacted by drought [107]. Lower molecular weight, more aliphatic DOM derived primarily from groundwater discharge to streams has been observed at lower flow periods, while higher molecular weight DOM has been observed at higher stream flows [108]. Accordingly, surface water DOM that is lower molecular weight, more aliphatic, and more hydrophilic, with lesser metal binding and hydrophobic organic carbon is expected during future summer drought periods [108]. While reduced connectivity between terrestrial and aquatic systems promotes greater influence from autochthonous sources of DOM, throughfall and decay processes remain relevant in the presence of vegetation, collectively leading to increased spatial heterogeneity in DOM sources [107]. In one investigation, higher levels of THM formation – associated with more hydrophobic and aromatic precursor materials – were observed during summer [109]. More frequent droughts may also shift dominant vegetation in peatlands to more drought-tolerant species; in one study, mild droughts contributed to an approximately 40% increase in production of DOC that was appreciably more difficult to remove by coagulation/flocculation [110]. Droughts can alter nutrient cycling and biota within both watersheds and reservoirs, thereby influencing water quality for periods of years or longer [111].

5.4 Wildfire

Wildfires are of particular concern because they rapidly consume biomass, alter the physical and chemical properties of upper soil layers, and transform vegetation into pyrogenic organic matter [112, 113]. Land cover is altered after severe wildfire (from forests and shrublands to barren and open areas) and more precipitation reaches the land surface [114], leading to increased erosion and solids runoff [64, 115]. Consequently, surface runoff regimes that strongly affect the quantity and quality of water exported from forested catchments are markedly altered [116]. Solid-associated metals [117, 118], nutrients [97, 119–121], and other contaminants (e.g., polycyclic aromatic hydrocarbons) [122–124] also can be elevated in wildfire-impacted waters. Wildfire transforms organic matter, resulting in the increased availability, mobilization, and delivery of lower molecular weight, more aromatic, and thus more difficult to coagulate post-fire DOC [97, 125–127]. In some cases, these effects can persist for a decade or longer [125, 128]. More aromatic DOC also tends to lead to greater formation of regulated and unregulated DBP precursors and increased potential to form THMs, HAAs, HANs, and chloropicrin, among others [113, 126, 129–131], with greater potential impacts from more severe wildfire [129]. In some areas, longer-term releases of sediment-associated bioavailable phosphorus to the water column [132, 133] and associated increases in primary productivity [119] have been observed after severe wildfires. Bench-scale investigations of interactions between wildfire ash and algae suggest that ash loads and bloom stage may affect DBP formation in source water [134].

5.5 Sea Level Rise

Sea level rise can lead to increases in saltwater intrusion into coastal aquifers and estuaries, resulting in higher salt concentrations in groundwater supplies [135]. In addition to aesthetic concerns, seawater contains higher concentrations of halides such as bromide and iodide than freshwater, resulting in increased bromide and iodide concentrations in affected source waters, increased potential to form DBPs, and changes in DBP speciation formed during drinking water treatment [73, 136, 137]. Increases in THM concentration and bromide substitution factor have been reported in association with increased saltwater intrusion [73]. Free halogen and reactive halogen species (RHSs) such as halogen radicals react with organic matter to produce brominated and iodinated DBPs [33, 34, 138]. DBP formation from NOM has been reported to relatively increase in the presence of iodide and decrease in the presence of bromide, while it decreases in the presence of halides when formed from AOM; however, DBP-associated cytotoxicity must also be considered as it is often highest in the presence of bromide [35].

6 Key Messages

- Warming temperatures and feedbacks between climate patterns and hydrological processes are leading to hydroclimatic change and climate shocks, including reduced glacial ice and snowpacks, increased heavy rainfall, increased drought, increased wildfire severity and extent, and sea level rise.
- Hydroclimatic change and climate shocks can alter or interact with key landscape characteristics, especially land cover, resulting in DBP precursor transport process responses that include increased erosion, increased slope failure, increased atmospheric deposition (dry and wet), and increased saltwater intrusion.
- By driving change in DBP precursor transport process responses, global warming alters the type and amount of DBP precursors (NOM, AnOM, halides, and AOM) delivered to water supplies and the associated potential for DBP formation during municipal drinking water treatment and distribution.

References

1. NOAA National Centers for Environmental Information (2016) What's the difference between climate and weather? Published online 9 Mar 2016, retrieved on 6 Mar 2024 from https://www.noaa.gov/explainers/what-s-difference-between-climate-and-weather
2. Kundzewicz ZW (2008) Climate change impacts on the hydrological cycle. Ecohydrol Hydrobiol 8(2–4):195–203
3. Emelko MB, Stone M (2024) Climate change threats to water quality and treatability. AWWA Water Sci. In review
4. Milly PCD, Betancourt J, Falkenmark M, Hirsch RM, Kundzewicz ZW, Lettenmaier DP, Stouffer RJ (2008) Stationarity is dead: whither water management? Science 319(5863): 573–574
5. AghaKouchak A, Chiang F, Huning LS, Love CA, Mallakpour I, Mazdiyasni O, Moftakhari H, Papalexiou SM, Ragno E, Sadegh M (2020) Climate extremes and compound hazards in a warming world. Annu Rev Earth Planet Sci 48:519–548
6. Clarke B, Otto F, Stuart-Smith R, Harrington L (2022) Extreme weather impacts of climate change: an attribution perspective. Environ Res Clim 1(1):012001
7. Trenberth KE (2011) Changes in precipitation with climate change. Climate Res 47(1–2): 123–138
8. Dai A, Zhao T, Chen J (2018) Climate change and drought: a precipitation and evaporation perspective. Curr Clim Chang Rep 4:301–312
9. Fernández-García V, Alonso-González E (2023) Global patterns and dynamics of burned area and burn severity. Remote Sens (Basel) 15(13):3401
10. Zhang X, Yu GQ, Li ZB, Li P (2014) Experimental study on slope runoff, erosion and sediment under different vegetation types. Water Resour Manag 28(9):2415–2433
11. Christopher SF, Mitchell MJ, McHale MR, Boyer EW, Burns DA, Kendall C (2008) Factors controlling nitrogen release from two forested catchments with contrasting hydrochemical responses. Hydrol Proces Int J 22(1):46–62
12. Groffman PM, Butterbach-Bahl K, Fulweiler RW, Gold AJ, Morse JL, Stander EK, Tague C, Tonitto C, Vidon P (2009) Challenges to incorporating spatially and temporally explicit phenomena (hotspots and hot moments) in denitrification models. Biogeochemistry 93:49–77

13. Ogden FL, Raj Pradhan N, Downer CW, Zahner JA (2011) Relative importance of impervious area, drainage density, width function, and subsurface storm drainage on flood runoff from an urbanized catchment. Water Resour Res 47(12)
14. Hatt BE, Fletcher TD, Walsh CJ, Taylor SL (2004) The influence of urban density and drainage infrastructure on the concentrations and loads of pollutants in small streams. Environ Manag 34:112–124
15. Croue JP, Korshin GV, Benjamin MM (2000) Characterization of natural organic matter in drinking water. American Water Works Association
16. Golea DM, Upton A, Jarvis P, Moore G, Sutherland S, Parsons SA, Judd SJ (2017) THM and HAA formation from NOM in raw and treated surface waters. Water Res 112:226–235
17. Westerhoff P, Mash H (2002) Dissolved organic nitrogen in drinking water supplies: a review. J Water Supply Res Technol AQUA 51(8):415–448
18. Bond T, Templeton MR, Graham N (2012) Precursors of nitrogenous disinfection by-products in drinking water – a critical review and analysis. J Hazard Mater 235–236:1–16
19. Shah AD, Mitch WA (2012) Halonitroalkanes, halonitriles, haloamides, and N-nitrosamines: a critical review of nitrogenous disinfection byproduct formation pathways. Environ Sci Technol 46(1):119–131
20. Hua LC, Lai CH, Wang GS, Lin TF, Huang C (2019) Algogenic organic matter derived DBPs: precursor characterization, formation, and future perspectives – a review. Crit Rev Environ Sci Technol 49(19):1803–1834
21. de Souza LL, Daniel LA, Bond T (2023) Algal organic matter as a disinfection by-product precursor during chlor(am)ination: a critical review. Environ Sci Water Res Technol 9(11): 2787–2802
22. Zhang Q, Liu B, Liu Y (2014) Effect of ozone on algal organic matters as precursors for disinfection by-products production. Environ Technol 35(14):1753–1759
23. Wang Z, Kim J, Seo Y (2012) Influence of bacterial extracellular polymeric substances on the formation of carbonaceous and nitrogenous disinfection byproducts. Environ Sci Technol 46(20):11361–11369
24. Pellerin BA, Wollheim WM, Hopkinson CS, McDowell WH, Williams MR, Vörösmarty CJ, Daley ML (2004) Role of wetlands and developed land use on dissolved organic nitrogen concentrations and DON/TDN in northeastern US rivers and streams. Limnol Oceanogr 49(4): 910–918
25. Liu Y, Evans SE, Friesen ML, Tiemann LK (2022) Root exudates shift how N mineralization and N fixation contribute to the plant-available N supply in low fertility soils. Soil Biol Biochem 165:108541
26. Villacorte LO, Ekowati Y, Neu TR, Kleijn JM, Winters H, Amy G, Schippers JC, Kennedy MD (2015) Characterisation of algal organic matter produced by bloom-forming marine and freshwater algae. Water Res 73:216–230
27. Paerl HW, Otten TG, Kudela R (2018) Mitigating the expansion of harmful algal blooms across the freshwater-to-marine continuum. Environ Sci Technol 52(10):5519–5529
28. Shon HK, Vigneswaran S, Snyder SA (2006) Effluent organic matter (EfOM) in wastewater: constituents, effects, and treatment. Crit Rev Environ Sci Technol 36(4):327–374
29. Krasner SW, Westerhoff P, Chen B, Rittmann BE, Nam SN, Amy G (2009) Impact of wastewater treatment processes on organic carbon, organic nitrogen, and DBP precursors in effluent organic matter. Environ Sci Technol 43(8):2911–2918
30. Liao Y, Ji W, Wang Z, Tian Y, Peng J, Li W, Pan Y, Li A (2024) Effects of alternative disinfection methods on the characteristics of effluent organic matter and the formation of disinfection byproducts. Environ Pollut 340:122796
31. Makepeace DK, Smith DW, Stanley SJ (1995) Urban stormwater quality: summary of contaminant data. Crit Rev Environ Sci Technol 25(2):93–139
32. Pamuru ST, Forgione E, Croft K, Kjellerup BV, Davis AP (2022) Chemical characterization of urban stormwater: traditional and emerging contaminants. Sci Total Environ 813:151887

33. Criquet J, Rodriguez EM, Allard S, Wellauer S, Salhi E, Joll CA, Von Gunten U (2015) Reaction of bromine and chlorine with phenolic compounds and natural organic matter extracts – electrophilic aromatic substitution and oxidation. Water Res 85:476–486
34. Criquet J, Allard S (2021) Influence of bromide and iodide on the formation of disinfection by-products in drinking water treatment. In: Comprehensive analytical chemistry 2021 Jan 1, vol 92. Elsevier, pp 117–138
35. Gao ZC, Lin YL, Xu B, Xia Y, Hu CY, Zhang TY, Qian H, Cao TC, Gao NY (2020) Effect of bromide and iodide on halogenated by-product formation from different organic precursors during UV/chlorine processes. Water Res 182:116035
36. Dillon PJ, Kirchner WB (1975) The effects of geology and land use on the export of phosphorus from watersheds. Water Res 9(2):135–148
37. Holloway JM, Dahlgren RA, Hansen B, Casey WH (1998) Contribution of bedrock nitrogen to high nitrate concentrations in stream water. Nature 395(6704):785–788
38. Poulsen DL, Simmons CT, Le Galle La Salle C, Cox JW (2006) Assessing catchment-scale spatial and temporal patterns of groundwater and stream salinity. Hydrgeol J 14:1339–1359
39. Lehmann J, Kleber M (2015) The contentious nature of soil organic matter. Nature 528(7580): 60–68
40. Wen H, Perdrial J, Abbott BW, Bernal S, Dupas R, Godsey SE, Harpold A, Rizzo D, Underwood K, Adler T, Sterle G, Li L (2020) Temperature controls production but hydrology regulates export of dissolved organic carbon at the catchment scale. Hydrol Earth Syst Sci 24(2):945–966
41. Kaiser K, Kalbitz K (2012) Cycling downwards – dissolved organic matter in soils. Soil Biol Biochem 52:29–32
42. Morison MQ, Higgins SN, Webster KL, Emilson EJS, Yao H, Casson NJ (2022) Spring coherence in dissolved organic carbon export dominates total coherence in boreal shield forested catchments. Environ Res Lett 17(1):014048
43. Vidon P, Wagner LE, Soyeux E (2008) Changes in the character of DOC in streams during storms in two Midwestern watersheds with contrasting land uses. Biogeochemistry 88(3): 257–270
44. Inamdar SP, Christopher SF, Mitchell MJ (2004) Export mechanisms for dissolved organic carbon and nitrate during summer storm events in a glaciated forested catchment in New York, USA. Hydrol Process 18(14):2651–2661
45. Minshall GW, Robinson CT, Lawrence DE (1997) Postfire responses of lotic ecosystems in Yellowstone National Park, USA. Can J Fish Aquat Sci 54(11):2509–2525
46. Kemp MJ, Dodds WK (2001) Spatial and temporal patterns of nitrogen concentrations in pristine and agriculturally-influenced prairie streams. Biogeochemistry 53:125–141
47. Follett RF, Delgado JA (2002) Nitrogen fate and transport in agricultural systems. J Soil Water Conserv 57(6):402–408
48. Bernhardt ES, Band LE, Walsh CJ, Berke PE (2008) Understanding, managing, and minimizing urban impacts on surface water nitrogen loading. Ann N Y Acad Sci 1134(1):61–96
49. Martin CW, Hornbeck JW, Likens GE, Buso DC (2000) Impacts of intensive harvesting on hydrology and nutrient dynamics of northern hardwood forests. Can J Fish Aquat Sci 57(S2): 19–29
50. Tetzlaff D, Malcolm IA, Soulsby C (2007) Influence of forestry, environmental change and climatic variability on the hydrology, hydrochemistry and residence times of upland catchments. J Hydrol 346(3–4):93–111
51. Issaka S, Ashraf MA (2017) Impact of soil erosion and degradation on water quality: a review. Geol Ecol Landscapes 1(1):1–11
52. van Beek R, Cammeraat E, Andreu V, Mickovski SB, Dorren L (2008) Hillslope processes: mass wasting, slope stability and erosion. In: Norris JE et al (eds) Slope stability and erosion control: ecotechnological solutions. Springer, Dordrecht
53. Greenwood JR, Norris JE, Wint J (2004) Assessing the contribution of vegetation to slope stability. Proc Inst Civ Eng Geotech Eng 157(4):199–207

54. Styczen ME, Morgan RP (2003) Engineering properties of vegetation. In: Slope stabilization and erosion control: a bioengineering approach. Taylor & Francis, pp 4–60
55. Fattet M, Fu Y, Ghestem M, Ma W, Foulonneau M, Nespoulous J, Le Bissonnais Y, Stokes A (2011) Effects of vegetation type on soil resistance to erosion: relationship between aggregate stability and shear strength. Catena 87(1):60–69
56. Chiarle M, Iannotti S, Mortara G, Deline P (2007) Recent debris flow occurrences associated with glaciers in the Alps. Global Planet Change 56(1–2):123–316
57. Kos A, Amann F, Strozzi T, Delaloye R, von Ruette J, Springman S (2016) Contemporary glacier retreat triggers a rapid landslide response, Great Aletsch Glacier, Switzerland. Geophys Res Lett 43(24):12–466
58. Deline P, Gruber S, Delaloye R, Fischer L, Geertsema M, Giardino M, Hasler A, Kirkbride M, Krautblatter M, Magnin F, McColl S (2015) Ice loss and slope stability in high-mountain regions. In: Snow and ice-related hazards, risks, and disasters. Academic Press, pp 521–561
59. Zhang LL, Zhang J, Zhang LM, Tang WH (2011) Stability analysis of rainfall-induced slope failure: a review. Proc Inst Civ Eng Geotech Eng 164(5):299–316
60. Legates DR, Mahmood R, Levia DF, DeLiberty TL, Quiring SM, Houser C, Nelson FE (2011) Soil moisture: a central and unifying theme in physical geography. Prog Phys Geogr 35(1): 65–86
61. Elosegi A, Díez J, Mutz M (2010) Effects of hydromorphological integrity on biodiversity and functioning of river ecosystems. Hydrobiologia 657:199–215
62. Neary DG, Ice GG, Jackson CR (2009) Linkages between forest soils and water quality and quantity. For Ecol Manage 258(10):2269–2281
63. Scott DF, Van Wyk DB (1990) The effects of wildfire on soil wettability and hydrological behaviour of an afforested catchment. J Hydrol 121(1–4):239–256
64. Moody JA, Shakesby RA, Robichaud PR, Cannon SH, Martin DA (2013) Current research issues related to post-wildfire runoff and erosion processes. Earth Sci Rev 122:10–37
65. Zhang Q, Huang J, Wang F, Mark L, Xu J, Armstrong D, Li C, Zhang Y, Kang S (2012) Mercury distribution and deposition in glacier snow over western China. Environ Sci Technol 46(10):5404–5413
66. Kang S, Zhang Y, Qian Y, Wang H (2020) A review of black carbon in snow and ice and its impact on the cryosphere. Earth Sci Rev 210:103346
67. Galloway JN, Thornton JD, Norton SA, Volchok HL, McLean RA (1982) Trace metals in atmospheric deposition: a review and assessment. Atmos Environ (1967) 16(7):1677–1700
68. Iavorivska L, Boyer EW, DeWalle DR (2016) Atmospheric deposition of organic carbon via precipitation. Atmos Environ 146:153–163
69. Ackerman D, Millet DB, Chen X (2019) Global estimates of inorganic nitrogen deposition across four decades. Global Biogeochem Cycles 33(1):100–107
70. Martin P, Bianchi TS (2023) Organic carbon cycling and transformation. Reference module in earth systems and environmental sciences. Elsevier. ISBN 9780124095489
71. Duce RA, Winchester JW, Van Nahl TW (1965) Iodine, bromine, and chlorine in the Hawaiian marine atmosphere. J Geophys Res 70(8):1775–1799
72. Li T, Chen H, Fung JC, Chan DH, Alfred LC, Leung KK, Yu JZ (2023) Large presence of bromine and toxic metals in ambient fine particles from urban fires. Atmos Environ 295: 119554
73. Kolb C, Pozzi M, Samaras C, VanBriesen JM (2017) Climate change impacts on bromide, trihalomethane formation, and health risks at coastal groundwater utilities. ASCE-ASME J Risk Uncertainty in Eng Syst A Civil Eng 3(3):04017006
74. Jeen SW, Kang J, Jung H, Lee J (2021) Review of seawater intrusion in western coastal regions of South Korea. Water 13(6):761
75. D'Andrilli J, Romero CM, Zito P, Podgorski DC, Payn RA, Sebestyen SD, Zimmerman AR, Rosario-Ortiz FL (2023) Advancing chemical lability assessments of organic matter using a synthesis of FT-ICR MS data across diverse environments and experiments. Org Geochem 184:104667

76. Pautler BG (2013) Novel analytical approaches for the characterization of natural organic matter in the cryosphere and its potential impacts on climate change. PhD thesis, University of Toronto, Chemistry Department
77. Zhai H, Cheng S, Zhang L, Luo W, Zhou Y (2022) Formation characteristics of disinfection byproducts from four different algal organic matter during chlorination and chloramination. Chemosphere 308:136171
78. Du Z, Ding S, Xiao R, Fang C, Song W, Jia R, Chu W (2022) Does snowfall introduce disinfection by-product precursors to surface water? Environ Sci Technol 56(20): 14487–14497
79. Aiken G, Cotsaris E (1995) Soil and hydrology: their effect on NOM. J Am Water Works Ass 87(1):36–45
80. Chen H, Uzun H, Chow AT, Karanfil T (2020) Low water treatability efficiency of wildfire-induced dissolved organic matter and disinfection by-product precursors. Water Res 184: 116111
81. Yuan Y, Li Q, Deng J, Ma X, Liao X, Zou J, Li G, Chen G, Dai H (2024) Rainwater extracting characteristics and its potential impact on DBPs generation: a case study. Sci Total Environ 906:167282
82. Altieri KE, Turpin BJ, Seitzinger SP (2009) Oligomers, organosulfates, and nitrooxy organosulfates in rainwater identified by ultra-high resolution electrospray ionization FT-ICR mass spectrometry. Atmos Chem Phys 9(7):2533–2542
83. Santos PSM, Santos EBH, Duarte AC (2013) Seasonal and air mass trajectory effects on dissolved organic matter of bulk deposition at a coastal town in South-Western Europe. Environ Sci Pollut Res 20(1):227–237
84. Mead RN, Mullaugh KM, Avery GB, Kieber RJ, Willey JD, Podgorski DC (2013) Insights into dissolved organic matter complexity in rainwater from continental and coastal storms by ultrahigh resolution Fourier transform ion cyclotron resonance mass spectrometry. Atmos Chem Phys 13(9):4829–4838
85. Mead RN, Felix JD, Avery GB, Kieber RJ, Willey JD, Podgorski DC (2015) Characterization of CHOS compounds in rainwater from continental and coastal storms by ultrahigh resolution mass spectrometry. Atmos Environ 105:162–168
86. Jung BJ, Lee JK, Kim H, Park JH (2014) Export, biodegradation, and disinfection byproduct formation of dissolved and particulate organic carbon in a forested headwater stream during extreme rainfall events. Biogeosciences 11(21):6119–6129
87. Pifer AD, Cousins SL, Fairey JL (2014) Assessing UV-and fluorescence-based metrics as disinfection byproduct precursor surrogate parameters in a water body influenced by a heavy rainfall event. J Water Supply Res Technol AQUA 63(3):200–211
88. Bao HY, Yi YY, Wang C, Spencer RGM, Deng X, Guo WD (2018) Dissolved organic matter in coastal rainwater – concentration, bioavailability and depositional flux to seawater in southeastern China. Mar Chem 205:48–55
89. Lee MH, Lee YK, Derrien M, Choi K, Shin KH, Jang KS, Hur J (2019) Evaluating the contributions of different organic matter sources to urban river water during a storm event via optical indices and molecular composition. Water Res 165:115006
90. Shi Y, Zhang L, Li Y, Zhou L, Zhou Y, Zhang Y, Huang C, Li H, Zhu G (2020) Influence of land use and rainfall on the optical properties of dissolved organic matter in a key drinking water reservoir in China. Sci Total Environ 699:134301
91. Hou M, Chu W, Wang F, Deng Y, Gao N, Zhang D (2018) The contribution of atmospheric particulate matter to the formation of CX3R-type disinfection by-products in rainwater during chlorination. Water Res 145:531–540
92. Liu Z, Lin YL, Chu WH, Xu B, Zhang TY, Hu CY, Cao TC, Gao NY, Dong CD (2020) Comparison of different disinfection processes for controlling disinfection by-product formation in rainwater. J Hazard Mater 385:121618

93. Kozak C, Leithold J, do Prado LL, Knapik HG, de Rodrigues Azevedo JC, Braga SM, Fernandes CV (2021) Adaptive monitoring approach to assess dissolved organic matter dynamics during rainfall events. Environ Monit Assess 193(7):423
94. Volk C, Wood L, Johnson B, Robinson J, Zhu HW, Kaplan L (2002) Monitoring dissolved organic carbon in surface and drinking waters. J Environ Monit 4(1):43–47
95. Delpla I, Rodriguez MJ (2017) Variability of disinfection by-products at a full-scale treatment plant following rainfall events. Chemosphere 166:453–462
96. Delpla I, Baurès E, Jung AV, Thomas O (2011) Impacts of rainfall events on runoff water quality in an agricultural environment in temperate areas. Sci Total Environ 409(9):1683–1688
97. Emelko MB, Silins U, Bladon KD, Stone M (2011) Implications of land disturbance on drinking water treatability in a changing climate: demonstrating the need for "source water supply and protection" strategies. Water Res 45(2):461–472
98. Liu C, Li Z, Berhe AA, Xiao H, Liu L, Wang D, Peng H, Zeng G (2019) Characterizing dissolved organic matter in eroded sediments from a loess hilly catchment using fluorescence EEM-PARAFAC and UV–visible absorption: insights from source identification and carbon cycling. Geoderma 334:37–48
99. Raymond PA (2005) The composition and transport of organic carbon in rainfall: insights from the natural (13C and 14C) isotopes of carbon. Geophys Res Lett 32(14)
100. Khamis K, Bradley C, Hannah DM (2020) High frequency fluorescence monitoring reveals new insights into organic matter dynamics of an urban river, Birmingham, UK. Sci Total Environ 710:135668
101. Kalev S, Toor GS (2020) Concentrations and loads of dissolved and particulate organic carbon in urban stormwater runoff. Water 12(4):1031
102. Majidzadeh H, Uzun H, Ruecker A, Miller D, Vernon J, Zhang H, Bao S, Tsui MTK, Karanfil T, Chow AT (2017) Extreme flooding mobilized dissolved organic matter from coastal forested wetlands. Biogeochemistry 136(3):293–309. https://doi.org/10.1007/s10533-017-0394-x
103. Paerl HW, Rudek J, Mallin MA (1990) Stimulation of phytoplankton production in coastal waters by natural rainfall inputs: nutritional and trophic implications. Mar Biol 107:247–254
104. Paerl HW (2009) Controlling eutrophication along the freshwater–marine continuum: dual nutrient (N and P) reductions are essential. Estuar Coasts 32:593–601
105. Watt C, Emelko MB, Silins U, Collins AL, Stone M (2021) Anthropogenic and climate-exacerbated landscape disturbances converge to alter phosphorus bioavailability in an oligotrophic river. Water 13(22):3151
106. Tranvik LJ, Downing JA, Cotner JB, Loiselle SA, Striegl RG, Ballatore TJ, Dillon P, Finlay K, Fortino K, Knoll LB, Kortelainen PL, Kutser T, Larsen S, Laurion I, Leech DM, McCallister SL, McKnight DM, Melack JM, Overholt E, Porter JA, Prairie Y, Renwick WH, Roland F, Sherman BS, Schindler DW, Sobek S, Tremblay A, Vanni MJ, Verschoor AM, von Wachenfeldt E, Weyhenmeyer GA (2009) Lakes and reservoirs as regulators of carbon cycling and climate. Limnol Oceanogr 54(6 part 2):2298–2314
107. Casas-Ruiz JP, Tittel J, von Schiller D, Catalán N, Obrador B, Gómez-Gener L, Zwirnmann E, Sabater S, Marcé R (2016) Drought-induced discontinuities in the source and degradation of dissolved organic matter in a Mediterranean river. Biogeochemistry 127:125–139
108. Maurice PA, Cabaniss SE, Drummond J, Ito E (2002) Hydrogeochemical controls on the variations in chemical characteristics of natural organic matter at a small freshwater wetland. Chem Geol 187(1–2):59–77
109. Kalankesh LR, Zazouli MA, Susanto H, Babanezhad E (2021) Variability of TOC and DBPs (THMs and HAA5) in drinking water sources and distribution system in drought season: the North Iran case study. Environ Technol 42(1):100–113
110. Ritson JP, Brazier RE, Graham NJ, Freeman C, Templeton MR, Clark JM (2017) The effect of drought on dissolved organic carbon (DOC) release from peatland soil and vegetation sources. Biogeosciences 14(11):2891–2902

111. Wright B, Stanford BD, Reinert A, Routt JC, Khan SJ, Debroux JF (2014) Managing water quality impacts from drought on drinking water supplies. J Water Supply Res Technol AQUA 63(3):179–188
112. Gonzalez-Perez JA, Gonzalez-Vila FJ, Almendros G, Knicker H (2004) The effect of fire on soil organic matter – a review. Environ Int 30(6):855–870
113. Chen H, Uzun H, Tolić N, Chu R, Karanfil T, Chow AT (2022) Molecular transformation of dissolved organic matter during the processes of wildfire, alum coagulation, and disinfection using ESI (−) and ESI (+) FT-ICR MS. ACS ES&T Water 3(8):2571–2580
114. Williams CH, Silins U, Spencer SA, Wagner MJ, Stone M, Emelko MB (2019) Net precipitation in burned and unburned subalpine forest stands after wildfire in the northern Rocky Mountains. Int J Wildland Fire 28(10):750–760
115. Vahedifard F, Abdollahi M, Leshchinsky BA, Stark TD, Sadegh M, AghaKouchak A (2024) Interdependencies between wildfire-induced alterations in soil properties, near-surface processes, and geohazards. Earth Space Sci 11(2):e2023EA003498
116. Cuo L (2016) Land use/cover change impacts on hydrology in large river basins: a review. In: Tang Q, Oki T (eds) Terrestrial water cycle and climate change: natural and human-induced impacts, vol 26, pp 103–134
117. Pinedo-Gonzalez P, Hellige B, West AJ, Sañudo-Wilhelmy SA (2017) Changes in the size partitioning of metals in storm runoff following wildfires: implications for the transport of bioactive trace metals. Appl Geochem 83:62–71
118. Magliozzi LJ, Matiasek SJ, Alpers CN, Korak JA, McKnight D, Foster AL, Ryan JN, Roth DA, Ku P, Tsui MT, Chow AT (2024) Wildland–urban interface wildfire increases metal contributions to stormwater runoff in paradise, California. Environ Sci Processes Impacts 26(4):667–685
119. Silins U, Bladon KD, Kelly EN, Esch E, Spence JR, Stone M, Emelko MB, Boon S, Wagner MJ, Williams CHS, Tichkowsky I (2014) Five-year legacy of wildfire and salvage logging impacts on nutrient runoff and aquatic plant, invertebrate, and fish productivity. Ecohydrology 7(6):1508–1523
120. Uzun H, Dahlgren RA, Olivares C, Erdem CU, Karanfil T, Chow AT (2020) Two years of post-wildfire impacts on dissolved organic matter, nitrogen, and precursors of disinfection by-products in California stream waters. Water Res 181:115891
121. Gustine RN, Hanan EJ, Robichaud PR, Elliot WJ (2022) From burned slopes to streams: how wildfire affects nitrogen cycling and retention in forests and fire-prone watersheds. Biogeochemistry 157(1):51–68
122. Crouch RL, Timmenga HJ, Barber TR, Fuchsman PC (2006) Post-fire surface water quality: comparison of fire retardant versus wildfire-related effects. Chemosphere 62(6):874–889
123. Mansilha C, Carvalho A, Guimarães P, Espinha MJ (2014) Water quality concerns due to forest fires: polycyclic aromatic hydrocarbons (PAH) contamination of groundwater from mountain areas. J Toxic Environ Health A 77(14–16):806–815
124. Chen H, Chow AT, Li XW, Ni HG, Dahlgren RA, Zeng H, Wang JJ (2018) Wildfire burn intensity affects the quantity and speciation of polycyclic aromatic hydrocarbons in soils. ACS Earth Space Chem 2(12):1262–1270
125. Chow AT, Tsai KP, Fegel TS, Pierson DN, Rhoades CC (2019) Lasting effects of wildfire on disinfection by-product formation in forest catchments. J Environ Qual 48(6):1826–1834
126. Hohner AK, Rhoades CC, Wilkerson P, Rosario-Ortiz FL (2019) Wildfires alter forest watersheds and threaten drinking water quality. Acc Chem Res 52(5):1234–1244
127. Blackburn EAJ, Dickson-Anderson SE, Anderson WB, Emelko MB (2023) Biological filtration is resilient to wildfire ash-associated organic carbon threats to drinking water treatment. ACS ES&T Water 3(3):639–649
128. Rhoades CC, Chow AT, Covino TP, Fegel TS, Pierson DN, Rhea AE (2019) The legacy of a severe wildfire on stream nitrogen and carbon in headwater catchments. Ecosystems 22:643–657

129. Wang JJ, Dahlgren RA, Erşan MS, Karanfil T, Chow AT (2016) Temporal variations of disinfection byproduct precursors in wildfire detritus. Water Res 99:66–73
130. Hickenbottom K, Pagilla K, Hanigan D (2023) Wildfire impact on disinfection byproduct precursor loading in mountain streams and rivers. Water Res 244:120474
131. Li Z, Samonte PR, Cao H, Miesel JR, Xu W (2023) Assess the formation of disinfection by-products from pyrogenic dissolved organic matter (pyDOM): impact of wildfire on the water quality of forest watershed. Sci Total Environ 898:165496
132. Stone M, Emelko MB, Droppo IG, Silins U (2011) Biostabilization and erodibility of cohesive sediment deposits in wildfire-affected streams. Water Res 45(2):521–534
133. Emelko MB, Stone M, Silins U, Allin D, Collins AL, Williams CHS, Martens AM, Bladon KD (2016) Sediment-phosphorus dynamics can shift aquatic ecology and cause downstream legacy effects after wildfire in large river systems. Glob Chang Biol 22(3):1168–1184
134. Tsai KP, Uzun H, Karanfil T, Chow AT (2017) Dynamic changes of disinfection byproduct precursors following exposures of Microcystis aeruginosa to wildfire ash solutions. Environ Sci Technol 51(15):8272–8282
135. Werner AD, Simmons CT (2009) Impact of sea-level rise on sea water intrusion in coastal aquifers. Groundwater 47(2):197–204
136. Ged EC, Boyer TH (2014) Effect of seawater intrusion on formation of bromine-containing trihalomethanes and haloacetic acids during chlorination. Desalination 345:85–93
137. Chowdhury S (2022) Effects of seawater intrusion on the formation of disinfection byproducts in drinking water. Sci Total Environ 827:154398
138. Wang JJ, Jiao Y, Rhew RC, Chow AT (2016) Haloform formation in coastal wetlands along a salinity gradient at South Carolina, United States. Environ Chem 13(4):745–756

Disinfection Byproducts in the Energy Sector

H. Larry Tang and Yuefeng F. Xie

Contents

1 DBPs Intertwined with Shale Gas Extraction 378
1.1 Environmental Risks of Shale Gas Development 380
1.2 Principles of DBP Formation as a Result of Shale Gas Development 380
1.3 DBP Precursors Research Intertwined with Shale Gas Development 382
1.4 DBP Formation Research Under the Influence of Fracking Wastewater Discharges 383
1.5 Advances in Fracking Wastewater Treatment Strategies for Reduction of DBP Formation 384
1.6 Outlook on DBPs in the Shale Gas Sector 386
2 DBPs Intertwined with Coal-Fired Power Plants 386
2.1 Coal Quality and Pollution Control Technologies 387
2.2 Origin of DBP Concerns in Coal-Fired Power Plant Sector 387
2.3 Advances in Reduction of DBP Formation 388
2.4 Outlook on DBPs in the Coal-Fired Power Plant Sector 389
3 Concluding Remarks 389
References 390

Abstract This chapter explores the intricate relationship between energy production activities, particularly in the shale gas and coal-fired power plant sectors, and the formation of disinfection byproducts (DBPs) in water treatment facilities. Highlighting recent studies and technological advancements, it delves into the sources of DBPs, emphasizing the role of both organic and inorganic DBP precursors from industrial discharges. Through a comprehensive analysis, the text offers insights into the environmental impact of these discharges, the challenges of managing them, and

H. Larry Tang (✉)
Environmental Engineering Program, Indiana University of Pennsylvania, Indiana, PA, USA
e-mail: htang@iup.edu

Y. F. Xie
Department of Civil, Construction and Environmental Engineering, The Pennsylvania State University, Middletown, PA, USA
e-mail: yxx4@psu.edu

Chao Chen, Susan Andrews, and Yuefeng Xie (eds.), *Disinfection By-Products in Water*, Hdb Env Chem (2024) 135: 377–392, DOI 10.1007/698_2024_1151,

Published online: 27 September 2024

the effectiveness of current and proposed treatment strategies to mitigate their presence in drinking water.

Keywords Bromide, Coal-fired power plant, DBP precursors, DBP reduction, Shale gas development

Water quality in the energy sector is a critical area of concern, encompassing various aspects of water utilization, contamination, and treatment within energy production processes. The energy sector, which includes oil and gas extraction, coal-fired power plants, and renewable energy sources, heavily relies on water for various operations, thus making water quality management an essential aspect of sustainable energy production. For instance, in hydraulic fracturing, water mixed with various chemicals is used to extract oil and gas, while in coal-fired power plants, water is crucial for cooling and emission control operations. The interaction of these processes with water often leads to changes in water quality, presenting challenges for both the industry and the environment. Key issues raised in this area include the sources and types of contaminants commonly associated with energy production, such as heavy metals, radioactive materials, and chemical additives. These contaminants can have significant impacts on the environment and human health, necessitating effective monitoring and management strategies. In this chapter, however, we will delve into an emerging area in the energy sector concerning the complexity intertwined with the formation and control of disinfection byproducts (DBPs).

1 DBPs Intertwined with Shale Gas Extraction

Shale gas, a natural gas trapped within shale formation, has transformed the energy landscape, with its exploration becoming a game-changer in the global energy sector. Although the first recorded shale gas well was in Fredonia, New York, the United States (U.S.) in 1821, the real breakthrough came much later in the late twentieth century. George P. Mitchell, often referred to as the "father of fracking," combined horizontal drilling with hydraulic fracturing, revolutionizing the extraction of shale gas.

Hydraulic fracturing, also known as fracking, involves injecting a high-pressure mixture of water, sand, and chemical additives into a well drilled deep into the earth. This well often reaches depths of 1 to 3 miles to the shale formation that contains natural gas or oil. The pressure applied exceeds the rock strength, causing fractures in the shale formation. The sand or other proppants in the fracturing fluids enter these fractures, preventing them from closing once the injection pressure is released. The internal pressure of the geological formation then forces the gas or oil to flow from the rock pores, through the fractures, into the wellbore. This gas or oil is then pumped out of the well and processed. Unlike completely vertical wells, which access only a small vertical section of the shale, horizontal drilling – initiated after reaching the shale formation – allows for a greater portion of the wellbore to be in contact with the gas or oil. This makes the extraction process more efficient and cost-effective. Figure 1 is a schematic illustration that helps to understand this process.

The U.S. currently leads globally in shale gas exploration, largely due to its extensive shale formations (Fig. 2). The most notable shale gas plays in the

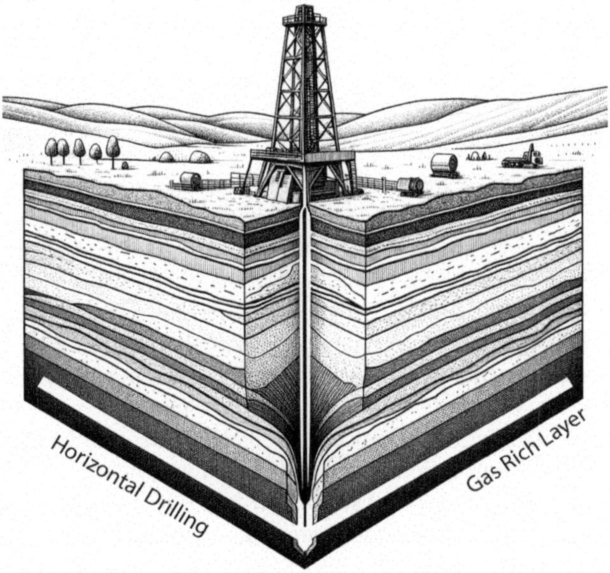

Fig. 1 Illustration of hydraulic fracturing that involves vertical drilling into a gas-rich layer followed by horizontal drilling

Fig. 2 Spatial distribution of shale plays in the lower 48 states of the U.S.

U.S. include the Marcellus Shale in the Northeast, the Barnett Shale in Texas, and the Bakken Formation in North Dakota. Each of these plays has significantly contributed to the U.S. natural gas supply. The Barnett Shale in Texas played a pivotal role in the early 2000s, serving as the testing ground for fracking. Its success led to a surge in exploration across various states. While shale gas exploration has contributed to job creation and economic growth, its environmental risks must be carefully weighed against the potential energy benefits.

1.1 Environmental Risks of Shale Gas Development

In fact, the development of shale gas has sparked considerable controversy. Concerns center on the potential risks to public health and the environment stemming from water use and water quality, air quality, seismic activity, land use, habitat disruption, and water and soil contamination, as well as broader community and health impacts [1, 2]. Should the sealing of pipes prove inadequate, gas and drilling fluids may be released into the environment, leading to groundwater contamination around fracking sites. Additionally, drinking water resources could be affected by the improper treatment and discharge of wastewater produced during and after fracking. The process itself demands significant water resources, averaging four million gallons per well, placing a heavy burden on water-scarce regions [3]. Methane gas leaks from ill-fitted pipes contribute to the greenhouse effect and raise the risk of dangerous explosions, adversely affecting air quality. Moreover, prevailing theories suggest a causal link between fracking and earthquakes, indicating that seismic activities near fracking sites are a non-negligible concern. Another pressing issue is the management of wastewater from fracking, which contains high concentrations of both natural and added chemicals that overwhelm existing treatment technologies, posing further risks to the environment and public health if not disposed of correctly [4–7]. In this chapter, we will focus on the environmental risks originating from water-related issues due to the formation of DBPs.

1.2 Principles of DBP Formation as a Result of Shale Gas Development

DBPs are a group of undesirable compounds that forms when disinfectants react with naturally occurring organic and inorganic matter in water. Their exposure has been linked to health issues, including bladder cancer, reproductive problems, and developmental defects. However, eliminating the disinfection step is not feasible, as it is a necessary public health measure to eliminate pathogenic microorganisms. The most common DBPs are trihalomethanes (THMs) and haloacetic acids (HAAs), along with others such as halonitromethanes and haloacetonitriles. These compounds differ in their chemical structures, formation pathways, and health implications. Given that wastewater generated from shale gas development is a source of groundwater and surface water contamination, it is critical to examine how fracking wastewater influences the formation of DBPs.

Fracking results in two types of wastewater: flowback water and produced water, differentiated mainly by the duration they spend in the well. Flowback water emerges in the initial weeks following fracking, whereas produced water is the wastewater that surfaces throughout the shale well's operational life. While a thorough chemical analysis can aid in differentiating them, pinpointing the exact transition from flowback to produced water is often challenging. Consequently, it is a typical practice to report the total volume of these waters combined, without distinct separation between the two types. In this chapter, we do not differentiate between these two types of wastewater and refer to them collectively as fracking wastewater.

FracFocus is a database where shale gas companies can report information about their fracking operations in the U.S. This platform provides public access to details such as water volume and chemicals added, which could end up in fracking wastewater, with disclosures from over 1,600 companies. Launched in 2010 as a joint project by the Ground Water Protection Council and the Interstate Oil and Gas Compact Commission, the system has been successful from the perspectives of both operators and consumers. As of this writing, 27 U.S. states now either require or allow companies to report their fracking data. The current version, FracFocus 3.0, released in 2016, features newly designed forms that improve the user experience in both reviewing and completing disclosures.

It is known that natural organic matter (NOM) commonly acts as an organic precursor in DBP formation. Its complex nature and reactivity with disinfectants require careful water treatment strategies to minimize the formation of DBPs. Thousands of chemicals, including volatile organic compounds (VOCs) and polycyclic aromatic hydrocarbons (PAHs), have been used in fracking or detected in fracking wastewater, and many of these chemicals are also organic precursors to DBPs [8–12].

While much attention has been given to organic precursors in DBP formation, inorganic precursors like bromide and iodide also play a crucial role [13]. Understanding their impact is vital for effective water treatment and minimizing potential health risks, as those brominated and iodinated byproducts are often more toxic than their chlorinated counterparts. Fracking wastewater generally has high levels of total dissolved solids (TDS), with values ranging up to hundreds of thousands of mg/L. Although TDS are typically dominated by sodium and chloride, elevated concentrations of bromide and iodide have also been noted, contributing to the formation of more cytotoxic and genotoxic brominated and iodinated DBPs.

1.3 DBP Precursors Research Intertwined with Shale Gas Development

Hoelzer et al. identified hydrocarbons (such as alkanes, alkenes, and aromatic compounds) as the most abundant compounds in their analysis of fracking wastewater [14]. This finding aligns with expectations, as these hydrocarbons are either openly disclosed as additives by shale gas companies or are naturally occurring in geological formations. Nevertheless, they also detected smaller quantities of substances with functional groups, including carbonyl compounds, alcohols, halogenated compounds, carboxylic acids, ethers, epoxides, and others like nitriles and siloxanes. Many of these compounds are not typically reported in shale formations or found in crude oil extracts, suggesting an anthropogenic origin. Given that groundwater and surface water contamination by fracking wastewater does not appear unlikely and considering that in Colorado alone there were 591 documented spills in 2013 [15], it is evident that fracking wastewater contributes to the proliferation of these organic DBP precursors [16, 17].

In addition to hydrocarbons, biocides are among the most common chemical additives used in fracking, with the total amount of biocides per fracking event often exceeding 1,000 gal [15]. Biocides are generally divided into two categories based on their methods of bacterial control: oxidizing and non-oxidizing compounds. Oxidizing biocides, such as sodium hypochlorite, chlorine dioxide, and peroxides, function by releasing free radical species that attack cellular components. These reactions with fracking chemicals may directly produce DBPs. For instance, peracetic acid, found in a small fraction of wells registered on FracFocus, is a stronger oxidizing biocide than H_2O_2 and has been identified as contributing to the formation of halogenated DBPs [18]. However, due to the short-lived nature of oxidizing biocides in controlling microbial growth, non-oxidizing biocides are more frequently used in fracking fluids. An example is cationic quaternary ammonium/amine compounds (QACs), a type of nonoxidizing biocide, which are injected into 22% of all wells registered on FracFocus and used in nearly every shale formation in the U.S. [15]. These compounds also act as precursors for DBP formation [19]. The potential for DBP formation due to the use of these biocides varies, and it has been suggested that they be categorized differently based on the outcomes of environmental risk assessments of DBPs [20]. Furthermore, as biocides are susceptible to various transformation and degradation processes, the impacts of the degradation products on DBP formation cannot be ignored. Under harsh conditions (i.e., high temperature, pressure, and salinity), the formation of DBPs could also be facilitated in underground reservoirs, and their concentrations could reach orders of magnitudes higher than in surface water.

Bromide and iodide, as inorganic DBP precursors of concern, prompt interest in their concentrations in fracking wastewater. Braun authored a report for the Ohio River Valley Water Sanitation Commission on the state of fracking in the Ohio River Basin, compiling the minimum and maximum concentrations of various geochemical constituents in fracking wastewater [21]. According to this report, bromide concentrations ranged from non-detectable to 10,600 mg/L. Hansen et al. also noted that bromide concentrations in Marcellus fracking wastewater varied from 185 to 1,190 mg/L in samples taken 5 days post-fracking, and from 15.8 to 1,600 mg/L in samples taken 14 days post-fracking [22]. Given that the USEPA's maximum contaminant level for bromide is 0.1 mg/L, a significant amount of water is theoretically required to dilute this pollution to meet drinking water standards. This theoretical requirement is referred to as the "grey water footprint," a term used in evaluating environmental sustainability. Additionally, Harkness et al. analyzed bromide and iodide in fracking wastewater from Marcellus and Fayetteville shales, finding iodide concentrations up to 56 mg/L [23]. A discharged effluent at a spill site in West Virginia even showed iodide levels as high as 28 mg/L. Clearly, the discharge and accidental spills of fracking wastewater elevate the levels of these inorganic DBP precursors, posing risks to both human health and the environment.

1.4 DBP Formation Research Under the Influence of Fracking Wastewater Discharges

After the pioneering research by Rossi et al. in 2011 [24] and Parker et al. in 2014 [25], subsequent notable research on the impact of fracking wastewater discharges as sources of DBPs in receiving streams was conducted by Hladik et al. in 2015 [26]. They collected water samples from four sites in Pennsylvania along a large river. These included an upstream site, a site downstream from a wastewater treatment plant that does not accept fracking wastewater, a site downstream from a commercial wastewater treatment plant that accepts fracking wastewater, and a site further downstream from both wastewater treatment plants. They discovered that the DBP profiles at the outfalls of the two treatment plants differed significantly, with brominated DBPs found at much higher concentrations at the commercial plant.

Pursuing the exploration of fracking wastewater contamination impacts on drinking water sources, Huang et al. visualized these effects by blending fracking wastewater with natural waters at varying ratios to simulate different degrees of contamination [27]. They found that as the blend ratio increased, the non-brominated DBP species exhibited a consistent decrease, while the fully brominated species showed a steady increase. The partially brominated species, however, displayed a fluctuating pattern. These findings confirmed the impact of increased bromide levels on DBP speciation. The study also highlighted the significance of background NOM on DBP formation, as the three different natural waters, when mixed with fracking wastewater at identical ratios, produced varied DBP profiles upon chlorination. Given that NOM is a complex mixture of hydrophilic and hydrophobic organic materials varying in size, functional group, and reactivity, its interplay with the constituents of fracking wastewater remains intricate.

While most studies focused on the significant impact of bromide due to its elevated concentrations in fracking wastewater, Huang et al. embarked on an investigation from a non-bromide perspective [28]. This approach specifically examined the impacts of metal ions, as ions such as calcium, magnesium, barium, and strontium are also key constituents of fracking wastewater and present in very high concentrations. They first removed bromide from the Fracking wastewater through ozonation and air purging to eliminate its known effects, and then spiked the debrominated fracking wastewater at different ratios into water samples simulated with individual organic precursors to simplify the analysis. They discovered that the complexation between these metal ions and the organic precursors played a significant role, enhancing DBP formation upon chlorination. A higher efficiency of Mg-Citrate chelates was inferred compared to Ca-Citrate in facilitating DBP yields. The impact of anionic species was also explored, revealing that sulfate played a more notable role in enhancing DBP formation than other anions.

1.5 Advances in Fracking Wastewater Treatment Strategies for Reduction of DBP Formation

While the above-mentioned ozonation following by air purging is available as a viable approach to reduce the impact of high bromide levels in fracking wastewater [27], more treatment strategies are present. Considering the promotional effects of metals and halides on DBP formation, salinity removal technologies such as ion exchange, lime-soda ash softening, electrodialysis, nanofiltration, and reverse osmosis should have positive effects on eliminating the deleterious effects of fracking wastewater.

If the target is to prevent bromide from entering waterways, an electrolysis method could be used to remove bromide from complex brines, which was proposed and tested in Sun et al.'s paper [29]. In this process, the bromide-containing brine undergoes electrolysis in a cell where the anode and cathode chambers are separated by a cation-permeable membrane. At the anode, negatively charged organics and bromide are oxidized. The produced bromine by bromide oxidation is then extracted from the electrolyte by air and collected in separate containers after being reduced by potassium iodide. Concurrently, hydrogen gas is formed at the cathode through the reduction of protons. This method could be effective for treating brines, as the high salinity naturally provides an electrolyte for the electrochemical reaction, thereby reducing energy consumption. However, when applying this selective bromide removal technique to brines containing chloride, a significant engineering challenge arises: the standard oxidizing potential for bromide is only 0.27 V lower than that for chloride. Given that chloride is present in high concentrations in brine but does not pose a health risk, precise control of the electrical potential is crucial to ensure the oxidation targets only bromide.

This electrolysis method, capable of modifying bromide and organics at the anode and influencing the formation of DBPs during chlorination of natural waters, led us to propose aerated electrolysis as an in-situ treatment strategy for controlling DBPs in fracking wastewater-affected water sources [30]. Each experiment began with 500 mL of fracking wastewater. We set the initial current at 600 mA, while maintaining the voltage at 4.5 V for synthetic fracking wastewater and 3.4 V for field-sourced fracking wastewater over an 8-h experimental period. The airflow was kept constant at 20 mL/min. After each experiment, the cathode was cleaned with 5% (v/v) sulfuric acid to eliminate precipitate debris. The results showed that aerated electrolysis treatment of fracking wastewater led to up to a 64% reduction in total DBP formation and up to a 79% reduction in the formation of the more toxic Br-DBPs after 8 h. We also considered softening the fracking wastewater through chemical precipitation to address cathode scaling issues. The aerated electrolysis process's energy consumption, at 5.9 kWh/m^3, suggested potential for improvement. Adjusting the applied voltage for selective oxidation of bromide could further control DBP formation in fracking wastewater-impacted sources.

While investigating electrochemical methods, Feng et al. shifted their focus from bromide to the transformation of dissolved organic matter (DOM) in fracking wastewater. Their research revealed that during electrochemical advanced oxidation processes, the removal effectiveness of different types of DOM varied, with terrestrial humic-like substances being most efficiently removed, followed by microbial humic-like and then protein-like substances. They also observed that as the applied current density and treatment duration increased, there was a corresponding rise in oxyhalides production, in the order of chlorate > bromate > perchlorate. Simultaneously, most of the DOM underwent mineralization, leaving behind residual products characterized by increased aromaticity. Although Feng et al. did not specifically examine DBP formation, their findings are significant, as many of the transformed products identified in their study are known precursors to DBPs.

Huang and Zhang presented a simpler alternative to setups involving electrodes, membranes, and regeneration components by introducing unactivated peroxymonosulfate (PMS) as an affordable and stable oxidant for the selective oxidation of bromide in mildly acidic conditions [31]. They achieved over 95% removal of bromide as bromine gas in under 10 min, without substantial chlorine gas formation. Additionally, the experiments revealed that the endothermic nature of this process makes higher temperatures of the fracking wastewater conducive to more effective bromide removal.

In addition to physical and chemical methods, biological processes have also been explored. Almaraz utilized biologically active filters (BAFs) as a pre-treatment method for fracking wastewater, aiming to eliminate organic matter and diminish fouling in subsequent membrane treatment processes used for desalination [32]. Considering the high halogen content in fracking wastewater, the study's findings were significant, revealing a notable presence of iodide oxidizing bacteria and the formation of iodinated DBPs during the biological treatment of fracking wastewater, a phenomenon not previously reported.

1.6 Outlook on DBPs in the Shale Gas Sector

While there have been significant technological advances in the shale gas sector, fracking wastewater continues to pose public health and environmental concerns. A typical fracking fluid composition is about 90% water, 9% proppants, and 1% chemical additives. The exact composition of these additives remains largely unknown, as each additive is a mixture of compounds with types and concentrations vary depending on the characteristics of the shale and operator specifications. Although the FracFocus database provides some information about these additives to the public, it is not comprehensive. Rogers et al. noted that out of 996 organic fracking fluid compounds identified by the US EPA and FracFocus in four states, data were only available for an initial screening of only 659 compounds [6]. Fifteen compounds, found in at least 50 FracFocus reports, were predicted to have a higher exposure potential. Among these, two (naphthalene and 2-butoxyethanol) appeared in more than 20% of FracFocus reports, four were found in more than 5% of reports, and three had health-based standards. There is still much work to be done in identifying the chemical additives in fracking wastewater and researching their transformation products as DBP precursors. The future direction in this area is promising, with a clear need for more comprehensive research and data collection to better understand and mitigate the associated risks.

In addition to the quality concerns surrounding fracking wastewater, its quantity also emerges as a critical area for future research. With millions of gallons of water injected into each well for fracking, approximately 10–70% of it returns to the surface as fracking wastewater. Recognizing the significant water footprint of fracking, reducing the amount of water used in the process could correspondingly decrease the volume of fracking wastewater needing treatment. In an unpublished study where we examined FracFocus disclosures for 222,472 wells fracked between 2011 and 2022, we observed a generally consistent increase in water usage per well over this 11 years. For example, in Webb County, Texas, USA, a key area in the East Ford Shale Play, water consumption escalated from 6.6 million gallons per well in 2012 to 18.5 million gallons per well in 2022, a nearly threefold increase. Similar patterns of increase were noted in other shale plays. These research findings underline the importance of efficient and sustainable fracking management. This involves developing technologies to reduce water and chemical usage, decrease wastewater generation, and optimize disposal methods. All these measures are crucial for safeguarding water resources and public health, and of course, they also contribute to the reduction of DBP formation.

2 DBPs Intertwined with Coal-Fired Power Plants

A coal-fired power plant is a type of thermal power station that generates electricity using coal as its primary fuel. The process begins with the delivery of coal to the plant, where it is often treated to enhance its quality and ensure efficient combustion.

This prepared coal is then burned in a furnace at high temperatures. It is important to note that coal-fired power plants have significant environmental impacts. They are major contributors to greenhouse gas emissions, exacerbating climate change, and also emit harmful pollutants like sulfur dioxide, nitrogen oxides, and particulate matter. In this section, however, we will focus on the DBPs intertwined with coal-fired power plants, a field that has garnered significant attention in recent years.

2.1 *Coal Quality and Pollution Control Technologies*

The operation of coal-fired power plants is relatively straightforward. Mitigating their environmental impact hinges on the quality of the coal used and the effectiveness of pollution control technologies. Similar to how the FracFocus database is a crucial resource in the shale gas sector, the CoalQual database is equally important for the power plant sector.

Developed and maintained by the United States Geological Survey (USGS), CoalQual, an abbreviation for "Coal Quality," is an extensive repository of coal quality data. This database was created to aid in the evaluation of coal resources, particularly with an emphasis on environmental compliance and technological developments. It includes a broad spectrum of information such as coal sulfur content, ash yield, trace element concentrations, and calorific value. These data points are vital for understanding the environmental impacts of coal combustion, including the emissions of sulfur dioxide, nitrogen oxides, and trace elements. Furthermore, CoalQual provides insights into the geographic distribution of different coal types, playing a crucial role in strategic planning and decision-making in the energy sector.

In terms of pollution control technologies, common methods employed at coal-fired power plants include electrostatic precipitators or baghouses for capturing fly ash from flue gas before it exits the smokestack, scrubbers for removing sulfur dioxide, and selective catalytic reduction systems for reducing nitrogen oxide emissions.

Moreover, controlling mercury emissions from coal-fired power plants is crucial for reducing environmental pollution and protecting public health. The USEPA has established specific limits under the Mercury and Air Toxics Standards (MATS). Technologies for mercury control include flue-gas desulfurization (FGD), activated carbon injection, and others.

2.2 *Origin of DBP Concerns in Coal-Fired Power Plant Sector*

In coal-fired power plants with FGD units, using brominated sorbents for mercury control is common. This practice leads to the discharge of bromide into FGD wastewater, which eventually reaches surface waters. As discussed earlier, bromide serves as an inorganic DBP precursor, hence the DBP concerns. The CoalQual database indicates that coal naturally contains bromide at trace levels (< 0.001%). In a typical coal-fired power plant, bromide is usually converted to bromine in the flue gas. However, plants with FGD units revert bromine back to bromide, which is then released in their wastewater. VanBriesen notes that FGD wastewater contains significant amounts of bromide, whereas other wastewaters from conventional coal-fired power plants have low bromide levels [33].

Bromide is also a byproduct of biocides used in power plant cooling towers [34]. Cooling towers produce concentrated wastewater that includes the original chemicals in the makeup water, plus those added for corrosion or biological fouling control. Bromine, used alone or with chlorine as a biocide due to its higher reactivity and stronger biocidal activity compared to chlorine, may increase oxidation rates. This leads to lower oxidant concentrations in the effluent at discharge, which is beneficial for ecological health. Consequently, DBPs may be directly formed in cooling water.

2.3 Advances in Reduction of DBP Formation

Recent concerns have emerged regarding the impact of refined coal usage on the formation of DBPs in downstream drinking water treatment facilities. U.S. power plants typically use feedstock coal of bituminous, subbituminous, or lignite rank, which contain halogens at trace levels. Refined coal, rich in halogens like bromide and iodide, has become popular due to tax incentives, as adding bromine to coal enhances mercury removal, aligning with the MATS and receiving tax credits for its use. A five-year study from 2016 to 2020 noted increases in refined coal consumption both spatially and temporally [35]. Despite the tax credit expiring in December 2021, the preference for refined coal usage shows no sign of declining, continuing to pose environmental challenges and complicating efforts to reduce DBP formation.

Reducing bromide loading into source water is the most effective strategy to minimize DBP formation in downstream drinking water facilities. Alternatives tend to be energy-intensive, costly, and less effective in complex natural water mixtures. Good and VanBriesen's 2016 analysis of the Allegheny River's bromide monitoring data attributed approximately 33% of bromide to coal-fired power plants with FGD, 49% to fracking wastewater discharges, and 18% to natural sources [36]. Their study highlighted the need for addressing future bromide increases.

In their 2017 follow-up, Good and VanBriesen corrected an oversight from their earlier work, arguing that assessing the impact of FGD discharges on surface water requires a watershed approach rather than a geographic buffer [37]. This approach better identifies drinking water intakes affected by upstream FGD discharges, focusing on dilution capacity rather than proximity. This confirmed power plants' significant impact on bromide levels at water treatment plants.

Good and VanBriesen's 2019 paper tackled bromide reduction methods, noting reverse osmosis (with over 99% removal efficiency) and anion exchange membranes (with 91% removal efficiency) and electrolysis and volatilization (with 40% removal efficiency) as effective [38]. They also promoted Zero Liquid Discharge (ZLD) as a feasible option for coal-fired plants to eliminate FGD-related bromide discharges, contingent on proper waste management.

2.4 *Outlook on DBPs in the Coal-Fired Power Plant Sector*

Reducing the concentrations of bromide in source waters will reduce the formation of DBPs. Addressing bromide at its source, through managing refined coal use and FGD processes, is more crucial than post-discharge treatments. Many drinking water utilities with high source water bromide concentrations switch from chlorination to chloramination to meet regulated DBP limits. Additionally, recently developed iodide-based coal additives have emerged as an alternative to mercury control. Iodide is also a precursor for DBP formation, particularly in chloramination disinfection processes. The unregulated iodinated DBPs, more toxic than their chlorine counterparts, and the potential role of iodide as a catalyst for the formation of N-nitrosodimethylamine (NDMA), listed by the EPA for future regulation, cannot be overlooked. Given the limited data on iodide content in coal and its addition for mercury control, future research is crucial to understand its potential impact on DBP formation in downstream drinking water utilities.

3 Concluding Remarks

In this chapter, we delve into the complexities of DBP formation within the energy sector, particularly focusing on shale gas extraction and coal-fired power plants. We highlight the significant roles of both organic and inorganic DBP precursors from fracking wastewater and power plant discharges in contributing to water system DBP levels, emphasizing the resulting environmental and health implications. Through analyzing pivotal research findings, the interplay between industrial practices and water quality management challenges is examined. We also evaluate various mitigation strategies, noting their benefits and limitations.

We need to strike a balance between energy needs and environmental protection by ongoing innovation to minimize pollutants from energy production, enforcing strict water quality regulations, and enhancing public awareness and community involvement. Future research directions might include focusing on the origins of DBPs to prevent their formation and developing more efficient treatment technologies. We advocate for a holistic approach, integrating policy, technology, and research, to sustain water resources amidst evolving energy sector developments.

References

1. Meegoda JN, Rudy S, Zou Z, Agbakpe M (2017) Can fracking be environmentally acceptable? J Hazard Toxic Radioact Waste 21:04016013. https://doi.org/10.1061/(ASCE)HZ.2153-5515.0000330
2. Deziel NC, Clark CJ, Casey JA, Bell ML, Plata DL, Saiers JE (2022) Assessing exposure to unconventional oil and gas development: strengths, challenges, and implications for epidemiologic research. Curr Environ Health Rep 9:436–450. https://doi.org/10.1007/s40572-022-00358-4
3. Harmon B, Logan LH, Spiese CE, Rahrig R (2023) Flow alterations in rivers due to unconventional oil and gas development in the Ohio River basin. Sci Total Environ 856:159126. https://doi.org/10.1016/j.scitotenv.2022.159126
4. Xiong B, Soriano MA, Gutchess KM, Hoffman N, Clark CJ, Siegel HG, Vera GADD, Li Y, Brenneis RJ, Cox AJ, Ryan EC, Sumner AJ, Deziel NC, Saiers JE, Plata DL (2022) Groundwaters in northeastern Pennsylvania near intense hydraulic fracturing activities exhibit few organic chemical impacts. Environ Sci Process Impacts 24:252–264. https://doi.org/10.1039/D1EM00124H
5. Shonkoff SBC, Hays J, Finkel ML (2014) Environmental public health dimensions of shale and tight gas development. Environ Health Perspect 122:787–795. https://doi.org/10.1289/ehp.1307866
6. Rogers JD, Burke TL, Osborn SG, Ryan JN (2015) A framework for identifying organic compounds of concern in hydraulic fracturing fluids based on their mobility and persistence in groundwater. Environ Sci Technol Lett 2:158–164. https://doi.org/10.1021/acs.estlett.5b00090
7. Trickey K, Hadjimichael N, Sanghavi P (2020) Public reporting of hydraulic fracturing chemicals in the USA, 2011–18: a before and after comparison of reporting formats. Lancet Planet Health 4:e178–e185. https://doi.org/10.1016/S2542-5196(20)30076-0
8. Luek JL, Harir M, Schmitt-Kopplin P, Mouser PJ, Gonsior M (2018) Temporal dynamics of halogenated organic compounds in Marcellus Shale flowback. Water Res 136:200–206. https://doi.org/10.1016/j.watres.2018.02.055
9. Clark CJ, Xiong B, Soriano Jr MA, Gutchess K, Siegel HG, Ryan EC, Johnson NP, Cassell K, Elliott EG, Li Y, Cox AJ, Bugher N, Glist L, Brenneis RJ, Sorrentino KM, Plano J, Ma X, Warren JL, Plata DL, Saiers JE, Deziel NC (2022) Assessing unconventional oil and gas exposure in the Appalachian Basin: comparison of exposure surrogates and residential drinking water measurements. Environ Sci Technol 56:1091–1103. https://doi.org/10.1021/acs.est.1c05081
10. Butkovskyi A, Bruning H, Kools SAE, Rijnaarts HHM, Van Wezel AP (2017) Organic pollutants in shale gas Flowback and produced waters: identification, potential ecological impact, and implications for treatment strategies. Environ Sci Technol 51:4740–4754. https://doi.org/10.1021/acs.est.6b05640
11. Hill CB, Yadav OP, Khan E (2022) Examining hydraulic fracturing chemicals: a temporal and comparative analysis. Water Res 208:117878. https://doi.org/10.1016/j.watres.2021.117878
12. USEPA (2015) Analysis of hydraulic fracturing fluid data from the FracFocus chemical disclosure registry 1.0
13. Landis MS, Kamal AS, Kovalcik KD, Croghan C, Norris GA, Bergdale A (2016) The impact of commercially treated oil and gas produced water discharges on bromide concentrations and modeled brominated trihalomethane disinfection byproducts at two downstream municipal drinking water plants in the upper Allegheny River, Pennsylvania, USA. Sci Total Environ 542:505–520. https://doi.org/10.1016/j.scitotenv.2015.10.074
14. Hoelzer K, Sumner AJ, Karatum O, Nelson RK, Drollette BD, O'Connor MP, D'Ambro EL, Getzinger GJ, Ferguson PL, Reddy CM, Elsner M, Plata DL (2016) Indications of transformation products from hydraulic fracturing additives in shale-gas wastewater. Environ Sci Technol 50:8036–8048. https://doi.org/10.1021/acs.est.6b00430

15. Kahrilas GA, Blotevogel J, Stewart PS, Borch T (2015) Biocides in hydraulic fracturing fluids: a critical review of their usage, mobility, degradation, and toxicity. Environ Sci Technol 49:16–32. https://doi.org/10.1021/es503724k
16. Abraham DG, Liberatore HK, Aziz MT, Burnett DB, Cizmas LH, Richardson SD (2023) Impacts of hydraulic fracturing wastewater from oil and gas industries on drinking water: quantification of 69 disinfection by-products and calculated toxicity. Sci Total Environ 882: 163344. https://doi.org/10.1016/j.scitotenv.2023.163344
17. Feng H, Zhang Z, Kuang Q, Chen S, Huang D, Zhou X (2023) The transformation of dissolved organic matter and formation of halogenated by-products during electrochemical advanced oxidation pretreatment for shale gas produced water. J Hazard Mater 455:131614. https://doi.org/10.1016/j.jhazmat.2023.131614
18. Farinelli G, Coha M, Vione D, Minella M, Tiraferri A (2022) Formation of halogenated byproducts upon water treatment with Peracetic acid. Environ Sci Technol 56:5123–5131. https://doi.org/10.1021/acs.est.1c06118
19. Krishnaswamy P (2021) Quaternary ammonium compounds as precursors to disinfection byproducts. University of Nevada
20. Usman M, Hüben M, Hahn S, Wieck S, Kehrer-Berger A, Linnemann V, Wintgens T (2023) Evaluation of the DBP formation potential of biocides and identification of knowledge gaps in environmental risk assessment. Environ Sci Eur 35:77. https://doi.org/10.1186/s12302-023-00781-w
21. Braun SM (2015) Hydraulic fracturing in the Ohio River basin. , Ohio River Valley Water Sanitation Commission
22. Hansen E, Mulvaney D, Betcher M (2013) Water resources reporting and water footprint from Marcellus shale development in West Virginia and Pennsylvania. Earthworks Oil & Gas Accountability Project
23. Harkness JS, Dwyer GS, Warner NR, Parker KM, Mitch WA, Vengosh A (2015) Iodide, bromide, and ammonium in hydraulic fracturing and oil and gas wastewaters: environmental implications. Environ Sci Technol 49:1955–1963. https://doi.org/10.1021/es504654n
24. Rossi N (2011) Effects of Marcellus shale gas drilling water discharge on Trihalomethane formation. Master's thesis, The Pennsylvania State University
25. Parker KM, Zeng T, Harkness J, Vengosh A, Mitch WA (2014) Enhanced formation of disinfection byproducts in shale gas wastewater-impacted drinking water supplies. Environ Sci Technol 48:11161–11169. https://doi.org/10.1021/es5028184
26. Hladik ML, Focazio MJ, Engle M (2014) Discharges of produced waters from oil and gas extraction via wastewater treatment plants are sources of disinfection by-products to receiving streams. Sci Total Environ 466–467:1085–1093. https://doi.org/10.1016/j.scitotenv.2013.08.008
27. Huang KZ, Xie YF, Tang HL (2019) Formation of disinfection by-products under influence of shale gas produced water. Sci Total Environ 647:741–751. https://doi.org/10.1016/j.scitotenv.2018.08.055
28. Huang KZ, Tang HL, Xie YF (2018) Impacts of shale gas production wastewater on disinfection byproduct formation: an investigation from a non-bromide perspective. Water Res 144: 656–664. https://doi.org/10.1016/j.watres.2018.07.048
29. Sun M, Lowry GV, Gregory KB (2013) Selective oxidation of bromide in wastewater brines from hydraulic fracturing. Water Res 47:3723–3731. https://doi.org/10.1016/j.watres.2013.04.041
30. Tang HL, Tang L, Xie YF (2018) Aerated electrolysis for reducing impacts of shale gas production wastewater on water sources regarding disinfection byproduct formation. Environ Sci Technol Lett 5:681–686. https://doi.org/10.1021/acs.estlett.8b00482
31. Huang KZ, Zhang H (2020) Highly efficient bromide removal from shale gas produced water by unactivated peroxymonosulfate for controlling disinfection byproduct formation in impacted water supplies. Environ Sci Technol 54:5186–5196. https://doi.org/10.1021/acs.est.9b06825

32. Almaraz N (2018) Emergence and fate of iodinated organic compounds and disinfection by-products during biological treatment of oil and gas produced water. Colorado School of Mines
33. VanBriesen JM (2014) Potential drinking water effects of bromide discharges from coal-fired electric power plants. EPA NPDES comments, pp 1–38
34. Saha P, Wang Y, Moradi M, Brüninghoff R, Moussavi G, Mei B, Mul G, Rijnaarts HM, Bruning H (2021) Advanced oxidation processes for removal of organics from cooling tower blowdown: efficiencies and evaluation of chlorinated species. Sep Purif Technol 278:119537. https://doi.org/10.1016/j.seppur.2021.119537
35. Good KD, Kolb C, VanBriesen JM (2023) Role of refined coal use in power plant effects on downstream drinking water quality. Environ Sci Technol Lett 10:260–266. https://doi.org/10.1021/acs.estlett.3c00026
36. Good KD, VanBriesen JM (2016) Current and potential future bromide loads from coal-fired power plants in the Allegheny River basin and their effects on downstream concentrations. Environ Sci Technol 50:9078–9088. https://doi.org/10.1021/acs.est.6b01770
37. Good KD, VanBriesen JM (2017) Power plant bromide discharges and downstream drinking water systems in Pennsylvania. Environ Sci Technol 51:11829–11838. https://doi.org/10.1021/acs.est.7b03003
38. Good KD, VanBriesen JM (2019) Coal-fired power plant wet flue gas desulfurization bromide discharges to U.S. watersheds and their contributions to drinking water sources. Environ Sci Technol 53:213–223. https://doi.org/10.1021/acs.est.8b03036

Part IV
Health effect of DBPs

Toxicity of DBPs

Haiying Du and Jinhua Li

Contents

1 In Vivo Toxicity of DBPs ... 396
  1.1 Carcinogenicity ... 396
  1.2 Reproductive and Developmental Toxicity ... 403
  1.3 Neurotoxicity ... 412
  1.4 Nephrotoxicity ... 413
  1.5 Hepatotoxicity ... 414
2 In Vitro Toxicity of DBPs ... 415
  2.1 Cytotoxicity ... 415
  2.2 Genotoxicity ... 425
3 Alternative Toxicity Tests Applied in DBPs Study ... 426
  3.1 Zebrafish as a Model for Toxicity Test of DBPs ... 426
  3.2 Nematodes as a Model for Toxicity Test of DBPs ... 428
  3.3 QSAR Models for Toxicity Test of DBPs ... 428
4 Perspectives and Limitations ... 430
References ... 432

Abstract Human is continuously exposed to DBPs, and there is currently a large body of evidence on the health effects of DBPs in addition to the health risks from epidemiological studies. This chapter summarizes the toxic effects of DBPs in terms of in vivo, in vitro, and alternative toxicity tests. The in vivo toxicity tests detailed the carcinogenicity, reproductive and developmental toxicity, neurotoxicity, hepatotoxicity, and nephrotoxicity of DBPs, the in vitro toxicity tests reported cytotoxicity and its currently identified mechanisms, as well as genotoxicity, and the alternative toxicity tests described the toxic effects of DBPs in terms of three models, the zebrafish, the nematode, and the QSAR model. This chapter provides a comprehensive, in-depth, and systematic analysis and evaluation of the toxic effects of DBPs, which provides new approaches and insights for drinking water safety.

H. Du and J. Li (✉)
School of Public Health, Jilin University, Changchun, China
e-mail: hydu@jlu.edu.cn; jinhua1@jlu.edu.cn

Chao Chen, Susan Andrews, and Yuefeng Xie (eds.), *Disinfection By-Products in Water*, Hdb Env Chem (2024) 135: 395–444, DOI 10.1007/698_2024_1147,

Published online: 16 September 2024

Keywords Alternative toxicity test, DBPs, In vitro toxicity, In vivo toxicity

Since the first DBPs were discovered in 1974, more than 700 DBPs have been identified, and these DBPs might only account for 30% of the total DBPs [1]. Recently, a database of DBPs was constructed, which included 6,310 DBPs reported between 1974 and 2022 [2]. However, very limited DBPs have been systematically evaluated for their toxic effects. Therefore, it is a big challenge to comprehensively evaluate the toxicities of DBPs. In vivo toxicity tests have been used to mainly examine the effects of DBPs on potential target organs, and in vitro toxicity tests have been applied to primarily assess the cytotoxicity and genotoxicity of DBPs, and to explore the potential mechanisms. In recent years, alternative methods have also been used to study the toxic effects of DBPs, such as zebrafish and quantitative structure-activity relationship (QSAR) analysis.

1 In Vivo Toxicity of DBPs

1.1 Carcinogenicity

Epidemiological studies have already suggested the correlation between exposure to DBPs and the increased risk of developing cancers, such as bladder, colon, and rectal cancer, while DBPs may lead to gastrointestinal, kidney, bladder, breast, liver, and thyroid cancers based on in vivo carcinogenicity tests. Among them, the liver and kidney are the most common target organs, which is different from the findings in epidemiological studies.

There were some short-term studies on the carcinogenicity of DBPs. Dichloroacetic acid (DCAA) and trichloroacetic acid (TCAA) could cause hepatocellular tumors in mice [3, 4]. Hooth et al. evaluated the carcinogenic effects of potassium bromate, 3-Chloro-4-dichloromethyl-5-hydroxy-2 (5H) – furanone (MX), chloroform, and bromodichloromethane (BDCM) using an Eker rat model of hereditary renal cell carcinoma. The findings suggested that most DBPs showed a dose response to renal tumors after 4 or 10 months [5]. Besides, high-dose MX exposure for 10 months brought about bladder transitional epithelial hyperplasia and cell proliferation in rats [6]. In addition, iodoacetic acid (IAA) demonstrated carcinogenic potential and induced the formation of transformed foci, and these transformed cells further resulted in the formation of fibrosarcoma after transplantation into nude mice for 15 days [7].

The long-term carcinogenicity studies, also called the two-year rodent carcinogenicity test, have been conducted through drinking water, mainly for the regulated DBPs, and the detailed information is listed in Table 1. Trihalomethanes (THMs) could cause liver, kidney, and intestinal tumors in vivo. Dichlorobromomethane (DCBM) could cause liver and kidney tumors in mice [8], while BDCM could cause kidney and intestinal tumors in male rats [8–10]. Dibromochloromethane (DBCM)

Table 1 The long-term carcinogenicity induced by DBPs

DBPs	Species	Sex	Dose	Exposure routes	Target organ	Reference
Bromodichloromethane	Mouse	Male	8.1, 27.2, 43.4 mg/kg	Oral administration	Kidney	[8, 10, 44]
		Female	9, 18, 36 mg/kg	Oral administration	Liver	
	Rat	Male	3.9, 20.6, 36.3 mg/kg 0, 6, 12, 25 mg/kg	Oral administration	Kidney, esophagus	
		Female	0, 0.014, 0.055, 0.22%	Oral administration	Kidney, large intestine	
Bromoform	Rat	Male	0, 100, 200 mg/kg	Oral administration	Large intestine	[45]
		Female		Oral administration	Large intestine	
Chlorodibromomethane	Mouse	Male	0, 50, 100 mg/kg	Oral administration	Liver	[11]
		Female		Oral administration	Liver	
Chloroform	Mouse	Male	138, 277 mg/kg	Oral administration	Liver	[12, 13, 18]
		Female	238, 477 mg/kg	Oral administration	Liver	
	Rat	Male	90, 180 mg/kg	Oral administration	Kidney	
		Male	0, 200, 400, 900, 1,800 mg/L	Drinking water	Kidney	
		Male	0, 25, 50, 100 ppm (v/v) + 1,000 ppm(w/w)	Inhalation and oral	Kidney	

(continued)

Table 1 (continued)

DBPs	Species	Sex	Dose	Exposure routes	Target organ	Reference
Dibromoacetic acid	Mouse	Male	0, 125, 250, 500, 1,000, 2,000 mg/L-2 weeks/3 months	Drinking water	Liver, lung, testes	[19]
		Female	0, 50, 500, 1,000 mg/L-2 years	Drinking water	Liver	
	Rat	Male	0, 125, 250, 500, 1,000, 2,000 mg/L-2 weeks/3 months	Drinking water	Mesotheliomas, mononuclear cell leukemia, liver, lung, testes	
		Female	0, 50, 500, 1,000 mg/L-2 years	Drinking water	Mononuclear cell leukemia, liver	
Dichloroacetic acid	Mouse	Male	1, 2 g/L or 0.05, 0.5, 1, 2, 3.5 g/L	Drinking water	Liver	[3, 4, 20]
	Rat	Male	0.05, 0.5, 1.6 g/L	Drinking water	Liver	
Trichloroacetic acid	Mouse	Male	1, 2 g/L	Drinking water	Liver	[3, 46]
		Female	2.0, 6.67, 20.0 mmol/L	Drinking water	Liver	
Bromochloroacetic acid	Rat	Male	250, 500, 1,000 mg/L	Drinking water	Mesotheliomas, pancreas, liver, large intestine	[21]
		Female		Drinking water	Large intestine, fibroadenomas	
Chloroacetonitrile	Mouse	Female		Oral administration	Lung	[35, 36]
				Skin application	Skin	
Trichloroacetonitrile	Mouse	Male		Oral administration	Lung	
				Skin application	Skin	

Dibromoacetonitrile	Mouse	Female		Skin application	Skin	
	Rat	Male	50, 100, 200 mg/L	Drinking water	Oral mucosa or tongue, stomach, liver	
		Female		Drinking water	Oral mucosa or tongue, stomach, skin	
Bromochloroacetonitrile	Mouse	Female		Oral administration	Lung	
				Skin application	Skin	
Formaldehyde	Mouse	Male	0, 2.0, 5.6, 14.3 ppm	Inhalation	Nasal	[28–31]
	Rat	Male	0, 2.0, 5.6, 14.3 ppm	Inhalation	Nasal	
			0, 10, 50, 100, 1,000, 1,500 mg/L or 0, 1.2, 15, 82 mg/kg	Drinking water	Lymphoma, stomach	
		Female	0, 2.0, 5.6, 14.3 ppm	Inhalation	Nasal	
			0, 10, 50, 100, 1,000, 1,500 mg/L or 0, 1.8, 21, 109 mg/kg	Drinking water	Mammary, lymphoma, stomach	
Acetaldehyde	Rat	Male	0, 0.75, 1.5, 3 ppm	Inhalation	Nasal	[31]
		Female		Inhalation	Nasal	
Chloral hydrate	Mouse	Male	1 g/L	Drinking water	Liver	[32, 33]
			13.5, 65.0, 146.6 mg/kg	Oral administration	Pituitary, liver	
Chloroacetaldehyde	Mouse	Male	0.1 g/L	Drinking water	Liver	[33]

(continued)

Table 1 (continued)

DBPs	Species	Sex	Dose	Exposure routes	Target organ	Reference
Bromate	Mouse	Male	0, 0.08, 0.4, 0.8 g/L	Drinking water	Kidney	[23, 24]
	Rat	Male	0, 0.02, 0.1, 0.2, 0.4 g/L or 0, 15, 30, 60, 125, 250, 500 ppm	Drinking water	Kidney, mesotheliomas, thyroid	
Chlorate	Rat	Male Female	125, 1,000, 2,000 mg/L	Drinking water	Thyroid	[27]
	Mouse	Male Female	500, 1,000, 2,000 mg/L	Drinking water	Thyroid	
3-Chloro-4-(dichloromethyl)-5-hydroxy-2(5*H*)-furanone (MX)	Rat	Male	0.4, 1.3, 5.0 mg/kg	Drinking water	Liver, adrenal, thyroid, lung	[34]
		Female	0.6, 1.9, 6.6 mg/kg	Drinking water	Mammary, liver, adrenal, thyroid, lymphoma, leukemia	
N-ethyl-N-nitrosourea	Mouse		60 μg/g	Inject	Lung, stomach, ovaries, lymphoma	[37–43]
N-nitrosodiethylamine	Mouse		25 μg/g	Inject	Gastrointestinal, skin, lung, lymphoma	
	Rat		0–16.896 ppm	Drinking water	Liver, lung, nasal	
	Syrian golden hamsters	Male	1.5, 7.5 mg	Intratracheal infusion	Lung	
N-nitrosodimethylamine	Syrian golden hamsters	Male	0.75, 1.5 mg	Intratracheal infusion	Liver, lung	
Nitrosodi-n-propylamine	Syrian golden hamsters	Male	1.5 mg	Intratracheal infusion	Lung	

N-nitrosamorphine	Syrian golden hamsters	Male	1.5 mg	Intratracheal infusion	Lung	
N-nitrosopyrrolidine	Syrian golden hamsters	Male	1.5 mg	Intratracheal infusion	Liver	
	MRC rats	Male	16 mg/kg	Drinking water	Liver, testes	
		Female				
	Rat	Male	0.3, 1, 3, 10 mg/kg	Drinking water	Liver, testes	
Nitrosodiethanolamine	Rat		3,900–31,250 ppm	Drinking water	Liver	

could also cause liver cell tumors in mice [11]. Chloroform caused hepatocellular tumors in male mice and kidney tumors in male rats [12, 13]. Another study was conducted in male rats using a combination of inhalation and drinking water, and the combination of these two exposure routes has significantly enhanced carcinogenicity and increased the incidence of kidney tumors in male rats [14–18].

Haloacetic acid (HAAs) have been confirmed to induce multi-organ tumors. Exposure to dibromoacetic acid (DBAA) via the drinking water route induced hepatocellular tumors and lung tumors in male and female mice, while mesotheliomas and leukemias were triggered in both male and female rats [19]. DCAA had carcinogenic effects on the liver in both rats and mice, and the rats were more sensitive to DCAA-induced hepatocellular carcinoma than mice [20]. Bromochloroacetic acid (BCAA) in drinking water caused mesothelioma in male rats, multiple fibroadenomas in female rats, colon adenomas in male and female rats, liver adenomas in male and female rats, and pancreatic adenomas in male rats [21]. In addition, exposure to bromodichloroacetic acid significantly increased breast fibroadenoma in female rats and led to colon adenoma in male rats [22].

Bromate and chlorate also have carcinogenic effects [23–26]. Bromate was found to induce renal tumors in mice and female rats, and renal tumors, mesothelioma, and thyroid follicular cell tumors in male rats. However, in studies of chlorate carcinogenesis, both male and female rats have developed follicular tumors of the thyroid gland [27].

Formaldehyde can cause cancer in rodents when administered either by inhalation or through drinking water. Formaldehyde demonstrated carcinogenic effects in different sites, such as nasal tumors, nasal squamous cell carcinoma, testicular Leytcell tumors, lymphoma, and leukemia [28–30]. Acetaldehyde has been shown to cause nasal tumors in rats when administered by inhalation, whereas when administered through drinking water, the ingested dose is very low so it is not carcinogenic [31]. Chloral hydrate may cause liver cell tumors in male mice, but this effect is not present in rats [32, 33]. Chloroacetaldehyde is a liver and pancreas carcinogen in mice [33].

MX has caused various cancers in both male and female rats, such as liver, adrenal, thyroid, lung, mammary, lymphoma, and leukemia [34]. And the tumor incidence increased in a dose-dependent manner.

Haloacetonitriles (HANs) have also caused various types of cancers. Dibromoacetonitrile exposure through drinking water can increase the incidence of oral squamous cell adenoma, or stomach and liver cancer in rats [35]. Chloroacetonitrile, trichloroacetonitrile, and bromochloroacetonitrile have induced skin tumors through skin contact, and have caused lung cancer through oral administration [36].

Nitrosamines have been detected as processing by-products in food, beer and cigarettes, impurities in cosmetics and rubber toys, and DBPs in drinking water. Various nitrosamines are classified as carcinogens by IARC. N-ethyl-N-nitrosourea (ENU) and N-nitrosodiethylamine (NDEA) can cause various tumors in mice after injection. Among them, ENU can cause tumors in the lungs, stomach, and ovaries, as well as lymphoma, while NDEA can cause tumors in the gastrointestinal tract, skin, lungs, and lymphatic cells. However, due to its metabolism in the liver, NDEA was more effective in inducing liver tumors [37]. A study has evaluated liver and esophageal tumors induced by lifelong exposure to NDEA or N-nitrosodimethylamine (NDMA) through drinking water in 4,080 rats, and the results indicated that exposure to concentrations as low as 0.01 ppm in drinking water can lead to 25% of animals developing liver tumors, and there were no signs of threshold effects [38]. Tanaka et al. showed that the carcinogenic rates of NDEA, NDMA, N-nitrosamorphine (NMOR), *N*-nitrosopyrrolidine (NPYR), and Nitrosodi-n-propylamine (NDPA) in respiratory organs of male Syrian golden hamsters were 100%, 6%, 43%, 0%, and 72%. The incidence of liver tumors was 19% in the NDMA group and 4% in the NPYR group, and liver tumors were not found in the other groups. The carcinogenic activity of nitrosamines on the respiratory organs was estimated to be in the following order: NDEA>NDPA>NMOR>NDMA = NPYR [39, 40]. NPYR can cause liver tumors in rats and testicular tumors (papillary mesothelioma) in male animals [41, 42]; while nitrosodiethanolamine can cause liver cancer and nasal cancer in rats [43].

1.2 Reproductive and Developmental Toxicity

Epidemiological and animal experimental evidence shows that drinking water DBPs pose potential threats to the reproductive and developmental health, affecting various stages of the entire life cycle, including adverse effects on parental reproductive function, such as germ cells, conception, pregnancy, parturition, and lactation, as well as adverse effects on fetal and postnatal development. The details of reproductive and developmental toxicities of DBPs, as well as neurotoxicity, nephrotoxicity, and hepatotoxicity, are summarized in Table 2.

DBPs demonstrated adverse effects on male reproductive function. Researchers found that BDCM could reduce serum levels of progesterone and luteinizing hormone (LH), leading to full-litter resorption [47–49]. In a study of four individual THMs (TCM, BDCM, DBCM, and TBM), it was found that they all caused a decrease in serum testosterone concentration in rats [50]. Bromoacetic acids can decrease the weight of male reproductive organs and semen parameters, reduce sperm activity and cause sperm acrosomal-nuclear malformations, promote the formation of atypical residual bodies, change the structure and/or function of supporting cells, and lower the serum testosterone levels [51–55]. HAAs mixture synergistically reduced the level of sperm membrane protein SP22, posing greater harm to male reproductive health [56, 57]. In addition, the mixture of THMs and HAAs was found to reduce sperm number and motility [58, 59]. Dibromoacetonitrile (DBAN) reduced the sperm count and motility of adult male albino mice, BDCM also decreased sperm motility in F344 rats [60, 61].

DBPs also could cause adverse effects on female reproductive function. DBAA disrupted the estrous cycle in female mammals, increased the circulating concentrations of estradiol (E2) and estrone (E1), and induced a dose-related increase in total

Table 2 The in vivo toxicities induced by DBPs

Types of toxicity	DBPs	Subjects	Exposure dose and time	Toxic manifestation	Reference
Nephrotoxicity	TCP, TBP, TIP	C57BL/6 mice	0.5, 10 and 200 μg/L for 12 weeks	At 0.5 μg/L showed renal injury, the rank order of renal toxicity was TIP > TBP > TCP	[101]
	DCBQ	ICR mice	31.25, 62.5, and 125 mg/kg for 28 days	Oxidative damage and renal inflammation	[103]
	Chloroform	Male mice	0, 50, 300, 700, 1,000, 1,500 mg/kg for 14 days	Kidney weight increase, induced renal cell proliferation and inflammatory reaction, and proximal renal tubular necrosis	[102]
Hepatotoxicity	TCP, TBP, TIP	C57BL/6 mice	0.5, 10 and 200 μg/L for 12 weeks	The antioxidant and detoxification capacity of the liver was impaired, and liver inflammation was triggered	[106]
	THMs, HAAs, HANs, HKs	C57BL/6 mice	Mixture of DBPs at 0, 200, and 500 ng/20 g, every other day for a week	The degree of liver injury increases with the increase of DBP concentration, and mixed DBPs might aggravate liver injury	[108]
	BANs	ICR mice	10, 100, 1,500 μg/L for 30 days	Oxidative stress and liver injury	[111]
	Chlorinated water	SD rats	Free chlorine levels were 1.4 to 1.6 mg/L and no free chlorine, swimming training performed 5 days a week for 12 weeks	Eyes and skin were more susceptible to chlorinated water stimulation, and the liver was the most likely target organ of DBPs	[104]
	TCAcAm	Mice	50, 500, and 5,000 μg/L for 90 days	TCAcAm exposure induced disruption of amino acid and lipid metabolism and led to hepatotoxicity	[109]
	MonoHAcAms, including IAcAm, BAcAm, CAcAm	ICR mice	0.75, 300, and 120,000 μg/L for 30 days	MonoHAcAms exposure induced oxidative stress in mice's liver and induced changes in metabolomics	[110]
	Chloroform	Male mice	0, 50, 300, 700, 1,000, 1,500 mg/kg for 14 days	Liver centrilobular necrosis and massive necrosis at 1,500 mg/kg	[102]

Reproductive and developmental toxicity	Bromoform (TBM), chloroform (TCM), BDCM, CDBM	SD rats	TCM (100, 200 and 400 mg/kg) TBM、BDCM and CDBM (50, 100 or 200 mg/kg), GD6-15	Depressed maternal weight gain, decreased fetal body weights, and caused fetal skeletal abnormalities	[70]
	THMs (TCM, BDCM, DBCM, TBM)	Male F344 rats	0.75, 1.5 mmol/kg for 7 days	All four THMs could decrease the serum testosterone concentration of rats	[50]
	BDCM	F344 rats	0, 75, or 100 mg/kg, GD0-PD6	BDCM-induced pregnancy loss and decreased serum progesterone and LH levels	[48]
	BDCM	F344 and SD rats	0, 75, and 100 mg/kg, GD6-15	On GD 6–10, F344 rats had a 62% incidence of full-litter resorption (FLR) at 75 mg/kg/day, SD rats maintained their litters; 24 h after a single dose, all dams with FLR reduced serum progesterone levels but LH levels were unaffected	[47]
	BDCM	F344 rats	0 or 100 mg/kg, GD6-9	Decreased corpora lutea and serum progesterone and LH levels	[49]
	DBAA	Male SD rats	0, 5, 50 mg/kg for 2 and 4 weeks, 250 mg/kg for 2 weeks	The atypical residual bodies and retention of Step 19 spermatids was observed in the high dose group, and retention of Step 19 spermatids was also observed in the low and middle-dose groups. DBAA had testicular toxicity	[51]
	DBAA	SD rats	0, 10, 30, 90, 270 mg/kg for 14 days 0、 2、 10、 50、 250 mg/kg for 79 days	DBAA may affect the reproductive results of rats. At high doses, the weight of testis epididymis decreased, sperm count decreased, sperm motility and morphology were affected, and so on	[52, 53]

(continued)

Table 2 (continued)

Types of toxicity	DBPs	Subjects	Exposure dose and time	Toxic manifestation	Reference
	DBAA	SD rats	0, 50, 250, 650 ppm; The exposure time of P generation was 92 days for males and 120 days for females (including pregnancy and lactation). Exposure of F1 generation rats began 29 days postpartum and lasted for at least 71 days	At 250 and 650 ppm, DBAA induced reproductive toxicity in P and F1 generation males, such as sperm count and epididymal tubules changes, and reproductive tract malformations in male rats, while no reproductive and developmental effects were observed in female rats exposed to high concentrations	[54]
	DBAA	Male Dutch-belted rabbits	1, 5, 50 mg/kg, from gestation day 15 to adulthood	Exposure to as low as 1 mg/kg could induce reproductive toxicity, such as unilateral cryptorchidism, destruction of spermatogenesis, etc.	[55]
	DBAA+BCAA	SD rats	2, 4 mg/kg DBAA; 1.6, 3.2 mg/kg BCAA; 2 mg/kg DBA + 1.6 mg/kg BCA	The mixture could reduce the level of sp22 and induce decreased fertility	[57]
	Concentrated and chlorinated drinking water	SD rats	Drinking water was concentrated 136-fold, P0 generation: exposure during pregnancy and lactation, F1 generation: continuous exposure	F1 female puberty delayed and reduced caput epidydimal sperm counts in F1 adult males	[58]
	DBAN	Adult male mice	50 mg/kg for 3 h	The number and motility of testicular sperm were decreased, the level of testicular lactate dehydrogenase-X (LDH-x) was decreased, and oxidative stress was induced in the testis of mice	[60]
	BDCM	F344 rats	22 and 39 mg/kg, drinking water exposure to 56 weeks of age	BDCM at 39 mg/kg decreased sperm motility vitality and caused reproductive toxicity	[61]

	DBAA	SD rats	0, 2, 10 and 50 mg/mL for 4, 8, 12, and 24 h	DBAA inhibited progesterone secretion under hCG-stimulated or baseline conditions; the pregnenolone abolished this decrease, with DBAA-exposed follicles exhibiting a linear increase in progesterone release	[81]
	DBAA	SD rats	0, 10, 30, 90, or 270 mg/kg for 14 days	DBAA exposure induced estrous cycles in female rats, with some animals showing a sustained period of vaginal keratosis, whereas others showed a sustained period of estrus	[80]
	DBAA	SD rats	0, 30, 60, 90, 120, or 270 mg/kg for 2 weeks	Increased luteinizing hormone and circulating estradiol	[62]
	DBAA	SD rats	0, 37.5, 75, or 150 mg/kg for 14 days	DBAA induced a dose-related increase in total estrogen concentrations	[63]
	DBAA	SD rats	0, 400, 600, 800 ppm, GD0- PD120	Delayed vaginal opening	[112]
	DBAA	Dutch-belted rabbits	0, 1, 5, or 50 mg/kg, GD15–24 weeks postpartum	50 mg/kg reduced primordial follicles and total healthy follicles in prepubertal animals, and 5 and 50 mg/kg reduced primordial follicles in adult animals	[65]
	DBAA	SD rats	0, 50, 150, and 300 ppm (5, 16, and 33 mg/kg) for 140 days	Increased circulating estradiol and disrupted estrous cyclicity	[64]
	TCAA	Long-Evans rats	0, 330, 800, 1,200, or 1,800 mg/kg, GD6-15	When ≥330 mg/kg, it inhibited maternal weight gain, increased embryonic mortality, and caused all soft tissue and skeletal malformations in live fetuses, which had developmental toxicity to pregnant rats	[72]
	TCAA	Charles Foster strain rats	1,000, 1,200, 1,400, 1,600, or 1,800 mg/kg, GD6-15	At ≥1,200 mg/kg, neuronal death increased and brain weight decreased with increasing dose	[71]

(continued)

Table 2 (continued)

Types of toxicity	DBPs	Subjects	Exposure dose and time	Toxic manifestation	Reference
	IAA	CD-1 mice	2, 5, and 10 μM	IAA exposure resulted in metaphase I (MI) arrest and polarbody-extrusion failure in mouse oocytes	[66]
	IAA	CD-1 mice	0.5, 10, 100, and 500 mg/L (3.48, 69, 490, and 1,700 μg per mouse per day), 35–40 days	Altered estrous cycle, ovarian gene expression, and estradiol levels	[68]
	IAA	CD-1 mice	10 or 500 mg/L (70 or 1,700 μg per dam per day), 35–40 days	Directly affecting the transcriptome of ovarian antral follicles	[67]
	IAA	CD-1 mice	10, 100, or 500 mg/L (40 to 1,500 μg per dam per day), 35 days	Exposure to IAA during pregnancy and lactation decreased the vaginal opening rate and the percentage of atretic follicles, and increased the absolute ovarian weight and anal reproductive index in female mice, and affect hormone levels	[73]
	CAN	CD-1 mice	12.5, 25, or 50 mg/kg, GD6-18	High-dose CAN (50 mg/kg) could increase the rate of fetal malformation and reduce the survival rate. Low levels of CAN (25 mg/kg) could reduce fetal weight and induce musculoskeletal malformations in mice	[76]
	CAN	Kunming mice	0, 12.5, 25 and 50 mg/kg for 28 days	CAN exposure altered oocyte gene expression, induced endoplasmic reticulum stress, and affected spindle assembly in oocytes, impaired follicular development	[69]

	CAN	CD-1 mice	25 mg/kg, GD6-18	Maternal body weight decreased, and fetal brain weight and body weight decreased, which caused oxidative stress and neuronal apoptosis in fetal brain tissue, and had adverse effects on the fetal liver in mice	[75, 78]
	DCAN	Long-Evans rats	0, 5, 15, 25, or 45 mg/kg, GD6-18	When≥15 mg/kg, it caused dams or fetuses death, and fetal soft tissue and skeletal malformations	[79]
	DCAN	BALB/c mice	7, 14 and 28 mg/kg, GD6-15	Maternal exposure to DCAN led to redox imbalance in the fetal cortex and cerebellum, induced apoptosis, and resulted in neuronal deformation	[77]
	9 DBP (4 THM + 5 HAA)	F344 rats	4THM (0, 307, 613, 920 μmol/kg), 5HAA (0, 308, 615, 1,231 μmol/kg), 4THM + 5HAA (0 + 0, 153 + 154, 307 + 308, 613 + 615 μmol/kg), GD6-20	When ≥613 μmol/kg/day THMs, HAAs, and DBPs caused pregnancy loss; at 615 μmol/kg/day HAAs and 307 μmol/kg/day DBPs increased resorption rates in surviving litters; at ≥308 μmol/kg/day HAAs caused eye malformations (anophthalmia, microphthalmia)	[74]
	Total THMs and total HAAs	SD rats	0, 500×, 1,000×, or 2,000× of the U.S. EPA's maximum contaminant levels (MCLs) for GD0-PD6	F1 pup weights were reduced at 2,000× on PND6 and at ≥1,000× on PND21; Postweaning F1 body weights were reduced at 2,000×; onset of puberty was delayed at 1,000× and 2,000×; F1 estrous cycles and fertility were unaffected, and F2 litters showed no effects on pup weight or survival	[59]

(continued)

Table 2 (continued)

Types of toxicity	DBPs	Subjects	Exposure dose and time	Toxic manifestation	Reference
Neurotoxicity	DBAA	F344rats	0, 0.2, 0.6, and 1.5 g/L for 6 months	DBAA-induced neurotoxicity in rats, mainly manifested as hypotonia, sensorimotor depression, and degeneration of spinal nerve fibers	[83]
	DCAA	F344 rats and LE rats	30, 100, 300, 1,000 mg/kg for 10 weeks; 137, 235 mg/kg for 24 months; 16–308 mg/kg for 8, 12 or 13 weeks	DCAA exposure produced neurotoxicity, which could occur when exposure levels are as low as 16 mg/kg/day, whereas high dose levels produced permanent neurotoxicity with prolonged exposure	[84]
	DCAA	Juvenile beagle dogs	Gelatin capsules orally, 0, 12.5, 39.5, and 72 mg/kg for 90 days	The dogs in the high-dose group showed partial paralysis of the hind limbs, and some of the dogs in the high, medium, and low-dose groups showed vacuolization of myelinated white fiber tracts in the brain, cerebellum, and spinal cord	[85]
	CAN	CD-1 mice	25 mg/kg, GD6-18	CAN could penetrate the placental and fetal blood-brain barrier and cause apoptotic neurodegeneration in fetal brain cells	[86]
	THMs+PCE	CD-1 mice (male and female)	10 and 100-fold of (THMs + PCE), PD0-30	Induced autism-like behavior in offspring male mice	[87]
	TCM, DCBM, DBCM, TBM	ICR mice	300 and 3,000 times of estimated ADI for 14 and 90 days, 100 mg/kg for 30 days, 100 and 400 mg/kg for 60 days	TCM, DBCM, and TBM at doses of 100,400 mg/kg had significant effects on the manipulation behavior of mice after 60 days of administration	[88]

	DBAA	SD rats	0–125 mg/kg for 4 weeks	DBAA has toxic effects on the hippocampus and prefrontal cortex of rats, and may also induce neurotoxicity through caspase-3-dependent apoptosis	[89, 90]
	DCAA	SD rats	DCAA (90 mg/kg) for 28 days	DCAA could induce neurotoxicity, such as oxidative stress, inflammation, and synaptic damage	[91]
	DCAA	SD rats	10, 30, 90 mg/kg twice a week for GD0-PD56	Exposure to DCAA-induced neurotoxicity in offspring mice, which is mainly manifested as oxidative stress, neuroinflammatory response, and synaptic damage in the hippocampus	[92]
	DBAN	Wistar rats	15, 30, or 60 mg/kg and treatment with taurine (200 mg/kg) for 7 days	The behavior of rats was impaired, which tends to cause significant oxidative damage in the brain of rats, and taurine could alleviate the damage caused by DBAN	[93]
	DCAN	SD rats and diabetic rats	11, 44, and 88 mg/kg for 28 days	DCAN altered motor behavior and induced anxiety behavior in normal rats, whereas diabetic rats were resistant to DCAN-induced neurotoxicity	[94]

estrogen [62–64]. Chronic exposure to DBAA had no effect on late follicular development or ovulation in rabbits, but DBAA reduced the total number of primordial and healthy follicles and delayed pubertal development [65]. IAA is an ovarian toxin that can cause metaphase I arrest and polar body extrusion failure in mouse oocytes, interfere with oocyte maturation [66], and significantly reduce estradiol levels and shorten proestrus [67, 68]. Chloroacetonitrile (CAN) also affected ovarian and follicular development [69].

In addition, DBP exposures also induce developmental toxicity. TCM, BDCM, and chlorodibromomethane (CDBM) inhibited maternal weight gain, TCM decreased fetal weight, and TCM, CDBM, and TBM caused fetal skeletal abnormalities [70]. Exposure to TCAA during pregnancy reduces maternal weight, resulting in embryonic death and skeletal malformations in living fetuses, reducing fetal brain weight and affecting the central nervous system [71, 72]. Exposure to IAA during gestation and lactation can affect F1 female vaginal opening, anogenital index, ovarian weight, atresia follicle percentage, and reproductive hormone levels [73]. The DBP mixtures including four THMs and five HAAs [Chloroacetic acid (CAA), DCAA, TCAA, Bromoacetic acid (BAA), and DBAA] would synergistically induce spontaneous abortion in rats and cause ocular deformities (anophthalmia, microphthalmia) in surviving litters. However, the presence of THMs in the complete mixture appeared to decline the incidence of HAAs-induced ocular defects [74]. Exposure to CAN and DCAN during pregnancy reduced the number of live births and fetal weight, affected the growth and development of fetal brain and liver tissues, and led to fetal behavioral learning and memory defects, as well as soft tissue and skeletal malformations [75–79].

Some DBPs such as BDCM, DBAA, and IAA could interfere with follicle-stimulating hormone (FSH), LH, testosterone, progesterone, estradiol hormone secretion, and disrupted estrous cyclicity. These HBQs could also act on the hypothalamic-pituitary gonad gland axis, thereby inhibiting the maturation of follicles and sperm, disrupting the estrous cycle and mating behavior in mammals [47, 48, 62–64, 67, 80–82]. Also, oxidative stress and apoptosis play important roles in the reproductive and developmental toxicities induced by DBPs. Maternal exposure to DBPs during pregnancy can affect the growth and development of the fetal brain and liver by inducing oxidative stress and neuronal apoptosis, resulting in adverse pregnancy outcomes and neurobehavioral deficits [75, 77–79]. In conclusion, DBPs produce reproductive and developmental toxicities in mammals mainly by interfering with hormone secretion and inducing oxidative stress and apoptosis.

1.3 Neurotoxicity

At present, there are relatively few studies on the neurotoxicity of DBPs. The neurotoxicity of DPBs in vivo is mainly manifested as neuromuscular toxicity and neurobehavioral toxicity. DBAA and DCAA can induce neuromuscular toxicity in rats, mainly including hypotonia, turning reflex defects, sensorimotor inhibition, and reduced response to tail-pinch and clicks [83, 84]. Additionally, DCAA causes paresis of the hind limb in dogs [85]. And CAN triggers apoptotic neurodegeneration in the fetal brain crossing the placenta and the fetal blood-brain barrier [86].

In addition to determining the neurotoxic changes caused by DBPs, the harmful effects of DBPs on the mammalian nervous system can also be determined by neurobehavioral test methods. The commonly used neurobehavioral testing methods mainly include the Morris water maze, open field test, rotarod test, etc. through which it can be observed and evaluated the spatial orientation and memory abilities, social ability, perseverance behavior, motor coordination, and exploratory learning behavior and other neurobehavioral performance. Maternal and offspring exposure to a mixture of two THMs (chloroform, bromoform) and PCE (perchloroethylene) led to autism-like behavior in offspring male CD-1 mice, comprising social withdrawal, repetitive behavior, communication difficulties, and sensory allergy as well as other symptoms [87]. TCM, DCBM, and TBM affected the operational behavior of the lever press in mice and tended to produce tolerance [88]. DBAA and DCAA exposures can impair cognitive function, learning, and memory capacity in rats [89–92]. DBAN had a noticeable negative impact on rat behavior, mainly manifested as reduced motor activity and coordination, and unable to maintain balance on the rotating rod [93]. Similarly, DCAN impaired rat locomotion and exploratory activities, induced anxiety behavior, as well as produced a significantly impaired effect on locomotion and exploratory activities [94].

Furthermore, it has been proved by in vivo studies that DBPs-impaired neurobehavioral development in mammals mainly by inducing oxidative stress, apoptosis, neuroinflammatory response, and diminished synaptic plasticity in the hippocampus and prefrontal cortical areas of the brain. Oxidative stress is one of the important toxic mechanisms of DBPs, and continuous oxidative stress leads to the development of inflammation and apoptosis [95]. Oxidative stress is further regulated by numerous molecular mechanisms, such as the PI3K/AKT/mTOR pathway, Nrf2 pathway, NF-κB pathway, and p38-MAPK pathway, which can induce inflammation and apoptosis and ultimately damage the development and function of the nervous system[96–100]. Next, it is necessary to further explore whether DBPs also have effects on other regions of the mammalian brain.

1.4 Nephrotoxicity

DBPs have raised increasing concerns due to their potential impact on human health, notably on the kidneys. The kidneys, responsible for filtering waste and maintaining fluid balance, are vulnerable to the accumulation of these DBPs. Continuous exposure to DBPs might lead to oxidative stress, inflammation, and impairment in renal tubular function, signifying nephrotoxicity.

Miao et al. investigated kidney damage induced by aromatic trihalogenated DBPs in mice [101]. Significant alterations, such as kidney impairment, proteinuria, and increased proliferation of glomerular mesangial cells, were observed following exposure to 2,4,6-trihalophenol, even at a low concentration of 0.5 μg/L in vivo. The ranking order for nephrotoxicity was determined as 2,4,6-triiodiophenol (TIP) > 2,4,6-tribromophenol (TBP) > 2,4,6-trichlorophenol (TCP). Furthermore, exposure to TCP in vivo resulted in noticeable alterations in urinary metabolic profiles. Ewaid et al. [102] found that exposure to 1,500 mg/kg chloroform resulted in significant kidney damage characterized by an increase in kidney weight, inflammation, renal cell proliferation, and proximal tubular necrosis. Liu et al. [103] pointed out that exposure to 2,6-dichloro-1,4-benzoquinone (DCBQ) caused nephrotoxicity in mice, and the potential mechanisms of nephrotoxicity induced by DCBQ might be through inflammation, oxidative stress, apoptosis, and gut microbiota. In addition, the altered gut microbiota displayed notable associations with the changed toxicological indicators following exposure to DCBQ. Therefore, various DBPs can induce renal toxicity in vivo, and DBPs can also alter the gut microbiota, thereby impacting renal function.

1.5 Hepatotoxicity

In addition to renal toxicity, the effects of DBPs on liver injury have also received attention. When individuals consume water containing elevated levels of DBPs over an extended period, the liver, a vital organ responsible for detoxification, could become susceptible to damage. The liver's intricate metabolic processes could be disrupted as DBPs infiltrate its tissues. This might lead to oxidative stress, inflammation, and disruption of liver enzyme activity, which are indicative of hepatotoxicity.

Exposure to DBPs has been confirmed to cause liver toxicity in vivo. Li et al. conducted a 12-week swimming training experiment on rats, and the level of free chlorine in chlorinated swimming pool water was controlled at 1.4–1.6 mg/L. The results showed that the liver was the most likely target organ of DBPs. Metabolomics analysis also proposed corresponding metabolic stress pathways and taurine-focused defense systems, and elucidated the potential toxicological mechanism of DBPs [104]. Epidemiological studies have indicated that elevated concentrations of urinary (THMs TCM, BDCM, DBCM, and TBM) are associated with the altered serum biomarkers of liver function, such as the positive associations between urinary DBCM and alanine aminotransferase, aspartate aminotransferase, total bilirubin, total protein, and albumin [105]. The results of a recent cross-sectional analysis of Chinese women of reproductive age showed that exposure to DCAA was associated with liver injury, which also extends the human evidence of hepatotoxicity of DBPs. In addition, similar findings were observed in trihalophenol-exposed mice, and after 12 weeks of exposure, trihalophenol caused a significant increase in serum levels of alkaline phosphatase and albumin [106]. Iodoform and chloroform also caused significant changes in red blood cells, white blood cells, hemoglobin (HGB), and other hematological indicators; glucose (GLU), total protein (TP), alanine aminotransferase (ALT), and other biochemical indicators [107].

The histopathological change in the liver was observed after DBP exposures. Ewaid et al. [102] conducted a study and found that exposure to 1,500 mg/kg chloroform for 14 days induced liver congestion, enlargement, and weight gain in mice. The primary feature of liver damage was centrilobular necrosis, with extensive necrosis observed at high doses. HANs, haloketones (HKs), and HAAs caused acute liver injury in mice along with the increase of hepatic macrophages [108]. In addition, liver inflammation was observed after trihalophenol and trichloroacetamide (TCAcAm) exposure [106, 109].

DBP exposures could induce oxidative stress in the liver, which might be the potential mechanism of liver toxicity. Twelve-week exposure to trihalophenols disrupted the glutathione metabolic pathway and impaired the antioxidant and detoxification capacity of the liver in mice according to integrated data of the metabolomic and transcriptomic analysis [106]. Exposure to monohaloacetamides (monoHAcAms) or bromoacetonitriles (BANs) also caused hepatic oxidative stress in mice [110, 111]. Additionally, exposure to DBPs, such as monoHAcAms and BANs, could disrupt amino acid, energy, and lipid metabolism in mouse liver.

The compounds within DBPs, such as THMs and HAAs, have been identified as potential culprits behind these toxic effects. In addition, certain individuals, such as those with existing liver diseases or impaired detoxification mechanisms, might be more vulnerable to DBPs-induced liver toxicity.

2 In Vitro Toxicity of DBPs

2.1 Cytotoxicity

In vitro bioassay is widely used to evaluate the quality of drinking water. The commonly used evaluation methods for DBPs-induced cytotoxicity include crystal violet staining, cell counting kit 8 (CCK-8), 3-(4,5-dimethylthiazol-2-yl)-2,5-diphenyl-2H-tetrazolium bromide (MTT) assay, 3-(4,5-dimethylthiazol-2-yl)-5-(3-carboxymethoxyphenyl) -2- (4-sulfophenyl) -2H tetrazolium (MTS) assay, the neutral red uptake (NRU) assay, and real-time cell analysis (RTCA). This section introduces the cytotoxicity of various DBPs on different cell lines based on different evaluation methods, and the details are listed in Table 3. Additionally, the cytotoxicity of DBPs on cells from target organs was also included.

Plewa et al. have evaluated the cytotoxicity of different DBPs on Chinese hamster ovary (CHO) cells using crystal violet staining. In these studies, CHO cells were exposed to a single DBP for 72 h, and finally, the minimum cytotoxic concentration and median lethal concentration (LC_{50}) were obtained. By this means, the rank order of DBPs was obtained based on their cytotoxicity, and a comparative database for the cytotoxicity of DBPs was generated. Plewa et al. have systematically analyzed the in vitro chronic cytotoxicity of 103 compounds from 11 DBP chemical classes that included halomethanes, HAAs, haloacids, halophenolics, HANs, cyanogen halides, nitrosamines and nitramines, haloacetamides, halonitromethanes,

Table 3 The comparison of cytotoxicity induced by DBPs

Methods	Cell lines	DBPs	Lowest cytotox. conc. (M)	LC_{50} (M)	Reference
Crystal violet staining	CHO	2,4,6-triiodophenol	5×10^{-6}	4.37×10^{-5}	[115]
		4-iodo-2-methylphenol	2.5×10^{-5}	1.63×10^{-4}	
		4-iodophenol	5×10^{-5}	2.16×10^{-4}	
		2-iodophenol	1.5×10^{-4}	6.01×10^{-4}	
		Bromoacetic acid	2.0×10^{-6}	9.60×10^{-6}	[113, 131–135]
		Chloroacetic acid	2.5×10^{-4}	8.10×10^{-4}	
		Iodoacetic acid	5.0×10^{-7}	2.95×10^{-6}	
		Dibromoacetic acid	2.0×10^{-4}	5.90×10^{-4}	
		Dichloroacetic acid	2.0×10^{-3}	7.30×10^{-3}	
		Diiodoacetic acid	1.0×10^{-4}	3.32×10^{-4}	
		Bromochloroacetic acid	3.0×10^{-4}	7.78×10^{-4}	
		Bromoiodoacetic acid	2.5×10^{-4}	8.97×10^{-4}	
		Chloroiodoacetic acid	1.7×10^{-4}	3.04×10^{-4}	
		Tribromoacetic acid	5.0×10^{-6}	8.50×10^{-5}	
		Trichloroacetic acid	4.0×10^{-4}	2.40×10^{-3}	
		Chlorodibromoacetic acid	1.0×10^{-4}	2.02×10^{-4}	
		Bromodichloroacetic acid	5.0×10^{-4}	6.85×10^{-4}	
		3,3-Dibromo-4-oxopentanoic acid	5.0×10^{-6}	1.64×10^{-5}	[113, 114, 135]
		3-Bromo-3-chloro-4-oxopentanoic acid	1.0×10^{-5}	2.89×10^{-5}	
		(Z)-3-Bromo-3-iodo-propenoic acid	7.5×10^{-5}	1.89×10^{-4}	
		(E)-3-Bromo-3-iodo-propenoic acid	2.5×10^{-5}	1.45×10^{-4}	
		(E)-3-Bromo-2-iodo-propenoic acid	1.8×10^{-5}	4.36×10^{-5}	
		3,3-Dibromopropenoic acid	2.5×10^{-5}	2.95×10^{-4}	
		(E)-2-Iodo-3-methylbutenedioic acid	7.0×10^{-4}	9.44×10^{-4}	
		2,3,3-Tribromopropenoic acid	7.5×10^{-4}	1.64×10^{-3}	
		2-Bromobutenedioic acid	1.0×10^{-3}	2.06×10^{-3}	
		2,3-Dibromopropenoic acid	1.0×10^{-3}	2.20×10^{-3}	
		2-Bromo-3-methyl-butenedioic acid	4.8×10^{-3}	5.27×10^{-3}	
		4-Hydroxy-3,5-diiodo-1-phenyl	3.3×10^{-5}	3.32×10^{-5}	[114]
		2,4,6-Triiodo-1-phenol	6.0×10^{-5}	7.01×10^{-5}	
		4-Hydroxy-3-iodo-1-phenolic acid	1.5×10^{-4}	3.18×10^{-4}	
		4-Hydroxy-3,5-diiodo-1-phenolic acid	1.1×10^{-4}	2.88×10^{-4}	
		4-Hydroxy-3-iodophenyl	1.0×10^{-4}	4.08×10^{-4}	

(continued)

Table 3 (continued)

Methods	Cell lines	DBPs	Lowest cytotox. conc. (M)	LC_{50} (M)	Reference
		Bromoacetonitrile	1.0×10^{-6}	3.21×10^{-6}	[136]
		Chloroacetonitrile	5.0×10^{-5}	6.83×10^{-5}	
		Iodoacetonitrile	1.0×10^{-7}	3.30×10^{-6}	
		Dibromoacetonitrile	1.0×10^{-6}	2.85×10^{-6}	
		Dichloroacetonitrile	1.0×10^{-5}	5.73×10^{-5}	
		Bromochloroacetonitrile	7.0×10^{-6}	8.46×10^{-6}	
		Trichloroacetonitrile	2.5×10^{-5}	1.60×10^{-4}	
Crystal violet staining	CHO	Cyanogen bromide	1.0×10^{-6}	2.09×10^{-5}	[114]
		Cyanogen chloride	3.0×10^{-3}	3.25×10^{-3}	
		Cyanogen iodide	1.0×10^{-6}	9.01×10^{-5}	
		Chlorodiiodomethane	1.0×10^{-4}	2.41×10^{-3}	[113, 135]
		Triiodomethane	1.0×10^{-5}	6.60×10^{-5}	
		Dibromoiodomethane	1.5×10^{-3}	1.91×10^{-3}	
		Bromochloroiodomethane	2.2×10^{-3}	2.42×10^{-3}	
		Tribromomethane	1.0×10^{-4}	3.96×10^{-3}	
		Chlorodibromomethane	7.5×10^{-4}	5.36×10^{-3}	
		Trichloromethane	6.0×10^{-3}	9.62×10^{-3}	
		Bromodichloromethane	4.0×10^{-3}	1.15×10^{-2}	
		Dichloroiodomethane	2.0×10^{-3}	4.13×10^{-3}	
		Bromodiiodomethane	1.5×10^{-3}	1.40×10^{-3}	
		Bromoacetaldehyde	8.0×10^{-6}	1.73×10^{-5}	[137]
		Chloroacetaldehyde	5.0×10^{-7}	3.51×10^{-6}	
		Iodoacetaldehyde	5.0×10^{-6}	6.00×10^{-6}	
		Dibromoacetaldehyde	2.0×10^{-6}	4.70×10^{-6}	
		Dichloroacetaldehyde	8.0×10^{-6}	2.92×10^{-5}	
		Bromochloroacetaldehyde	2.5×10^{-6}	5.34×10^{-6}	
		Tribromoacetaldehyde	2.0×10^{-6}	3.58×10^{-6}	
		Trichloroacetaldehyde	3.7×10^{-4}	1.16×10^{-3}	
		Dibromochloroacetaldehyde	4.0×10^{-6}	5.15×10^{-6}	
		Bromodichloroacetaldehyde	1.0×10^{-5}	2.04×10^{-5}	
		Bromonitromethane	NA	7.08×10^{-6}	[132]
		Chloronitromethane	NA	5.29×10^{-4}	
		Dibromonitromethane	NA	6.09×10^{-6}	
		Dichloronitromethane	NA	3.73×10^{-4}	
		Bromochloronitromethane	NA	4.05×10^{-5}	
		Tribromonitromethane	NA	8.57×10^{-6}	
		Trichloronitromethane	NA	5.36×10^{-4}	
		Bromodichloronitromethane	NA	1.32×10^{-5}	
		Dibromochloronitromethane	NA	6.88×10^{-6}	
		Bromoacetamide	5.0×10^{-7}	1.89×10^{-6}	[138]
		Chloroacetamide	7.5×10^{-5}	1.48×10^{-4}	
		Iodoacetamide	5.0×10^{-7}	1.42×10^{-6}	

(continued)

Table 3 (continued)

Methods	Cell lines	DBPs	Lowest cytotox. conc. (M)	LC_{50} (M)	Reference
		Dibromoacetamide	2.5×10^{-6}	1.22×10^{-5}	
		Dichloroacetamide	8.0×10^{-4}	1.92×10^{-3}	
		Diiodoacetamide	2.5×10^{-8}	6.78×10^{-7}	
		Bromoiodoacetamide	2.0×10^{-6}	3.81×10^{-6}	
		Chloroiodoacetamide	2.0×10^{-6}	5.97×10^{-6}	
		Tribromoacetamide	2.0×10^{-6}	3.14×10^{-6}	
		Bromochloroacetamide	1.0×10^{-6}	1.71×10^{-5}	
		Dibromochloroacetamide	1.0×10^{-6}	4.75×10^{-6}	
		Bromodichloroacetamide	2.0×10^{-6}	8.68×10^{-6}	
		Trichloroacetamide	5.0×10^{-4}	2.05×10^{-3}	
		Bromate	NA	9.63×10^{-4}	[131, 139, 140]
		MX (3-chloro-4 (dichloromethyl)-5-hydroxy-2[5H]-furanone	NA	2.75×10^{-4}	
		4-Hydroxy-3,5-diiodo-1-nitrobenzine	2.0×10^{-5}	3.35×10^{-5}	
		Tribromopyrrole	1.0×10^{-5}	6.10×10^{-5}	
		2,6-Dichloro-p-benzoquinone	5.0×10^{-6}	1.12×10^{-5}	
		2,6-Dibromo-p-benzoquinone	1.0×10^{-5}	1.44×10^{-5}	
		5-Amino-N1, N3 -bis (1,3-dihydroxypropan-2-yl)-2,4,6-triiodoisophthalamide	9.0×10^{-4}	1.44×10^{-3}	
		N1, N3 -Bis(1,3-dihydroxy-propan-2-yl)-2,4,6-triiodo-5-nitroisophthalamide	9.0×10^{-4}	9.34×10^{-4}	
		4-Chloro-N1, N3 -bis (1,3-dihydroxypropan-2-yl)-2,6-diiodo-5-nitroiso-phthalamide	1.0×10^{-3}	1.30×10^{-3}	
		4,6-Dichloro-N1, N3 -bis (1,3-dihydroxypropan-2-yl)-2-iodo-5-nitroisophthalamide	2.5×10^{-4}	8.23×10^{-4}	
RTCA	CHO-K1	OH-DBBQ	NA	5.04×10^{-5}	[122]
		OH-DCBQ	NA	9.06×10^{-5}	
		OH-DCMBQ	NA	2.40×10^{-5}	
		DCMBQ	NA	1.59×10^{-5}	
		OH-TCBQ	NA	6.98×10^{-5}	
		2,6-DCBQ	NA	3.22×10^{-5}	[120]
		2,5-DCBQ	NA	2.41×10^{-5}	
		2,6-DBBQ	NA	2.60×10^{-5}	
		2,5-DBBQ	NA	2.09×10^{-5}	

(continued)

Table 3 (continued)

Methods	Cell lines	DBPs	Lowest cytotox. conc. (M)	LC_{50} (M)	Reference
		2,3-DIBQ	NA	1.07×10^{-5}	
		2-CBQ	NA	1.31×10^{-5}	
		TCBQ	NA	3.51×10^{-5}	
		TetraCBQ	NA	2.40×10^{-5}	
		3-cl-Tyr	NA	1.15×10^{-2}	[121]
		3-Br-Tyr	NA	1.16×10^{-2}	
		3-I-Tyr	NA	1.06×10^{-2}	
		3,5-di-cl-Tyr	NA	4.20×10^{-3}	
		3,5-di-Br-Tyr	NA	1.90×10^{-3}	
		3,5-di-I-Tyr	NA	5.50×10^{-4}	
		3-I-Tyr-Ala	NA	5.00×10^{-3}	
		3,5-di-I-Tyr-Ala	NA	7.80×10^{-4}	
		NDMA	NA	3.10×10^{-2}	[123]
		NPyr	NA	3.20×10^{-2}	
		NPip	NA	1.10×10^{-2}	
		NDPhA	NA	7.20×10^{-4}	
	A549	NDMA	NA	2.20×10^{-2}	
		NPyr	NA	5.00×10^{-2}	
		NPip	NA	1.50×10^{-2}	
		NDPhA	NA	7.50×10^{-4}	
	T24	NDMA	NA	1.50×10^{-2}	
		NPyr	NA	1.40×10^{-2}	
		NPip	NA	4.70×10^{-3}	
		NDPhA	NA	5.90×10^{-4}	
	HepG2	NDMA	NA	9.50×10^{-2}	
		NPyr	NA	5.20×10^{-2}	
		NPip	NA	1.60×10^{-2}	
		NDPhA	NA	1.90×10^{-3}	
CCK-8/ MTT/ MTS	HepG2	CAN	NA	7.34×10^{-4}	[127]
		DCAN	NA	2.34×10^{-3}	
		TCAN	NA	1.09×10^{-3}	
		BAN	NA	6.97×10^{-6}	
		DBAN	NA	8.08×10^{-6}	
	T24	2,6-DCBQ	NA	9.45×10^{-5}	[141]
		DCMBQ	NA	1.10×10^{-4}	
		TCBQ	NA	1.51×10^{-4}	
		2,6-DBBQ	NA	1.42×10^{-4}	
	CHO-K1	2-ClBZAM	1.00×10^{-4}	1.04×10^{-3}	[119]
		2-BrBZAM	1.00×10^{-4}	1.73×10^{-3}	
		2-IBZAM	8.00×10^{-5}	0.68×10^{-3}	
		2,4-DCBZAM	8.00×10^{-4}	2.79×10^{-3}	
		2,6-DCBZAM	4.00×10^{-4}	2.17×10^{-3}	
		3,5-DBBZAM	1.00×10^{-5}	1.4×10^{-4}	

(continued)

Table 3 (continued)

Methods	Cell lines	DBPs	Lowest cytotox. conc. (M)	LC_{50} (M)	Reference
		5-Br-2-OH-BZAM	1.00×10^{-4}	1.66×10^{-3}	
		2-Cl-4-OH-BZAM	1.00×10^{-3}	1.66×10^{-2}	
		2-Chlorophenyl-acetonitrile	3.20×10^{-5}	1.33×10^{-4}	[142]
		3,4-Dichlorophenyl-acetonitrile	5.00×10^{-6}	8.30×10^{-5}	
		2-CPAcAm	2.50×10^{-4}	4.97×10^{-3}	[143]
		3-CPAcAm	2.50×10^{-4}	3.07×10^{-3}	
		2,4-DCPAcAm	5.00×10^{-4}	1.40×10^{-3}	
		2,6-DCPAcAm	5.00×10^{-4}	1.14×10^{-3}	
		3,4-DCPAcAm	1.00×10^{-3}	1.42×10^{-3}	
		2-BPAcAm	2.00×10^{-3}	3.69×10^{-3}	
		3-BPAcAm	2.00×10^{-3}	2.80×10^{-3}	
		2,6-DBPAcAm	1.88×10^{-4}	7.17×10^{-4}	
		2-IPAcAm	1.00×10^{-3}	1.98×10^{-3}	
		2,6-BCPAcAm	1.25×10^{-4}	1.00×10^{-3}	
		3,5-dichloro-4-hydroxybenzonitrile	6.30×10^{-5}	3.21×10^{-4}	[144]
		3,5-dichloro-2-hydroxybenzonitrile	1.00×10^{-4}	2.33×10^{-4}	
		5-bromo-3-chloro4-hydroxybenzonitrile	3.80×10^{-5}	1.87×10^{-4}	
CCK-8/ MTT/ MTS	Caco-2	Bromoacetic acid	NA	3.18×10^{-5}	[99, 145]
		2,4,6-Tribromophenol	NA	2.15×10^{-4}	
		3,5-Dibromo-4-hydroxybenzoic acid	NA	1.78×10^{-3}	
		3,5-Dibromo-4-hydroxybenzaldehyde	NA	6.22×10^{-4}	
	SH-SY5Y	Bromoacetic acid	NA	1.91×10^{-5}	
		2,4,6-Tribromophenol	NA	1.55×10^{-4}	
		3,5-Dibromo-4-hydroxybenzoic acid	NA	1.97×10^{-3}	
		3,5-Dibromo-4-hydroxybenzaldehyde	NA	5.87×10^{-4}	
		2,6-DCBQ	NA	2.69×10^{-5}	
	CHO-K1	5-bromo-3-chloro2-hydroxybenzonitrile	7.50×10^{-5}	1.93×10^{-4}	[144]
		3,5-dibromo-4-hydroxybenzonitrile	3.80×10^{-5}	1.58×10^{-4}	
		3,5-dibromo-2-hydroxybenzonitrile	1.00×10^{-4}	1.63×10^{-4}	
		3,5-diiodo-4-hydroxybenzonitrile	1.90×10^{-5}	9.50×10^{-4}	
		2,4,6-tribromophenol	7.56×10^{-5}	9.99×10^{-5}	[146]

(continued)

Table 3 (continued)

Methods	Cell lines	DBPs	Lowest cytotox. conc. (M)	LC_{50} (M)	Reference
		2,6-dibromo-4-chlorophenol	1.75×10^{-5}	1.33×10^{-4}	
		2,6-dichloro-4-bromophenol	2.07×10^{-5}	1.75×10^{-4}	
		2,4,6-trichlorophenol	1.52×10^{-4}	2.44×10^{-4}	
		2,6-dibromo-4-nitrophenol	2.36×10^{-5}	5.07×10^{-5}	
		2-bromo-4-chloro-6-nitrophenol	5.94×10^{-5}	8.01×10^{-5}	
		2,6-dichloro-4-nitrophenol	4.81×10^{-5}	6.73×10^{-5}	
		3,5-dibromo-4-hydroxybenzaldehyde	1.79×10^{-5}	1.30×10^{-4}	
		3-bromo-5-chloro-4-hydroxybenzalde hyde	1.27×10^{-4}	2.04×10^{-4}	
		3,5-dichloro-4-hydroxybenzaldehyde	1.57×10^{-4}	3.2×10^{-4}	
		3,5-dibromo-2-hydroxybenzoic acid	3.38×10^{-5}	2.21×10^{-4}	
		3-bromo-5-chloro-2-hydroxybenzoic acid	2.39×10^{-4}	3.56×10^{-4}	
		3,5-dichloro-2-hydroxybenzoic acid	4.83×10^{-5}	4.77×10^{-4}	
		3,5-dibromo-4-hydroxybenzoic acid	3.38×10^{-5}	1.96×10^{-3}	
		3,5-dichloro-4-hydroxybenzoic acid	4.83×10^{-5}	2.16×10^{-3}	
	GES-1	DIAcAm	5.00×10^{-7}	7.29×10^{-7}	[118]
		IAcAm	5.00×10^{-7}	1.06×10^{-6}	
		TBAcAm	5.00×10^{-7}	1.40×10^{-6}	
		BAcAm	1.00×10^{-6}	2.13×10^{-6}	
		DBCAcAm	1.00×10^{-6}	2.24×10^{-6}	
		BIAcAm	4.00×10^{-6}	4.52×10^{-6}	
		BDCAcAm	2.00×10^{-6}	4.92×10^{-6}	
	GES-1	CIAcAm	8.00×10^{-6}	9.64×10^{-6}	[118]
		BCAcAm	6.40×10^{-5}	7.32×10^{-5}	
		DBAcAm	6.40×10^{-5}	8.96×10^{-5}	
		CAcAm	6.40×10^{-5}	1.04×10^{-4}	
		TCAcAm	2.50×10^{-3}	4.52×10^{-3}	
		DCAcAm	2.03×10^{-4}	8.93×10^{-3}	
	HaCaT	DIAcAm	1.00×10^{-6}	1.61×10^{-6}	
		IAcAm	1.00×10^{-6}	1.49×10^{-6}	
		TBAcAm	2.50×10^{-7}	1.42×10^{-6}	
		BAcAm	1.00×10^{-6}	3.34×10^{-6}	
		DBCAcAm	2.00×10^{-6}	3.09×10^{-6}	
		BIAcAm	1.56×10^{-6}	5.55×10^{-6}	
		BDCAcAm	4.00×10^{-6}	6.10×10^{-6}	

(continued)

Table 3 (continued)

Methods	Cell lines	DBPs	Lowest cytotox. conc. (M)	LC_{50} (M)	Reference
		ClAcAm	3.20×10^{-5}	3.17×10^{-5}	
		BCAcAm	6.4×10^{-5}	1.54×10^{-4}	
		DBAcAm	1.28×10^{-4}	1.52×10^{-4}	
		CAcAm	6.40×10^{-5}	1.40×10^{-4}	
		TCAcAm	6.25×10^{-4}	2.14×10^{-3}	
		DCAcAm	2.03×10^{-4}	2.36×10^{-2}	
CCK-8/MTT/MTS	HEK 293T	IAcAm	NA	3.49×10^{-6}	[125]
		DIAcAm	NA	7.90×10^{-7}	
		BIAcAm	NA	3.96×10^{-6}	
		ClAcAm	NA	1.18×10^{-5}	
		BAcAm	NA	3.80×10^{-6}	
		DBAcAm	NA	1.55×10^{-4}	
		TBAcAm	NA	1.27×10^{-5}	
		BCAcAm	NA	1.38×10^{-4}	
		DBCAcAm	NA	3.11×10^{-6}	
		BDCAcAm	NA	1.92×10^{-5}	
		CAcAm	NA	1.11×10^{-4}	
		DCAcAm	NA	2.37×10^{-2}	
		TCAcAm	NA	9.78×10^{-3}	
		IAN	NA	5.04×10^{-6}	
		BAN	NA	1.11×10^{-5}	
		DBAN	NA	2.55×10^{-5}	
		BCAN	NA	2.00×10^{-5}	
		CAN	NA	2.80×10^{-4}	
		DCAN	NA	4.44×10^{-4}	
		TCAN	NA	1.44×10^{-3}	
		BNM	NA	3.73×10^{-5}	
		DBNM	NA	3.13×10^{-5}	
		TBNM	NA	2.18×10^{-5}	
		CNM	NA	2.37×10^{-4}	
		DCNM	NA	1.59×10^{-3}	
		TCNM	NA	1.61×10^{-4}	
		BCNM	NA	4.56×10^{-5}	
		BDCNM	NA	1.54×10^{-4}	
		DBCNM	NA	6.64×10^{-5}	
		MCAA	NA	1.21×10^{-3}	[126]
		MBAA	NA	1.61×10^{-5}	
		MIAA	NA	6.17×10^{-6}	
	HEK	2,6-DCBQ	NA	5.80×10^{-5}	[147]
		DCMBQ	NA	3.70×10^{-5}	
		TCBQ	NA	1.20×10^{-4}	
		2,6-DBBQ	NA	6.30×10^{-5}	

haloacetaldehydes, and bromate [113, 114]. With CHO cell cytotoxicity, the rank order for DBPs could be achieved. The cytotoxicity of most DBPs on CHO is at the μM level, while the cytotoxicity of nitrosamines and halomethanes is at the mM level. The cytotoxicity of the same class of DBPs generally follows the rule of iodo- > bromo-> > chloro-DBPs. In addition, compared to regulated carbonaceous DBPs (such as HAAs), nitrogenous DBPs (such as HANs) have stronger cytotoxicity to CHO cells [115].

CCK-8, MTT assay, and MTS assay are similar colorimetric assays that examine the formation of formazan, which is proportional to the number of live cells. Many DBPs have been examined for cytotoxicity using these colorimetric assays. In the cytotoxicity study of aliphatic halogenated DBPs and aromatic halogenated DBPs on human hepatoma cells (HepG2 cells) [116], the cytotoxicity ranking was as follows: iodo-DBPs>bromo-DBPs>chloro-DBPs, which were consistent with the findings from cytotoxicity in CHO cells. The toxicity of different aliphatic monohalogenated DBPs to HepG2 was usually higher than that of dihalogenated and trihalogenated DBPs with the same halogen substitution, while the aromatic DBPs with more halogen exhibited higher cytotoxicity under the same category [117]. Ding et al. exposed 13 HAcAms to human gastric epithelial GES-1 and immobilized keratinocytes HaCaT, and the findings revealed that different cell lines had different sensitivities to the same DBPs [118]. In the cytotoxicity study of eight halobenzamides (HBZAMs) on CHO-K1 cells, seven HBZAMs exhibited higher cytotoxicity than the regulated DCAA [119].

RTCA is a cellular microelectronic sensing technology that measures cell viability by monitoring impedance changes between cells and microelectrodes in real-time, providing a sensitive, convenient, and multi-endpoints method. Therefore, RTCA has been developed as an alternative technique for chemical cytotoxicity testing. Researchers used CHO-K1 cells to evaluate the cytotoxicity of a variety of DBPs. In the study of HBQs, the impact of isomeric structure and halogen substitution on cytotoxicity was highlighted [120]. Studies of newly emerging DBPs halogenated amino acids and peptides have shown that dihalogenated tyrosine compounds were more cytotoxic than their monohalogenated analogs [121]. However, studies on HBQs and their transformation products halo-hydroxyl-benzoquinones (OH-HBQs) found that both HBQs and OH-HBQs were much more toxic than regulated DBPs [122]. With RTCA, Boyd et al. examined the cytotoxicity of four nitrosamines [NDMA, N-nitrosodiphenylamine (NDPhA), N-nitrosopiperidine (NPip), and NPyr using four cell lines, including T24 bladder, HepG2 liver, A549 lung carcinoma cells, and CHO-K1 cells. The in vitro cytotoxicity values (IC_{50}) for nitrosamines were in a range of 0.6–95 mM in all four cell lines, and T24 cells were the most sensitive to nitrosamine exposure among the four cell lines tested (T24 > CHO > A549 > HepG2) [123].

Besides the comparison for rank order of cytotoxicity of DBPs, the cytotoxicity study focusing on cells from target organs of DBPs was studied as well, including neural cells, kidney cells, liver cells, bladder cells, etc. In vitro studies have shown that various DBPs are cytotoxic to neural cells, and DBPs could reduce cellular activity and inhibit the proliferation and differentiation of neural cells [124]. Fu et al.

examined the effects of four DBPs on the proliferation and differentiation of human neural stem cells (hNSCs) and found that BAA and CAA (0.5 μM) affected neural differentiation, while DCBQ and DBBQ (1 μM) influenced the proliferation of hNSCs [124]. In addition, Liu et al. found that DCBQ decreased cell viability and neural differentiation of SH-SY5Y cells, and the IC_{50} values were in a range of 20–30 μM after 24–96 h exposure [99]. Different DBPs can cause nephrotoxicity in vitro, and the mechanisms underlying renal toxicity encompass DNA damage, oxidative stress, apoptosis, inflammation, and more. Miao et al. also used the mouse glomerular mesangial cell line MES-13 for in vitro cellular experiments, an increase in cell proliferation was noted in MES-13 cells following exposure to 2,4,6-trihalophenols, particularly in the cases of TBP and TIP [101]. Haloacetaldehydes (HALs) and iodo-THMs could cause toxicological and pathological damage to human embryonic kidney (HEK293) cell. N-DBPs have a higher cytotoxicity on HEK 293T cells [125]. HBQs including DCBQ, 2,6-dibromo-1,4-benzoquinone (DBBQ), and TCBQ, also exhibited cytotoxic effects on human embryonic kidney cells (HEK293 cells), and led to an elevation in reactive oxygen species (ROS) levels. The toxic effect of MCAA on HEK293 cells may be related to the response of cellular molecules [126]. Human Hepatocellular (Hep3B) cells display greater susceptibility to HANs and HKs compared to THMs. The cytotoxicity order of five types of HANs [MCAN, DCAN, trichloroacetonitrile (TCAN), MBAN, DBAN] to liver cancer cell lines (HepG2) was MBAN > DBAN > MCAN > TCAN > DCAN; exposure to DCAN, TCAN, and DBAN notably inhibited HepG2 cell viability, induced DNA damage and promoted ROS generation [127]. In addition [128], DCBQ led to an elevation in ROS levels and a decrease in cell viability in HepG2 cells. In the study of the toxicity of HBQs to bladder cancer T24 cells, the MTT assay showed that the IC_{50} values of phenazine at 24 h and 48 h were 368 μM and 82 μM, respectively [129]. In addition, the IC_{50} values of six HBQS in human uroepithelial cells (SV-HUC-1) at 24–72 h were determined to be at 20–50 μM levels [130].

Many mechanisms cause cytotoxicity. Previous studies have shown that the mechanism by which most DBPs produce cytotoxicity might be related to oxidative stress [148, 149]. Stalter et al. tested the cytotoxic pathways of 50 DBPs and found that 49 of them activated Nrf2-mediated oxidative stress response [150]. Jiao et al. revealed IAA exposure increased ROS levels in oocytes in a dose-dependent manner, while the addition of antioxidant N-acetyl-L-cysteine (NAC) reduced ROS levels and restored meiotic maturation, indicating that IAA-induced oocyte maturation failure was mainly mediated by oxidative stress [66]. Studies on the cytotoxicity and mechanism of HBQs have shown that HBQs induced ROS generation in cells and affected the cellular antioxidant system, further triggering oxidative damage to cell proteins and DNA, leading to cytotoxicity, potential genotoxicity, and carcinogenicity [151]. A study indicated that oxidative stress-induced mitochondrial-dependent apoptosis regulated by the p38 MAPK activation pathway plays an important role in CAA-induced neuronal cell death [96]. Similarly, studies on N-DBPs have found that DNA methylation and oxidative stress might be the reasons for the cytotoxic effects induced by NDMA [152]. In the study by Xue et al. on the toxicity of seven HANs to HaCaT cells, it was found that HANs induced oxidative stress and activated Nrf2-ARE-mediated antioxidant responses [153]. In addition to oxidative stress, Zhou et al. found that glutathione methoxy HBQ (SG Me O-HBQ) was the notable metabolite significantly associated with abnormal cytotoxic changes of HBQs. By hydroxylation and glutathione methylation, detoxifying hydroxyl HBQ (OH-HBQ) and SG-HBQ were generated, followed by methylation to produce SG-Me-O-HBQ with enhanced toxicity, leading to cytotoxicity [154].

Overall, the cytotoxicity of most DBPs may be related to oxidative stress, oxidative damage, electrophilic/nucleophilic reactions, and methylation. The diverse structure of DBPs could also influence their cytotoxicity, and the discussion will be included in the QSAR models.

2.2 *Genotoxicity*

There are several methods to determine genotoxicity, with the Ames test, single cell gel electrophoresis (SCGE) test, and micronucleus test as the most commonly used methods for DBPs-induced genotoxicity. These three methods evaluate the genotoxicity of DBPs from different genetic endpoints. In addition, there were also several other methods used for genotoxicity, such as the examination of γH2AX, 8-OHdG.

SCGE, also known as comet assay, quantitatively measures the level of induced genomic DNA damage in a single nucleus through the tail length and fluorescence intensity of DNA fragments. Plewa et al. set up an acute genotoxicity database for 103 DBPs from 11 DBP chemical classes, which were the same DBPs as in the cytotoxicity database [114]. Here, a single DBP was exposed to CHO cells for 4 h, followed by an SCGE test, and 50% tail DNA value and lowest genotoxic concentration were generated for comparison of genotoxicity. The rank order of genotoxicity also followed the pattern of iodo - > bromo -> > chloro-DBPs, which was consistent with the cytotoxicity pattern. Compared to regulated carbonaceous DBPs, nitrogenous DBPs have higher genotoxicity to CHO cells. In addition, the SCGE test was also used to determine genotoxicity by other researchers. Manasfi et al. studied the genotoxicity of chloral hydrate (CH) and bromohydrate (BH) in CHO cells, and only BH demonstrated statistically significant DNA damage even without S9 rat liver microsome [155]. However, in the presence or absence of the S9 mixture, CH did not induce the DNA strand. Zhang et al. compared the DNA damage induced by 15 DBPs in HepG2 cells, and found that HepG2 cells combined with SCGE are suitable to evaluate the genotoxicity of DBPs and the genotoxicity also followed the order as iodinated DBPs > brominated DBPs > chlorinated DBPs [156]. Dong et al. revealed that the nitrosamine mixture was treated at 1,000-fold actual drinking-water concentration induced DNA damage in HepG2 cells, while a single nitrosamine at the same concentration did not cause DNA damage[157]. However, a study found that there was no DNA damage in T24 cells after chronic exposure to low doses of HAAs for 8 weeks [158].

Micronucleus test is a genotoxicity test with chromosome damage as the endpoint. Li et al. observed chromosome damage of human gastric cancer cells, bladder cancer cells, and human colon cancer cells by four HBQs [159]. Tu et al. found that HBQs also caused significant chromosomal damage in human hepatocytes (L-02), and the rank order of HBQ isomers on genotoxicity was 2,5-HBQs > their corresponding 2,6-HBQs [160]. The nitrosamine mixtures at 1,000-fold actual water level could induce chromosomal damage [157]. Both BAA and CAA at 10 μM induce micronucleus formation in A549 cells, while diiodoacetic acid (DIAA) requires more than 100 μM to produce significant micronuclei [161]. However, Liviac et al. found that IAA, BAA, and CAA could not significantly increase micronucleus frequency in human lymphoblastic TK6 cells, and this phenomenon may be due to that micronucleus test only showed fixed DNA damage, while primary DNA damage is easily repaired [162]. Additionally, BH did not exhibit chromosomal damage in CHO cells, but BH demonstrated mutagenic activity in Salmonella strain TA100 and DNA damage in comet assay in CHO cells, indicating the genotoxic mechanism of BH involves inducing mutations and DNA damage rather than chromosomal damage or aberrations [155].

Ames test, also named as Salmonella typhimurium reverse mutation assay, determines the genetic mutations as endpoints. For the regulated DBPs, most of them are mutagenic by Ames test in the presence or absence of S9. THMs except for chloroform were mutagenic in the presence of GSTT1-1 enzymes, and their rank order of mutagenic potency was chloroform > BDCM > DBCM [163]. HAAs have the mutagenic potential except for TCAA. Chlorite may form ROS, however, there is little data on its toxicity. Nitrosamines are more sensitive to show toxicity during the Ames test with *Salmonella typhimurium YG7108* than with SCGE, with 1,4-dinitropiperazine, N-nitrosodiethanolamine, N-nitrodiethanolamine, and N-nitromonoethanolamine demonstrating weak direct mutagenic effects, while 1-nitrosopiperazine and 1-nitropiperazine were not mutagenic [164]. However, very few data with the Ames test were available for the unregulated DBPs. Halonitromethanes (HNMs) were found to be mutagenic with or without S9 [165]. As a strong mutagen, MX had a mutagenic ability three orders of magnitude higher than HNMs, with a minimum effective concentration of 50 nM[166]. CH did not exhibit mutagenic effects in any of the four standard *Salmonella* strains, while BH produced mutagenicity in the TA100 strain [155]. Within the HANs, the chlorinated acetonitriles are mutagenic in the presence or absence of S9, whereas brominated analogs are generally not mutagenic in *Salmonella*.

3 Alternative Toxicity Tests Applied in DBPs Study

3.1 Zebrafish as a Model for Toxicity Test of DBPs

Zebrafish is a widely distributed ornamental fish in the world and is currently listed as the third largest model organism by the National Institutes of Health (NIH). It has many advantages, such as strong reproductive ability, in vitro fertilization and development, embryo transparency, short sexual maturity cycle, etc. Due to these merits, zebrafish has been widely applied in studies of DBP toxicities. There are multiple toxic effects of various DBPs on zebrafish, mainly including the effects on adult fish and the effects on embryos.

In the case of adult zebrafish, NDEA were lethal to zebrafish and further accumulated in their bodies, with the highest NDEA level in the kidney, followed liver and gills [167]. DCBQ demonstrated oxidative damage to the heart and brain of adult zebrafish after exposure for 96 h [168]. Exposure to 2,2-dichloroacetamide for 30 days resulted in apparent changes in the abundance and diversity of zebrafish gut microbiota [169]. Whereas, in an endocrine disruption study comparing the effects of nine DBPs on human and zebrafish estrogen receptors (hERα and zERα), IAA, CAN, and BAN triggered estrogenic activity in hERα, while IAA, chloroacetamide (CAM), and bromoacetamide (BAM) inhibited estrogenic activity in E2 in zERα, making zebrafish more susceptible to endocrine disruption than humans [170].

With zebrafish embryos, the developmental toxicity of DBPs was examined. Some DBPs could inhibit the growth of zebrafish embryos, induce embryo death, decrease hatching rate, affect locomotor behavior, and cause abnormal development of zebrafish embryos. THM exposures resulted in adverse developmental effects and significantly shortened tail length, while tribromoacetic acid and DCAA, as well as sodium bromate, expressed a notable increase in the probability of malformation [171]. Chloroacetamide (CAcAm), bromoacetamide (BAcAm), and iodoacetamide (IAcAm) could cause zebrafish morphological abnormalities such as body bending, pericardium and yolk sac edema, and also affect the locomotor behavior of zebrafish such as embryonic mobility, speed and relative movement distance [172]. A study has found that 17 typical halogenated DBPs caused abnormal development of zebrafish embryos, and affected the swimming ability of adult zebrafish [173].

Some DBPs demonstrated neurotoxicity to zebrafish embryos. Zhang et al. found that low-dose NDMA was not developmentally toxic but potentially neurobehaviorally toxic [174]. HBQ exposures induced neurotoxicity in zebrafish embryos, accompanied by changes in gene expression related to neurodevelopment and abnormal locomotor behavior, decreased levels of neurotransmitters, and inhibition of acetylcholinesterase (AchE) enzyme activity [175].

Cardiovascular developmental toxicity of DBPs was also observed in zebrafish embryos. DCBQ affected cardiovascular development in zebrafish embryos. Pericardial edema and significant atrial-ventricular segregation occurred after 72 h post -fertilization (hpf) of exposure to DCBQ (0, 2, 4, 8 μM), and the rate of atrial-ventricular segregation aberrations increased significantly with increasing dosage, following by a decrease in blood flow activity [176].

Some studies compared the toxicity of DBPs using zebrafish embryos. A study evaluated the developmental toxicity of seven unregulated DBPs (including nitrosamines, aldehydes, alcohols, and ketones) using zebrafish embryos, and showed yolk sac and pericardial edema, tail abnormalities, and spinal scoliosis at specific developmental stages of zebrafish (24, 48, 72, and 96 hpf), demonstrating higher toxicity compared to regulated DBPs [177]. The developmental toxicities of 15 chlorinated, brominated, and iodinated DBPs in zebrafish embryos have also been investigated, and the DBPs followed a ranked order of toxicity: monohalogenated acetamides > monohalogenated acetic acids > monohalogenated acetonitriles > nitrosamines, which was in good agreement with previous in vitro mammalian data. Dihalogenated and trihalogenated acetonitriles were more toxic than their monohalogenated analogs, while bromo- and iodo-DBPs were more toxic than their chlorinated analogs [178]. Whereas in the toxicity comparisons between novel cyclic DBPs and aliphatic DBPs in zebrafish embryos, cyclic DBPs typically exhibited higher developmental and acute toxicity than aliphatic DBPs [117].

3.2 Nematodes as a Model for Toxicity Test of DBPs

The nematode is the first animal whose genome has been completely sequenced, and these characteristics have made it become an important model animal in various fields such as biology, pharmacology, genetics, and toxicology. Particularly, it can be used in mechanism studies, high-throughput screening, and toxicity assessment. *Caenorhabditis elegans (C. elegans)* is a microscopic nematode with a transparent body, short life span, self-fertilization, and easy cultivation. In addition, *C. elegans* is sensitive to various environmental toxicants, and has proven to be an important model animal model in toxicology. Therefore, *C. elegans* has been applied in studies of DBP toxicities.

Zuo et al. revealed that DCBQ exposure increased the mortality of *C. elegans* and inhibited respiration, and exposure to DCBQ and NDMA resulted in DNA damage, further suggesting that DCBQ had a greater ability to induce general toxicity than some regulated DBPs and revealing the genotoxic effects of DCBQ in vivo [179]. DBAA could stimulate nematode growth and longevity at 50 μM but delay the onset of reproduction, and microarray analysis showed that DBAA also inhibited signal transduction and neural processes [180]. Additionally, DBAA displayed neural stimulation on nematodes and temporary effects on defecation intervals [181]. It has also been found that low concentrations of DBAA and NDMA did not exhibit toxic effects on nematodes but regulate resource allocation by extending nematode lifespan and increasing body length [182].

3.3 QSAR Models for Toxicity Test of DBPs

In recent years, researchers have applied the QSAR model to assess the cytotoxicity of DBPs. Log P, the log value of the partition coefficient of a solute between octanol and water, has been proven to be effective in properly predicting DBPs. Chen and Zhang [183] used Log P of 173 DBPs to construct QSAR models to evaluate DBPs toxicity. The models were developed using three molecular descriptors: energy of the lowest unoccupied molecular orbital (E_{LUMO}), number of chlorines, and number of carbons. Given the intricate characteristics of DBPs, the existing QSAR models demonstrated commendable performance in terms of their accuracy, reliability, and capacity to make predictions. Researchers have successfully developed a robust model for accurately predicting the cytotoxic effects of DBPs on CHO cells [184]. Jiang et al. [185]. employed a robust dimensionless parameter scaling approach to accurately establish the correlation between halogen substitution and the cytotoxicity of different groups of DBPs across three cell lines, namely, human breast carcinoma (MVLN), CHO, and HepG2. This method allows for a quantitative analysis without the requirement to consider absolute values or other confounding factors. The researchers discovered that MVLN cells demonstrated the highest sensitivity in evaluating the influence of halogen substitution on cyclic DBPs, while the CHO cells presented the highest sensitivity in assessing the impact of halogen substitution on aliphatic DBPs. In addition, the comparative cytotoxicity of 15 halogenated aromatic DBPs from four groups (HPs, HNPs, halohydroxybenzaldehydes, and halohydroxybenzoic acids) was assessed using CHO cells [146]. This study evaluated the probable causes of toxicity associated with the antioxidant system by molecular docking analysis, and the structural descriptors included physicochemical, quantum-chemical, and topological descriptors. The results obtained from QSAR analysis demonstrated the significance of the interactions involving aromatic DBPs in relation to catalase, as well as the electrophilic and nucleophilic reactivity of these aromatic DBPs in the mechanisms underlying their toxicity. In the cytotoxicity analysis of 29 N-DBPs on HEK 293T cells, two distinct QSAR models have been constructed to estimate the IC_{50} values. The first model was based on that of HANs and HAcAms, while the second model was based on that of HNMs [125]. The cytotoxicity mechanisms of halophenylacetamides (HPAcAms) were shown to be influenced significantly by their cellular absorption efficiency and the polarized distributions of electrons [143]. The QSAR analysis also found that these aromatic DBPs generally exhibited much greater toxicity compared to haloaliphatic DBPs [186]. Li et al. have set up a QSAR model for eight HBQs in CHO-K1 cells, and found E_{LUMO} and dipole moment are the two driven descriptors for HBQs' cytotoxicity [120]. Here, E_{LUMO} characterizes the electron affinity and indicates the ability of HBQs to produce free radicals, while dipole moment reflects the average electron distribution on a molecule, indicating the reactivity of HBQs.

QSAR analysis has also been applied to investigate the thiol reactivity of HANs in human cells. The order of reactivity for thiols could be approximated as follows: tribromoacetonitrile (TBAN) demonstrated the highest reactivity, followed by bromodichloroacetonitrile (BDCAN) and chlorodibromoacetonitrile (CDBAN), which exhibited similar reactivity levels, and TCAN had the lowest reactivity among the thiols. The observed toxicity measurements exhibited correlations with both membrane permeability and chemical reactivity [187]. Wang et al. found that HBQs, due to their electron-deficient structure, were the reactive electrophilic reagents that can easily react with various binucleophiles, leading to a decrease in GSH [188].

QSAR analysis can also be used to evaluate the genotoxicity and developmental toxicities of DBPs. Quantitative toxicogenomic methods were used to evaluate the genotoxicity of 20 DBPs, and found that the genetic toxicity of DBPs was positively correlated with topology and quantum chemical descriptors [189]. A study combined GC × GC-qMS along with QSAR Toolbox for the identification and prioritization of volatile and semi-volatile DBPs, and finally 58 DBPs were found to exhibit genotoxic effects or potentially genotoxicity [190]. A QSAR model of HBQ isomers on DNA reactivity and genotoxicity was constructed and dipole moment was demonstrated to be the key structural descriptor [160]. The developmental toxicity of aromatic halogenated DBPs in chlorinated saline sewage effluent was evaluated by Yang et al. [191] utilizing QSAR analysis. Among 19 DBPs, 2,5-dibromohydroquinone exhibited the highest developmental toxicity, followed by 2,6-diiodo-4-nitrophenol and 2,4,6-triiodophenol, and the QSAR indicated that the descriptors including E_{LUMO}, the energy of the highest occupied molecular orbital (E_{HOMO}), and pKa play important roles in the developmental toxicity. In addition, Qin et al. [192] have constructed a predicted QSAR models to assess the reactive toxicities of 50 DBPs across five bioassays, namely, X-Microtox, GSH+, GSH-, DNA+, and DNA-. With the selected descriptors, the developed QSAR models could be used to predict the toxicities of DBPs.

In addition, the researchers applied QSAR to assess the toxicity of DBPs to zebrafish and sewage bacteria. Li et al. [193] assessed the toxicity with 10 halogenated aromatic DBPs by utilizing the model organism zebrafish. The findings of the molecular docking analysis showed that the ten DBPs manifested binding affinity towards zebrafish's catalase (CAT), cytochrome P450 (CYP450), p53, and AChE, and the toxicity of halogenated aromatic DBPs to zebrafish has shown to be primarily influenced by three factors: hydrophobicity, interaction with the catalase enzyme, and hydrogen bonding acidity, as determined by QSAR models. Wu et al. [117] applied QSAR analysis to estimate the toxicity of 11 cyclic DBPs and nine aliphatic DBPs in both the HepG2 cells and zebrafish embryos. Based on the toxicity data and the descriptors of DBPs, two QSAR models were constructed to predict the toxicity rank order of other in vitro and in vivo assays. These models aid in the elucidation of the toxicity mechanisms of DBPs at several endpoints.

4 Perspectives and Limitations

Toxicological data provide evidence for the agencies to regulate some DBPs. Therefore, toxicological researches are very fundamental for DBP research and treatment. In the face of the large number of DBPs, it is a big challenge to examine the toxicity of every single DBP. QSAR models help to prioritize the emerging DBPs with toxicological importance, so these prioritized DBPs could be studied for toxicity first. With the development of artificial intelligence and machine learning, the QSAR analysis could be much more helpful and predictive for DBP toxicities. In addition, the animal models especially the zebrafish are useful tools to facilitate screening for toxicity of DBPs and to set up the database for the comparison of DBP toxicities. However, a standard procedure or method of toxicity examination is needed for these animal models.

As indicated, Dr. Plewa and his colleagues have contributed to the quantitative database of both chronic cytotoxicity and acute genotoxicity of >100 DBPs in CHO cells, which enables the comparison and ranking of their toxicity. However, compared with the over 6,000 reported DBPs, there is still a lot of work to do in the future, and a high-put method is needed to accelerate the speed of toxicity screening. Although in vitro toxicity tests are very helpful, cultured cells lack metabolic ability and interactions within the bodies, highlighting the need for in vivo toxicity testing. In addition, some in vitro toxicity tests use cells originating from tumors, which are not suitable for toxicity studies. Instead, the cells from normal tissues or organs are recommended in toxicological research. Also, it is needed to consider using the cells from the target organs of DBPs, such as the liver, kidney, and bladder.

In vitro toxicity tests can also be useful to examine the potential toxic mechanisms. Most of the DBPs demonstrated the ability to induce oxidative stress and DNA damage, therefore a fast-screening method and comparison database on the mechanisms of action should be established. With the help of CRISPR-Cas 9 gene-editing technology, this purpose could be achieved by using engineered cell models. Additionally, understanding the mechanisms of action can lead to the identification of biomarkers for DBP exposures, which could be used in epidemiology studies of DBPs.

In vivo toxicity tests are much more relevant to humans in comparison with in vitro tests, animal models, and QSAR analysis. Given the potential targets of DBPs in epidemiological studies and bladder cancer, in vivo carcinogenicity data suggested that the liver and kidney are the target organs of DBPs. In vivo toxicity tests, especially the two-year rodent carcinogenicity test, are quite time-consuming and expensive, and only 26 DBPs have been demonstrated to have carcinogenic potential within the testing DBPs. Additionally, the doses used in these in vivo tests are much higher than the levels in real drinking water, and the exposure time is around 28–90 days, which represents the subacute and subchronic toxicity; however, this is still not long enough to cover the exposure time of DBPs. Therefore, the combination of both in vivo and in vitro toxicity tests is recommended, with the screening of in vitro tests or other tests to first assess the rank order of toxicity of DBPs, and with in vivo tests to examine the toxicity of the prioritized DBPs. Also, with the development of omics technology, multiple-omics research should be used in DBPs toxicity tests, especially toxicogenomics, proteomics, microbiomics, and metabonomics, which help screen the biomarkers for epidemiology studies and understand the mechanisms of actions of toxicity.

Although DBPs occur within complex mixtures in drinking water, the toxicity tests mainly focus on the toxicity of single DBPs. This might be one of the reasons for the inconsistency between the results of toxicity tests and observations of epidemiological studies. There are several toxicity studies on DBP mixtures with the combination of two single DBPs or more, but more studies are still needed on the effects of real drinking water samples to better understand the toxicity of DBP mixtures. In addition, more studies are needed to understand the combined toxicity of DBP mixtures in the aspect of additive, synergistic, or antagonistic effects.

References

1. Srivastav AL, Patel N, Chaudhary VK (2020) Disinfection by-products in drinking water: occurrence, toxicity and abatement. Environ Pollut 267:115474. https://doi.org/10.1016/j.envpol.2020.115474
2. Chen HC, Xie JD, Huang CX, Liang YN, Zhang YL, Zhao XY, Ling YH, Wang L, Zheng Q, Yang XQ (2024) Database and review of disinfection by-products since 1974: constituent elements, molecular weights, and structures. J Hazard Mater 462:132792. https://doi.org/10.1016/j.jhazmat.2023.132792
3. Bull RJ, Sanchez IM, Nelson MA, Larson JL, Lansing AJ (1990) Liver tumor induction in B6C3F1 mice by dichloroacetate and trichloroacetate. Toxicology 63(3):341–359. https://doi.org/10.1016/0300-483x(90)90195-m
4. DeAngelo AB, George MH, House DE (1999) Hepatocarcinogenicity in the male B6C3F1 mouse following a lifetime exposure to dichloroacetic acid in the drinking water: dose-response determination and modes of action. J Toxicol Environ Health A 58(8):485–507. https://doi.org/10.1080/009841099157115
5. Hooth MJ, McDorman KS, Hester SD, George MH, Brooks LR, Swank AE, Wolf DC (2002) The carcinogenic response of Tsc2 mutant Long-Evans (Eker) rats to a mixture of drinking water disinfection by-products was less than additive. Toxicol Sci 69(2):322–331. https://doi.org/10.1093/toxsci/69.2.322
6. McDorman KS, Chandra S, Hooth MJ, Hester SD, Schoonhoven R, Wolf DC (2003) Induction of transitional cell hyperplasia in the urinary bladder and aberrant crypt foci in the colon of rats treated with individual and a mixture of drinking water disinfection by-products. Toxicol Pathol 31(2):235–242. https://doi.org/10.1080/01926230390183733
7. Wei X, Wang S, Zheng W, Wang X, Liu X, Jiang S, Pi J, Zheng Y, He G, Qu W (2013) Drinking water disinfection byproduct iodoacetic acid induces tumorigenic transformation of NIH3T3 cells. Environ Sci Technol 47(11):5913–5920. https://doi.org/10.1021/es304786b
8. National Toxicology Program (2006) NTP toxicology and carcinogenesis studies of bromodichloromethane (CAS No. 75-27-4) in male F344/N rats and female B6C3F1 mice (Drinking Water Studies). Natl Toxicol Program Tech Rep Ser 532:1–248
9. george-et-al-2002-carcinogenicity-of-bromodichloromethane-administered-in-drinking-water-to-male-f344-n-rats-and-b6c3f1-290.pdf. https://doi.org/10.1080/1091581029009635
10. Aida Y, Yasuhara K, Takada K, Kurokawa Y, Tobe M (1992) Chronic toxicity of microencapsulated BROMODICHLOROMETHANE administered in the diet to WISTAR rats. J Toxicol Sci 17(2):51–68. https://doi.org/10.2131/jts.17.51
11. National Toxicology Program (1985) NTP toxicology and carcinogenesis studies of chlorodibromomethane (CAS No. 124-48-1) in F344/N rats and B6C3F1 mice (Gavage Studies). Natl Toxicol Program Tech Rep Ser 282:1–174
12. National Toxicology Program (1976) Report on the carcinogenesis bioassay of chloroform (CAS No. 67-66-3). Natl Cancer Inst Carcinog Tech Rep Ser 1976:1–60
13. Jorgenson TA, Meierhenry EF, Rushbrook CJ, Bull RJ, Robinson M (1985) Carcinogenicity of chloroform in drinking water to male Osborne-Mendel rats and female B6C3F1 mice. Fundam Appl Toxicol 5(4):760–769. https://doi.org/10.1016/0272-0590(85)90200-3
14. DeAngelo AB, Geter DR, Rosenberg DW, Crary CK, George MH (2002) The induction of aberrant crypt foci (ACF) in the colons of rats by trihalomethanes administered in the drinking water. Cancer Lett 187(1–2):25–31. https://doi.org/10.1016/s0304-3835(02)00356-7

15. Geter DR, George MH, Moore TM, Kilburn S, Huggins-Clark G, DeAngelo AB (2004) Vehicle and mode of administration effects on the induction of aberrant crypt foci in the colons of male F344/N rats exposed to bromodichloromethane. J Toxicol Environ Health A 67(1):23–29. https://doi.org/10.1080/15287390490253642
16. Geter DR, George MH, Moore TM, Kilburn SR, Huggins-Clark G, DeAngelo AB (2004) The effects of a high animal fat diet on the induction of aberrant crypt foci in the colons of male F344/N rats exposed to trihalomethanes in the drinking water. Toxicol Lett 147(3):245–252. https://doi.org/10.1016/j.toxlet.2003.11.006
17. Geter DR, Moore TM, George MH, Kilburn SR, Allen JW, Nelson GM, Winkfield E, DeAngelo AB (2005) Tribromomethane exposure and dietary folate deficiency in the formation of aberrant crypt foci in the colons of F344/N rats. Food Chem Toxicol 43(9):1405–1412. https://doi.org/10.1016/j.fct.2005.03.015
18. Nagano K, Kano H, Arito H, Yamamoto S, Matsushima T (2006) Enhancement of renal carcinogenicity by combined inhalation and oral exposures to chloroform in male rats. J Toxicol Environ Health A 69(20):1827–1842. https://doi.org/10.1080/15287390600630146
19. Melnick RL, Nyska A, Foster PM, Roycroft JH, Kissling GE (2007) Toxicity and carcinogenicity of the water disinfection byproduct, dibromoacetic acid, in rats and mice. Toxicology 230(2–3):126–136. https://doi.org/10.1016/j.tox.2006.11.006
20. DeAngelo AB, Daniel FB, Most BM, Olson GR (1996) The carcinogenicity of dichloroacetic acid in the male Fischer 344 rat. Toxicology 114(3):207–221. https://doi.org/10.1016/s0300-483x(96)03510-x
21. National Toxicology Program (2009) Toxicology and carcinogenesis studies of bromochloroacetic acid (CAS No. 5589-96-8) in F344/N rats and B6C3F1 mice (drinking water studies). Natl Toxicol Program Tech Rep Ser 549:1–269
22. National Toxicology Program (2018) RoC monograph series. In: Report on carcinogens monograph on Haloacetic acids found as water disinfection by-products: RoC monograph, vol 12. National Toxicology Program
23. DeAngelo AB, George MH, Kilburn SR, Moore TM, Wolf DC (1998) Carcinogenicity of potassium bromate administered in the drinking water to male B6C3F1 mice and F344/N rats. Toxicol Pathol 26(5):587–594. https://doi.org/10.1177/019262339802600501
24. Kurokawa Y, Aoki S, Matsushima Y, Takamura N, Imazawa T, Hayashi Y (1986) Dose-response studies on the carcinogenicity of potassium bromate in F344 rats after long-term oral administration. J Natl Cancer Inst 77(4):977–982
25. Kurokawa Y, Hayashi Y, Maekawa A, Takahashi M, Kokubo T, Odashima S (1983) Carcinogenicity of potassium bromate administered orally to F344 rats. J Natl Cancer Inst 71(5): 965–972
26. Kurokawa Y, Takayama S, Konishi Y, Hiasa Y, Asahina S, Takahashi M, Maekawa A, Hayashi Y (1986) Long-term in vivo carcinogenicity tests of potassium bromate, sodium hypochlorite, and sodium chlorite conducted in Japan. Environ Health Perspect 69:221–235
27. National Toxicology Program (2005) Toxicology and carcinogenesis studies of sodium chlorate (Cas No. 7775-09-9) in F344/N rats and B6C3F1 mice (drinking water studies). Natl Toxicol Program Tech Rep Ser 517:1–255
28. Kerns WD, Pavkov KL, Donofrio DJ, Gralla EJ, Swenberg JA (1983) Carcinogenicity of formaldehyde in rats and mice after long-term inhalation exposure. Cancer Res 43(9): 4382–4392
29. Soffritti M, Belpoggi F, Lambertin L, Lauriola M, Padovani M, Maltoni C (2002) Results of long-term experimental studies on the carcinogenicity of formaldehyde and acetaldehyde in rats. Ann N Y Acad Sci 982:87–105. https://doi.org/10.1111/j.1749-6632.2002.tb04926.x
30. Til HP, Woutersen RA, Feron VJ, Hollanders VH, Falke HE, Clary JJ (1989) Two-year drinking-water study of formaldehyde in rats. Food Chem Toxicol 27(2):77–87. https://doi.org/10.1016/0278-6915(89)90001-x

31. Woutersen RA, Appelman LM, Van Garderen-Hoetmer A, Feron VJ (1986) Inhalation toxicity of acetaldehyde in rats. III. Carcinogenicity study. Toxicology 41(2):213–231. https://doi.org/10.1016/0300-483x(86)90201-5
32. George MH, Moore T, Kilburn S, Olson GR, DeAngelo AB (2000) Carcinogenicity of chloral hydrate administered in drinking water to the male F344/N rat and male B6C3F1 mouse. Toxicol Pathol 28(4):610–618. https://doi.org/10.1177/019262330002800415
33. Daniel FB, DeAngelo AB, Stober JA, Olson GR, Page NP (1992) Hepatocarcinogenicity of chloral hydrate, 2-chloroacetaldehyde, and dichloroacetic acid in the male B6C3F1 mouse. Fundam Appl Toxicol 19(2):159–168. https://doi.org/10.1016/0272-0590(92)90147-a
34. Komulainen H, Kosma VM, Vaittinen SL, Vartiainen T, Kaliste-Korhonen E, Lötjönen S, Tuominen RK, Tuomisto J (1997) Carcinogenicity of the drinking water mutagen 3-chloro-4-(dichloromethyl)-5-hydroxy-2(5H)-furanone in the rat. J Natl Cancer Inst 89(12):848–856. https://doi.org/10.1093/jnci/89.12.848
35. National Toxicology Program (2010) Toxicology and carcinogenesis studies of dibromoacetonitrile (CAS No. 3252-43-5) in F344/N rats and B6C3F1 mice (drinking water studies). Natl Toxicol Program Tech Rep Ser 544:1–193
36. Bull RJ (1985) Carcinogenic and mutagenic properties of chemicals in drinking water. Sci Total Environ 47:385–413. https://doi.org/10.1016/0048-9697(85)90344-4
37. Gurley KE, Moser RD, Kemp CJ (2015) Induction of liver tumors in mice with N-ethyl-N-nitrosourea or N-nitrosodiethylamine. Cold Spring Harb Protoc 2015(10):941–942. https://doi.org/10.1101/pdb.prot077438
38. Peto R, Gray R, Brantom P, Grasso P (1991) Effects on 4080 rats of chronic ingestion of N-nitrosodiethylamine or N-nitrosodimethylamine: a detailed dose-response study. Cancer Res 51(23 Pt 2):6415–6451
39. Tanaka A, Hisanaga A, Inamasu T, Hirata M, Ishinishi N (1988) A comparison of the carcinogenicity of N-nitrosodiethylamine and N-nitrosodimethylamine after intratracheal instillation into Syrian golden hamsters. Food Chem Toxicol 26(10):847–850. https://doi.org/10.1016/0278-6915(88)90025-7
40. Ishinishi N, Tanaka A, Hisanaga A, Inamasu T, Hirata M (1988) Comparative study on the carcinogenicity of N-nitrosodiethylamine, N-nitrosodimethylamine, N-nitrosomorpholine, N-nitrosopyrrolidine and N-nitrosodi-n-propylamine to the lung of Syrian golden hamsters following intermittent instillations to the trachea. Carcinogenesis 9(6):947–950. https://doi.org/10.1093/carcin/9.6.947
41. Preussmann R, Schmähl D, Eisenbrand G (1977) Carcinogenicity of N-nitrosopyrrolidine: dose-response study in rats. Z Krebsforsch Klin Onkol Cancer Res Clin Oncol 90(2):161–166. https://doi.org/10.1007/bf00285322
42. Greenblatt M, Lijinsky W (1972) Nitrosamine studies: neoplasms of liver and genital mesothelium in nitrosopyrrolidine-treated MRC rats. J Natl Cancer Inst 48(6):1687–1696
43. Lijinsky W, Reuber MD, Manning WB (1980) Potent carcinogenicity of nitrosodiethanolamine in rats. Nature 288(5791):589–590. https://doi.org/10.1038/288589a0
44. George MH, Olson GR, Doerfler D, Moore T, Kilburn S, DeAngelo AB (2002) Carcinogenicity of bromodichloromethane administered in drinking water to male F344/N rats and B6C3F1 mice. Int J Toxicol 21(3):219–230. https://doi.org/10.1080/10915810290096351
45. National Toxicology Program (1989) NTP toxicology and carcinogenesis studies of tribromomethane (Bromoform) (CAS No. 75-25-2) in F344/N rats and B6C3F1 mice (Gavage Studies). Natl Toxicol Program Tech Rep Ser 350:1–194
46. Pereira MA (1996) Carcinogenic activity of dichloroacetic acid and trichloroacetic acid in the liver of female B6C3F1 mice. Fundam Appl Toxicol 31(2):192–199. https://doi.org/10.1006/faat.1996.0091
47. Bielmeier SR, Best DS, Guidici DL, Narotsky MG (2001) Pregnancy loss in the rat caused by bromodichloromethane. Toxicol Sci 59(2):309–315. https://doi.org/10.1093/toxsci/59.2.309

48. Bielmeier SR, Best DS, Narotsky MG (2004) Serum hormone characterization and exogeneous hormone rescue of bromodichloromethane-induced pregnancy loss in the F344 rat. Toxicol Sci 77(1):101–108. https://doi.org/10.1093/toxsci/kfh017
49. Bielmeier SR, Murr AS, Best DS, Harrison RA, Pegram RA, Goldman JM, Narotsky MG (2007) Effects of bromodichloromethane on ex vivo and in vitro luteal function and bromodichloromethane tissue dosimetry in the pregnant F344 rat. Toxicol In Vitro 21(5): 919–928. https://doi.org/10.1016/j.tiv.2007.01.017
50. Potter CL, Chang LW, DeAngelo AB, Daniel FB (1996) Effects of four trihalomethanes on DNA strand breaks, renal hyaline droplet formation and serum testosterone in male F-344 rats. Cancer Lett 106(2):235–242. https://doi.org/10.1016/0304-3835(96)04331-5
51. Tsuchiya T, Ooyama N, Murakami T, Sano F, Sugimoto J, Mutai M (2000) Collaborative work to evaluate toxicity on male reproductive organs by repeated dose studies in rats 25. Effects of 2- and 4-week repeated-dosing of dibromoacetic acid. J Toxicol Sci:25. Spec No. 241–249. https://doi.org/10.2131/jts.25.specialissue_241
52. Linder RE, Klinefelter GR, Strader LF, Suarez JD, Roberts NL, Dyer CJ (1994) Spermatotoxicity of dibromoacetic acid in rats after 14 daily exposures. Reprod Toxicol 8(3):251–259. https://doi.org/10.1016/0890-6238(94)90009-4
53. Linder RE, Klinefelter GR, Strader LF, Narotsky MG, Suarez JD, Roberts NL, Perreault SD (1995) Dibromoacetic acid affects reproductive competence and sperm quality in the male rat. Fundam Appl Toxicol 28(1):9–17. https://doi.org/10.1006/faat.1995.1140
54. Christian MS, York RG, Hoberman AM, Frazee J, Fisher LC, Brown WR, Creasy DM (2002) Oral (drinking water) two-generation reproductive toxicity study of dibromoacetic acid (DBA) in rats. Int J Toxicol 21(4):237–276. https://doi.org/10.1080/10915810290096432
55. Veeramachaneni DNR, Palmer JS, Klinefelter GR (2007) Chronic exposure to low levels of dibromoacetic acid, a water disinfection by-product, adversely affects reproductive function in male rabbits. J Androl 28(4):565–577. https://doi.org/10.2164/jandrol.107.002550
56. Monetti C, Vigetti D, Gornati R, Prati M, Klinefelter GR, Bernardini G (2002) Identification and molecular cloning of Xenopus laevis SP22, a protein associated with fertilization in mammals. Comp Biochem Physiol B Biochem Mol Biol 132 (4):761–767. Pii S1096-4959 (02)00098-2. https://doi.org/10.1016/S1096-4959(02)00098-2
57. Kaydos EH, Suarez JD, Roberts NL, Bobseine K, Zucker R, Laskey J, Klinefelter GR (2004) Haloacid induced alterations in fertility and the sperm biomarker SP22 in the rat are additive: validation of an ELISA. Toxicol Sci 81(2):430–442. https://doi.org/10.1093/toxsci/kfh218
58. Narotsky MG, Klinefelter GR, Goldman JM, Best DS, McDonald A, Strader LF, Suarez JD, Murr AS, Thillainadarajah I, Hunter ES et al (2013) Comprehensive assessment of a chlorinated drinking water concentrate in a rat multigenerational reproductive toxicity study. Environ Sci Technol. https://doi.org/10.1021/es402646c
59. Narotsky MG, Klinefelter GR, Goldman JM, DeAngelo AB, Best DS, McDonald A, Strader LF, Murr AS, Suarez JD, George MH et al (2015) Reproductive toxicity of a mixture of regulated drinking-water disinfection by-products in a multigenerational rat bioassay. Environ Health Perspect 123(6):564–570. https://doi.org/10.1289/ehp.1408579
60. Abdel-Wahab MH (2003) Testicular toxicity of dibromoacetonitrile and possible protection by tertiary butylhydroquinone. Pharmacol Res 47(6):509–515. https://doi.org/10.1016/S1043-6618(03)00039-2
61. Klinefelter GR, Suarez JD, Roberts NL, DeAngelo AB (1995) Preliminary screening for the potential of drinking water disinfection byproducts to alter male reproduction. Reprod Toxicol 9(6):571–578. https://doi.org/10.1016/0890-6238(95)02007-1
62. Goldman JM, Murr AS (2003) Dibromoacetic acid-induced elevations in circulating estradiol: effects in both cycling and ovariectomized/steroid-primed female rats. Reprod Toxicol 17(5): 585–592. https://doi.org/10.1016/S0890-6238(03)00068-6
63. Goldman J, Murr A, Buckalew A, Ferrell J, Cooper R (2007) Moderating influence of the drinking water disinfection by-product dibromoacetic acid on a dithiocarbamate-induced

suppression of the luteinizing hormone surge in female rats☆. Reprod Toxicol 23(4):541–549. https://doi.org/10.1016/j.reprotox.2007.03.001
64. Murr AS, Goldman JM (2005) Twenty-week exposures to the drinking water disinfection by-product dibromoacetic acid: reproductive cyclicity and steroid concentrations in the female Sprague-Dawley rat. Reprod Toxicol 20(1):73–80. https://doi.org/10.1016/j.reprotox.2004.12.006
65. Bodensteiner KJ (2004) Chronic exposure to dibromoacetic acid, a water disinfection byproduct, diminishes primordial follicle populations in the rabbit. Toxicol Sci 80(1):83–91. https://doi.org/10.1093/toxsci/kfh135
66. Jiao X, Gonsioroski A, Flaws JA, Qiao H (2021) Iodoacetic acid disrupts mouse oocyte maturation by inducing oxidative stress and spindle abnormalities. Environ Pollut:268. https://doi.org/10.1016/j.envpol.2020.115601
67. Gonsioroski A, Laws M, Mourikes VE, Neff A, Drnevich J, Plewa MJ, Flaws JA (2022) Iodoacetic acid exposure alters the transcriptome in mouse ovarian antral follicles. J Environ Sci 117:46–57. https://doi.org/10.1016/j.jes.2022.01.018
68. Gonsioroski A, Meling DD, Gao L, Plewa MJ, Flaws JA (2021) Iodoacetic acid affects estrous cyclicity, ovarian gene expression, and hormone levels in mice. Biol Reprod 105(4): 1030–1042. https://doi.org/10.1093/biolre/ioab108
69. Qiao F-X, Sun M-X, Xu Z-R, Liu Y-C, Chen Y-Z, Wang H-L, Qi Z-Q, Xu C-L, Liu Y (2023) Chloroacetonitrile exposure induces endoplasmic reticulum stress and affects spindle assembly in mouse oocytes. Food Chem Toxicol:176. https://doi.org/10.1016/j.fct.2023.113736
70. Ruddick JA, Villeneuve DC, Chu I, Valli VE (1983) A teratological assessment of four trihalomethanes in the rat. J Environ Sci Health B 18(3):333–349. https://doi.org/10.1080/03601238309372373
71. Singh R (2006) Neuroembryopathic effect of trichloroacetic acid in rats exposed during organogenesis. Birth Defects Res B Dev Reprod Toxicol 77(1):47–52. https://doi.org/10.1002/bdrb.20064
72. Smith MK, Randall JL, Read EJ, Stober JA (1989) Teratogenic activity of trichloroacetic acid in the rat. Teratology 40(5):445–451. https://doi.org/10.1002/tera.1420400506
73. Gonsioroski A, Plewa MJ, Flaws JA (2022) Effects of prenatal and lactational exposure to iodoacetic acid on the F1 generation of mice. Biol Reprod 107(2):650–663. https://doi.org/10.1093/biolre/ioac079
74. Narotsky MG, Best DS, McDonald A, Godin EA, Hunter ES, Simmons JE (2011) Pregnancy loss and eye malformations in offspring of F344 rats following gestational exposure to mixtures of regulated trihalomethanes and haloacetic acids. Reprod Toxicol 31(1):59–65. https://doi.org/10.1016/j.reprotox.2010.08.002
75. Ahmed AE, Jacob S, Campbell GA, Harirah HM, Perez-Polo JR, Johnson KM (2005) Fetal origin of adverse pregnancy outcome: the water disinfectant by-product chloroacetonitrile induces oxidative stress and apoptosis in mouse fetal brain. Dev Brain Res 159(1):1–11. https://doi.org/10.1016/J.Devbrainres.2005.05.008
76. Ahmed AE, El-Mazar HM, Nagy AA, Abdel-Naim AB (2008) Chloroacetonitrile induces intrauterine growth restriction and musculoskeletal toxicity in fetal mouse. Toxicol Ind Health 24(8):511–518. https://doi.org/10.1177/0748233708098129
77. Esmat A, Ghoneim AI, El-Demerdash E, Khalifa AE, Abdel-Naim AB (2012) Dichloroacetonitrile induces oxidative stress and developmental apoptotic imbalance in mouse fetal brain. Environ Toxicol Pharmacol 33(1):78–84. https://doi.org/10.1016/J.Etap.2011.11.004
78. Abdel-Naim AB, Nagy AA, Mohamadin AM, El-Mazar HM, Ahmed AE (2009) Chloroacetonitrile induces oxidative stress and apoptosis in mouse fetal liver. Toxicol Lett 190(2):123–127. https://doi.org/10.1016/j.toxlet.2009.06.867
79. Smith MK, Randall JL, Stober JA, Read EJ (1989) Developmental toxicity of dichloroacetonitrile: a by-product of drinking water disinfection. Fundam Appl Toxicol 12(4):765–772. https://doi.org/10.1016/0272-0590(89)90008-0

80. Balchak SK, Hedge JM, Murr AS, Mole ML, Goldman JM (2000) Influence of the drinking water disinfection by-product dibromoacetic acid on rat estrous cyclicity and ovarian follicular steroid release in vitro. Reprod Toxicol 14(6):533–539. https://doi.org/10.1016/S0890-6238(00)00104-0
81. Goldman JM, Murr AS (2002) Alterations in ovarian follicular progesterone secretion by elevated exposures to the drinking water disinfection by-product dibromoacetic acid: examination of the potential site(s) of impact along the steroidogenic pathway. Toxicology 171(2–3):83–93. Pii S0300-483x(01)00565-0. https://doi.org/10.1016/S0300-483x(01)00565-0
82. Gonzalez RVL, Weis KE, Gonsioroski AV, Flaws JA, Raetzman LT (2021) Iodoacetic acid, a water disinfection byproduct, disrupts hypothalamic, and pituitary reproductive regulatory factors and induces toxicity in the female pituitary. Toxicol Sci 184(1):46–56. https://doi.org/10.1093/toxsci/kfab106
83. Moser VC (2004) Neurotoxicity produced by dibromoacetic acid in drinking water of rats. Toxicol Sci 79(1):112–122. https://doi.org/10.1093/toxsci/kfh081
84. Moser VC, Phillips PM, McDaniel KL, MacPhail RC (1999) Behavioral evaluation of the neurotoxicity produced by dichloroacetic acid in rats. Neurotoxicol Teratol 21(6):719–731. https://doi.org/10.1016/S0892-0362(99)00029-X
85. Cicmanec JL, Condie LW, Olson GR, Wang SR (1991) 90-Day toxicity study of dichloroacetate in dogs. Fundam Appl Toxicol 17(2):376–389. https://doi.org/10.1016/0272-0590(91)90227-u
86. Ahmed AE, Campbell GA, Jacob S (2005) Neurological impairment in fetal mouse brain by drinking water disinfectant byproducts. Neurotoxicology 26(4):633–640. https://doi.org/10.1016/j.neuro.2004.11.001
87. Guariglia SR, Jenkins EC, Chadman KK, Wen GY (2011) Chlorination byproducts induce gender specific autistic-like behaviors in CD-1 mice. Neurotoxicology 32(5):545–553. https://doi.org/10.1016/j.neuro.2011.06.008
88. Balster RL, Borzelleca JF (1982) Behavioral toxicity of trihalomethane contaminants of drinking water in mice. Environ Health Perspect. https://doi.org/10.1289/ehp.8246127
89. Jiang W, Li B, Chen Y, Gao S (2017) The toxic influence of dibromoacetic acid on the hippocampus and pre-frontal cortex of rat: involvement of neuroinflammation response and oxidative stress. Metab Brain Dis 32(6):2009–2019. https://doi.org/10.1007/s11011-017-0095-0
90. Jiang W, Chen Y, Li B, Gao S (2017) DBA-induced caspase-3-dependent apoptosis occurs through mitochondrial translocation of cyt-c in the rat hippocampus. Mol Biosyst 13(9):1863–1873. https://doi.org/10.1039/c7mb00246g
91. Wei W, Dong Q, Jiang W, Wang Y, Chen Y, Han T, Sun C (2021) Dichloroacetic acid-induced dysfunction in rat hippocampus and the protective effect of curcumin. Metab Brain Dis 36(4):545–556. https://doi.org/10.1007/s11011-020-00657-5
92. Wang Y, Jiang WB, Dong QY, Zhao Y, Chen YY, Sun CH, Sun GL (2020) Fetal exposure to dichloroacetic acid and impaired cognitive function in the adulthood. Brain Behav 10(10) ARTN e01801. https://doi.org/10.1002/brb3.1801
93. Sayed RH, Salem HA, El-Sayeh BM (2012) Potential protective effect of taurine against dibromoacetonitrile-induced neurotoxicity in rats. Environ Toxicol Pharmacol 34(3):849–857. https://doi.org/10.1016/j.etap.2012.08.015
94. Li F, Dong Y, Shen H, Lu R, Yin S, Tian W, Wang S, Xing G (2017) Tolerance to dichloroacetonitrile-induced neurotoxicity in streptozotocin-induced diabetic rats. Environ Toxicol Pharmacol 56:61–67. https://doi.org/10.1016/j.etap.2017.08.037
95. Liu Y, Tong C, Xu Y, Cong P, Liu Y, Shi L, Shi X, Zhao Y, Bi G, Jin H et al (2019) CD28 deficiency ameliorates blast exposure-induced lung inflammation, oxidative stress, apoptosis, and T cell accumulation in the lungs via the PI3K/Akt/FoxO1 signaling pathway. Oxid Med Cell Longev 2019:1–15. https://doi.org/10.1155/2019/4848560

96. Chen CH, Chen SJ, Su CC, Yen CC, Tseng TJ, Jinn TR, Tang FC, Chen KL, Su YC, Leek I et al (2013) Chloroacetic acid induced neuronal cells death through oxidative stress-mediated p38-MAPK activation pathway regulated mitochondria-dependent apoptotic signals. Toxicology 303:72–82. https://doi.org/10.1016/j.tox.2012.10.008
97. Li F, Zhou J, Zhu X, Lu R, Ye Y, Wang S, Xing G, Shen H (2022) Oxidative injury induced by drinking water disinfection by-products dibromoacetonitrile and dichloroacetonitrile in mouse hippocampal neuronal cells: the protective effect of N-acetyl-L-cysteine. Toxicol Lett 365:61–73. https://doi.org/10.1016/j.toxlet.2022.06.005
98. Liu J, Wu Y, Liu S, Gao Z, Fan W, Ding C, Tang Z, Zhang S, Shi X, Tan L et al (2022) Bromoacetic acid induces neurogenic injury in the chicken brain by activating oxidative stress and NF-κB inflammatory pathway. Chem Biol Interact:365. https://doi.org/10.1016/j.cbi.2022.110115
99. Liu T, Chen X, Li WL, Zhang X, Wang GH, Wang J, Liang LQ, Yang F, Li J, Li JH (2023) Oxidative stress as a key event in 2,6-dichloro-1,4-benzoquinone-induced neurodevelopmental toxicity. Ecotoxicol Environ Saf:263. ARTN 115357. https://doi.org/10.1016/j.ecoenv.2023.115357
100. Lu T-H, Su C-C, Tang F-C, Chen C-H, Yen C-C, Fang K-M, Lee K-I, Hung D-Z, Chen Y-W (2015) Chloroacetic acid triggers apoptosis in neuronal cells via a reactive oxygen species-induced endoplasmic reticulum stress signaling pathway. Chem Biol Interact 225:1–12. https://doi.org/10.1016/j.cbi.2014.10.022
101. Miao T, Li M, Shao T, Jiang X, Jiang L, Zhou Q, Pan Y, Wang Y, Qiu J (2022) The involvement of branched-chain amino acids (BCAAs) in aromatic trihalogenated DBP exposure-induced kidney damage in mice. Chemosphere 305., Article. https://doi.org/10.1016/j.chemosphere.2022.135351
102. Ewaid SH, Abed SA; Al-Ansari N (2020) Acute toxicity of the water chlorination byproduct (chloroform) in male mice. In: 8th international conference on applied science and technology (ICAST), Univ Kerbala, Coll Sci, Karbala, IRAN, 2020 Apr 15–16, 2020; vol 2290. https://doi.org/10.1063/5.0027353
103. Liu T, Wang J, Dang X, Wan S, Luo X, Tang W, Du H, Jin M, Li J, Li J (2023) Investigation of the nephrotoxicity of 2,6-dichloro-1,4-benzoquinone disinfection by-product in mice through a 28-day toxicity test. Toxicology 487:153459. https://doi.org/10.1016/j.tox.2023.153459
104. Li J-H, Wang Z-H, Zhu X-J, Deng Z-H, Cai C-X, Qiu L-Q, Chen W, Lin Y-J (2015) Health effects from swimming training in chlorinated pools and the corresponding metabolic stress pathways. PLoS One 10(3). https://doi.org/10.1371/journal.pone.0119241
105. Yang L, Chen L, Hao Y, Zhou R, Zhu J, Zhu X, Wang Q, Li X, Ding X, Qian Y (2023) Urinary trihalomethane concentrations and liver function indicators: a cross-sectional study in China. Environ Sci Pollut Res 30(14):39724–39732. https://doi.org/10.1007/s11356-022-25072-4
106. Jiang X, Shi P, Jiang L, Qiu J, Xu B, Pan Y, Zhou Q (2022) In vivo toxicity evaluations of halophenolic disinfection byproducts in drinking water: a multi-omics analysis of toxic mechanisms. Water Res 218., Article. https://doi.org/10.1016/j.watres.2022.118431
107. Perveen S, Hashmi I, Khan R (2019) Evaluation of genotoxicity and hematological effects in common carp (Cyprinus carpio) induced by disinfection by-products. J Water Health 17(5): 762–776. https://doi.org/10.2166/wh.2019.261
108. Dong F, Chen J, Li C, Ma X, Jiang J, Lin Q, Lin C, Diao H (2019) Evidence-based analysis on the toxicity of disinfection byproducts in vivo and in vitro for disinfection selection. Water Res 165., Article. https://doi.org/10.1016/j.watres.2019.114976
109. Zhang Y, Zhang Z, Zhao Y, Cheng S, Ren H (2013) Identifying health effects of exposure to trichloroacetamide using transcriptomics and metabonomics in mice (Mus musculus). Environ Sci Technol 47(6):2918–2924., Article. https://doi.org/10.1021/es3048976
110. Deng Y, Zhang Y, Zhang R, Wu B, Ding L, Xu K, Ren H (2014) Mice in vivo toxicity studies for monohaloacetamides emerging disinfection byproducts based on metabolomic methods. Environ Sci Technol 48(14):8212–8218., Article. https://doi.org/10.1021/es502010v

111. Deng Y, Zhang Y, Lu Y, Lu K, Bai H, Ren H (2017) Metabolomics evaluation of the in vivo toxicity of bromoacetonitriles: one class of high-risk nitrogenous disinfection byproducts. Sci Total Environ 579:107–114., Article. https://doi.org/10.1016/j.scitotenv.2016.11.003
112. Klinefelter GR (2004) Continuous exposure to dibromoacetic acid delays pubertal development and compromises sperm quality in the rat. Toxicol Sci 81(2):419–429. https://doi.org/10.1093/toxsci/kfh217
113. Plewa MJ, Wagner ED (2009) Mammalian cell cytotoxicity and genotoxicity of disinfection by-products. Water Research Foudation
114. Wagner ED, Plewa MJ (2017) CHO cell cytotoxicity and genotoxicity analyses of disinfection by-products: an updated review. J Environ Sci 58:64–76. https://doi.org/10.1016/j.jes.2017.04.021
115. Liberatore HK, Plewa MJ, Wagner ED, VanBriesen JM, Burnett DB, Cizmas LH, Richardson SD (2017) Identification and comparative mammalian cell cytotoxicity of new lodo-phenolic disinfection byproducts in chloraminated oil and gas wastewaters. Environ Sci Technol Lett 4(11):475–480. https://doi.org/10.1021/acs.estlett.7b00468
116. Wang J, Zhang M, Hu S, Xian Q, Chen H, Gong T (2022) Occurrence and cytotoxicity of aliphatic and aromatic halogenated disinfection byproducts in indoor swimming pool water and their incoming tap water. Environ Sci Technol 56(24):17763–17775. https://doi.org/10.1021/acs.est.2c07175
117. Wu Y, Wei W, Luo J, Pan Y, Yang M, Hua M, Chu W, Shuang C, Li A (2022) Comparative toxicity analyses from different endpoints: are new cyclic disinfection byproducts (DBPs) more toxic than common aliphatic DBPs? Environ Sci Technol 56(1):194–207., Article. https://doi.org/10.1021/acs.est.1c03292
118. Ding X, Zhu J, Wang X, Zhou W, Wu K, Zhou Z, Zhou K, Wu D, Jiao J, Xia Y et al (2019) Different cytotoxicity of disinfection by-product haloacetamides on two exposure pathway-related cell lines: human gastric epithelial cell line GES-1 and immortalized human keratinocyte cell line HaCaT. Sci Total Environ 692:1267–1275. https://doi.org/10.1016/j.scitotenv.2019.07.279
119. Lou J, Yin L, Meng Z, Fang S, Pan X (2023) Occurrence, stability and cytotoxicity of halobenzamides: a new group of nitrogenous disinfection byproducts in drinking water. Water Res 245:120670. https://doi.org/10.1016/j.watres.2023.120670
120. Li J, Moe B, Vemula S, Wang W, Li X-F (2016) Emerging disinfection byproducts, halobenzoquinones: effects of isomeric structure and halogen substitution on cytotoxicity, formation of reactive oxygen species, and genotoxicity. Environ Sci Technol 50(13):6744–6752. https://doi.org/10.1021/acs.est.5b05585
121. Tian D, Moe B, Huang G, Jiang P, Ling ZC, Li XF (2020) Cytotoxicity of halogenated tyrosyl compounds, an emerging class of disinfection byproducts. Chem Res Toxicol 33(4):1028–1035. https://doi.org/10.1021/acs.chemrestox.0c00049
122. Wang W, Qian YC, Li JH, Moe B, Huang RF, Zhang HQ, Hrudey SE, Li X-F (2014) Analytical and toxicity characterization of halo-hydroxyl-benzoquinones as stable halobenzoquinone disinfection byproducts in treated water. Anal Chem 86(10):4982–4988. https://doi.org/10.1021/Ac5007238
123. Boyd JM, Huang L, Xie L, Moe B, Gabos S, Li XF (2008) A cell-microelectronic sensing technique for profiling cytotoxicity of chemicals. Anal Chim Acta 615(1):80–87. https://doi.org/10.1016/j.aca.2008.03.047
124. Fu KZ, Li J, Vemula S, Moe B, Li X-F (2017) Effects of halobenzoquinone and haloacetic acid water disinfection byproducts on human neural stem cells. J Environ Sci 58:239–249. https://doi.org/10.1016/j.jes.2017.02.006
125. Hong H, Lu Y, Zhu X, Wu Q, Jin L, Jin Z, Wei X, Ma G, Yu H (2023) Cytotoxicity of nitrogenous disinfection byproducts: a combined experimental and computational study. Sci Total Environ 856., Article. https://doi.org/10.1016/j.scitotenv.2022.159273
126. Weng H, Wang C, Ye T, Xu Z, Sun H, Lin H, Deng W-J, Wu F, Hong H (2022) Precursor characteristics of mono-HAAs during chlorination and cytotoxicity of mono-HAAs on

HEK-293T cells. Chemosphere 301., Article. https://doi.org/10.1016/j.chemosphere.2022.134689
127. Lu G, Qin D, Wang Y, Liu J, Chen W (2018) Single and combined effects of selected haloacetonitriles in a human-derived check for hepatoma line. Ecotoxicol Environ Saf 163: 417–426., Article. https://doi.org/10.1016/j.ecoenv.2018.07.104
128. Hung S, Mohan A, Reckhow DA, Pollitt KJG (2019) Assessment of the in vitro toxicity of the disinfection byproduct 2,6-dichloro-1,4-benzoquinone and its transformed derivatives. Chemosphere 234:902–908., Article. https://doi.org/10.1016/j.chemosphere.2019.06.086
129. McGuigan CF, Li XF (2014) Cytotoxicity and genotoxicity of phenazine in two human cell lines. Toxicol In Vitro 28(4):607–615. https://doi.org/10.1016/J.Tiv.2013.12.007
130. Li J, Moe B, Liu Y, Li X-F (2018) Halobenzoquinone-induced alteration of gene expression associated with oxidative stress signaling pathways. Environ Sci Technol 52(11):6576–6584. https://doi.org/10.1021/acs.est.7b06428
131. Plewa MJ, Kargalioglu Y, Vankerk D, Minear RA, Wagner ED (2002) Mammalian cell cytotoxicity and genotoxicity analysis of drinking water disinfection by-products. Environ Mol Mutagen 40(2):134–142. https://doi.org/10.1002/em.10092
132. Plewa MJ, Wagner ED, Jazwierska P, Richardson SD, Chen PH, McKague AB (2004) Halonitromethane drinking water disinfection byproducts: chemical characterization and mammalian cell cytotoxicity and genotoxicity. Environ Sci Technol 38(1):62–68. https://doi.org/10.1021/es030477l
133. Plewa MJ, Wagner ED, Richardson SD, Thruston Jr AD, Woo YT, McKague AB (2004) Chemical and biological characterization of newly discovered iodoacid drinking water disinfection byproducts. Environ Sci Technol 38(18):4713–4722. https://doi.org/10.1021/es049971v
134. Plewa MJ, Simmons JE, Richardson SD, Wagner ED (2010) Mammalian cell cytotoxicity and genotoxicity of the haloacetic acids, a major class of drinking water disinfection by-products. Environ Mol Mutagen 51(8–9):871–878. https://doi.org/10.1002/em.20585
135. Richardson SD, Fasano F, Ellington JJ, Crumley FG, Buettner KM, Evans JJ, Blount BC, Silva LK, Waite TJ, Luther GW et al (2008) Occurrence and mammalian cell toxicity of iodinated disinfection byproducts in drinking water. Environ Sci Technol 42(22):8330–8338. https://doi.org/10.1021/es801169k
136. Muellner MG, Wagner ED, McCalla K, Richardson SD, Woo YT, Plewa MJ (2007) Haloacetonitriles vs. regulated haloacetic acids: are nitrogen-containing DBPs more toxic? Environ Sci Technol 41(2):645–651. https://doi.org/10.1021/es0617441
137. Jeong CH, Postigo C, Richardson SD, Simmons JE, Kimura SY, Mariñas BJ, Barcelo D, Liang P, Wagner ED, Plewa MJ (2015) Occurrence and comparative toxicity of haloacetaldehyde disinfection byproducts in drinking water. Environ Sci Technol 49(23): 13749–13759. https://doi.org/10.1021/es506358x
138. Plewa MJ, Muellner MG, Richardson SD, Fasano F, Buettner KM, Woo YT, McKague AB, Wagner ED (2008) Occurrence, synthesis, and mammalian cell cytotoxicity and genotoxicity of haloacetamides: an emerging class of nitrogenous drinking water disinfection byproducts. Environ Sci Technol 42(3):955–961. https://doi.org/10.1021/es071754h
139. Richardson SD, Thruston Jr AD, Rav-Acha C, Groisman L, Popilevsky I, Juraev O, Glezer V, McKague AB, Plewa MJ, Wagner ED (2003) Tribromopyrrole, brominated acids, and other disinfection byproducts produced by disinfection of drinking water rich in bromide. Environ Sci Technol 37(17):3782–3793. https://doi.org/10.1021/es030339w
140. Prochazka E, Escher BI, Plewa MJ, Leusch FD (2015) In vitro cytotoxicity and adaptive stress responses to selected haloacetic acid and Halobenzoquinone water disinfection byproducts. Chem Res Toxicol 28(10):2059–2068. https://doi.org/10.1021/acs.chemrestox.5b00283
141. Du HY, Li JH, Moe B, McGuigan CF, Shen SW, Li X-F (2013) Cytotoxicity and oxidative damage induced by halobenzoquinones to T24 bladder cancer cells. Environ Sci Technol 47(6):2823–2830. https://doi.org/10.1021/Es303762p

142. Zhang D, Chu WH, Yu Y, Krasner SW, Pan Y, Shi J, Yin DQ, Gao NY (2018) Occurrence and stability of chlorophenylacetonitriles: a new class of nitrogenous aromatic DBPs in chlorinated and chloraminated drinking waters. Environ Sci Technol Lett 5(6):394–399. https://doi.org/10.1021/acs.estlett.8b00220
143. Hu S, Kaw HY, Zhu L, Wang W (2022) Formation and cytotoxicity of halophenylacetamides: a new group of nitrogenous aromatic halogenated disinfection byproducts in drinking water. Environ Sci Technol 56(5):3181–3192. https://doi.org/10.1021/acs.est.1c08419
144. Hu S, Kaw HY, Zhu L, Wang W (2022) Halohydroxybenzonitriles as a new group of halogenated aromatic DBPs in drinking water: are they of comparable risk to halonitrophenols? Water Res 219:118547. https://doi.org/10.1016/j.watres.2022.118547
145. Liu J, Gibb M, Pradhan SH, Sayes CM (2022) Synergistic cytotoxicity of bromoacetic acid and three emerging bromophenolic disinfection byproducts against human intestinal and neuronal cells. Chemosphere:287. https://doi.org/10.1016/j.chemosphere.2021.131794
146. Zhang Z, Zhu Q, Huang C, Yang M, Li J, Chen Y, Yang B, Zhao X (2020) Comparative cytotoxicity of halogenated aromatic DBPs and implications of the corresponding developed QSAR model to toxicity mechanisms of those DBPs: binding interactions between aromatic DBPs and catalase play an important role. Water Res 170:115283. https://doi.org/10.1016/j.watres.2019.115283
147. Li J, Bauer M, Moe B, Leslie EM, Li X-F (2017) Multidrug resistance protein 4 (MRP4/ABCC4) protects cells from the toxic effects of halobenzoquinones. Chem Res Toxicol 30(10):1815–1822., Article. https://doi.org/10.1021/acs.chemrestox.7b00156
148. Uchimiya M, Stone AT (2009) Reversible redox chemistry of quinones: impact on biogeochemical cycles. Chemosphere 77(4):451–458. https://doi.org/10.1016/j.chemosphere.2009.07.025
149. Wang C, Yang X, Zheng Q, Moe B, Li XF (2018) Halobenzoquinone-induced developmental toxicity, oxidative stress, and apoptosis in Zebrafish embryos. Environ Sci Technol 52(18):10590–10598. https://doi.org/10.1021/acs.est.8b02831
150. Stalter D, O'Malley E, von Gunten U, Escher BI (2016) Fingerprinting the reactive toxicity pathways of 50 drinking water disinfection by-products. Water Res 91:19–30. https://doi.org/10.1016/j.watres.2015.12.047
151. Li J, Wang W, Moe B, Wang H, Li XF (2015) Chemical and toxicological characterization of halobenzoquinones, an emerging class of disinfection byproducts. Chem Res Toxicol 28(3):306–318. https://doi.org/10.1021/tx500494r
152. Erkekoglu P, Baydar T (2010) Evaluation of the protective effect of ascorbic acid on nitrite- and nitrosamine-induced cytotoxicity and genotoxicity in human hepatoma line. Toxicol Mech Methods 20(2):45–52. https://doi.org/10.3109/15376510903583711
153. Xue P, Wang H, Yang L, Jiang Z, Li H, Liu Q, Zhang Q, Andersen ME, Crabbe MJC, Hao L et al (2022) NRF2-ARE signaling is responsive to haloacetonitrile-induced oxidative stress in human keratinocytes. Toxicol Appl Pharmacol 450:116163. https://doi.org/10.1016/j.taap.2022.116163
154. Zhou M, Li J, Du M, Wang J, Kaw HY, Zhu L, Wang W (2023) Methoxylated modification of glutathione-mediated metabolism of halobenzoquinones in vivo and in vitro. Environ Sci Technol 57(9):3581–3589. https://doi.org/10.1021/acs.est.2c06765
155. Manasfi T, De Méo M, Di Giorgio C, Coulomb B, Boudenne JL (2017) Assessing the genotoxicity of two commonly occurring byproducts of water disinfection: chloral hydrate and bromal hydrate. Mutat Res Genet Toxicol Environ Mutagen 813:37–44. https://doi.org/10.1016/j.mrgentox.2016.11.009
156. Zhang L, Xu L, Zeng Q, Zhang SH, Xie H, Liu AL, Lu WQ (2012) Comparison of DNA damage in human-derived hepatoma line (HepG2) exposed to the fifteen drinking water disinfection byproducts using the single cell gel electrophoresis assay. Mutat Res 741(1–2):89–94. https://doi.org/10.1016/j.mrgentox.2011.11.004
157. Dong L, Jiang Z, Yang L, Hu F, Zheng W, Xue P, Jiang S, Andersen ME, He G, Crabbe MJC et al (2022) The genotoxic potential of mixed nitrosamines in drinking water involves

oxidative stress and Nrf2 activation. J Hazard Mater 426:128010. https://doi.org/10.1016/j.jhazmat.2021.128010
158. Marsà A, Cortés C, Hernández A, Marcos R (2018) Hazard assessment of three haloacetic acids, as byproducts of water disinfection, in human urothelial cells. Toxicol Appl Pharmacol 347:70–78. https://doi.org/10.1016/j.taap.2018.04.004
159. Li J, Zhang H, Han Y, Chao H, Ma M, Yang M (2020) Cytotoxicity and genotoxicity assays of halobenzoquinones disinfection byproducts using different human cell lines. Environ Mol Mutagen 61(5):526–533. https://doi.org/10.1002/em.22369
160. Tu NN, Liu H, Li WL, Yao S, Liu JL, Guo ZY, Yu R, Du HY, Li JH (2022) Quantitative structure-toxicity relationships of halobenzoquinone isomers on DNA reactivity and genotoxicity. Chemosphere 309. https://doi.org/10.1016/j.chemosphere.2022.136763
161. Pérez-Albaladejo E, Pinteño R, Aznar-Luque MDC, Casado M, Postigo C, Porte C (2023) Genotoxicity and endocrine disruption potential of haloacetic acids in human placental and lung cells. Sci Total Environ 879. https://doi.org/10.1016/j.scitotenv.2023.162981
162. Liviac D, Creus A, Marcos R (2010) Genotoxicity testing of three monohaloacetic acids in TK6 cells using the cytokinesis-block micronucleus assay. Mutagenesis 25(5):505–509. https://doi.org/10.1093/mutage/geq034. geq034 [pii]
163. Pegram RA, Andersen ME, Warren SH, Ross TM, Claxton LD (1997) Glutathione S-transferase-mediated mutagenicity of trihalomethanes in Salmonella typhimurium: contrasting results with bromodichloromethane off chloroform. Toxicol Appl Pharmacol 144(1):183–188. https://doi.org/10.1006/taap.1997.8123
164. Wagner ED, Osiol J, Mitch WA, Plewa MJ (2014) Comparative in vitro toxicity of nitrosamines and nitramines associated with amine-based carbon capture and storage. Environ Sci Technol 48(14):8203–8211. https://doi.org/10.1021/es5018009
165. Kundu B, Richardson SD, Swartz PD, Matthews PP, Richard AM, DeMarini DM (2004) Mutagenicity in Salmonella of halonitromethanes: a recently recognized class of disinfection by-products in drinking water. Mutat Res 562(1–2):39–65. https://doi.org/10.1016/j.mrgentox.2004.05.007
166. DeMarini DM, Abu-Shakra A, Felton CF, Patterson KS, Shelton ML (1995) Mutation spectra in salmonella of chlorinated, chloraminated, or ozonated drinking water extracts: comparison to MX. Environ Mol Mutagen 26(4):270–285
167. Zheng J, Lin T, Chen W (2018) Removal of the precursors of N-nitrosodiethylamine (NDEA), an emerging disinfection byproduct, in drinking water treatment process and its toxicity to adult zebrafish (Danio rerio). Chemosphere 191:1028–1037. https://doi.org/10.1016/j.chemosphere.2017.10.059
168. Xiao C, Wang C, Zhang Q, Yang X, Huang S, Luo Y, Feng Y, Zheng Q (2021) Transcriptomic analysis of adult zebrafish heart and brain in response to 2, 6-dichloro-1, 4-benzoquinone exposure. Ecotoxicol Environ Saf:226. https://doi.org/10.1016/j.ecoenv.2021.112835
169. Xue B, Li C, Wang S, Zhao C, Dai K, Li W, Xi Z, Wang J, Qiu Z, Shen Z (2019) Effects of 2,2-dichloroacetamide (DCAcAm), an emerging disinfection by-product in drinking water, on the intestinal microbiota of adult zebrafish. J Water Health 17(5):683–690., Article. https://doi.org/10.2166/wh.2019.081
170. Kim YJ, Yang JH, Ryu CS, Choi I, Esterhuizen M, Park CG, Lee S-A (2023) Investigating endocrine disrupting impacts of nine disinfection byproducts on human and Zebrafish Estrogen receptor alpha. Front Biosci Landmark 28(3). https://doi.org/10.31083/j.fbl2803048
171. Gómez-Catalán J, Llobet JM, Gonzalez-Linares J, Piqué E, Teixidó E (2015) Developmental effects and genotoxicity of 10 water disinfection by-products in zebrafish. J Water Health 13(1):54–66. https://doi.org/10.2166/wh.2014.006
172. Ding X, Zhu J, Zhang J, Dong T, Xia Y, Jiao J, Wang X, Zhou W (2020) Developmental toxicity of disinfection by-product monohaloacetamides in embryo-larval stage of zebrafish. Ecotoxicol Environ Saf:189. https://doi.org/10.1016/j.ecoenv.2019.110037

173. Cui H, Chen B, Jiang Y, Tao Y, Zhu X, Cai Z (2021) Toxicity of 17 disinfection by-products to different trophic levels of aquatic organisms: ecological risks and mechanisms. Environ Sci Technol 55(15):10534–10541. https://doi.org/10.1021/acs.est.0c08796
174. Zhang Y, Zhang B, Xu T, Yang X, Wei S, Yin D (2021) Developmental and neurobehavioral assessment of low-dose N-nitrosodimethylamine (NDMA) using zebrafish embryo bioassay. Sci Total Environ:770. https://doi.org/10.1016/j.scitotenv.2020.144748
175. Yang X, Wang C, Yang L, Zheng Q, Liu Q, Wawryk NJP, Li X-F (2022) Neurotoxicity and transcriptome changes in embryonic zebrafish induced by halobenzoquinone exposure. J Environ Sci 117:129–140. https://doi.org/10.1016/j.jes.2022.03.042
176. Yang X, Wang C, Zheng Q, Liu Q, Wawryk NJP, Li X-F (2022) Emerging disinfection byproduct 2,6-dichlorobenzoquinone-induced cardiovascular developmental toxicity of embryonic Zebrafish and larvae: imaging and transcriptome analysis. ACS Omega 7(49): 45642–45653. https://doi.org/10.1021/acsomega.2c06296
177. Chaves RS, Guerreiro CS, Cardoso VV, Benoliel MJ, Santos MM (2020) Toxicological assessment of seven unregulated drinking water disinfection by-products (DBPs) using the zebrafish embryo bioassay. Sci Total Environ:742. https://doi.org/10.1016/j.scitotenv.2020.140522
178. Hanigan D, Truong L, Simonich M, Tanguay R, Westerhoff P (2017) Zebrafish embryo toxicity of 15 chlorinated, brominated, and iodinated disinfection by-products. J Environ Sci 58:302–310. https://doi.org/10.1016/j.jes.2017.05.008
179. Zuo Y-T, Hu Y, Lu W-W, Cao J-J, Wang F, Han X, Lu W-Q, Liu A-L (2017) Toxicity of 2,6-dichloro-1,4-benzoquinone and five regulated drinking water disinfection by-products for the Caenorhabditis elegans nematode. J Hazard Mater 321:456–463. https://doi.org/10.1016/j.jhazmat.2016.09.038
180. Saul N, Baberschke N, Chakrabarti S, Stürzenbaum SR, Lieke T, Menzel R, Jonáš A, Steinberg CEW (2014) Two organobromines trigger lifespan, growth, reproductive and transcriptional changes in Caenorhabditis elegans. Environ Sci Pollut Res 21(17): 10419–10431. https://doi.org/10.1007/s11356-014-2932-6
181. Ju J, Lieke T, Saul N, Pu Y, Yin L, Kochan C, Putschew A, Baberschke N, Steinberg CEW (2014) Neurotoxic evaluation of two organobromine model compounds and natural AOBr-containing surface water samples by a Caenorhabditis elegans test. Ecotoxicol Environ Saf 104:194–201. https://doi.org/10.1016/j.ecoenv.2014.03.009
182. Baberschke N, Steinberg CEW, Saul N (2015) Low concentrations of dibromoacetic acid and N-nitrosodimethylamine induce several stimulatory effects in the invertebrate model Caenorhabditis elegans. Chemosphere 124:122–128. https://doi.org/10.1016/j.chemosphere.2014.12.002
183. Chen RZ (2009) Development, validation and uncertainty analysis of quantitative structure and activity relationship models for Log P of disinfection by-products
184. Qin L-T, Zhang X, Chen Y-H, Mo L-Y, Zeng H H, Liang Y P, Lin H, Wang D-Q (2019) Predicting the cytotoxicity of disinfection by-products to Chinese hamster ovary by using linear quantitative structure-activity relationship models. Environ Sci Pollut Res 26(16): 16606–16615., Article. https://doi.org/10.1007/s11356-019-04947-z
185. Jiang L, Luo J, Wei W, Song M, Shi W, Li A, Zhou Q, Pan Y (2023) Comparative cytotoxicity analyses of disinfection byproducts in drinking water using dimensionless parameter scaling method: effect of halogen substitution type and number. Water Res 240., Article. https://doi.org/10.1016/j.watres.2023.120087
186. Liu J, Zhang X, Li Y, Li W, Hang C, Sharma VK (2019) Phototransformation of halophenolic disinfection byproducts in receiving seawater: kinetics, products, and toxicity. Water Res 150: 68–76., Article. https://doi.org/10.1016/j.watres.2018.11.059
187. Wei X, Yang M, Zhu Q, Wagner ED, Plewa MJ (2020) Comparative quantitative toxicology and QSAR modeling of the haloacetonitriles: forcing agents of water disinfection byproduct toxicity. Environ Sci Technol 54(14):8909–8918., Article. https://doi.org/10.1021/acs.est.0c02035

188. Wang W, Qian Y, Li J, Aljuhani N, Siraki AG, Le XC, Li XF (2018) Characterization of mechanisms of glutathione conjugation with halobenzoquinones in solution and HepG2 cells. Environ Sci Technol 52(5):2898–2908. https://doi.org/10.1021/acs.est.7b05945
189. Lan J, Rahman SM, Gou N, Jiang T, Plewa MJ, Alshawabkeh A, Gu AZ (2018) Genotoxicity assessment of drinking water disinfection byproducts by DNA damage and repair pathway profiling analysis. Environ Sci Technol 52(11):6565–6575., Article. https://doi.org/10.1021/acs.est.7b06389
190. Li C, Wang D, Li N, Luo Q, Xu X, Wang Z (2016) Identifying unknown by-products in drinking water using comprehensive two-dimensional gas chromatography-quadrupole mass spectrometry and *in silico* toxicity assessment. Chemosphere 163:535–543., Article. https://doi.org/10.1016/j.chemosphere.2016.08.053
191. Yang M, Zhang X (2013) Comparative developmental toxicity of new aromatic halogenated DBPs in a chlorinated saline sewage effluent to the marine polychaete *Platynereis dumerilii*. Environ Sci Technol 47(19):10868–10876., Article. https://doi.org/10.1021/es401841t
192. Qin L, Zhang X, Chen Y, Mo L, Zeng H, Liang Y (2017) Predictive QSAR models for the toxicity of disinfection byproducts. Molecules 22(10) Article. https://doi.org/10.3390/molecules22101671
193. Li JJ, Yue YX, Shi SJ, Xue JZ (2023) Investigation on toxicity mechanism of halogenated aromatic disinfection by-products to zebrafish based on molecular docking and QSAR model. Chemosphere 341:139916–139916. https://doi.org/10.1016/j.chemosphere.2023.139916

Health Risk of DBPs

Min Zhang and Qiang Zeng

Contents

1 Introduction 446
2 Internal Exposure Levels 446
 2.1 Exhaled THM Concentrations 447
 2.2 Blood THM Concentrations 447
 2.3 Urinary HAA Concentrations 448
3 Cancer Risks 448
 3.1 Bladder Cancer Risk 449
 3.2 Colon and Rectal Cancer Risks 450
 3.3 Other Cancer Risks 452
4 Adverse Reproductive and Developmental Outcomes 453
 4.1 Male Reproductive Health Outcomes 453
 4.2 Female Reproductive Health Outcomes 456
 4.3 Fetal Developmental Outcomes 458
5 Other Adverse Health Outcomes 464
References 465

Abstract Human is exposed to disinfection byproducts (DBPs) everyday through the disinfected drinking water. The potential health risk of exposure to DBPs has been an increasing global concern in field of drinking water safety. In this chapter, we reported the internal levels of DBP exposures in human's exhaled air, blood, and urine, indicating widespread DBP exposures across populations. Furthermore, we summarized human evidence on exposure to DBPs associated with cancer risks and reproductive and developmental outcomes, as well as other adverse health outcomes (e.g., asthma). Overall, epidemiological studies support the association between exposure to DBPs and bladder cancer risk; however, no consistent relationships are observed for colon, rectal, and other cancers. Existing epidemiological evidence

M. Zhang and Q. Zeng (✉)
Department of Occupational and Environmental Health, School of Public Health, Tongji Medical College, Huazhong University of Science and Technology, Wuhan, Hubei, People's Republic of China
e-mail: zengqiang506@hust.edu.cn

Chao Chen, Susan Andrews, and Yuefeng Xie (eds.), *Disinfection By-Products in Water*, Hdb Env Chem (2024) 135: 445–472, DOI 10.1007/698_2024_1153,

Published online: 27 September 2024

remains insufficient to support the associations of DBP exposures with adverse reproductive and developmental outcomes, but emerging human studies relying on blood and urinary biomarkers have shown that DBP exposures are associated with declined semen quality, impaired ovarian function, and adverse birth outcomes. Human evidence concerning other adverse health outcomes is limited and inconclusive.

Keywords Asthma, Birth outcomes, Cancer, Internal exposure, Reproductive health

1 Introduction

Disinfection byproducts (DBPs) are widely present in drinking water and humans are exposed to such chemicals in daily water-use activities. Based on the measurements of trihalomethanes (THMs) in expired air/blood and haloacetic acids (HAAs) in urine, there have been ubiquitous DBP exposures across different populations. Exposure to DBPs resulting in potential health risks has been a public health concern worldwide. During the past several decades, a large number of epidemiological investigations have focused on the effects of exposure to DBPs on cancer risks such as bladder, colon, and rectal cancer, as well as adverse reproductive and developmental outcomes such as semen quality and fetal growth. However, there is paucity of human evidence on exposure to DBPs associated with other adverse health outcomes such as liver and kidney injury, thyroid dysfunction, respiratory diseases, and type II diabetes.

2 Internal Exposure Levels

Humans can be continuously exposed to DBPs via ingestion, inhalation, and dermal absorption from daily water-use activities, such as drinking, bathing, swimming, laundry, and washing. After exposure, about 60–70% of DBPs are excreted as the prototypes from the lungs, feces, or urine, while the remaining are metabolized in the body. The metabolism of DBPs is complex and depends on the extent of biotransformation and elimination between compounds and between species. For example, THMs are mainly metabolized and bioactivated in the liver via the cytochrome P450 (CYP450), generating free radicals, oxidation intermediates, and ultimately carbon dioxide, which is excreted by the expired air. HAAs are generally metabolized in the liver via CYP450-catalyzed reductive dechlorination followed by glutathione-S-transferases or further reductive dechlorination, and then excreted through urine. Currently, THMs in expired air/blood and HAAs in urine have been primarily utilized to reflect human internal DBP exposure levels.

2.1 Exhaled THM Concentrations

Exhaled THMs are commonly used to represent individual short-term exposures after water-use activities. Swimming, showering, or bathing can significantly increase the exhaled THM concentrations within a short time. An increase in the median chloroform/trichloromethane (TCM) concentration in exhaled breath was observed from 5 μg/m^3 before showing to 24 μg/m^3 after a 13-min-hot water showing among four adults from North Carolina, USA [1]. In a study among 116 non-smoker American adults after 40-min-swimming, the median concentrations of TCM, brominated THMs [(Br-THMs, the sum of bromodichloromethane (BDCM), dibromochloromethane (DBCM), and bromoform/tribromomethane (TBM)], and total THMs (TTHMs, the sum of DBCM, BDCM, TCM and TBM) in exhaled breath increased from 0.4 to 11.5 μg/m^3, 0.1 to 2.7 μg/m^3, and 0.5 to 14.4 μg/m^3, respectively [2]. However, the collection of exhaled breath needs a special collection device and THM concentrations in such samples are also generally non-detectable in the general population, which limits the use of exhaled THM biomarkers to assess human internal exposure levels.

2.2 Blood THM Concentrations

Blood THM concentrations can indicate an integrative measure from multiple exposure routes and are generally more sensitive to low levels of exposure. Daily water-use activities such as drinking, showering, bathing, and washing can increase blood THM levels. Generally, TCM has the highest detectable frequency and concentration in blood samples, followed by BDCM, DBCM, and TBM. Among 6,924 American adults in the National Health and Nutrition Examination Survey (NHANES) 1999–2006, the detection frequencies of blood TCM, BDCM, DBCM, and TBM were 94.9%, 79.1%, 56.3%, and 48.8%, respectively, with the median concentrations of 12.9 ng/L, 1.6 ng/L, 0.6 ng/L, and 0.8 ng/L, respectively [3]. In China, the detection frequencies of blood TCM, BDCM, DBCM, and TBM were 92.4%, 79.4%, 42.6%, 48.5% from 1760 pregnant women, respectively, with the median concentrations of 10.1 ng/L, 0.8 ng/L, less than the limit for detection (LOD, 0.68 ng/L), and less than LOD (2.0 ng/L), respectively [4]. Similar results were also detected among 1,199 Chinese men who screened as potential sperm donors [5]. However, blood THM concentrations are readily influenced by recent water-use activities, while the THM measurements in the morning blood sample collected before any major water-use activities can produce baseline internal exposure levels.

2.3 Urinary HAA Concentrations

Ingestion is the main route of exposure to HAAs due to their non-volatility and low skin permeability. Studies have demonstrated significant correlations between urinary trichloroacetic acid (TCAA) levels and ingestion of TCAA and THMs from drinking water [6–8]. As a result, urinary TCAA has been considered a leading internal exposure biomarker for ingestion of drinking water DBPs. Among 402 American adults from NHANES III 1988–1994, TCAA was detectable in 76% of the urine samples, with the median concentration of 3.3 μg/L [9]. In a study of 2,009 Chinese men seeking semen analysis, the median concentration of urinary TCAA was 7.97 μg/L, with a detection frequency of 98.6% [10]. Recently, among 956 Chinese women undergoing assisted reproductive technology treatment from the Tongji Reproductive and Environmental (TREE) cohort, TCAA and dichloroacetic acid (DCAA) were detected in more than 99% of urine samples, with the median concentrations of 4.30 and 4.61 μg/L, respectively [11]. Due to the short biological half-lives of HAAs in the human body, a single urinary measurement can produce misclassification of exposure. In a panel study of 11 healthy Chinese men, high within-person variability was observed for urinary DCAA and TCAA concentrations over three months [12]. The authors further found that the probabilities of correctly classifying the men into the lowest group (three-month average urinary DCAA and TCAA levels below 67%) using a single-spot urinary measurement were high (specificities were 0.79 and 0.78 for DCAA and TCAA, respectively), but the probabilities of correctly classifying the men into the highest group (three-month averaged urinary DCAA and TCAA levels above 33%) were relatively low sensitivity (sensitivities were 0.47 and 0.50 for DCAA and TCAA, respectively) [12].

3 Cancer Risks

Some DBPs are carcinogenic in rodents. The World Health Organization International Agency for Research on Cancer (IARC) classifies certain DBPs including TCM, BDCM, DCAA, TCAA, dibromoacetic acid (DBAA), dibromoacetonitrile, N-nitrosodimethylamine (NDMA), N-nitrosodiethylamine (NDEA), 3-chloro-4-(dichloromethyl)-5-hydroxy-2(5H)-furanone (MX), and potassium bromate as possible human carcinogens [13, 14]. The effects of exposure to drinking water DBPs on human cancer risk have attracted increasing attention around the world. Early epidemiological studies primarily utilized the disinfection methods of drinking water as individual exposure assessment. The subsequent epidemiological investigations have enhanced exposure assessment based on the monitoring of DBP concentrations in drinking water and/or combining the individual's daily water activities. The differences in study designs, DBP exposure assessments, study populations, and residual confounding factors may contribute to the heterogeneous results among epidemiological studies of exposure to DBPs and cancer risks. However, increasing epidemiological evidence suggests an association between exposure to DBPs and bladder cancer.

3.1 Bladder Cancer Risk

Bladder cancer is the most frequently evaluated cancer associated with exposure to drinking water DBPs. Early ecological studies (ecological study is an observational study that compares exposures and disease rates across populations or over time to examine their associations) found that exposure to chlorinated water was associated with an increased risk of bladder cancer. For instance, a relative risk of 1.21-fold [95% confidence interval (CI): 1.10, 1.32] for bladder cancer was reported in an ordinary municipality (a local administrative unit) using the chlorinated surface water via a public water supply system compared with those not using the public water supply (e.g., private well) from 56 Finnish municipalities [15]. However, a study conducted in 261 municipalities from Valencia Province of Spain did not support the association between the use of chlorinated surface water and bladder cancer [16]. Recently, a study using tumor registration information and DBP monitoring data from 46 cities in China reported that the cities with higher median DBP concentrations in tap water (103.44 μg/L) were characterized by higher incidence rates of bladder cancer than those with lower median DBP concentrations in tap water (71.46 μg/L) [17].

Most case-control studies have reported the associations between the exposure to DBPs (specifically THMs) and the increased risk of bladder cancer. In a 1998–2001 case-control study in Spain among 1,219 cases and 1,271 controls, average household THM levels of >49 μg/L vs. ≤8 μg/L were associated with an adjusted odds ratio (OR) of 2.10 (95% CI: 1.09, 4.02) for bladder cancer occurrence [18]. In a study from New England, USA, among 1,213 cases and 1,418 controls, the adjusted ORs for bladder cancer in the exposure group with the highest average daily intake (>103.89 vs. <8.69 μg/day for TTHMs; >7.04 vs. <0.5 μg/day for Br-THMs) were 1.53 (95% CI: 1.01, 2.32) for TTHMs and 1.98 (95% CI: 1.19, 3.29) for Br-THMs [19].

Several cohort studies have generated inconsistent results of the associations between exposure to DBPs and bladder cancer. A statistically significant excess risk of bladder cancer [relative risk (RR) = 1.48, 95% CI: 1.01, 2.18] was found for women living in a town using chlorinated surface water compared to those not exposed to chlorinated surface water in a historical cohort of 621,431 persons living in 56 towns from Finland [20]. However, in a prospective cohort study among 28,237 postmenopausal women in Iowa, USA, women with municipal surface water sources (geometric mean of water TTHM concentrations: 56.16 μg/L) were at a similar risk of bladder cancer in comparison with women who used municipal ground-water sources (geometric mean of water TTHM concentrations: 0.52 μg/L) [21]. Similarly, a large-scale prospective cohort study from Sweden found no overall association of THM exposures with the risk of bladder cancer, with a hazard ratio (HR) of 0.90 (95% CI: 0.73, 1.11) for the highest THM exposure (≥15 μg/L) compared to the non-exposure [22].

Meta-analysis and pooled analysis have consistently shown that DBP exposures are associated with an increased risk of bladder cancer, and such an association seems to be more evident in men. The meta-analysis of six case-control and two cohort studies showed that consumption of chlorinated drinking water was associated with an increased risk of bladder cancer in men (combined OR = 1.40, 95% CI: 1.10, 1.90), but not in women (combined OR = 1.20, 95% CI: 0.70, 1.80) [23]. A pooled analysis of six case-control studies found that there was an adjusted OR of 1.24 (95% CI: 1.09, 1.41) in men exposed to an average of more than 1 μg/L of THMs compared with those who had lower or no exposure [24]. In a pooled and two-stage random-effect meta-analyses of three European case-control studies from France, Finland, and Spain, an increased bladder cancer risk (OR = 1.47, 95% CI:1.05, 2.05) was estimated for men exposed to an average residential TTHM levels >50 μg/L compared to those exposed to levels ≤5 μg/L, but null association was found among women [25].

Genetic polymorphisms, particularly the heritable variation in metabolic enzymes involved in the metabolic pathways of DBPs, may modify the associations between DBP exposures and bladder cancer risk. A case-control study in Spain showed that the associations between THMs and bladder cancer were stronger among individuals who carried present glutathione transferase theta-1 (GSTT1), glutathione transferase zeta-1(GSTZ1) rs1046428 CT/TT, or cytochrome P450 2E1 (CYP2E1) rs2031920 CC genotypes, and those with the highest quartile of average THM levels were at a 5.9-fold (95%CI: 1.8. 19.0) increased risk of bladder cancer compared with those carriers with the lowest quartile [26]. Another two large case-control studies from the New England Bladder Cancer Study in the USA and the Spanish Bladder Cancer Study reported that rs907611 at 11p15.5 [leukocyte-specific protein 1 (*LSP1* region)] exhibited the strongest modification among the identified 16 genetic variants for DBP exposures and bladder cancer [27]. Among subjects with the rs907611-GG genotype, those with the highest quartile of THM level had a 1.66-fold (95% CI: 1.17, 2.34) increased risk of bladder cancer compared with those the lowest exposure quartile [27]. Apart from gene-environment interactions, long interspersed elements-1 (LINE-1) methylation status may also modify the association between THM levels and bladder cancer risk. Among subjects with low levels of LINE-1 5-methylcytosine level, those with average residential THM levels >26 μg/L were at a 2.14-fold (95%CI: 1.39, 3.30) increased risk of bladder cancer compared with those with average residential THM levels ≤26 μg/L [28].

3.2 Colon and Rectal Cancer Risks

Human evidence concerning the associations of DBP exposures with colon and rectal cancer is inconsistent. The ecological studies in Iowa, USA reported the positive associations between the use of chlorinated surface water and the incidence of colon and rectal cancers [29, 30]. In an ecological study based on Norwegian national cancer registries, the chlorination of drinking water was associated with a 35% (OR = 1.35, 95%CI: 1.17, 1.57) increase in the incidence of colorectal cancer among men and a 17% (OR = 1.17, 95%CI: 1.05, 1.30) increase among women compared to those using unchlorinated water [31]. A positive association between the consumption of chlorinating drinking water and rectum cancer was also observed in another ecological study in Taiwan, China [32].

Several case-control studies have explored DBP exposures in association with colon and/or rectal cancer, but the results are inconclusive. A matched case-control study in Louisiana, USA observed that subjects using chlorinated water had an increased risk of rectal cancer (OR = 1.68, 95% CI: 1.17, 2.42) compared with those using unchlorinated water but not for colon cancer risk [33]. Another study in New South Wales, Australia reported that an interquartile range increase in TBM concentrations was associated with an increased RR of 2.5% (95%CI:1.01, 1.04) in colon and rectal cancers among men, and this effect was more pronounced for colon cancer (RR = 1.04, 95%CI: 1.02, 1.05), but null associations were observed for other THMs [34]. Another case-control study involving 767 colon cases, 661 rectal cases, and 1,545 controls in southern Ontario, Canada observed that males exposed to an estimated THM level of 75 μg/L for ≥35 years were at 2.1-fold (95% CI: 1.21, 3.66) risk for colon cancer compared to those exposed for <10 years, but no associations with rectal cancer risk were found [35]. However, a study among white women teachers in New York State, USA reported that individual cumulative TCM exposure was not associated with the deaths of colorectal cancer [36]. Similarly, another study included 65 municipalities in Taiwan, China observed that there was a null statistically significant association between TTHM levels in drinking water and the risk of death from colon cancer [37]. Recently, a multicenter case-control study in Spain and Italy in 2008–2013 observed that TTHM concentrations in residential tap water were not associated with colorectal cancer risk, but TCM concentrations were inversely associated with colorectal cancer risk (OR = 0.31, 95% CI: 0.24, 0.41 for highest vs. lowest quartile) [38].

Several cohort studies have also examined the relationships between exposure to DBPs and the risks of colon and rectal cancer. A historical cohort study of 621,431 persons living in 56 towns in Finland reported a statistically significant excess risk of rectum cancer (RR = 1.48, 95% CI: 1.01, 2.18) for women living in a town using chlorinated surface water compared to those not using chlorinated surface water [20]. A prospective cohort study among 28,237 postmenopausal women in Iowa, USA found that TCM levels in drinking water were associated with an increased risk of colon cancer (HR_{Q4vsQ1} = 1.68, 95% CI:1.11, 2.53), but not for rectal cancer [21]. The Iowa Women's Health Study cohort of 15,910 postmenopausal women observed that TTHM (HR_{Q5vsQ1} = 1.71, 95% CI:1.00, 2.92), BDCM (HR_{Q4vsQ1} = 1.89, 95% CI:1.17, 3.00), and TCAA (HR_{Q4vsQ1} = 1.92, 95% CI:1.20, 3.09) levels in drinking water were associated with increased risk of rectal cancer, but null associations were found between TTHMs, TCM, BDCM, DCAA, TCAA, and bromochloroacetic acid exposures and colon cancer risk

[39]. Recently, one large-scale prospective cohort study from Sweden found a 1.26-fold (95% CI: 1.05, 1.51) increased risk of colorectal cancer among men exposed to THMs in drinking water at concentrations ≥15 μg/L compared to those not exposed to THMs, and this association was more pronounced for proximal colon cancer (HR = 1.59, 95% CI: 1.11, 2.27) but not for distal colon cancer or rectal cancer [40].

Evidence with respect to the meta-analysis studies is inconclusive. An analysis of ten case-control studies found that rectal cancer yielded a pooled relative risk estimate for exposure to DBPs (i.e., consumption of chlorinated water, surface water, or water with high levels of chloroform) of 1.38 (95% CI: 1.01, 1.87) [41]. The meta-analysis of ten case-control and three cohort studies used a random effects approach to estimate the RRs or ORs of colon and rectal cancers for the highest exposure of DBP levels compared with the lowest exposure [42]. For colon cancer, the pooled RR/OR estimates were 1.11 (95% CI: 0.73, 1.70), 1.33 (95% CI: 1.12, 1.57), and 1.27 (95% CI: 1.08, 1.50) for cohort studies, case-control studies and combining both study types, respectively; the corresponding RR/OR estimates for the rectal cancer were 0.88 (95% CI: 0.57, 1.35), 1.40 (95% CI: 1.15, 1.70), and 1.30 (95% CI: 1.06, 1.59), respectively [42].

3.3 Other Cancer Risks

There has been sparse human evidence of DBP exposures in associations with other cancers (e.g., liver, pancreas, kidney, brain, breast, esophagus, stomach, and lung cancer, in addition to leukemia, melanoma, and non-melanoma skin cancer). A study conducted in Erie County, New York, USA found a significantly positive association between THM levels and pancreatic cancer incidence, but no significant associations with esophageal and stomach cancer [43]. Another study among white females in Wisconsin, USA reported that chlorinated drinking water was not associated with mortality from liver, kidney, esophagus, stomach, pancreas, lung, brain, or breast cancers [44].

As for leukemia, an increased risk of chronic myeloid leukemia was observed in Canadian adults for more than 31 years exposed to TTHM levels >20 μg/L, to TTHM levels >40 μg/L, and to BDCM level >5 μg/L, with adjusted OR of 1.70 (95% CI: 1.00, 3.03), 1.72 (95% CI: 1.01, 3.08), and 1.63 (95% CI: 1.00, 3.10), respectively, whereas a protective against chronic lymphatic leukemia was estimated for more than 31 years exposed to THMs >40 μg/L (OR = 0.60, 95% CI: 0.41, 0.87) [45].

In the Iowa Women's Health Study cohort of postmenopausal women, researchers used historical monitoring and treatment data to estimate levels of long-term TTHM and/or five haloacetic acids (HAA5) and reported no association between average TTHM levels and pancreatic cancer risk [46], and null associations were also found between long-term average TTHMs or HAA5 and kidney cancer risk [47]. However, significant associations between average concentrations of DBPs in drinking water and endometrial cancer risk were observed among those postmenopausal women (TTHM: HR = 2.19, 95% CI: 1.41, 3.40; HAA5: HR = 1.84, 95% CI: 1.19, 2.83 for the ≥95th percentile vs. first quartile) [48].

4 Adverse Reproductive and Developmental Outcomes

DBPs are reproductive and developmental toxicities in toxicological studies, though inconsistent results have also been reported. Many human studies have focused on the effects of exposure to DBPs on interest in reproductive and developmental outcomes. The available epidemiological evidence has indicated that exposure to DBPs is associated with a wide range of reproductive and developmental outcomes, including reduced semen quality, altered reproductive hormones, intrauterine growth retardation, low birth weight, small for gestational age, preterm birth, and stillbirth, but the findings are inconclusive and insufficient to support such conclusion. One of the main challenges is the different exposure assessments that present difficulties in making comparisons among DBP exposures in relation to adverse reproductive and developmental outcomes. However, based on blood THMs and urinary HAAs as internal biomarkers to enhance exposure assessment, emerging human studies have fairly suggested evidence of associations of DBP exposures with decreased semen quality, impaired ovarian function, and adverse birth outcomes.

4.1 Male Reproductive Health Outcomes

Toxicological studies have suggested that DBPs, particularly HAAs, are male reproductive toxicants. However, human studies have focused on DBP exposures in relation to semen quality and reproductive hormones, with inconsistent results (Table 1). Previous epidemiological studies have primarily utilized DBP concentrations in tap water serving the subject's residential address or combined individual water-use activities as exposure surrogates. Among healthy males from couples without known risk factors for infertility in Santa Clara, California, USA, exposure to TTHMs was not associated with decreased semen quality, but exposure to BDCM was related to decreased sperm linearity. Higher levels of TTHM ingestion metric (>160 μg/L× glasses/day, which is equivalent to more than two glasses/day of water containing 80 μg/L) were associated with decreased percent normal morphology and increased percent abnormal head [49]. In a presumed American fertile male population with different DBP profiles, there were no associations between exposure to THMs or HAAs at levels approaching regulated limits and semen quality, but a negative association was observed between sperm concentration and total organic halides, a global measure of halogen-containing DBPs [50]. In a large-scale study of fertility clinics from multiple centers in England and Wales, there was no evidence of associations between exposure to THMs and semen quality [51]. However, THM exposure via ingestion was observed to be associated with decreased sperm

Table 1 Epidemiological studies on exposure to DBPs and male reproductive health

Author (year)	Study design	Study population	Exposure	Outcomes[a]
Fenster et al. (2003) [49]	Prospective study	Healthy male population in California, US	TTHMs in tap water	Semen parameters ×
			Highest levels of TTHM ingestion metric	Percent normal morphology ↓ Percent head defects ↑
			BDCM in tap water	Sperm linearity ↓
Luben et al. (2007) [50]	Prospective study	Presumed fertile male population in North Carolina, US	THMs in tap water	Semen parameters ×
			HAAs in tap water	Semen parameters ×
			total organic halides in tap water	Sperm concentration ↓
Iszatt et al. (2013) [51]	Cross-sectional study	Men from 13 fertility clinics in England and Wales	THMs in tap water	Semen parameters ×
Zeng et al. (2014) [52]	Prospective study	Men from an infertility clinic in China	THM exposure via ingestion	Sperm concentration ↓ Sperm count ↓
			THM exposure via showering/bathing	Semen parameters ×
Xie et al. (2011) [53]	Cross-sectional study	Men from an infertility clinic in China	Urinary TCAA levels	Sperm motility ↓
Zeng et al. (2014) [10]	Cross-sectional study	Men from an infertility clinic in China	Urinary TCAA levels	Sperm concentration ↓ Sperm motility ↓ Sperm count ↓
Zhang et al. (2023) [54]	Prospective study	Men seeking assisted reproductive technology treatment in China	Urinary DCAA levels	Sperm count ↓ Sperm progressive motility ↓ Sperm motility ↓
			Urinary TCAA levels	Sperm progressive motility ↓ Sperm motility ↓
Chen et al. (2022) [55]	Cross-sectional study	Men from an infertility clinic in China	Urinary TCAA levels	Percentage of tail DNA (tail%) ↑ Tail-distributed moment (TDM) ↑ Percentage of Annexin V+/PI- spermatozoa ↑
Zeng et al. (2013) [56]	Cross-sectional study	Men from an infertility clinic in China	Blood TTHM levels	Sperm concentration ↓ Serum total testosterone×
			Blood TCM levels	Sperm concentration ↓ Serum total testosterone×

(continued)

Table 1 (continued)

Author (year)	Study design	Study population	Exposure	Outcomes[a]
			Blood BDCM levels	Sperm count ↓ Serum total testosterone×
			Blood DBCM levels	Sperm linearity ↓ Serum total testosterone ↓
Zeng et al. (2016) [57]	Cross-sectional study	Men from an infertility clinic in China	Co-exposure to blood Br-THMs and urinary TCAA levels	Sperm count ↓
Chen et al. (2020) [5]	Prospective study	Healthy men screened as potential sperm donors in China	Blood TTHM, Br-THM, and DBCM levels	Sperm count ↓ Sperm concentration ↓
			Blood TCM levels	Sperm count ↓ Sperm motility ↓ Sperm progressive motility ↓
Nickmilder et al. (2011) [58]	Cross-sectional study	School male adolescents aged 14–18 years in Belgium	Cumulative indoor pool attendance	Serum inhibin B ↓ Serum total testosterone ↓
Tian et al. (2024) [59]	Cross-sectional study	Men seeking semen examination	Urinary TCAA and DCAA levels	Urinary testosterone ↑ Urinary progesterone ↑ Urinary estradiol ↑ Urinary testosterone/estradiol ↓

Note: TTHMs: sum of TCM, BDCM, DBCM and TBM; Br-THMs: sum of BDCM, DBCM and TBM. ↓ represents a negative association; ↑ represents a positive association

[a] × represents a null association

concentration and sperm count among men from an infertility clinic in China, though THM exposures via showering/bathing was not associated with semen quality [52].

Emerging human studies using urinary HAAs and blood THMs as internal biomarkers have fairly reported negative associations between DBP exposures and semen quality parameters such as sperm motility, sperm concentration, and sperm count. Among adult men from an infertility clinic in Wuhan, China, an inverse association was firstly found between urinary TCAA concentrations and sperm motility, but this association became non-significant after adjustment for potential confounding factors [53]. The relatively small sample sizes in this study may produce the imprecise finding. In a subsequent large-scale study involving 2,009 Chinese men, higher quartiles of urinary TCAA levels were observed to be associated with decreased sperm concentration, sperm motility, and sperm count [10]. Furthermore, given that single measurement of exposure biomarkers and/or semen quality parameters at a one-time point due to their high within-individual variability over time may bias the observed results, a recent study among 1,230 Chinese reproductive-aged men undergoing assisted reproductive technology treatment from the TREE cohort assessed the associations between repeated measures of urinary TCAA and DCAA and semen quality parameters [54]. Consistent with previous studies, the repeated longitudinal analysis observed that urinary TCAA was negatively associated with sperm progressive motility and total motility, and also firstly found inverse associations of urinary DCAA with sperm count, progressive motility, and total motility [54]. Moreover, a study reported positive exposure-response associations of urinary TCAA with sperm DNA damage, as well as a positive association between urinary TCAA and spermatozoa apoptotic cells among men from an infertility clinic in Wuhan, China [55].

Among 401 Chinese men seeking semen examination, the increased baseline blood TCM and TTHM levels were associated with decreased sperm concentration, and the moderate levels of baseline blood BDCM (1.02–2.35 ng/L) and DBCM (0.68–1.00 ng/L) were associated with decreased sperm count and sperm linearity, respectively [56]. Notably, the authors further firstly found that the co-exposure to baseline blood Br-THMs and urinary TCAA had additive effects on decreased sperm count [57]. Recently, a cohort study among 1,199 healthy Chinese men screened as potential sperm donors also reported inverse dose-response relationships of blood TCM with sperm count, total motility, and progressive motility, as well as of blood DBCM, Br-THM, and TTHM with sperm count and concentration in the cross-sectional analysis [5]. The authors further confirmed that the increased concentrations of blood TCM, DBCM, Br-THM, and TTHM were inversely associated with the sperm count in the longitudinal and repeated measure analysis.

Although DBPs have also been found to alter male reproductive hormones in animal models, the human evidence is limited. Swimming in indoor chlorinated pools during childhood has been reported to be significantly associated with lower levels of serum inhibin B and total testosterone [58]. Among 401 Chinese men seeking semen examination, an inverse dose-response relationship was also observed between the baseline blood DBCM and serum total testosterone, but there were null associations of the baseline blood TCM, TBM, BDCM, Br-THM, and TTHM with the serum total testosterone [56]. A recent study among 449 Chinese men seeking semen examination reported that urinary concentrations of DCAA and TCAA were positively associated with urinary levels of testosterone, progesterone, and estradiol [59].

4.2 Female Reproductive Health Outcomes

Toxicological studies have demonstrated that DBPs have endocrine-disrupting effects and can cause ovarian toxicity. However, there have been a lack of epidemiological studies on exposure to DBPs and female reproductive health (Table 2). In a prospective cohort study of 403 females in Northern California, USA, elevated tap-water TTHM concentrations were associated with decrements in cycle and

Table 2 Epidemiological studies on exposure to DBPs and female reproductive health

Author (year)	Study design	Study population	Exposure	Outcomes[a]
Windham et al. (2003) [60]	Prospective cohort study	Premenopausal women in Northern California	TTHM levels in tap-water	Cycle and follicular phase length ↓ Luteal phase length, menses length, or cycle variability ×
MacLehose et al. (2008) [61]	Prospective cohort study	Women in the United States	Ingested THMs	Time to pregnancy ↓
Deng et al. (2022a) [11]	Cross-sectional study	Women seeking assisted reproductive technology treatment in China	Urinary DCAA levels	Total antral follicle count ↓ Reproductive hormone ×
			Urinary TCAA levels	Total antral follicle count ↓ Serum anti-müllerian hormone ↓
Liu et al. (2023) [62]	Case-control study	Women seeking assisted reproductive technology treatment in China	Urinary DCAA levels	Diminished ovarian reserve ↑
			Urinary TCAA levels	Diminished ovarian reserve ×
Deng et al. (2022b) [63]	Cross-sectional study	Women seeking assisted reproductive technology treatment in China	Urinary DCAA levels	Irregular menstrual cycle ↑ Menstrual cycle ↑ Variation in cycle length ↑
			Urinary TCAA levels	Bleeding duration ↑
Liu et al. (2022a) [64]	Cross-sectional study	Women seeking assisted reproductive technology treatment in China	Urinary DCAA levels	Serum progesterone levels ↑
			Urinary TCAA levels	Reproductive hormones ×

Note: TTHMs: sum of TCM, BDCM, DBCM and TBM; ↓ represents a negative association; ↑ represents a positive association
[a] × represents a null association

follicular phase length, but there were little associations with luteal phase length, menses length, or cycle variability [60]. In a large prospective cohort study of 1,315 women conducted in the United States, high levels of ingested THMs was found to be associated with a decrease in fecundability (time to pregnancy) [61].

Recently, based on the epidemiological study on couples undergoing assisted reproductive technology treatment from the TREE cohort in China, the researchers have reported that urinary DCAA and TCAA were associated with diminished ovarian function, disrupted menstrual cycle, and altered hormone homeostasis. There were consistently negative associations of urinary TCAA levels with total antral follicle count and anti-Mullerian hormone, two well-known indicators of ovarian reserve function [11]. The authors also found that urinary DCAA levels were inversely associated with total antral follicle count, and this negative association was more evident among the right ovary [11]. Furthermore, based on a case-control design within the TREE cohort, the elevated urinary DCAA levels were observed to be associated with a higher risk of diminished ovarian reserve [62]. Women with higher urinary DCAA concentrations were also reported to suffer an increased risk of irregular menstrual cycle and long menstrual cycle [63], as well as prolonged variation in cycle length [63], and increased serum progesterone [64], whereas the women with higher urinary TCAA concentrations were associated with prolonged bleeding duration [63].

4.3 Fetal Developmental Outcomes

Accumulating toxicological studies have shown the developmental toxicities of DBPs, which can cause various adverse pregnancy outcomes. Early epidemiological studies utilized different disinfection methods as external surrogates to investigate the effects of exposure to DBPs during pregnancy on fetal growth and development outcomes (Table 3). For example, a case-control study in Massachusetts, USA found that pregnant women exposed to chlorinated surface water during the entire pregnancy had a slightly increased risk of stillbirth compared to those exposed to non-chlorinated drinking water [65]. Another case-control study in Liguria, Italy found a higher frequency of small body length ($\leq$49.5 cm) and small cranial circumference ($\leq$35.0 cm) in infants born to mothers who drank chlorinated water compared to those who drank non-chlorinated water [66]. In a case-control study from Taiwan, China, a higher risk of preterm birth was also observed in areas where mothers consumed chlorinated water than in areas where mothers consumed non-chlorinated water, but there was no association with an increased risk of low birth weight [67].

Later epidemiological studies applied DBP concentrations in tap water serving the subject's residential address or combined individual water-use activities as exposure surrogates (Table 3). One large-scale retrospective cohort study in Canada found that maternal exposure to TTHMs in drinking water during pregnancy was associated with an increased risk of fetal stillbirth, but there were little associations with fetal weight or gestational age [68]. Another retrospective cohort study in the USA also found that maternal exposure to DCAA and TCAA in drinking water during late pregnancy was associated with an increased risk of intrauterine growth retardation, and that exposure to DBAA during late pregnancy was associated with an increased risk of term low birth weight [69]. A prospective cohort study in Lithuania found that there were statistically significant dose-response relationships between maternal TTHM, DBCM, and TCM exposures during whole pregnancy and an increased risk of low birth weight, and between maternal TCM exposures during late pregnancy and an increased risk of small for gestational age [70]. A case-control study of birth defects in Massachusetts, USA with complete quarterly 1999–2004 THM and HAA monitoring data found that maternal exposure to DCAA and total HAAs (the sum of DCAA, TCAA, DBAA, and monochloroacetic acid, monobromoacetic acid) during early pregnancy were associated with an increased

Table 3 Epidemiological studies on external surrogates of exposure to DBPs and fetal developmental outcomes

Author (year)	Study design	Study population	Exposure period	Exposure	Fetal developmental outcomes[a]
Aschengrau et al. (1993) [65]	Case-control study	Pregnant women in Massachusetts, USA	Entire pregnancy	Use of chlorinated surface water	Stillbirth ↑
Kanitz et al. (1996) [66]	Case-control study	Pregnant women in Liguria, Italy	Entire pregnancy	Use of water treated with chlorine compounds	Small body length ↑ Small cranial circumference ↑
Yang et al. (2000) [67]	Case-control study	Pregnant women in Taiwan, China	Entire pregnancy	Use of chlorinated water	Preterm birth ↑ Low birth weight ×
Dodds et al. (1999) [68]	Retrospective cohort study	Pregnant women in Nova Scotia, Canada	Pre-pregnancy and Entire pregnancy	TTHMs in drinking water	Stillbirth ↑ Fetal weight or gestational age ×
Hinckley et al. (2005) [69]	Retrospective cohort study	Pregnant women in a Arizona community, USA	The third trimester of pregnancy	DCAA and TCAA in drinking water	Intrauterine growth retardation ↑
				Dibromoacetic acid in drinking water	Term low birth weight ↑
Grazuleviciene et al. (2011) [70]	Prospective cohort study	Pregnant women in Kaunas, Lithuania	Entire pregnancy, and each trimester of pregnancy	TTHM, DBCM, and TCM in drinking water during whole pregnancy	Low birth weight ↑
				TCM in drinking water during late pregnancy	Small for gestational age ↑
Wright et al. (2017) [71]	Case-control study	Pregnant women in Massachusetts, USA	The first trimester of pregnancy	highest dose group of total HAAs in tap water	Tetralogy of Fallot ↑
				Highest dose group of TBM in tap water	Atrial septal defect ↑ Ventricular septal defect ↑ Pulmonary stenosis ↑

(continued)

Table 3 (continued)

Author (year)	Study design	Study population	Exposure period	Exposure	Fetal developmental outcomes[a]
Säve-Söderbergh et al. (2020) [72]	Prospective cohort study	Pregnant women in Sweden	The third trimester of pregnancy	TTHMs concentrations in drinking water	Small for gestational age ↑
Aggazzotti et al. (2004) [74]	Case-control study	Pregnant women in Italy	The third trimester of pregnancy	THMs in drinking water	Preterm birth × Small for gestational age ×
Toledano et al. (2005) [75]	Cross-sectional study	Pregnant women in England	Entire pregnancy	Areas with high TTHM concentrations in drinking water	Low birth weight × Very low birth weight ×
Kogevinas et al. (2016) [76]	Prospective cohort study	Pregnant women from France, Greece, Lithuania, Spain, and the UK	Entire pregnancy	THMs in drinking water	Preterm birth × Low birth weight × Small for gestational age ×

Note: HAAs: haloacetic acids; THMs: trihalomethanes; TTHMs: sum of TCM, BDCM, DBCM and TBM; ↑ represents a positive association
[a] × represents a null association

risk of tetralogy of Fallot in newborns, and that maternal exposure to TBM during early pregnancy was associated with an increased risk of atrial septal defect, ventricular septal defect, and pulmonary valve stenosis [71]. Based on a nationwide register-based prospective study of more than 500,000 newborns from Sweden, a dose-response relationship was observed between maternal exposure to TTHMs in drinking water during mid- and late-pregnancy and an increased risk for small for gestational age [72]. A meta-analysis also found that maternal exposures to TCM, DBCM, Br-THMs, TTHMs, DCAA, TCAA, and total HAAs in drinking water were associated with an increased risk of small for gestational age [73].

However, some epidemiological studies have also reported no evidence of associations between exposure to DBPs during pregnancy and fetal growth and development outcomes. For example, a case-control study in Italy found null associations of maternal exposure to drinking water THMs during pregnancy with preterm birth and small for gestational age [74]. A cross-sectional investigation in England, UK found no risk for low birth weight (OR = 1.09; 95% CI, 0.93–1.27) and very low birth weight (OR = 1.05; 95% CI: 0.82, 1.34) in areas with higher TTHM concentrations compared to areas with lower TTHM concentrations [75]. In a large-scale European cohort study from France, Greece, Lithuania, Spain, and the United Kingdom, null statistically significant associations were also reported between maternal exposure to drinking water THMs during pregnancy and the risk of preterm birth, low birth weight, and small for gestational age [76]. A systematic review and meta-analysis involving 37 studies showed little or no evidence for associations between maternal exposure to TTHMs during late pregnancy and preterm birth, low birth weight, and low birth weight at term [77]. A recent prospective cohort study also did not find a statistically significant association between maternal exposure to TTHMs in drinking water during mid- and late-pregnancy with preterm birth based on more than 500,000 newborns registered in the Swedish national registry [72].

Emerging human studies have used urinary HAAs and blood THMs as internal biomarkers to examine the associations between maternal exposure to DBPs and birth outcomes (Table 4). A nested case-control study conducted in France found that women with urinary TCAA levels >10 μg/L had a higher risk of fetal growth restriction (OR = 1.8, 95% CI: 0.9, 3.7) than those with TCAA levels below the detection limit (≤ 10 μg/L) [78]. In a hospital-based cross-sectional study in Wuhan, China, prenatal higher urinary TCAA concentrations were associated with decreased birth weight [79]. Prenatal urinary TCAA levels were also negatively associated with biparietal diameter, head circumference, and femur length in boys but not in girls [80]. Consistently, negative associations of maternal urinary TCAA levels during late pregnancy with birth weight and ponderal index were observed in a large-scale study [81]. A prospective cohort study conducted in Xiaogan and Wuhan, China reported that blood TTHMs during late pregnancy were associated with lower mean birth weight and increased risk of small for gestational age, and blood BDCM and DBCM concentrations during late pregnancy were associated with smaller birth length [82]. Another prospective cohort study conducted in Xiaogan, China found that blood TCM concentrations in the second and third trimester of pregnancy were associated with an increased risk of small for gestational age, and the 23–34 week

Table 4 Epidemiological studies on internal biomarkers of exposure to DBPs and fetal developmental outcomes

Author (year)	Study design	Study population	Exposure period	Exposure	Outcomes[a]
Costet et al (2012) [78]	Nested case-control study	Pregnant women in France	Entire pregnancy and early pregnancy	Urinary TCAA levels	Fetal growth restriction ↑
Zhou et al. (2012) [79]	Prospective cohort study	Pregnant women in Wuhan, China	The late pregnancy	Urinary TCAA levels	Birth weight ↓
Deng et al. (2019) [80]	Prospective cohort study	Pregnant women in Wuhan, China	The late pregnancy	Urinary TCAA levels	Biparietal diameter in boys ↓ Head circumference in boys↓ Femur length in boys ↓
Yang et al. (2019) [81]	Prospective cohort study	Pregnant women in Wuhan, China	The late pregnancy	Urinary TCAA levels	Birth weight ↓ Ponderal index ↓
Cao et al. (2016) [82]	Prospective cohort study	Pregnant women in Wuhan and Xiaogan, China	The late pregnancy	blood TTHM levels	Birth weight ↓ Small for gestational age ↑
				blood BDCM and DBCM levels	Birth length ↓
Sun et al. (2020) [83]	Prospective cohort study	Pregnant women in Xiaogan, China	Entire pregnancy, and the first, second and third trimester of pregnancy	Blood TCM levels	Small for gestational age ↑
Liu et al. (2021b) [84]	Prospective cohort study	Pregnant women in Xiaogan, China	Entire pregnancy, and the first, second and third trimester of pregnancy	Blood TTHM and Br-THM levels	Fetal abdominal circumference ↓
Chen et al. (2019) [85]	Prospective cohort study	Pregnant women in Xiaogan, China	Early pregnancy	Blood BDCM levels	Neonatal behavioral neurological assessment (NBNA) score↓
Liu et al. (2021a) [86]	Prospective cohort study	Pregnant women in Xiaogan, China	Entire pregnancy, and the first, second and third trimester of pregnancy	Blood BDCM levels and urinary TCAA levels	Neonatal behavioral neurological assessment (NBNA) score ↓

(continued)

Table 4 (continued)

Author (year)	Study design	Study population	Exposure period	Exposure	Outcomes[a]
Deng et al. (2023) [87]	Prospective cohort study	Women undergoing assisted reproductive technology treatment	–	Urinary DCAA levels	Proportion of best-quality embryos ↓ Proportion of successful implantation ↓ Proportion of clinical pregnancy ↓ Proportion of live birth ↓

Note: HAAs: haloacetic acids; THMs: trihalomethanes; TTHMs: sum of TCM, BDCM, DBCM and TBM; Br-THMs: sum of BDCM, DBCM and TBM; ↓ represents a negative association; ↑ represents a positive association

[a] × represents a null association

gestation was identified as the windows of vulnerability for exposure [83]. Similarly, prenatal blood TTHM and Br-THM levels during the first trimester was associated with reduced fetal abdominal circumference [84]. Furthermore, inverse associations of prenatal blood BDCM and urinary TCAA concentrations with neonatal neurobehavioral development were observed, and these associations were more evident in male infants [85, 86]. Recently, based on the TREE cohort among women undergoing assisted reproductive technology treatment, the researchers first found that urinary DCAA concentrations were inversely associated with a lower proportion of best-quality embryos, successful implantation, clinical pregnancy, and live birth [87], suggesting that preconception exposure to DBPs may exert adverse effects on early pregnancy outcomes.

Studies have also shown that the associations between prenatal DBP exposures and birth outcomes may be modified by maternal or neonatal genetic polymorphisms in metabolic enzymes. A case-control study from Canada showed that maternal exposure to TTHMs in tap water in positive association with intrauterine growth restriction only was observed among newborns with the presence of a variant in the CYP2E1 gene [88]. One study conducted in 426 Chinese pregnant women reported that elevated maternal blood Br-THMs during late pregnancy in relation to reduced birth weight was only present in neonates carrying the CYP2E1 rs2031920 CT/TT genotype [89]. The subsequent cohort study based on a larger sample from Wuhan and Xiaogan, China found that maternal urinary TCAA during late pregnancy in association with decreased birth weight was observed only among subjects with CYP2E1 rs3813867 GC/CC, and the associations with decreased birth length, ponderal index, and gestational age were observed only among subjects with GSTZ1 rs7975 GA/AA [81]. These findings highlight that individual genetic susceptibility plays an important role in the effects of exposure to DBPs on fetal growth.

5 Other Adverse Health Outcomes

DBPs are cytotoxic, hepatotoxic, nephrotoxic, haematotoxic, coagulotoxic, or immunotoxic in toxicological studies. However, epidemiological studies of exposure to DBPs associated with such adverse health effects/outcomes are sporadic.

Swimmers can be exposed to higher DBP levels, and exposure to such irritant chemicals may affect respiratory health. A higher prevalence of certain respiratory symptoms such as asthma has been observed among swimming pool workers or elite swimmers from the UK, the Netherlands, France, Italy, and the USA. In a meta-analysis, an increased frequency of asthma (meta-ORs ranged from 2.3 to 2.6 with all 95% CIs excluding 1.0) was estimated for elite swimmers compared to other athletes [90]. However, the results among children are conflicting. Cross-sectional studies conducted in Belgium reported that chlorinated pool attendance was associated with an increased risk of asthma among children [91, 92], whereas several studies in other European countries such as the Netherlands and the United Kingdom reported a null association between swimming and children's asthma [93–95]. Furthermore, the evidence from the general population has been still lacking. A cross-sectional study among 2,359 adolescents aged 12–19 years from NHANES 2005–2012 found that blood DBCM concentrations were associated with a higher risk of asthma (OR = 1.54, 95% CI:1.07, 2.21, comparing the extreme exposure categories), and this positive association was more pronounced among those exposed to tobacco smoke (OR = 3.96, 95% CI: 1.89, 8.30, comparing the extreme exposure categories) [96]. Another cross-sectional study based on NHANES 2005–2006 also reported a positive association between blood THM concentrations and a greater risk of allergic sensitization among adolescents [97].

Several studies have investigated the associations between DBP exposures and impaired liver function. A case-control study among 2,781 participants from NHANES 1999–2006 found that the cases with liver injury were 1.35-fold (95% CI: 1.02, 1.79) more likely to have circulating DBCM concentrations exceeding the median values compared to the controls [98]. A pilot case-control study in Nicosia, Cyprus observed a significant increase in the odds of having above the median serum alanine aminotransferase (ALT) levels (OR = 6.38, 95% CI: 1.11, 42.84) for each unit increase in creatinine-unadjusted urinary TBM levels [99]. One cross-sectional study among 182 Chinese adults also reported positive associations of urinary DBCM Br-THMs and TTHMs with ALT, aspartate aminotransferase (AST), total bilirubin, total protein, and albumin [100]. Another cross-sectional study among 922 Chinese women from the TREE cohort, China observed that urinary DCAA concentrations were positively associated with serum ALT levels and inversely associated with serum AST/ALT, whereas urinary TCAA concentrations showed null associations with liver function indicators [101]. Furthermore, urinary DCAA concentrations were found to be positively associated with risk of liver injury [101].

Few epidemiological evidence has suggested that DBP exposures may affect thyroid function. One early study among three glucose-6-phosphate dehydrogenase deficient adults reported reduced serum levels of T4 after daily ingestion of drinking water with a sodium chlorite concentration of 5 mg/L for 12 weeks [102]. Another study from Quebec, Canada found that newborns with low birth weight had higher blood TSH levels in municipalities where drinking water was disinfected by chlorine dioxide instead of chlorine [103]. Recently, a cross-sectional study among 2,233 adults from NHANES 2007–2008 showed that blood THM concentrations were associated with altered serum indicators of thyroid function including free thyroxine, total triiodothyronine, and free thyroxine/free triiodothyronine ratio) but not with thyroid autoimmunity [104]. Similarly, one cross-sectional study among 1,190 Chinese women from the TREE cohort found that urinary TCAA concentrations had positive associations with serum thyrotropin and thyrotropin/free thyroxine ratio and negative associations with serum free triiodothyronine and free thyroxine, and that urinary DCAA concentrations were inversely associated with serum free thyroxine/free triiodothyronine ratio [105].

Limited human studies have examined the associations of exposure to DBPs with blood parameters, renal injury, and type II diabetes. One cross-sectional study among 505 Chinese adults observed an inverse association of urinary DCAA levels with platelet distribution width, as well as positive associations of urinary DCAA and TCAA levels with systolic blood pressure, diastolic blood pressure, and pulse pressure [106]. Cross-sectional studies from the TREE cohort found higher levels of urinary DCAA and TCAA in associations with prolonged prothrombin time [107] and a negative dose-response relationship between urinary DCAA concentrations and estimated glomerular filtration rate [108]. Furthermore, a pilot case-control study conducted in Nicosia, Cyprus reported that urinary THM concentrations were not associated with the risk of type II diabetes [99]. Similarly, one nested case-control study also reported that neither urinary Br-THM nor TCM levels were associated with the risk of type II diabetes [109].

References

1. Nuckols JR, Ashley DL, Lyu C et al (2005) Influence of tap water quality and household water use activities on indoor air and internal dose levels of trihalomethanes. Environ Health Perspect 113(7):863–870. https://doi.org/10.1289/ehp.7141
2. Font-Ribera L, Marco E, Grimalt JO et al (2019) Exposure to disinfection by-products in swimming pools and biomarkers of genotoxicity and respiratory damage - the PISCINA2 study. Environ Int 131:104988. https://doi.org/10.1016/j.envint.2019.104988
3. Riederer AM, Dhingra R, Blount BC et al (2014) Predictors of blood trihalomethane concentrations in NHANES 1999-2006. Environ Health Perspect 122(7):695–702. https://doi.org/10.1289/ehp.1306499
4. Wang YX, Liu C, Chen YJ et al (2019) Profiles, variability and predictors of concentrations of blood trihalomethanes and urinary haloacetic acids along pregnancy among 1760 Chinese women. Environ Res 172:665–674. https://doi.org/10.1016/j.envres.2019.03.017
5. Chen YJ, Duan P, Meng TQ et al (2020) Associations of blood trihalomethanes with semen quality among 1199 healthy Chinese men screened as potential sperm donors. Environ Int 134:105335. https://doi.org/10.1016/j.envint.2019.105335

6. Smith RB, Nieuwenhuijsen MJ, Wright J et al (2013) Validation of trichloroacetic acid exposure via drinking water during pregnancy using a urinary TCAA biomarker. Environ Res 126:145–151. https://doi.org/10.1016/j.envres.2013.05.004
7. Zhang W, Gabos S, Schopflocher D et al (2009) Validation of urinary trichloroacetic acid as a biomarker of exposure to drinking water disinfection by-products. J Water Health 7(3): 359–371. https://doi.org/10.2166/wh.2009.009
8. Zhang W, Gabos S, Schopflocher D et al (2009) Reliability of using urinary and blood trichloroacetic acid as a biomarker of exposure to chlorinated drinking water disinfection byproducts. Biomarkers 14(6):355–365. https://doi.org/10.1080/13547500903079186
9. Calafat AM, Kuklenyik Z, Caudill SP et al (2003) Urinary levels of trichloroacetic acid, a disinfection by-product in chlorinated drinking water, in a human reference population. Environ Health Perspect 111(2):151–154. https://doi.org/10.1289/ehp.5644
10. Zeng Q, Wang YX, Xie SH et al (2014) Drinking-water disinfection by-products and semen quality: a cross-sectional study in China. Environ Health Perspect 122(7):741–746. https://doi.org/10.1289/ehp.1307067
11. Deng YL, Luo Q, Liu C et al (2022) Urinary biomarkers of exposure to drinking water disinfection byproducts and ovarian reserve: a cross-sectional study in China. J Hazard Mater 421:126683. https://doi.org/10.1016/j.jhazmat.2021.126683
12. Wang YX, Zeng Q, Wang L et al (2014) Temporal variability in urinary levels of drinking water disinfection byproducts dichloroacetic acid and trichloroacetic acid among men. Environ Res 135:126–132. https://doi.org/10.1016/j.envres.2014.05.035
13. IARC (1991) IARC monographs on the evaluation of carcinogenic risks to humans. Chlorinated drinking-water; chlorination by-products; some other halogenated compounds; cobalt and cobalt compounds, IARC (ed) IARC
14. IARC, International Agency for Research on Cancer (1999) Monographs on the evaluation of carcinogenic risks to humans. Some chemicals that cause tumors of the kidney or urinary bladder in rodents and some other substances, I.A.f.R.o. Cancer (ed) International Agency for Research on Cancer: IARC, pp 131–182
15. Koivusalo M, Jaakkola JJ, Vartiainen T et al (1994) Drinking water mutagenicity and gastrointestinal and urinary tract cancers: an ecological study in Finland. Am J Public Health 84(8):1223–1228. https://doi.org/10.2105/ajph.84.8.1223
16. Morales Suarez-Varela MM, Llopis Gonzalez A, Tejerizo Perez ML et al (1994) Chlorination of drinking water and cancer incidence. J Environ Pathol Toxicol Oncol 13(1):39–41
17. Liu M, Graham N, Wang W et al (2022) Spatial assessment of tap-water safety in China. Nat Sustain 5(8):689–698. https://doi.org/10.1038/s41893-022-00898-5
18. Villanueva CM, Cantor KP, Grimalt JO et al (2007) Bladder cancer and exposure to water disinfection by-products through ingestion, bathing, showering, and swimming in pools. Am J Epidemiol 165(2):148–156. https://doi.org/10.1093/aje/kwj364
19. Beane Freeman LE, Cantor KP, Baris D et al (2017) Bladder cancer and water disinfection by-product exposures through multiple routes: a population-based case-control study (New England, USA). Environ Health Perspect 125(6):067010. https://doi.org/10.1289/ehp89
20. Koivusalo M, Pukkala E, Vartiainen T et al (1997) Drinking water chlorination and cancer-a historical cohort study in Finland. Cancer Causes Control 8(2):192–200. https://doi.org/10.1023/a:1018420229802
21. Doyle TJ, Zheng W, Cerhan JR et al (1997) The association of drinking water source and chlorination by-products with cancer incidence among postmenopausal women in Iowa: a prospective cohort study. Am J Public Health 87(7):1168–1176. https://doi.org/10.2105/ajph.87.7.1168
22. Helte E, Säve-Söderbergh M, Ugge H et al (2022) Chlorination by-products in drinking water and risk of bladder cancer – a population-based cohort study. Water Res 214:118202. https://doi.org/10.1016/j.watres.2022.118202
23. Villanueva CM, Fernández F, Malats N et al (2003) Meta-analysis of studies on individual consumption of chlorinated drinking water and bladder cancer. J Epidemiol Community Health 57(3):166–173. https://doi.org/10.1136/jech.57.3.166

24. Villanueva CM, Cantor KP, Cordier S et al (2004) Disinfection byproducts and bladder cancer: a pooled analysis. Epidemiology 15(3):357–367. https://doi.org/10.1097/01.ede.0000121380.02594.fc
25. Costet N, Villanueva CM, Jaakkola JJ et al (2011) Water disinfection by-products and bladder cancer: is there a European specificity? A pooled and meta-analysis of European case-control studies. Occup Environ Med 68(5):379–385. https://doi.org/10.1136/oem.2010.062703
26. Cantor KP, Villanueva CM, Silverman DT et al (2010) Polymorphisms in GSTT1, GSTZ1, and CYP2E1, disinfection by-products, and risk of bladder cancer in Spain. Environ Health Perspect 118(11):1545–1550. https://doi.org/10.1289/ehp.1002206
27. Beane Freeman LE, Kogevinas M, Cantor KP et al (2022) Disinfection by-products in drinking water and bladder cancer: evaluation of risk modification by common genetic polymorphisms in two case-control studies. Environ Health Perspect 130(5):57006. https://doi.org/10.1289/ehp9895
28. Salas LA, Villanueva CM, Tajuddin SM et al (2014) LINE-1 methylation in granulocyte DNA and trihalomethane exposure is associated with bladder cancer risk. Epigenetics 9(11): 1532–1539. https://doi.org/10.4161/15592294.2014.983377
29. Isacson P, Bean JA, Lynch C (1983) Relationships of cancer incidence rates in Iowa municipalities to chlorination status of drinking water. Environ Impact Health Effects 2:1353–1364
30. Isacson P, Bean JA, Splinter R et al (1985) Drinking water and cancer incidence in Iowa. III. Association of cancer with indices of contamination. Am J Epidemiol 121(6):856–869. https://doi.org/10.1093/oxfordjournals.aje.a114056
31. Flaten TP (1992) Chlorination of drinking water and cancer incidence in Norway. Int J Epidemiol 21(1):6–15. https://doi.org/10.1093/ije/21.1.6
32. Yang CY, Chiu HF, Cheng MF et al (1998) Chlorination of drinking water and cancer mortality in Taiwan. Environ Res 78(1):1–6. https://doi.org/10.1006/enrs.1997.3823
33. Gottlieb MS, Carr JK (1982) Case-control cancer mortality study and chlorination of drinking water in Louisiana. Environ Health Perspect 46:169–177. https://doi.org/10.1289/ehp.8246169
34. Rahman MB, Cowie C, Driscoll T et al (2014) Colon and rectal cancer incidence and water trihalomethane concentrations in New South Wales, Australia. BMC Cancer 14:445. https://doi.org/10.1186/1471-2407-14-445
35. King WD, Marrett LD, Woolcott CG (2000) Case-control study of colon and rectal cancers and chlorination by-products in treated water. Cancer Epidemiol Biomarkers Prev 9(8): 813–818
36. Lawrence CE, Taylor PR, Trock BJ et al (1984) Trihalomethanes in drinking water and human colorectal cancer. J Natl Cancer Inst 72(3):563–568
37. Kuo HW, Tiao MM, Wu TN et al (2009) Trihalomethanes in drinking water and the risk of death from colon cancer in Taiwan. J Toxicol Environ Health A 72(20):1217–1222. https://doi.org/10.1080/15287390903129176
38. Villanueva CM, Gracia-Lavedan E, Bosetti C et al (2017) Colorectal cancer and long-term exposure to Trihalomethanes in drinking water: a multicenter case-control study in Spain and Italy. Environ Health Perspect 125(1):56–65. https://doi.org/10.1289/ehp155
39. Jones RR, DellaValle CT, Weyer PJ et al (2019) Ingested nitrate, disinfection by-products, and risk of colon and rectal cancers in the Iowa Women's Health Study cohort. Environ Int 126: 242–251. https://doi.org/10.1016/j.envint.2019.02.010
40. Helte E, Säve-Söderbergh M, Larsson SC et al (2023) Disinfection by-products in drinking water and risk of colorectal cancer: a population-based cohort study. J Natl Cancer Inst. https://doi.org/10.1093/jnci/djad145
41. Morris RD, Audet AM, Angelillo IF et al (1992) Chlorination, chlorination by-products, and cancer: a meta-analysis. Am J Public Health 82(7):955–963. https://doi.org/10.2105/ajph.82.7.955
42. Rahman MB, Driscoll T, Cowie C et al (2010) Disinfection by-products in drinking water and colorectal cancer: a meta-analysis. Int J Epidemiol 39(3):733–745. https://doi.org/10.1093/ije/dyp371

43. Carlo GL, Mettlin CJ (1980) Cancer incidence and trihalomethane concentrations in a public drinking water system. Am J Public Health 70(5):523–525. https://doi.org/10.2105/ajph.70.5.523
44. Young TB, Kanarek MS, Tsiatis AA (1981) Epidemiologic study of drinking water chlorination and Wisconsin female cancer mortality. J Natl Cancer Inst 67(6):1191–1198
45. Kasim K, Levallois P, Johnson KC et al (2006) Chlorination disinfection by-products in drinking water and the risk of adult leukemia in Canada. Am J Epidemiol 163(2):116–126. https://doi.org/10.1093/aje/kwj020
46. Quist AJL, Inoue-Choi M, Weyer PJ et al (2018) Ingested nitrate and nitrite, disinfection by-products, and pancreatic cancer risk in postmenopausal women. Int J Cancer 142(2): 251–261. https://doi.org/10.1002/ijc.31055
47. Jones RR, Weyer PJ, DellaValle CT et al (2017) Ingested nitrate, disinfection by-products, and kidney cancer risk in older women. Epidemiology 28(5):703–711. https://doi.org/10.1097/ede.0000000000000647
48. Medgyesi DN, Trabert B, Sampson J et al (2022) Drinking water disinfection byproducts, ingested nitrate, and risk of endometrial cancer in postmenopausal women. Environ Health Perspect 130(5):57012. https://doi.org/10.1289/ehp10207
49. Fenster L, Waller K, Windham G et al (2003) Trihalomethane levels in home tap water and semen quality. Epidemiology 14(6):650–658. https://doi.org/10.1097/01.ede.0000077980.52697.01
50. Luben TJ, Olshan AF, Herring AH et al (2007) The healthy men study: an evaluation of exposure to disinfection by-products in tap water and sperm quality. Environ Health Perspect 115(8):1169–1176. https://doi.org/10.1289/ehp.10120
51. Iszatt N, Nieuwenhuijsen MJ, Bennett J et al (2013) Chlorination by-products in tap water and semen quality in England and Wales. Occup Environ Med 70(11):754–760. https://doi.org/10.1136/oemed-2012-101339
52. Zeng Q, Chen YZ, Xu L et al (2014) Evaluation of exposure to trihalomethanes in tap water and semen quality: a prospective study in Wuhan, China. Reprod Toxicol 46:56–63. https://doi.org/10.1016/j.reprotox.2014.03.005
53. Xie SH, Li YF, Tan YF et al (2011) Urinary trichloroacetic acid levels and semen quality: a hospital-based cross-sectional study in Wuhan, China. Environ Res 111(2):295–300. https://doi.org/10.1016/j.envres.2010.12.010
54. Zhang M, Liu XY, Deng YL et al (2023) Associations between urinary biomarkers of exposure to disinfection byproducts and semen parameters: a repeated measures analysis. J Hazard Mater 461:132638. https://doi.org/10.1016/j.jhazmat.2023.132638
55. Chen YJ, Liu C, Tu ZZ et al (2022) Associations of urinary trichloroacetic acid concentrations with spermatozoa apoptosis and DNA damage in a Chinese population. Environ Sci Technol 56(10):6491–6499. https://doi.org/10.1021/acs.est.1c07725
56. Zeng Q, Li M, Xie SH et al (2013) Baseline blood trihalomethanes, semen parameters and serum total testosterone: a cross-sectional study in China. Environ Int 54:134–140. https://doi.org/10.1016/j.envint.2013.01.016
57. Zeng Q, Zhou B, He DL et al (2016) Joint effects of trihalomethanes and trichloroacetic acid on semen quality: a population-based cross-sectional study in China. Environ Pollut 212:544–549. https://doi.org/10.1016/j.envpol.2016.02.032
58. Nickmilder M, Bernard A (2011) Associations between testicular hormones at adolescence and attendance at chlorinated swimming pools during childhood. Int J Androl 34(5 Pt 2):e446–e458. https://doi.org/10.1111/j.1365-2605.2011.01174.x
59. Tian M, Li H, Wu S et al (2024) Exposure to haloacetic acid disinfection by-products and male steroid hormones: an epidemiological and in vitro study. J Hazard Mater 468:133796. https://doi.org/10.1016/j.jhazmat.2024.133796
60. Windham GC, Waller K, Anderson M et al (2003) Chlorination by-products in drinking water and menstrual cycle function. Environ Health Perspect 111(7):935–941.; discussion A409. https://doi.org/10.1289/ehp.5922

61. MacLehose RF, Savitz DA, Herring AH et al (2008) Drinking water disinfection by-products and time to pregnancy. Epidemiology 19(3):451–458. https://doi.org/10.1097/EDE.0b013e31816a23eb
62. Liu XY, Zhang M, Gu XL et al (2023) Urinary biomarkers of drinking-water disinfection byproducts in relation to diminished ovarian reserve risk: a case-control study from the TREE cohort. Sci Total Environ 912:168729. https://doi.org/10.1016/j.scitotenv.2023.168729
63. Deng YL, Luo Q, Yuan XQ et al (2022) Associations between drinking water disinfection byproducts and menstrual cycle characteristics: a cross-sectional study among women attending an infertility clinic. Int J Hyg Environ Health 241:113931. https://doi.org/10.1016/j.ijheh.2022.113931
64. Liu C, Deng YL, Yuan XQ et al (2022) Exposure to disinfection by-products and reproductive hormones among women: results from the Tongji reproductive and environmental (TREE) study. Environ Res 209:112863. https://doi.org/10.1016/j.envres.2022.112863
65. Aschengrau A, Zierler S, Cohen A (1993) Quality of community drinking water and the occurrence of late adverse pregnancy outcomes. Arch Environ Health 48(2):105–113. https://doi.org/10.1080/00039896.1993.9938403
66. Kanitz S, Franco Y, Patrone V et al (1996) Association between drinking water disinfection and somatic parameters at birth. Environ Health Perspect 104(5):516–520. https://doi.org/10.1289/ehp.96104516
67. Yang CY, Cheng BH, Tsai SS et al (2000) Association between chlorination of drinking water and adverse pregnancy outcome in Taiwan. Environ Health Perspect 108(8):765–768. https://doi.org/10.1289/ehp.00108765
68. Dodds L, King W, Woolcott C et al (1999) Trihalomethanes in public water supplies and adverse birth outcomes. Epidemiology 10(3):233–237
69. Hinckley AF, Bachand AM, Reif JS (2005) Late pregnancy exposures to disinfection by-products and growth-related birth outcomes. Environ Health Perspect 113(12):1808–1813. https://doi.org/10.1289/ehp.8282
70. Grazuleviciene R, Nieuwenhuijsen MJ, Vencloviene J et al (2011) Individual exposures to drinking water trihalomethanes, low birth weight and small for gestational age risk: a prospective Kaunas cohort study. Environ Health 10:32. https://doi.org/10.1186/1476-069x-10-32
71. Wright JM, Evans A, Kaufman JA et al (2017) Disinfection by-product exposures and the risk of specific cardiac birth defects. Environ Health Perspect 125(2):269–277. https://doi.org/10.1289/ehp103
72. Säve-Söderbergh M, Toljander J, Donat-Vargas C et al (2020) Exposure to drinking water chlorination by-products and fetal growth and prematurity: a Nationwide register-based prospective study. Environ Health Perspect 128(5):57006. https://doi.org/10.1289/ehp6012
73. Summerhayes RJ, Rahman B, Morgan GG et al (2021) Meta-analysis of small for gestational age births and disinfection byproduct exposures. Environ Res 196:110280. https://doi.org/10.1016/j.envres.2020.110280
74. Aggazzotti G, Righi E, Fantuzzi G et al (2004) Chlorination by-products (CBPs) in drinking water and adverse pregnancy outcomes in Italy. J Water Health 2(4):233–247
75. Toledano MB, Nieuwenhuijsen MJ, Best N et al (2005) Relation of trihalomethane concentrations in public water supplies to stillbirth and birth weight in three water regions in England. Environ Health Perspect 113(2):225–232. https://doi.org/10.1289/ehp.7111
76. Kogevinas M, Bustamante M, Gracia-Lavedán E et al (2016) Drinking water disinfection by-products, genetic polymorphisms, and birth outcomes in a European mother-child cohort study. Epidemiology 27(6):903–911. https://doi.org/10.1097/ede.0000000000000544
77. Grellier J, Bennett J, Patelarou E et al (2010) Exposure to disinfection by-products, fetal growth, and prematurity: a systematic review and meta-analysis. Epidemiology 21(3):300–313. https://doi.org/10.1097/EDE.0b013e3181d61ffd
78. Costet N, Garlantézec R, Monfort C et al (2012) Environmental and urinary markers of prenatal exposure to drinking water disinfection by-products, fetal growth, and duration of

gestation in the PELAGIE birth cohort. Am J Epidemiol 175(4):263–275. https://doi.org/10.1093/aje/kwr419
79. Zhou WS, Xu L, Xie SH et al (2012) Decreased birth weight in relation to maternal urinary trichloroacetic acid levels. Sci Total Environ 416:105–110. https://doi.org/10.1016/j.scitotenv.2011.10.073
80. Deng YL, Yang P, Cao WC et al (2019) Urinary biomarker of late pregnancy exposure to drinking water disinfection by-products and ultrasound measures of fetal growth in Wuhan, China. Environ Res 170:128–133. https://doi.org/10.1016/j.envres.2018.12.018
81. Yang P, Cao WC, Zhou B et al (2019) Urinary biomarker of prenatal exposure to disinfection byproducts, maternal genetic polymorphisms in CYP2E1 and GSTZ1, and birth outcomes. Environ Sci Technol 53(20):12026–12034. https://doi.org/10.1021/acs.est.9b03847
82. Cao WC, Zeng Q, Luo Y et al (2016) Blood biomarkers of late pregnancy exposure to trihalomethanes in drinking water and fetal growth measures and gestational age in a Chinese cohort. Environ Health Perspect 124(4):536–541. https://doi.org/10.1289/ehp.1409234
83. Sun Y, Wang YX, Liu C et al (2020) Trimester-specific blood trihalomethane and urinary haloacetic acid concentrations and adverse birth outcomes: identifying windows of vulnerability during pregnancy. Environ Health Perspect 128(10):107001. https://doi.org/10.1289/ehp7195
84. Liu C, Sun Y, Mustieles V et al (2021) Prenatal exposure to disinfection byproducts and intrauterine growth in a Chinese cohort. Environ Sci Technol 55(23):16011–16022. https://doi.org/10.1021/acs.est.1c04926
85. Liu C, Messerlian C, Chen YJ et al (2021) Trimester-specific associations of maternal exposure to disinfection by-products, oxidative stress, and neonatal neurobehavioral development. Environ Int 157:106838. https://doi.org/10.1016/j.envint.2021.106838
86. Chen YJ, Liu C, Huang LL et al (2019) First-trimester blood concentrations of drinking water trihalomethanes and neonatal neurobehavioral development in a Chinese birth cohort. J Hazard Mater 362:451–457. https://doi.org/10.1016/j.jhazmat.2018.09.040
87. Deng YL, Liu C, Yuan XQ et al (2023) Associations between urinary concentrations of disinfection byproducts and in vitro fertilization outcomes: a prospective cohort study in China. Environ Health Perspect 131(9):97003. https://doi.org/10.1289/ehp12447
88. Infante-Rivard C (2004) Drinking water contaminants, gene polymorphisms, and fetal growth. Environ Health Perspect 112(11):1213–1216. https://doi.org/10.1289/ehp.7003
89. Zhou B, Yang P, Gong YJ et al (2018) Effect modification of CPY2E1 and GSTZ1 genetic polymorphisms on associations between prenatal disinfection by-products exposure and birth outcomes. Environ Pollut 243(Pt B):1126–1133. https://doi.org/10.1016/j.envpol.2018.09.083
90. Goodman M, Hays S (2008) Asthma and swimming: a meta-analysis. J Asthma 45(8): 639–647. https://doi.org/10.1080/02770900802165980
91. Bernard A, Carbonnelle S, de Burbure C et al (2006) Chlorinated pool attendance, atopy, and the risk of asthma during childhood. Environ Health Perspect 114(10):1567–1573. https://doi.org/10.1289/ehp.8461
92. Bernard A, Carbonnelle S, Dumont X et al (2007) Infant swimming practice, pulmonary epithelium integrity, and the risk of allergic and respiratory diseases later in childhood. Pediatrics 119(6):1095–1103. https://doi.org/10.1542/peds.2006-3333
93. Font-Ribera L, Kogevinas M, Zock JP et al (2009) Swimming pool attendance and risk of asthma and allergic symptoms in children. Eur Respir J 34(6):1304–1310. https://doi.org/10.1183/09031936.00180608
94. Font-Ribera L, Villanueva CM, Nieuwenhuijsen MJ et al (2011) Swimming pool attendance, asthma, allergies, and lung function in the Avon longitudinal study of parents and children cohort. Am J Respir Crit Care Med 183(5):582–588. https://doi.org/10.1164/rccm.201005-0761OC
95. Jacobs JH, Fuertes E, Krop EJ et al (2012) Swimming pool attendance and respiratory symptoms and allergies among Dutch children. Occup Environ Med 69(11):823–830. https://doi.org/10.1136/oemed-2011-100621

96. Sun Y, Xia PF, Xie J et al (2022) Association of blood trihalomethane concentrations with asthma in US adolescents: nationally representative cross-sectional study. Eur Respir J 59(5). https://doi.org/10.1183/13993003.01440-2021
97. Sun Y, Wang YX, Mustieles V et al (2023) Blood trihalomethane concentrations and allergic sensitization: a nationwide cross-sectional study. Sci Total Environ 871:162100. https://doi.org/10.1016/j.scitotenv.2023.162100
98. Burch JB, Everson TM, Seth RK et al (2015) Trihalomethane exposure and biomonitoring for the liver injury indicator, alanine aminotransferase, in the United States population (NHANES 1999-2006). Sci Total Environ 521–522:226–234. https://doi.org/10.1016/j.scitotenv.2015.03.050
99. Makris KC, Andrianou XD, Charisiadis P et al (2016) Association between exposures to brominated trihalomethanes, hepatic injury and type II diabetes mellitus. Environ Int 92-93: 486–493. https://doi.org/10.1016/j.envint.2016.04.012
100. Yang L, Chen L, Hao Y et al (2023) Urinary trihalomethane concentrations and liver function indicators: a cross-sectional study in China. Environ Sci Pollut Res Int 30(14):39724–39732. https://doi.org/10.1007/s11356-022-25072-4
101. Deng YL, Lu TT, Hao H et al (2024) Association between urinary haloacetic acid concentrations and liver injury among women: results from the Tongji reproductive and environmental (TREE) study. Environ Health Perspect 132(1):17006. https://doi.org/10.1289/ehp13386
102. Lubbers JR, Chauan S, Bianchine JR (1982) Controlled clinical evaluations of chlorine dioxide, chlorite and chlorate in man. Environ Health Perspect 46:57–62. https://doi.org/10.1289/ehp.824657
103. Ouhoummane N, Levallois P, Gingras S (2004) Thyroid function of newborns and exposure to chlorine dioxide by-products. Arch Environ Health 59(11):582–587. https://doi.org/10.1080/00039890409603437
104. Sun Y, Xia PF, Korevaar TIM et al (2021) Relationship between blood trihalomethane concentrations and serum thyroid function measures in U.S. Adults. Environ Sci Technol 55(20):14087–14094. https://doi.org/10.1021/acs.est.1c04008
105. Wu Y, Deng YL, Zhang M et al (2024) Urinary haloacetic acid concentrations and thyroid function among women: results from the TREE study. Sci Total Environ 927:172368. https://doi.org/10.1016/j.scitotenv.2024.172368
106. Zhang SH, Guo AJ, Wei N et al (2021) Associations of urinary dichloroacetic acid and trichloroacetic acid exposure with platelet indices: exploring the mediating role of blood pressure in the general population. J Hazard Mater 402:123452. https://doi.org/10.1016/j.jhazmat.2020.123452
107. Cui FP, Miao Y, Deng YL et al (2023) Associations of exposure to disinfection by-products with blood coagulation parameters among women: results from the Tongji reproductive and environmental (TREE). Ecotoxicol Environ Saf
108. Li CR, Deng YL, Miao Y et al (2023) Exposures to drinking water disinfection byproducts and kidney function in Chinese women. Environ Res 244:117925. https://doi.org/10.1016/j.envres.2023.117925
109. Gängler S, Waldenberger M, Artati A et al (2019) Exposure to disinfection byproducts and risk of type 2 diabetes: a nested case-control study in the HUNT and lifelines cohorts. Metabolomics 15(4):60. https://doi.org/10.1007/s11306-019-1519-0

Future Perspectives

Susan Andrews, Chao Chen, and Yuefeng Xie

Contents

1 Detection and Quantification of DBPs and Their Precursors 474
2 Water and Wastewater Treatment 475
3 Considerations for Climate Change and Energy Generation 477
4 Health Risks/DBP Toxicity Considerations (Collaboration, Modeling for Predictions...) 477
5 Some Final Messages 479

Abstract This chapter attempts to take the information conveyed throughout this book and predict future activities concerning disinfection byproduct (DBP) research, regulation, and management. A lot has happened in the past 50 years since Johannes Rook's influential 1974 paper, and the DBP realm will continue to evolve. Hopefully, it is evident from this book that it is difficult to be specific about what will be the research and practical DBP perspectives in the future, except to say that there will be more.

It is also important to remain mindful that water is disinfected so that it is safe for human consumption, and that the practice of disinfection has saved countless lives by limiting the incidence and spread of waterborne disease. As long as disinfection will be required into the future, we will have to manage the unintended consequence of the formation of DBPs. To this end, it will not be surprising to see more research into essentially all of the aspects that surround DBPs that are described in this book,

S. Andrews (✉)
University of Toronto, Toronto, ON, Canada
e-mail: susan.andrews@utoronto.ca

C. Chen
Tsinghua University, Beijing, China
e-mail: chen_water@tsinghua.edu.cn

Y. Xie
Pennsylvania State University, Middletown, PA, USA
e-mail: yxx4@psu.edu

Chao Chen, Susan Andrews, and Yuefeng Xie (eds.), *Disinfection By-Products in Water*, Hdb Env Chem (2024) 135: 473–480, DOI 10.1007/698_2024_1165,

Published online: 16 October 2024

as well as others that might have been missed. As such, this chapter attempts to summarize many of the efforts that likely will be continued.

Keywords Artificial intelligence, Climate, Detection, Disinfection by-products (DBPs), Energy, Future, Health, Models, Regulation, Toxicity, Treatment

A lot has happened in the past 50 years since Johannes Rook's influential 1974 paper, and the disinfection byproduct (DBP) realm will continue to evolve. Hopefully, it is evident from this book that it is difficult to be specific about what will be the research and practical DBP perspectives in the future, except to say that there will be more.

It is also important to be mindful that water is disinfected so that it is safe for human consumption, and that the practice of disinfection has saved countless lives by limiting the incidence and spread of waterborne disease. As long as disinfection will be required into the future, we will have to manage the unintended consequence of the formation of DBPs. To this end, it will not be surprising to see more research into essentially all of the aspects that surround DBPs that are described in this book, as well as others that might have been missed. As such, the following passages attempt to summarize many of the efforts that likely will be continued.

1 Detection and Quantification of DBPs and Their Precursors

Hundreds of DBPs have been identified and quantified, and yet this is still understood to be a small fraction of the total. Analytical methods will continue to develop to be more specific and sensitive, so more DBPs will be identified at lower and lower concentrations. The early development of gas and liquid chromatographic methods (GC, LC) to separate compounds so they could be measured with relatively simple but non-specific detectors will likely continue to be a standard procedure for measuring DBPs in water.

However, high resolution mass spectrometric detection methods are being refined to rely less on pre-separation and/or derivatization techniques to result in faster and more specific analytical detection and quantification. These advances enhance DBP detection, minimize the workload to see them (i.e., optimize and/or minimize need for separation from background matrix), and reduce costs of measurement, making monitoring and response efforts more accessible to greater amounts of the population, helping to ensure safer water for more people.

There will be a continued increase in focus on DBPs containing more complex structures and/or heteroatoms (N, S, P, I, metals. . .etc.), especially those known to be associated with chemicals of human health concern. Even within the relatively simple category of halogenated DBPs, the interest has come to include not only chlorinated and brominated species, but iodinated species as well. Interestingly, studies have shown I-DBPs to be formed under different conditions to the chloro or bromo species, making them not just a simple extension of the halogenated DBP series but a more unique group for which different strategies for their control are being studied. The species of iodine involved, the types of disinfection/oxidation methods used, and the structure of organic precursors all have important effects on I-DBP formation.

Similarly, new DBP classes will continue to be discovered and studied, considering both organic and inorganic DBPs. While it seems that halogenated DBPs of interest will be of relatively low molecular weight (generally <300 g/mol), the structures that are being identified with toxicity are becoming more complex (e.g., halogenated furanones and halobenzoquinones).

When concerning various possible heteroatoms in DBPs, there has already been much research into DBPs that contain nitrogen atoms, largely in response to toxicity testing that showed nitrogenous DBPs (e.g., nitrosamines, haloacetonitriles or HANs, haloacetamides or HAMs, halonitromethanes or HNMs, cyanogen halides) generally to be more of a human health concern than the early-detected halogenated DBPs (e.g., trihalomethanes or THMs, haloacetic acids or HAAs).

As for inorganic DBPs, interest will continue in chlorite, chlorate, and bromate, but it is expected that this list will also lengthen with advances in analytical and toxicological methods.

Several methods are being developed to aid in the identification of relevant possible DBPs. Non-targeted mass spectrometric (MS) analysis such as precursor ion scanning (PIS) shows promise in this area.

Due to the likely continuing complexity and cost of measuring individual DBP species as this list gets longer, an understanding of DBP precursors is crucial for effectively controlling the harmful DBPs in water treatment. New developments will continue to take advantage of the fundamental characteristics of DBP precursors, such as the aromaticity, molecular weight, and hydrophobicity, that significantly influence DBP formation.

Still, DBP precursor materials are not generally simple molecules. Due to the complex nature of water matrices, research on DBP precursors still faces challenges in effective detection and characterization, as well as relationships between DBP generation and properties of DBP precursors. Innovations in water treatment technology are essential for removing these precursors and ensuring water safety. Additionally, it is vital to study the impact of environmental factors such as climate change and urbanization on the distribution of DBP precursors to identify affordable and reliable surrogates of DBPs and their precursors to facilitate monitoring and operations optimization.

2 Water and Wastewater Treatment

In regard to minimizing DBP concentrations via water and wastewater treatment, work will continue to identify and remove DBPs and DBP precursor materials. The aim will always be to strive to prevent DBP formation rather than remove DBPs from treated water, so removal of precursors and identification of alternative disinfectants will be ongoing. We already see significant shifts away from primary disinfection with free chorine to alternatives like chloramine, chlorine dioxide, UV, and ozone. The challenge is to provide equivalent or better disinfection while avoiding unintended consequences of switching disinfectants (i.e., avoiding simply causing different but still potentially toxic DBPs to be formed).

There always will be challenges for optimizing existing treatment processes and process trains to minimize DBP formation while maintaining disinfection targets. Levels of DBPs were higher when pre-chlorination was used compared to water treatment plants that chlorinated toward the end of the treatment train. The removal of precursors varies according to the precursors present and the type of treatment implemented and varies seasonally.

New treatment processes will continue to be developed which avoid or minimize DBP formation. Increased removal of natural organic matter, both NOM as DBP precursors (to reduce DBP formations) and as biodegradable organic matter (to minimize microbial regrowth and DBP formation in distribution systems) will possibly involve the use of more "natural" processes (e.g., bio-treatment).

As fresh water supplies become less abundant, maintaining adequate water quantity as well as quality will require employing water reuse strategies, requiring enhanced vigilance concerning all aspects of water quality prior to reuse by humans.

There will be continued efforts to improve the design and operation of small and remote treatment systems, including point-of-use (POU) and point-of-entry (POE) systems, to provide good disinfection while minimizing DBPs. Lessons can be learned from POU and recreational water systems for comparison to conventional drinking water systems. Various methods for removing DBPs through household treatment processes, also known as POU processes, include boiling, activated carbon adsorption, membrane filtration, and ultraviolet disinfection and photolysis. Work will continue to optimize the performance of these processes as the understanding of similarities and differences of conventional and POU processes becomes more advanced.

We need to recognize that drinking water distribution systems are an integral component of safe water and they must be designed, operated, maintained, and optimized with the same priorities afforded to water treatment plants. There will be a continuing and increased effort to optimize distribution system management to enhance the delivery of treated drinking water with minimal DBP formation. Secondary disinfection has already been moving away from the use of free chlorine to less reactive chloramines with some success. Consideration is also given to removing biodegradable organic carbon from treated water before its distribution.

3 Considerations for Climate Change and Energy Generation

As with many aspects of modern life, we will need to consider the impacts of climate change on water treatment practices. Hydroclimatic changes and climate shocks can alter or interact with key landscape characteristics, especially land cover, resulting in DBP precursor transport process responses that include increased erosion, increased slope failure, increased atmospheric deposition (dry and wet), increased forest fires, increased algae growth, and increased saltwater intrusion. As well, by driving change in DBP precursor transport process responses, global warming alters the type and amount of DBP precursors (natural organic matter (NOM), man-made or "anthropogenic" organic matter (AnOM, things like pharmaceuticals and personal care products), halides, and algal-related organic matter (AOM) [resulting from nutrients, algae, and other microorganisms]) delivered to water supplies and the associated potential for DBP formation during municipal drinking water treatment and distribution.

We will also need to consider the interplay of various energy generation practices on the possible DBP formation of impacted source waters. For example, for shale gas extraction and coal-fired power plants, there are significant roles for both organic and inorganic DBP precursors from discharged water that contribute to DBPs, especially brominated and iodinated DBPs, emphasizing the resulting environmental and health implications. Recently developed iodide-based coal additives have emerged as an alternative to mercury control. However, iodide is also a precursor for DBP formation, particularly in chloramination disinfection processes. Given the limited data on iodide content in coal and its addition for mercury control, future research is crucial to understand its potential impact on DBP formation in downstream drinking water utilities. By continuing to analyze pivotal research findings, the interplay between industrial practices and water quality management challenges will help to evaluate various mitigation strategies.

4 Health Risks/DBP Toxicity Considerations (Collaboration, Modeling for Predictions. . .)

Although comments on the toxicity and potential health effects of DBPs are discussed last in this epilogue, the whole area of DBPs really both starts and ends with their toxicity. Without that aspect, there would be no need for interest, concern, or awareness of DBPs. . . .or this book. Efforts to identify and mitigate human health effects are the primary drivers of DBP regulation and, as such, could be discussed at the beginning of this chapter. However, they are mentioned here as a reminder that they will continue to be key in evolving our responses to DBPs in water. These efforts will likely involve many or most of the elements listed below.

For example, since most of DBPs demonstrated the ability in inducing oxidative stress and DNA damage, therefore a fast screening method and comparison database on the mechanisms of action should be established. In general, it would be good to if we could develop toxicity tests to identify potentially toxic water, regardless of the source of the toxicity.

Toxicological data provide evidence for various agencies to regulate DBPs. Therefore, toxicological research is fundamental for DBP research and treatment. Considering the large number of possible DBPs, it is a big challenge to examine the toxicity of every single DBP. Quantitative structure-activity relationship (QSAR) models help to prioritize emerging DBPs with toxicological importance, so these prioritized DBPs might be studied for toxicity first. With the development of artificial intelligence and machine-learning, the QSAR analysis could be much more helpful and predictive for DBP toxicities. In addition, the animal models, especially zebrafish, are useful tools to facilitate screening for toxicity of DBPs and to set up database for the comparison of DBP toxicities. However, standard procedures or methods for toxicity testing are needed for these animal models, and a combination of both in vivo and in vitro toxicity tests is recommended, with the screening of in vitro tests or other tests to firstly assess the rank order of toxicity of DBPs, and with in vivo tests to examine the toxicity of the prioritized DBPs. Also, with the development of omics technology, the multiple-omics research should be used in DBPs toxicity test, especially the toxicogenomics, proteomics, microbiomics, and metabonomics, which are helpful to screen the biomarkers for epidemiology studies and to understand the mechanisms of actions of toxicity.

It will be important to consider possible synergistic and antagonistic effects of DBPs with each other and with other water quality parameters when assessing possible toxicity outcomes, and to continue to be aware of possible exposures to DBPs from both drinking water and non-drinking activities (cooking, bathing, recreation) and their contribution to the overall body burden of DBPs to toxicity. Awareness that exposure to DBPs in drinking water is only one route, exposure via food, air, and other contacts is also important (sometimes more important). Regardless, these multiple routes of exposure complicate the challenge to attribute causes and effect for DBPs and toxicity.

More research into possible impacts of adsorption of DBPs onto solids that might be ingested or inhaled (e.g., microplastics, but also natural materials) is also likely. These solids can act as reservoirs of DBPs to produce possible pinpoint exposures of higher concentrations than may be seen via chronic exposures to dissolved DBPs.

Considering the caveats associated with available toxicological methodologies and the large number of possible DBPs, we should further develop DBP-related mathematical models to help with predicting the results of changes in treatment conditions on DBP concentrations for both known and likely DBPs (e.g., QSARs). We should also improve and enhance efforts concerning the epidemiology of drinking water as it relates to DBPs. Overall, epidemiological studies fairly support the association between exposure to DBPs and bladder cancer risk; however, no consistent relationships are observed for colon, rectal, and other cancers. Existing epidemiological evidence remains insufficient to support the associations of DBP exposures with adverse reproductive and developmental outcomes, but emerging human studies relying on blood and urinary biomarkers have shown that DBP exposures are associated with declined semen quality, impaired ovarian function, and adverse birth outcomes. Human evidence concerning other adverse health outcomes is limited and inconclusive.

For DBP regulations, regulating more individual DBPs may not be the most cost-effective way to protect public health. With more information on the speciation, formation, stability, occurrence, removal, and toxicities of individual DBPs, regulating DBP mixtures, DBP surrogates, aggregate halogen constituents, and/or overall water toxicity may be an effective way to minimize the health risks of DBPs. As well, given that the relationship between the toxicology, treatment, and regulation of DBPs adds further levels of complexity to the DBP story, managing all of the data and possible calculated outcomes to provide informed advice could very well involve the use of artificial intelligence (AI) techniques that are only beginning to be developed. Even in these early stages of development, however, it seems that a single AI technique may not be realistic and that hybrid AI technologies will be required.

DBP research began with the identification of chloroform as a byproduct of chlorine disinfection, and the knowledge at the time that chloroform was harmful to human health. Knowledge of human health impacts of DBPs will always play a role, and likely be a significant driver, in determining means to regulate and manage them appropriately. As such, there will continue to be increased collaboration with toxicologists to help identify DBPs of potential human health concern. . . . .toxicology is advancing in parallel with (but separately from) advances in DBP research.

5 Some Final Messages

While progress is being made concerning DBP identification and management, water practitioners can feel confident in providing the safest water possible by making use of documented standard methods and regulations that have been adopted by the scientific and regulatory organizations that are active in virtually every nation that produces drinking water. They will continue to provide design and operational guidance to improve drinking water and help educate not only water providers and design engineers but also the political bodies that make decisions concerning spending on safe water, and the public.

Finally, it is important to reiterate that water is disinfected so that it is safe for human consumption, and that the practice of disinfection has saved countless lives by limiting the incidence and spread of waterborne disease. As long as disinfection will be required into the future, we will have to manage the unintended consequence of the formation of DBPs. To this end, it will not be surprising to see more research into essentially all of the aspects that surround DBPs that are described in this book, as well as others that might have been missed, and with climate change and energy challenges being at the forefront of concerns it will be interesting to see how these issues develop or are solved over the next 50 years.

Correction to: Climate Change Implications for DBP Occurrence in Municipal Drinking Water

Monica B. Emelko, Alex Chow, Fariba Amiri, Huan Chen, and Micheal Stone

Correction to:
Chapter 1168 in: Chao Chen, Susan Andrews, and Yuefeng Xie (eds.), *Disinfection By-Products in Water*, Hdb Env Chem, https://doi.org/10.1007/698_2024_1168

The original version of the chapter was inadvertently published with incorrect author affiliations. The correct author affiliations are as follows:

M. B. Emelko
Department of Civil & Environmental Engineering, University of Waterloo, Waterloo, ON, Canada
e-mail: mbemelko@uwaterloo.ca

A. Chow
The Chinese University of Hong Kong, EESC, Hong Kong, People's Republic of China
e-mail: achow@cuhk.edu.hk

The updated version of this chapter can be found at
https://doi.org/10.1007/698_2024_1168

The manufacturer's authorised representative in the EU is Springer Nature Customer Service Centre GmbH, Europaplatz 3, 69115 Heidelberg, Germany. If you have any concerns regarding our products, please contact ProductSafety@springernature.com

Printed and bound by CPI Group (UK) Ltd, Croydon, CR0 4YY
01/10/2026
02225965-0015